Surveying for engineers

J. Uren
School of Civil Engineering
University of Leeds

W. F. Price
School of the Environment
University of Brighton

4th edition

palgrave
macmillan

First edition 1978
Second edition 1985
Third edition 1994
This edition 2006
Published by
PALGRAVE MACMILLAN
Houndmills, Basingstoke, Hampshire RG21 6XS and
175 Fifth Avenue, New York, N. Y. 10010
Companies and representatives throughout the world

PALGRAVE MACMILLAN is the global academic imprint of the Palgrave Macmillan division of St. Martin's Press LLC and of Palgrave Macmillan Ltd. Macmillan® is a registered trademark in the United States, United Kingdom and other countries. Palgrave is a registered trademark in the European Union and other countries.

ISBN-13: 978–1–4039–2054–6
ISBN-10: 1–4039–2054–0

This book is printed on paper suitable for recycling and made from fully managed and sustained forest sources. Logging, pulping and manufacturing processes are expected to conform to the environmental regulations of the country of origin.

A catalogue record for this book is available from the British Library.

10   9   8   7   6   5   4
14 13 12 11 10 09 08 07

Printed and bound in China

Contents

Preface

During the decade or so since the publication of the previous edition of *Surveying for Engineers*, many of the techniques and most of the equipment used in modern site surveying have undergone considerable changes and improvements. Virtually all of these have evolved due to the ever-increasing pace of developments in computers, electronics and communications. As a result, much of the instrumentation now commonplace in engineering surveying is very sophisticated and offers portable computing power, speed and data storage capabilities that were only dreamt of a few years ago. These, when linked to the immense capabilities of easily affordable plotters and desktop computers, have had a significant effect on many of the techniques used in surveying, particularly those associated with the processing and output of survey data.

All of these changes and improvements are covered in detail in the fourth edition while some of the traditional but still very relevant techniques are retained. For example, although considerable emphasis is given to new technologies, especially satellite and airborne surveying systems, these have not been included at the expense of taping, which still has a place in this edition.

As well as describing the improvements that have occurred over the last few years in the hardware and software used in everyday surveying, this edition also brings the reader up to date with topics such as the latest highway design standards, the changes in surveying brought about by GPS, the services provided by the Ordnance Survey and the latest information on emerging technologies.

To reflect the continuing evolution of surveying, this new edition represents a complete revision and rewriting of the previous one and contains new illustrations throughout. In response to student demand, exercises are included at the end of each chapter and much use is made of the Internet, with many references being given to this in the text. In addition, each chapter and section now begins with a statement of aims, and to reinforce these reflective summaries are also provided. It is intended that the aims and reflective summaries, together with the easy-to-read format and improved layout, will help the reader understand each topic better as it is developed.

In keeping with the emphasis of previous editions, this edition has been written with civil engineering, building and construction students in mind. However, it will also be found useful by any other students who undertake surveying as an elective subject and it is anticipated that practising engineers and those engaged in site surveying and construction will use *Surveying for Engineers* as a reference text.

John Uren and Bill Price
August 2005

Acknowledgements

The authors wish to thank all those who have contributed in any way to the preparation of this book and, in particular, the following persons and organisations:

Anthony Mills, Thales Navigation
Barbara Molloy, PV Publications Ltd
Colin Prior and Bob Seago, University of Brighton
British Standards Institution
Construction Industry Research and Information Association (CIRIA)
David Partridge, QinetiQ Ltd
Department of Transport
Gareth Brookman, Stuart Hartley, Steve Kennedy and Emma Powell, Pentax (UK) Ltd
Garmin Ltd
Gemma Illidge and Rachel Scallan, Sokkia Ltd
Henrietta Flynn and Paula Innes, Leica Geosystems
Institution of Civil Engineering Surveyors
Institution of Civil Engineers
International Organization for Standardization (ISO)
Jenny Clark and Duncan Lees, Plowman Craven & Associates
John Strodachs, AiC Ltd
John van den Berg, The National Swedish Institute for Building Research
Karen Myhill, Fisco Tools Ltd
Katsumi Kaji, Ushikata Mfg. Co. Ltd, Tokyo, Japan
Lilian Bew, Andrew Cowling, Christine Martin, Roy Trembath and David Walsh, University of Leeds
Lucy Abrahams, Kay Clarke and Martin Smith, Survey Supplies
Mariam Alston and Neil Sutherland, Positioning Resources Ltd
Martin Smith, IESSG, University of Nottingham
Nigel Lorriman, McCarthy Taylor Systems Ltd
Paul Cruddace, Ordnance Survey
Paul Finney, Zenith Surveys Ltd
Royal Institution of Chartered Surveyors
Saxon Stockwell-Brown, York Survey Supply Centre
Scott Ryder, Itronix (UK) Ltd
Steven Ramsey, APR Services Ltd
Stuart Edwards and Matt King, School of Civil Engineering and Geosciences, University of Newcastle upon Tyne
Sylvie Leiss, Marikki Honkala, Simon Hyatt and Lucy Parry, Trimble Navigation
Tanja Eigenhuis and Jim Pelham, Topcon
Trevor Burton, BKS Surveys Ltd

Special thanks are also due to Ian Kingston for typesetting the book and to Jenna Steventon, our editor.

chapter one

Introduction

 ## Aims

After studying this chapter you should be able to:

- Define what surveying is and what its various disciplines are
- Explain that engineering surveying is that part of surveying used mostly for civil engineering, building and construction projects
- State the main purposes of engineering surveying
- Discuss the reasons why geospatial engineering and geomatics have been introduced to describe the activities of surveyors
- Describe, in outline, the methods by which engineering surveying is carried out and the equipment and methods that are used for this
- Give reasons why engineering surveyors now play a major role in data management for engineering projects
- Recognise those areas of surveying that will develop in the near future and appreciate why engineering surveyors have an important part to play in this
- Obtain information about surveying from a variety of sources, including the main institutions that promote surveying

This chapter contains the following sections:

1.1   Engineering surveying

1.2   Survey institutions and organisations

Exercises

Further reading and sources of information

1.1 Engineering surveying

After studying this section you should be able to explain what geospatial engineering and geomatics are and why these terms have been introduced. You should be aware that engineering surveying is used extensively in building and construction, but that this can also involve many other different specialist areas of surveying. You should have an outline knowledge of the equipment and methods used in engineering surveys and have some appreciation of the way in which these are expected to develop. You should also have a clear idea of the aims of this book.

This section includes the following topics:

● Engineering surveying, geospatial engineering or geomatics?

● What is engineering surveying?

● How are engineering surveys carried out?

● What will be the role of engineering surveyors in future?

● What are the aims of this book?

Engineering surveying, geospatial engineering or geomatics?

To many, the traditional role of a surveyor has been to determine the position of features in both the natural and built environment on or below the surface of the Earth and to represent these on a map. Even though this view of surveying is still true in some respects, in an age where the acquisition, processing and presentation of data are paramount, surveyors today will be familiar with many different methods for collecting spatial data about the Earth and its environment, they will be able to process this data in various formats and they will be able to present this in an assortment of media.

Although this gives an idea of what contemporary surveying is, to the majority of engineers working on construction sites, surveying is the process of measuring angles, distances and heights to help in the design and construction of civil engineering projects. This gives rise to the term *engineering surveying*, which describes any survey work carried out in connection with construction and building. This also involves all of the different methods of data acquisition, processing and presentation now available in surveying. Engineering surveying is one of the most important areas of expertise in surveying and to reflect this, it is the main subject of this book.

Many on site think that engineering surveying is a labour-intensive method that uses old-fashioned instruments for taking measurements and requires never-ending calculations to be done. Although theodolites, levels and tapes are still used and engineering surveying will always require some calculations to be done on site, the way in which surveys are carried out for civil engineering and construction projects has been transformed in recent years. For example, most measurements of distance,

Total station   Digital level

Figure 1.1 ● Total station and digital level (courtesy Pentax UK Ltd and Trimble Navigation Ltd).

angle and height are now recorded and processed electronically using total stations and digital levels similar to those shown in Figure 1.1. Satellite surveying systems such as *GPS (the Global Positioning System)* are in everyday use and new airborne technologies such as *LiDAR (Light Detection and Ranging)* can now be used by engineers for mapping purposes. These are shown in Figure 1.2.

The large amounts of data that can be collected by these measuring systems is easily processed by computers, which are capable of handling data and performing calculations in a fraction of the time taken to do this a few years ago. Another benefit

GPS equipment on site   LiDAR survey of railway

Figure 1.2 ● GPS and LiDAR technology in surveying (courtesy Survey Supplies and Fugro Inpark).

Figure 1.3 ● Using mobile phone for data transfer (courtesy Survey Supplies).

of the digital age in data recording and processing is that it can be transmitted between instrument and office using the Internet and mobile phone technology, as can be seen in Figure 1.3.

Not surprisingly, all of these new technologies have resulted in some changes to the way in which engineering surveying is carried out. Up till now, the main purposes of engineering surveying have been to supply the survey data required for preparing maps and plans for site surveys together with all aspects of dimensional control and setting out on site. However, even though these are still relevant, there is now much more emphasis on providing survey data and managing this for both an engineering and a built environment.

With all these developments in mind, one of the institutions that regulates the activities of surveyors working in civil engineering in the UK has redefined engineering surveying as *geospatial engineering*. According to the Institution of Civil Engineering Surveyors (the ICES):

> Geospatial engineers work within construction on the measurement and monitoring of projects as well as producing maps, plans and charts of different features.

In their definition, the ICES state that the main profession of a geospatial engineer in civil engineering is engineering surveying. However, to reflect the changes in the way in which survey data is collected and processed for civil engineering projects today, geospatial engineers may also be involved in other specialist areas such as photogrammetry and remote sensing, geographic information systems as well as cartography and visualisation. It is the addition of these to the engineering surveyor's role and the change of emphasis towards data management that has given rise to the term *geospatial engineer*.

Another institution that regulates surveying in the UK is the Royal Institution of Chartered Surveyors (RICS). It takes a much broader view of surveying which is organised into faculties covering areas ranging from arts and antiques to valuation. The faculty that is responsible for land surveying is known as the *geomatics faculty*.

Geomatics is a new word that is used to describe surveying as it is today and not only covers the traditional work of the surveyor in mapping and on site but also reflects the changing role of the surveyor in data management. As discussed above, this has arisen because of the advances made in surveying which make it possible to collect, process and display large amounts of spatial data with relative ease using digital technology. This in turn has created an enormous demand for this data from a wide variety of sources such as geology, geophysics, hydrology, forestry, transportation, government and human resources. For all of these, data is collected and processed by a computer in a Geographic Information System (GIS). These are databases that can integrate the spatial data provided by surveyors with environmental, geographic and social information layers (see Figure 1.4) which can be combined, processed and displayed in any format according to the needs of the end user. Without any doubt, the most important part of a GIS is the spatial data on which all other information is based and the provision of this has been a huge growth area in surveying. Because of the different emphasis in surveying and other advances made in instrumentation for data collection and processing, it is now felt that a change of name from surveying to geomatics reflects the nature of the profession as practised today in the same way that geospatial engineering describes surveying in a construction environment. Throughout *Surveying for Engineers*, the term *surveying* will continue to be used, but the use of the term *geomatics* to replace this is noted here.

Bearing in mind the reasons for the use of geomatics in surveying, the geomatics faculty of the RICS gives the following definition:

> Geomatics is the science and study of spatially related information and is particularly concerned with the collection, manipulation and presentation of the natural, social and economic geography of the natural and built environments.

Within the faculty, *engineering surveys* and the *monitoring of structures* are identified as specialist areas but others such as mapping and positioning, GIS, photogrammetry and remote sensing, together with cartography are also included.

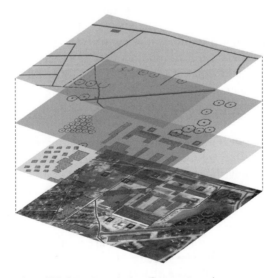

Figure 1.4 ● Layers in a GIS (courtesy Leica Geosystems).

What is engineering surveying?

Both the ICES and RICS include engineering surveying in their definitions of geospatial engineering and geomatics. Taking the ICES definition, engineering surveying involves:

- Investigating land, using computer-based measuring instruments and geographical knowledge, to work out the best position to construct bridges, tunnels and roads
- Producing up-to-date plans which form the basis for the design of a project
- Setting out a site, so that a structure is built in the correct location and to the correct size
- Monitoring the construction process to make sure that the structure remains in the right position and recording the final as-built position
- Providing control points by which the future movement of structures such as dams and bridges can be monitored

Both institutions also identify some of the other types of survey that might be used on civil engineering projects as the following.

- *Hydrographic surveying.* This is surveying in a marine environment where the traditional role for centuries was to map the coastlines and sea bed to produce navigational charts. More recently, a lot of hydrographic surveys have been carried out for offshore oil and gas exploration and production. Hydrographic surveys are also used in the design, construction and maintenance of harbours, inland water routes, river and sea defences, in control of pollution and in scientific studies of the ocean.
- *Photogrammetry.* This is the technique of acquiring measurements from photographic images. The use of this in topographic mapping for engineering is well established and is carried out today using digital aerial photography and computers with a high-resolution display in a soft copy workstation similar to that shown in Figure 1.5. The photographs are taken with special cameras mounted in fixed wing aircraft or helicopters. Because it is a non-contact technique, photogrammetry is particularly useful in hazardous situations; another of its advantages is that it produces data in a digital format, which makes it ideal for use in GIS and CAD.
- *Remote sensing.* This technique is closely allied to photogrammetry because it also uses imagery to collect information. In this case, information is gathered about the ground surface without coming into contact with it. Remote sensing can be carried out for engineering projects using satellite imagery, spectral imaging (in which different colour images are analysed) and more recently with airborne platforms such as LiDAR and IFSAR.
- *Geographic Information Systems (GISs).* These are computer-based systems which allow spatial information to be stored and integrated with many other different types of data. As far as geospatial engineering is concerned, they involve

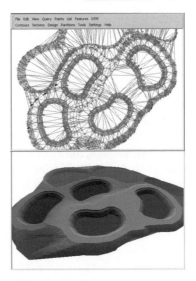

Figure 1.5 ● Photogrammetric soft copy workstation (courtesy BKS).

Figure 1.6 ● Computer visualisation produced from survey data (courtesy AiC)

obtaining, compiling, input and manipulation of geographic and related data and the presentation of this in ways and formats specifically required by a user.

● *Cartography and visualisation.* This is the art and technique of making maps, plans and charts accurately and representing three dimensions on a variety of media. Cartography and map making can be considered to be the traditional role of the surveyor and anyone who uses a map to find their way round town or countryside is using information gathered and presented by surveyors. Compared to this, visualisation is a new technology that uses spatial data to show computer-generated views of landscapes, as shown in Figure 1.6. These could be used for undertaking environmental impact assessments.

As can be seen, engineering surveying involves a number of specialist areas, all of which will overlap from time to time. Although geospatial engineering and geomatics encompass these, this book concentrates on engineering surveying. Guidance on how to obtain information on the other specialist areas in geospatial engineering and geomatics is given at the end of this chapter.

How are engineering surveys carried out?

Plane surveying, geodetic surveying and setting out

Engineering surveys are usually based on horizontal and vertical control, which consist of a series of fixed points located throughout a site whose positions must be determined on some coordinate system. For most construction work, *horizontal control* defines points on a two-dimensional horizontal plane which covers the site, whereas *vertical control* is the third dimension added to the chosen horizontal datum. The measurements for a control survey can be taken with a variety of equipment. For a small building site, theodolites and tapes could be used to observe horizontal control in the form of a traverse with levelling providing the vertical control. As the site gets larger, total stations and digital levels may be used. In all of these surveys, horizontal angles and distances are measured for horizontal control and vertical angles and distances for vertical control.

What has been described so far is known as *plane surveying*, in which a flat horizontal surface is used to define the local shape of the Earth, with the vertical always taken to be perpendicular to this. The reason for adopting a flat rather than curved surface for surveying is to simplify the calculation of horizontal position by plane trigonometry. Heights are easily defined to be vertically above (or below) the chosen horizontal datum. Another advantage of plane surveying is that all dimensions obtained from drawings and other design data will be the same as those on the ground. Of course, as a site gets bigger, there comes a point when the assumptions made in plane surveying are no longer valid and the curvature of the Earth has to be accounted for. This limit occurs when a site is greater than 10–15 km in extent in any direction. The type of surveying that accounts for the true shape of the Earth is known as *geodetic surveying*. This can be carried out over very large areas, but can be quite complicated because measurements are often taken over long distances and computations are based on a curved surface instead of a flat one. Fortunately, GPS is capable of performing geodetic surveys for even the biggest sites with relative ease compared to previous methods. Consequently, GPS is the predominant method in use today for providing three-dimensional survey control over large areas. Unlike total station and other terrestrial systems, GPS equipment determines position using data transmitted from orbiting satellites. Even though position is fixed on a regional or global datum and is based on an assumed shape of the Earth in satellite surveying systems, the coordinates obtained are usually transformed to another coordinate system for use on site.

Although satellite surveying systems have provided a practical solution to the problem of surveying over large areas on curved surfaces, they are also often the preferred method for taking measurements in plane surveys.

In practice, nearly all engineering surveys are carried out by plane surveying methods. *Surveying for Engineers* reflects this and concentrates mostly on plane surveying techniques, but geodetic surveying is discussed in those chapters that cover GPS.

The positions of control points are usually established at the start of a project for a topographic or detail survey. This is carried out to produce maps or plans of the site. Without a topographic survey of some sort, the construction project could not proceed. For large projects, photogrammetry will be used for producing the plans, but for smaller projects, ground methods using total stations and GPS are more cost-effective. Whatever method is used for data capture in a topographic survey, it is all processed by computer to produce a digital map of some sort which can then be exported to a CAD system for use by the design team. Figure 1.7 shows a detail survey being carried out with a total station and subsequent data processing in the office. As well as producing maps, a *Digital Terrain Model (DTM)*, which is a three-

Figure 1.7 ● Detail survey by total station with computer processing (courtesy Leica Geosystems).

dimensional representation of the natural ground surface stored as spatial data in a computer, is also produced for most construction projects. This is used to generate a *design surface*, and with most instruments now relying on electronic data this is probably used more than drawings or any other media as it can be downloaded or transmitted to a total station or GPS receiver ready for use on site.

Although the emphasis in engineering surveying is changing towards the management and processing of data, a lot of time on site is devoted to *setting out*. Sometimes called dimensional control, this is the surveying carried out to establish all the marks, lines and levels needed for construction purposes. It can vary from the measurement of angles and distances by theodolite and tape to the most sophisticated machine control systems using GPS and computer visualisation as shown in Figure 1.8. One of the responsibilities of surveyors and engineers involved in setting out is to choose the right equipment for the job from the array of instruments and systems currently available. For some sites, theodolites, levels and tapes are quite sufficient, but as the work gets larger and more complex, total stations, GPS and lasers may be more appropriate. When accuracy is important, the choice of equipment and methods used for setting out is crucial. All of this requires anyone engaged in setting out to have an understanding of the precision of the equipment and methods they use and the way in which errors can propagate through a survey. This is reinforced by the greater emphasis now placed on quality control and specifications for building and construction. When setting out, it is essential that good practice is always followed. This requires a systematic approach to be developed in which equipment is regularly checked and established procedures are followed in order that mistakes can be detected at an early stage.

Modern survey equipment

Although traditional theodolites, levels and tapes are still used on site, many new technologies are now available for engineering surveys.

Theodolite and tape

Grader with GPS antennae mounted on blade

Display unit in cab

Figure 1.8 ● Setting out by theodolite and tape in comparison to machine control by GPS (courtesy Leica Geosystems, Trimble Navigation Ltd and Topcon).

These include *total stations*, which are capable of measuring angles and distances with a high degree of precision in a single instrument. Today, these are high-performance opto-electrical instruments with many different measuring functions built into them. They can be used with data recorders, field computers and controllers to read, store and edit data on site and can use the latest wireless technology for data transfer. The *digital level* is a similar instrument, but this reads and records height information electronically by using digital image processing techniques.

The equipment that has had the biggest impact on engineering surveying in recent years is *GPS*. This is well known by most people as something they might use in their cars for navigation or for map reading when out walking. However, it has also found widespread use in surveying for control surveys, mapping and setting out, and in GIS (see Figure 1.9). With GPS, it is possible to obtain a position at any time anywhere by receiving and processing signals from orbiting satellites. This can be done with an accuracy at the metre level for GIS, but at the centimetre level for control surveys and setting out. Without computers, GPS would not be possible because the enormous amount of data a receiver generates needs to be processed in real time to be of use for

Figure 1.9 ● GPS receivers and data collectors available for GIS (courtesy Positioning Resources Ltd).

many survey applications. Because GPS is used extensively on site today, the engineering surveyor needs to study geodesy, which leads to a better understanding of the way in which GPS works. *Geodesy* is the study of the shape and size of the Earth and these are of some importance when defining position on a regional or global scale.

Although photogrammetry using film-based cameras and analytical plotters has been in use for a long time, a development that has revolutionised this is the use of digital cameras and plotters. *Digital photogrammetry* uses workstations for mapping in which digital images are displayed on a computer screen. These are integrated hardware and software systems for photogrammetric processing, data capture and analysis of softcopy images. They are particularly suitable for generating digital terrain models, orthophotography and perspective views for visualisation. *Digital orthophotos* look like conventional aerial photographs but they have the accuracy of a map. These are used extensively for GIS and in engineering to create fly-throughs and for showing as-built images at the design stage of a project.

Even though photogrammetry is still the preferred method for producing digital terrain models of large areas, other emerging technologies are challenging this. These include *LiDAR* and *IFSAR* (*InterFerometric Synthetic Aperture Radar*). LiDAR involves measuring the time it takes laser pulses to travel from an aircraft to the ground and back to determine elevation data. This is capable of capturing data at very fast rates at sub-metre accuracies. In contrast, IFSAR uses radar imagery obtained from an aircraft for producing terrain data. Both of these technologies have the advantage of being remote sensing techniques.

Another new technology that is gaining in popularity for engineering surveys is *laser scanning*. These instruments use a rotating laser distance and angle measuring system to automatically record the three-dimensional coordinates of the surface of an object. They are very good for producing as-built CAD models of complex structures, for calculating areas and volumes and for generating images for visualisations. Being a non-contact method of measurement, laser scanners can be used in areas where access is difficult or restricted for safety reasons.

Data collection and communications

For most surveying, data collection using pencil and paper is a thing of the past and a number of different methods of recording data electronically have evolved to replace these. Today, data collectors and controllers (handheld computers rather like a large calculator) have replaced pencil and paper. As shown in Figure 1.10, these are connected to a survey instrument and then programmed to ask the operator to record data and perform calculations. Field computers are also available, and these are laptop computers adapted to survey data collection. Data collectors, controllers and field computers are interactive devices but data can also be stored onto memory cards which are plugged into an instrument. These are similar to the compact flash cards used in digital cameras and most survey instruments will also have an internal memory capable of bulk storage. These are controlled using a microprocessor and keyboard mounted on the instrument. All of these methods enable large amounts of data to be stored much more easily than before and without these, most of the new surveying technologies would not be possible. For example, GPS, laser scanning and airborne terrain modelling all rely on enormous amounts of data being collected at very fast rates.

Advanced handheld or onboard computer technology makes it possible for surveyors to have as much computing power on site as in the office, and it provides the hardware needed to run the sophisticated applications and data collecting software now available in surveying. The hardware platform that is gaining in popularity is the *Windows CE* operating system, marketed by Microsoft. This provides a single platform from which it is possible to run several application programs and it allows different instruments to be used with the same data collector. Windows CE also provides Internet access on site. This makes it possible for data to be emailed to the office for further processing and it enables data held in the office to be accessed (useful if this has been left behind and is needed on site). Parallel developments in hardware and software have made it possible for survey systems such as GPS to be able to perform complex survey adjustments and compute coordinates in real time. These can then be used immediately for data collection in mapping or, with design data already stored in an instrument, for setting out.

Many survey instruments and data collectors now have large graphic screens that make it possible for the data collected to be edited, checked and verified in real time before it is stored or transmitted (see Figure 1.11). This can be done in full colour using touch screen technology.

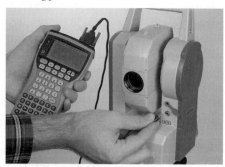

Figure 1.10 ● Connecting data recorder to total station (courtesy Pentax UK Ltd).

Figure 1.11 ● Large graphics screen on total station (courtesy Leica Geosystems).

Alongside these developments in data collection and storage, data communications have also improved dramatically in recent years. As mentioned above, with Internet access, data can be sent or received from the office while surveying on site. This has been made possible by improvements in mobile phone technology. The most recent development in data communication has been the introduction of short-range systems such as *Bluetooth* wireless technology. This new technology makes it possible for a computer to have a keyboard, mouse and Internet connection without using any cables. In surveying, wireless technology enables survey instruments and data collectors to communicate with each other and transfer data at very high speeds, also without using cables. This is especially useful for GPS equipment, where a high-speed cable-free data link between receiver and controller is essential and for other equipment when data is to be transferred between instruments away from the office without having to carry cables for this.

Developments in surveying technology

Taking account of all the above, major advances have been made in surveying technology in recent years in total stations, satellite and airborne surveying systems, in electronic data collection, in data communications as well as in applications and CAD software. Putting all of these together, the trend in surveying is definitely towards *integrated surveying*. An integrated surveying system will be made up of the following components, all of which will work together instead of separately:

● A surveying sensor (for example, total station, GPS receiver, laser scanner)

● Data collection hardware and software

● Data communications

● Processing and design software

In practical terms, integrated surveying means that all survey instruments (sensors) will be interchangeable and the data flow across all surveying disciplines will be seamless. The data for a control survey could be taken by GPS and then transferred directly on site to a total station (as in Figure 1.12) or to a laser scanner for mapping purposes. Several instruments and systems may be involved in collecting

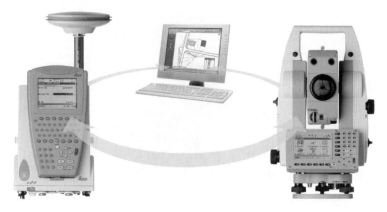

Figure 1.12 ● Data integration between total station and GPS (courtesy Leica Geosystems).

the data for a site survey, but each dataset will be combined by one software application to produce a map or DTM. The data generated by a CAD system for a project design will be in a format compatible with all the equipment and computer control systems used on site for setting out.

If we may predict the future in specific areas, it is expected that most sensor technology will continue to improve, especially laser scanning and those systems based on airborne mapping technologies. Some marked changes to satellite surveying systems are also planned, with the introduction of the European system known as Galileo and modifications to GPS signals. An area where much development is expected is in wireless communications, which will become commonplace for survey instruments. For long-range data transfer, which is especially important for GPS, data communication based on wireless mobile (cellular) phone technology will become increasingly important. Because of the capability of most survey sensors to produce huge amounts of data, data integration is an area where improvements are being planned and it is expected that an industry standard will be introduced allowing data to be transferred between the software supplied by different manufacturers. Proper data integration will enable surveyors and engineers to combine all kinds of information without experiencing complicated conversions and data loss. Another area where data integration will play an important role is in the upgrading of the spatial data in existing geographic information systems.

What will be the role of engineering surveyors in future?

Even though site surveying can involve using the most high-tech systems, the ability to observe and measure angles, distances and heights using fairly basic equipment and methods will always be required on site. All those engaged in engineering surveying must be able to use a conventional theodolite for measuring angles, or be able to use a steel tape for measuring distances and should be able to determine heights by levelling. The ability to use a calculator to process these observations by hand is also required.

During the construction phase of a project, most time in engineering surveying is devoted to setting out. This is the practical application of routine surveying techniques to construction and requires a knowledge and understanding of these. Setting out could be carried out using an optical theodolite and tape or the most sophisticated GPS methods might be used. Data could be calculated and recorded by hand in field books or by using a state-of-the-art field computer. The engineering surveyor needs to be conversant with all of these and appreciate the limitations they have as well as knowing the best application for each method when setting out. This knowledge will only be obtained through training and experience.

Beside the role of providing data on site for control surveys, mapping and the expertise required for setting out, engineering surveyors will be involved in all the developments currently taking place to improve data integration. The ultimate aims of these are to enable everyone working on a construction project to have access to all the spatial data needed on site for control surveys, mapping, setting out, design and quality control. This will only be possible when proper real-time data infrastructures have been developed. When these systems are finally on line, engineering surveyors will become data managers overseeing the seamless transfer of data between site and office from the start to the finish of a construction project.

What are the aims of this book?

As can be seen, engineering surveyors are expected to combine the traditional methods used on site with new IT based technologies. This requires a wider range of experience and ability than was needed in the past. Consequently, the main aims of *Surveying for Engineers* are not only to provide a thorough grounding in the basic surveying techniques required in civil engineering but also to make the reader aware of developments in engineering surveying technology and their impact. As noted earlier, the text concentrates on plane surveying because this is the norm for most engineering projects. However, some of the concepts of geodetic surveying are introduced to help the understanding of satellite surveying systems.

Since engineers use surveying as a means by which they can carry out their work, there is a limit to the surveying knowledge required by them beyond which specialist surveyors should be consulted to deal with unusual and complicated problems. Another aim of this book is to set this limit so that an engineer knows when to call in the specialist.

Although the methods involved in engineering surveying can be studied in textbooks, such is the practical nature of the subject that no amount of reading will turn a student into a competent engineering surveyor. Only by combining some hands-on surveying with a textbook such as *Surveying for Engineers* will a student become a useful engineering surveyor.

Reflective summary

With reference to engineering surveying, remember:

— Engineering surveying describes any survey work carried out in connection with civil engineering and building projects.

— Because most civil engineering projects are complex, it would not be possible to construct these to the required specifications and costs today without the experience and knowledge provided by engineering surveyors.

— The traditional role of the engineering surveyor has been to provide the survey data required for producing site plans and to provide the expertise required on site for all dimensional control (setting out, measurement of quantities, monitoring and so on).

— Nowadays, the engineering surveyor's responsibilities have changed, and as well as their traditional role they are seen to be data managers that oversee the continuous acquisition, processing and transfer of data between site and office.

— Because of the changing role of engineering surveyors, they are now expected to have some knowledge of other specialist areas in surveying, such as geodesy, photogrammetry, remote sensing, terrain modelling and visualisation, GIS and spatial data management.

— Engineering surveys are now carried out using a wide range of equipment and methods. It is the site engineer's responsibility to be aware of these and to know which instrument or method is best suited to each task on site.

— Surveying technology is changing at a very fast rate – another responsibility of the engineering and site engineer is to keep up-to-date with this.

— To become a competent engineering surveyor, it is necessary to acquire some practical experience on site – no amount of reading and study will achieve this.

— Although *Surveying for Engineers* is a textbook mostly concerned with engineering surveying, introductions to other specialist areas of surveying are included where appropriate.

1.2 Survey institutions and organisations

After studying this section you should be aware that there are many national and international institutions and organisations promoting surveying all of which provide high-quality sources of information through publications and the Internet.

This section includes the following topics:

- The Royal Institution of Chartered Surveyors (RICS)
- The Institution of Civil Engineering Surveyors (ICES)
- The Ordnance Survey
- The International Federation of Surveyors (FIG)
- Other UK organisations
- North American survey organisations

The Royal Institution of Chartered Surveyors (RICS)

This is the largest and oldest of the UK institutions that endorse surveying and has 110,000 members in 120 countries. The main aims of the RICS are to promote the knowledge and skills of its members and the services they offer, to maintain high standards of professional conduct and to ensure the continuing development of surveyors.

To represent the many specialist areas in surveying, the RICS is currently organised into 16 faculties. Each one of these has the responsibility for promoting, reviewing and updating the educational, training and professional standards of their area. They also play a significant role in providing information by sponsoring research and by organising seminars and conferences. One of their most important roles is to monitor the quality of the profession by making sure members follow RICS codes of practice.

Some of the faculties that are related to engineering surveying include Building Surveying, Construction, Environment, Minerals and Waste Management, Planning and Development, Project Management and, of course, the Geomatics faculty mentioned earlier.

The Geomatics faculty itself identifies many areas of specialist knowledge in surveying and those that involve engineering surveying include:

- Land and hydrographic surveying
- Mapping and positioning
- Global and local navigation systems
- Geographic information systems
- Cartography

- Photogrammetry and remote sensing
- Spatial data and metadata management, interpretation and manipulation
- Land, coastal and marine information management
- Ocean bed and resource surveys
- Monitoring of structures

As well as promoting surveying worldwide, the RICS has a library and also publishes promotional material, guidelines and a number of journals. One of the more important of these is *Geomatics World*, which is a bi-monthly journal published for the RICS that includes articles, technical advice, reports on seminars, lectures and conferences, book reviews as well as information on forthcoming events in surveying and mapping. A lot of information about the RICS is now available on their web site at http://www.rics.org/. The primary source of information on geomatics is the faculty web site based at http://www.rics.org/geo/.

The Institution of Civil Engineering Surveyors (ICES)

Founded in the UK in 1969 as the Association of Surveyors in Civil Engineering, the ICES was formed from this in 1972. This is a professional institution representing those employed as quantity and land surveyors in geospatial engineering and commercial management in the civil engineering industry. This institution aims to support, encourage, regulate and promote the interests of all its members and to support the profession by encouraging research, education and training.

In 1992, the ICES became an Associated Institution of the Institution of Civil Engineers (ICE) and formed the Geospatial Engineering Board (GEB) with them. The GEB aims to publicise engineering surveying knowledge and expertise within civil engineering in collaboration with the ICE through seminars and publications. One of the most important of these is the ICE design and practice guide *The management of setting out in construction*, which was published by the Joint Engineering Survey Board (which is now the GEB) in 1997. This guide is referred to throughout *Surveying for Engineers*.

Like the RICS, the ICES organises an extensive programme of meetings, workshops, lectures and conferences in order to promote surveying. In addition, they publish a monthly journal entitled the *Civil Engineering Surveyor* together with annual reviews on electronic surveying and GIS with information on the latest geospatial technology and instrumentation.

The web site for the ICES is http://www.ices.org.uk/.

The Ordnance Survey

The Ordnance Survey is the national mapping agency responsible for the official, definitive surveying and topographic mapping of Great Britain. As the importance of geographic information systems increases, it is also responsible for maintaining consistent national coverage of other geospatial datasets. The stated aims of the Ordnance Survey are to satisfy the national interest and customer need for accurate and readily available geospatial data and maps of the whole of Great Britain in the most effective and efficient way.

The Ordnance Survey is the most important source of geographic information in Great Britain and this is provided in an extensive range of products. The most familiar of these are paper maps such as the 1:50 000 Landranger and 1:25 000 Explorer maps from their range of leisure products. In contrast, OS MasterMap is a definitive digital map of Great Britain that has been produced for use with geographic information systems and other spatial databases. This includes topographic information on every landscape feature and represents a significant evolution from traditional cartography. Other products that are particularly useful in building and construction are Siteplan and Superplan, which can be supplied on paper or in a digital format at scales varying from 1:100 to 1:10 000.

All of the mapping products and services currently supplied by the Ordnance Survey are described in catalogues and other promotional literature as well as on their web site at http://www.ordnancesurvey.co.uk/. They are also described in Chapter 10, which deals with detail surveying and mapping.

Another very important service provided by the Ordnance Survey is the *National GPS Network* (http://www.gps.gov.uk/). According to the introduction given by the Ordnance Survey, this web site is an essential resource for the precise GPS user in Great Britain and also contains useful information for all users of GPS, both recreational and professional. It is intended for GPS-equipped surveyors (land, hydrographic, and engineering surveyors) and for GIS developers who work with Ordnance Survey mapping. Using the free services provided by it, GPS surveyors can obtain precise coordinates in the European standard GPS coordinate system ETRS89 and instantly convert these to British National Grid coordinates and heights above mean sea level (Newlyn datum). This is done by using Ordnance Survey high-precision transformation models. Surveyors and GIS developers can also convert two- and three-dimensional spatial datasets from GPS coordinates to Ordnance Survey coordinates and vice versa with high accuracy. The National GPS Network is described in more detail in Chapter 8, which explains how GPS coordinates are obtained as ETRS89 coordinates, what the Ordnance Survey National Grid and Newlyn datum are and how GPS coordinates are transformed to these.

Similar mapping and GPS services to those offered by the Ordnance Survey in Great Britain are also available from the Ordnance Survey of Northern Ireland (OSNI) and from the Ordnance Survey Ireland (OSi). Their web sites are http://www.osni.gov.uk/ and http://www.osi.ie/.

The International Federation of Surveyors (FIG)

The FIG (Fédération Internationale des Géomètres) is a federation of national associations and is the only international body that represents all surveying disciplines. As a non-governmental organisation, its main aim is to ensure that all who practise surveying meet the needs of their clients. This aim is realised by promoting all aspects of surveying and by encouraging the development of professional standards. What makes the FIG different to all the institutions and organisations that deal with surveying and geomatics, is that it carries out all of its activities through international collaboration.

At present, the FIG is made up of 10 commissions whose membership is drawn from surveyors all over the world. The commission of most interest here is *Commission 6: Engineering Surveys*, but others dealing with Spatial Information Management, Hydrography, Positioning and Measurement in addition to Construction Economics and Management are also of interest in civil engineering. The members of each commission serve a four-year term during which they meet regularly in working groups. These present their work through technical programmes and formal publications but working groups are also responsible for organising seminars and conferences on technical and professional topics. A special remit of each commission is that they try to assist the professional development of surveying in developing countries and those going through an economic transition.

To make known its activities, the FIG produces newsletters, information leaflets and a number of different publications, the most useful of which are the proceedings of their various working weeks, seminars and conferences. The biggest of these is the *FIG Congress*, which takes place once every four years.

For further information on the FIG, visit their web site on `http://www.fig.net/`.

Other UK organisations

- *The Survey Association (TSA)*, is a trade association that brings together businesses engaged in land and hydrographic surveying. Within the association, companies benefit by having members available to advise them on the best way to collect, interpret and apply spatial data. Their web site is `http://www.tsa-uk.org.uk/`.

- The *Remote Sensing and Photogrammetric Society* (`http://www.rspsoc.org/`) publish a number of journals that include many articles related to engineering surveying.

- The *Association for Geographic Information* (`http://www.agi.org.uk/`) represents the interests of the geographic information industry including users in both the public and private sectors, suppliers of software, hardware and data services, consultants and academics.

- The *Hydrographic Society* (`http://www.hydrographicsociety.org/`) aims to promote the science of surveying over water. Members, both individual and institutional, represent the fields of hydrography, oceanography, geophysics, civil engineering and associated disciplines at all levels of expertise.

- The *Institution of Civil Engineers* (`http://www.ice.org.uk/`), although promoting civil engineering across all fields of interest, also provides services and information related to engineering surveying and geospatial engineering.

North American survey organisations

Some of the professional survey organisations in the USA provide very useful information for surveyors and engineers engaged in engineering surveying. Some of these, with their web sites, are listed below.

- *American Congress on Surveying and Mapping* (http://www.survmap.org/).
- *American Society for Photogrammetry and Remote Sensing* (http://www.asprs.org/).
- The *Geomatics Division* of the *American Society of Civil Engineers (ASCE)* publishes a number of journals including the *Journal of Surveying Engineering*. This can be searched on http://www.pubs.asce.org/.

Reflective summary

With reference to survey institutions and organisations, remember:

— There are many societies and organisations that are useful sources of survey information – these can all be accessed on the Internet.

— The largest surveying institution in the UK is the RICS. It addresses all aspects of surveying including such diverse areas as arts, antiques and waste management.

— The geomatics faculty of the RICS is responsible for promoting engineering surveying and all of its allied subjects.

— The ICES is an institution that represents those employed as quantity and engineering surveyors in the civil engineering industry.

— Both the RICS and ICES promote surveying through meetings, lectures, seminars and conferences as well as publications such as their regular journals *Geomatics World* and *Civil Engineering Surveyor*.

— The Ordnance Survey is the national mapping agency in the UK. It is responsible for the topographic mapping of Great Britain and for providing services such as the National GPS Network that help realise this.

— Ordnance Survey products include paper maps and plots but also include an extensive range of digital mapping products – many of these are used in engineering surveying.

— The FIG is a non-governmental international organisation representing all surveying disciplines – their Commission 6 is responsible for engineering surveying.

Exercises

1.1 What are geospatial engineering and geomatics?

1.2 Using information on the RICS web site, make a list of all the specialist areas of surveying given by their Geomatics faculty. Which ones are relevant to engineering surveying?

1.3 Using the definition given on the ICES web site, explain what the main aims of engineering surveying are.

1.4 What are photogrammetry and remote sensing? Why are they important in engineering surveying?

1.5 Explain how plane surveying differs from geodetic surveying.

1.6 In engineering surveying, what do the following terms mean?

horizontal and vertical control
topographic survey
digital terrain model
setting out

1.7 Describe the different methods that are used in surveying today for electronic data collection. What impact will improved data communications have on these?

1.8 What is integrated surveying?

1.9 Discuss the ways in which engineering surveying is expected to develop in the near future. Why will engineering surveyors become data managers?

1.10 Visit the Ordnance Survey web site and make a list of all the products and services they offer. Which of these are the most useful for civil engineering and construction projects?

1.11 Give details of the way in which the FIG promotes surveying. Using their web site, make a list of recent symposia and conferences they have organised for those interested in engineering surveying.

1.12 By obtaining information from web sites, describe the activities of the ACSM and ASCE.

Further reading and sources of information

For convenience, all of the institutions and organisations mentioned in this chapter are listed below with their web sites.

- Royal Institution of Chartered Surveyors (RICS):
  http://www.rics.org/

- Institution of Civil Engineering Surveyors (ICES):
  http://www.ices.org.uk/

- Ordnance Survey:
  http://www.ordnancesurvey.co.uk/

- International Federation of Surveyors (FIG):
  http://www.fig.net/
- The Survey Association (TSA):
  http://www.tsa-uk.org.uk/
- Remote Sensing and Photogrammetric Society:
  http://www.rspsoc.org/
- Association for Geographic Information (AGI):
  http://www.agi.org.uk/
- Hydrographic Society:
  http://www.hydrographicsociety.org/
- Institution of Civil Engineers:
  http://www.ice.org.uk/
- American Congress on Surveying and Mapping:
  http://www.survmap.org/
- American Society for Photogrammetry and Remote Sensing:
  http://www.asprs.org/
- American Society of Civil Engineers:
  http://www.asce.org/

The following books offer introductions to the specialist subjects that are included in the definition of engineering surveying:

Egels, Y. and Kasser, M. (2001) *Digital Photogrammetry*. Taylor & Francis, London.
Heywood, I., Cornelius, S. and Carver, S. (2002) *An Introduction to Geographical Information Systems*. Pearson, London.
Lillesand, T. M., Kiefer, R. W. and Chipman, J. (2003) *Remote Sensing and Image Interpretation*. John Wiley & Sons, Chichester.
Slocum, T. A., McMaster, R. B., Kessler, F. C. and Howard, H. H. (2004) *Thematic Cartography and Geographic Visualization*. Prentice Hall, London.
Wolf, P. R. (2000) *Elements of Photogrammetry with Applications in GIS*. McGraw-Hill, London.

The survey journals published by the RICS and ICES are:

- *Geomatics World*, which is published bi-monthly for the RICS by GITC in the Netherlands. GITC also publish a range of surveying journals including the *Engineering Surveying Showcase*, which is a twice yearly look at the UK's surveying industry and *GIM International*. For further information on Geomatics World and GIM International, visit http://www.gitc.nl/ and for information on the Engineering Surveying Showcase, visit http://www.pvpubs.com/.
- The *Civil Engineering Surveyor* is published monthly by the ICES. It also publishes two annual supplements to this known as the *Supplement on Electronic Surveying* and the *GIS/GPS Supplement*. For further information on these see the ICES web site.

The following document is essential reading for those engaged in surveying:

RICS (2002) *Surveying Safely: A commitment to personal safety*. RICS Construction Faculty. Also available on http://www.rics.org/.

chapter two

Levelling

 ## Aims

After studying this chapter you should be able to:

- Discuss the various types of datum and bench marks that can be used in levelling including Ordnance Datum Newlyn (ODN)
- Describe how automatic, digital and tilting levels work
- Describe the field procedures that are used for determining heights when levelling
- Perform all the necessary calculations and checks for determining heights by levelling including an assessment of the quality of the results obtained
- Appreciate that levelling is subject to many sources of error and that it is possible to manage these
- Outline some methods used in levelling to obtain heights at difficult locations

This chapter contains the following sections:

2.1   Heights, datums and bench marks

2.2   Levelling equipment

2.3   Field procedure for levelling

2.4   Calculating reduced levels

2.5   Precision of levelling

2.6   Sources of error in levelling

2.7   Other levelling methods

Exercises

Further reading and sources of information

2.1 Heights, datums and bench marks

After studying this section you should understand the differences between horizontal and vertical lines or surfaces and why these are important in levelling. You should know what a levelling datum is and be aware that a national levelling datum has been set up by the Ordnance Survey and is known as Ordnance Datum Newlyn (ODN). You should be aware that bench marks are points of known height specified on a chosen datum and that these can either be Temporary Bench Marks (TBMs) or Ordnance Survey Bench Marks (OSBMs). In addition, you should appreciate the difficulties of using an OSBM for any levelling and why these are being replaced by GPS bench marks.

Levelling and how heights are defined

In surveying, three basic quantities are measured – heights, angles and distances – levelling is the name given to one of the methods available for determining heights.

When levelling, it is possible to measure heights within a few millimetres and this order of precision is more than adequate for height measurement on the majority of civil engineering projects. As well as levelling, it is worth noting that heights can also be measured by using total stations, handheld laser distance meters and GPS – these are described in subsequent chapters of the book. In comparison to these, levelling offers a versatile yet simple, accurate and inexpensive field procedure for measuring heights and this is the reason for its continued use on construction sites in competition with other methods.

The equipment required to carry out levelling is an optical, digital or laser level. This chapter deals with optical and digital levelling – the techniques used with laser levels to determine and process heights on site are given in Chapter 11 (this covers setting out).

All methods of height measurement determine the heights of points above (or below) an agreed datum. On site or in the office, surveyors, builders and engineers all use, on a daily basis, horizontal and vertical datums as references for all types of measurement including levelling. To illustrate what is meant by horizontal and vertical, Figure 2.1 shows a plumb-bob (a weight on a length of string or cord) suspended over a point P. The direction of gravity along the plumb-line defines the vertical at P and a horizontal line is a line taken at right angles to this. Any horizontal line can be chosen as a datum and the height of a point is measured along a vertical above or below the chosen horizontal line. On most survey and construction sites, a permanent feature of some sort is usually chosen as a datum for levelling and this is given an arbitrary height to suit site conditions. The horizontal line or surface passing through this feature, with its assigned height, then becomes the levelling

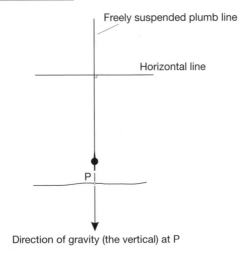

Figure 2.1 ● Horizontal line.

datum. Although it may seem logical to assign a height of 0 m to such a datum, a value often used is 100 m and this is chosen to avoid any negative heights occurring as these can lead to mistakes if the minus sign was omitted for some reason. The heights of points relative to a datum are known as *reduced levels*.

Any permanent reference point which has an arbitrary height assigned to it or has had its height accurately determined by levelling is known as a *bench mark*. For most surveys and construction work, it is usual to establish the heights of several bench marks throughout a site and if these have heights based on an arbitrary datum, they are known as *Temporary Bench Marks* (*TBMs*).

The positions of TBMs have to be carefully chosen to suit site conditions and various suggestions for the construction of these are given in Chapter 11 and in BS 5964:1990 *Building Setting Out and Measurement*.

The definition of a levelling datum given above is a horizontal or level line or surface that is always at right angles to the direction of gravity. As might be expected, the direction of gravity is generally towards the centre of the Earth and over large areas, because the Earth is curved, a level surface will become curved as shown in Figure 2.2. On this diagram, the height of A above B is measured along a vertical between the level surfaces through A and B.

If heights are to be based on the same datum for the whole of a large area such as the UK, a curved level surface of zero height has to be defined. For mainland Great Britain, this has been established by the Ordnance Survey and is known as the Ordnance Datum Newlyn (ODN) vertical coordinate system. This corresponds to the mean sea level measured at Newlyn, Cornwall and heights which refer to this partic-ular level surface as zero height are known as ODN heights, but are often called heights above mean sea level. Mean sea level is represented by a surface known as the *Geoid* which is the level surface to which all height measurements are referenced, whether these are national or local. For more information about Ordnance Datum Newlyn, the Geoid and what mean sea level represents, refer to Chapter 8.

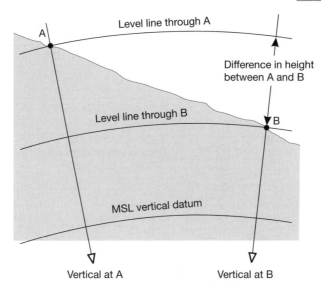

Figure 2.2 ● Level surfaces showing difference of height between two points A and B.

All heights and contours marked on Ordnance Survey maps and plans covering mainland Great Britain are ODN heights and across the country, the Ordnance Survey have established, by levelling, about seven hundred thousand bench marks, known as Ordnance Survey Bench Marks (OSBMs) all of which have a quoted ODN height. The most common of these are cut into stone or brick at the base of buildings as shown in Figure 2.3 and the positions and heights of these are shown on Ordnance Survey plans at scales of 1:1250 and 1:2500. Bench mark lists, which can be purchased from the Ordnance Survey or approved agents, give map references, a description and also height values for these bench marks.

Some caution must be exercised when using OSBMs as they have not been maintained by the Ordnance Survey since 1989. In fact, many of these have not had their heights revised since the 1970s and they may have been affected by local subsidence or some other physical disturbance since the date they were last levelled. For these reasons, if the heights given on maps or in bench mark lists are to be used for any survey work, it is essential to include two or more OSBMs in levelling schemes so that their height values can be checked for errors by levelling between them. The Ordnance Survey have not been maintaining OSBMs for a number of years because ODN heights of reference points are now determined using GPS methods. These enable the height of any point at almost any location to be determined so there is now no need to visit an Ordnance Survey bench mark in order to obtain heights relative to mean sea level. The use of GPS in height measurement is described in Chapters 7 and 8.

Despite the existence of ODN heights and GPS means of realising these, it is not always necessary to use a mean sea level datum and many construction projects use an arbitrary datum for defining heights.

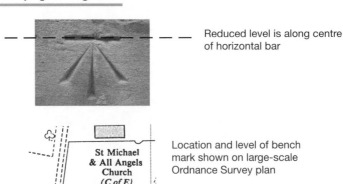

— — — — — Reduced level is along centre of horizontal bar

Location and level of bench mark shown on large-scale Ordnance Survey plan

Staff in place for taking reading on bench mark

Figure 2.3 ● Ordnance Survey Bench Mark (part reproduced by permission of Ordnance Survey on behalf of the Controller of Her Majesty's Stationery Office, © Crown Copyright 100024463).

Reflective summary

With reference to datums and bench marks, remember:

— The Ordnance Survey have established a national vertical datum at mean sea level in mainland Great Britain called Ordnance Datum Newlyn (ODN) – if heights are to be based on this an OSBM has to be used.

— Be careful when using OSBMs as they have not been maintained by the Ordnance Survey for many years – if you need to have ODN heights on site you should use GPS to do this.

— For most construction work and civil engineering projects, levelling is based on an arbitrary datum and involves using TBMs.

2.2   Levelling equipment

After studying this section you should be able to describe the differences between tilting, automatic and digital levels. In practical terms, you should know how to set up and use a level, how to handle and read a levelling staff and how to carry out a two-peg test on a level. As well as this, you should understand the operation of a survey telescope, what parallax is and how to remove this.

This section includes the following topics:

- The levelling staff
- Automatic levels
- Tilting levels
- Digital levels
- Adjustment of the level

The levelling staff

Levelling involves the measurement of vertical distance with reference to a horizontal plane or surface. To do this, a levelling staff is needed to measure vertical distances and an instrument known as a level is required to define the horizontal plane, both of which are shown in Figure 2.4.

A *levelling staff* is the equivalent of a long ruler and it enables distances to be measured vertically from the horizontal plane established by a level to points where heights are required. Many types of staff are in current use and these can have

Figure 2.4 ● Levelling equipment (courtesy Leica Geosystems).

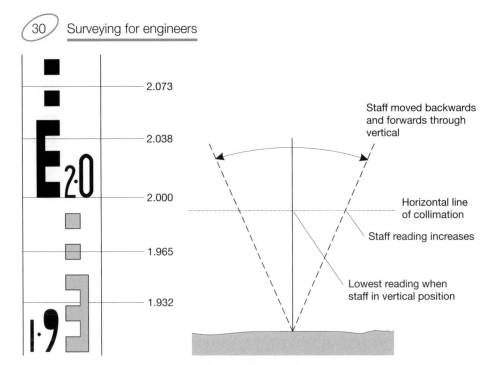

Figure 2.5 ● Level staff with example readings   **Figure 2.6** ● Taking vertical reading on staff

lengths of up to 5 m. The staff is usually telescopic but can be socketed in as many as five sections for ease of carrying and use and it is made of aluminium or non-conductive fibreglass. The staff markings can take various forms but the E-type staff face recommended in BS 4484 is used in the UK. This is shown in Figure 2.5 and can be read directly to 0.01 m and by estimation to 0.001 m.

Since the staff must be held vertically at each point where a height is to be measured most staffs are fitted with a circular bubble to help do this. If no bubble is available, the staff should be *slowly* moved back and forth through the vertical and the lowest reading noted – this will be the reading when the staff is vertical, as shown in Figure 2.6.

Automatic levels

The general features of the automatic level are shown in Figure 2.7. These instruments establish a horizontal plane at each point where they are set up and consist of a telescope and compensator. The *telescope* provides a magnified line of sight for taking measurements and the *compensator*, built into the telescope, ensures that the line of sight viewed through the telescope is horizontal even if the optical axis of the telescope is not exactly horizontal.

Surveying telescopes

The type of telescope used in automatic and other levels is very similar to that used in other surveying instruments such as theodolites and total stations and is shown in

Figure 2.7 ● Automatic level: 1 focusing screw; 2 eyepiece; 3 footscrew; 4 horizontal circle; 5 base plate; 6 tangent screw (slow motion screw); 7 circular bubble; 8 collimator (sight); 9 object lens (courtesy Trimble Navigation Ltd and Topcon).

Figure 2.8. The following section refers mostly to the telescope of a level but is also applicable to measurements taken using other surveying instruments.

When looking through the *eyepiece* of the telescope, a set of lines can be seen in the field of view and these provide a reference against which measurements are taken. This part of the telescope is called the *diaphragm* (or *reticule*) and this consists of a circle of plane glass upon which lines are etched, as shown in Figure 2.9. Conventionally, the pattern of vertical and horizontal lines is called the *cross hairs*.

The *object lens*, *focusing lens*, diaphragm and eyepiece are all mounted on the same optical axis and the imaginary line passing through the centre of the cross hairs and the optical centre of the object lens is called the *line of sight* or the *line of collimation* of the telescope. The diaphragm is held in the telescope by means of four adjusting screws so that the position of the line of sight can be moved within the telescope (see *Adjustment of the level*, p. 39).

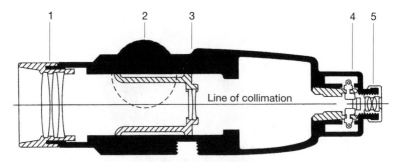

Line of collimation

Figure 2.8 ● Surveying telescope: 1 object lens; 2 focusing screw; 3 focusing lens; 4 diaphragm; 5 eyepiece (courtesy Sokkia Ltd).

Figure 2.9 ● Diaphragm patterns.

The action of the telescope is as follows. Light rays from the levelling staff (or target) pass through the object lens and are brought to a focus in the plane of the diaphragm by rotating the focusing screw. Rotating the focusing screw moves the focusing lens along the axis of the telescope. When the eyepiece is rotated, this also moves axially along the telescope and since it has a fixed focal point that lies outside the lens combination, its focal point can also be made to coincide with the plane of the diaphragm. Since the image of the levelling staff has already been focused on the diaphragm, an observer will see in the field of view of the telescope the levelling staff focused against the cross hairs. The image of the staff will also be highly magnified (see Figure 2.10) making accurate measurement of vertical distances possible over long distances.

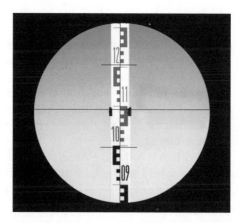

Figure 2.10 ● View of levelling staff through telescope (courtesy Leica Geosystems).

A problem often encountered with outdoor optical instruments is water and dust penetration. In order to provide protection from these, the telescope and compensator compartment of some levels are sealed and filled, under pressure, with dry nitrogen gas. This is known as *nitrogen purging*, and since the gas is pressurised, water and dust are prevented from entering the telescope. The use of dry nitrogen also prevents lens clouding and moisture condensation inside the telescope.

Typical specifications for a surveying telescope for use in construction work are a magnification of up to about 30, a field of view of between 1 and 2° and a minimum focusing distance of 0.5–1.0 m. Some telescopes use *autofocusing*, where focusing is achieved by pressing a button in a similar manner to a camera (see also Section 5.3 under *Useful accessories* and Figure 5.19).

Parallax

For a surveying telescope to work correctly, the focusing screw has to be adjusted so that the image of the staff falls exactly in the plane of the diaphragm and the eyepiece must be adjusted so that its focal point is also in the plane of the diaphragm. Failure to achieve either of these settings results in a condition called *parallax*, and this can be a source of error when using levels, theodolites and total stations. Parallax can be detected by moving the eye to different parts of the eyepiece when viewing a levelling staff – if different parts of the staff appear against the cross hairs the telescope has not been properly focused and parallax is present, as shown in Figure 2.11.

It is difficult to take accurate staff readings under these conditions, since the position of the cross hairs alters for different positions of the eye. For these reasons, *it is essential that parallax is removed* before any readings are taken when using a level or any optical instrument with an adjustable eyepiece.

To remove parallax, a piece of white paper or page from a field book is held in front of the objective and the eyepiece is adjusted so the cross hairs are in focus. The paper or field book is removed from in front of the objective, and the staff at which readings

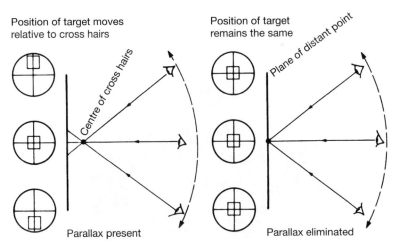

Figure 2.11 ● Parallax.

are required is now sighted and brought into focus using the focusing screw. Next, the staff is observed whilst moving the eye up and down, and if it does not appear to move relative to the cross hairs then parallax has been eliminated (see Figure 2.11). If there is apparent movement then the procedure should be repeated. Once adjusted, it is not usually necessary to adjust the eyepiece again until a new set of readings is taken, say, on another day. For all levelling, the focusing screw has to be adjusted for each staff reading, as focus depends on the sighting distance.

The compensator

In an automatic level, the function of the compensator is to deviate a horizontal ray of light at the optical centre of the object lens through the centre of the cross hairs. This ensures that the line of sight (or collimation) viewed through the telescope is horizontal even if the telescope is tilted.

Whatever type of automatic level is used it must be levelled to within about 15' of the vertical through the level to allow the compensator to work. This is achieved by using the three footscrews together with the circular bubble.

Figure 2.12 shows a compensator and the position in which it is usually mounted in the telescope. The action of the compensator is shown in Figure 2.13, which has been exaggerated for clarity. The main component of the compensator is a prism which is assumed to be freely suspended within the telescope tube when the instrument has been levelled and which takes up a position under the influence of gravity according to the angle of tilt of the telescope. Provided the tilt is within the working range of the compensator, the prism moves to a position to counteract this and a horizontal line of sight (collimation) is always observed at the centre of the cross hairs.

The wires used to suspend a compensator are made of a special alloy to ensure stability and flexibility under rapidly changing atmospheric conditions, vibration and shock. The compensator is also screened against magnetic fields and uses some form of damping, otherwise it might be damaged when the level is in transit and might be affected by wind and vibration preventing readings from being taken.

Figure 2.12 ● Compensator (courtesy Sokkia Ltd).

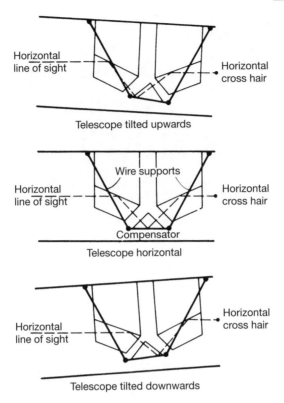

Horizontal line of sight

Horizontal cross hair

Telescope tilted upwards

Wire supports

Horizontal line of sight

Horizontal cross hair

Compensator

Telescope horizontal

Horizontal cross hair

Horizontal line of sight

Telescope tilted downwards

Figure 2.13 ● Action of compensator.

Use of the automatic level

The first part of the levelling process is to set the tripod in position for the first reading, ensuring that the top of the tripod is levelled by eye after the tripod legs have been pushed firmly into the ground. Following this, the level is attached to the tripod using the clamp provided and the circular bubble is centralised using the three footscrews.

When an automatic level has been set up and levelled in this way, the compensator automatically moves to a position to establish a horizontal line of sight at the centre of the cross hairs. Therefore, at each set up, no further levelling is required after the circular bubble has been set.

As with all types of level, parallax must be removed before any readings are taken.

In addition to the levelling procedure and parallax removal, a test should be made to see if the compensator is working before readings commence. One of the levelling footscrews should be moved slightly off level and, if the reading to the levelling staff remains constant, the compensator is working. If the reading changes, it may be necessary to gently tap the telescope tube to free the compensator. When using some automatic levels, this procedure does not have to be followed as a small lever is attached to the level which is pressed when a staff has been sighted (see Figure 2.14).

Figure 2.14 ● Compensator check (courtesy Leica Geosystems).

If the compensator is working this time, the horizontal cross hair is seen to move and then return immediately to the horizontal line of sight. Some levels incorporate a warning device that gives a visual indication to an observer, in the field of view of the telescope, when the instrument is not level.

A problem sometimes encountered with levels that use a compensator is that machinery operating nearby will cause the compensator to vibrate, which in turn causes the image of the staff to appear to vibrate so that readings become very difficult to take. This problem is sometimes encountered on construction sites, particularly where the site is narrow or constricted.

Tilting levels

Because of the popularity of the automatic and digital levels, the tilting level is rarely used on site these days. However, brief details are included here so that it can be compared with the others.

On this instrument, the telescope is not fixed to the base of the level and can be tilted a small amount in the vertical plane about a pivot placed below the telescope. The amount of tilt is controlled by a *tilting screw* which is usually directly underneath or next to the telescope eyepiece. Instead of a compensator, a tilting level will have a *spirit level* tube fixed to its telescope to enable a horizontal line of sight to be set. The spirit level tube (Figure 2.15) is a short barrel-shaped glass tube, sealed at both ends, that is partially filled with purified synthetic alcohol. The remaining space in the tube is an air bubble and there are a series of graduations marked on the glass top of the tube that are used to locate the relative position of the air bubble within the spirit level. The imaginary tangent to the surface of the glass tube at the centre of these graduations is known as the axis of the spirit level. When the bubble is centred in its run and takes up a position with its ends an equal number of graduations (or divisions) either side of the centre, the axis of the spirit level will be horizontal, as shown in Figure 2.15. By attaching a spirit level to a telescope such that its axis is parallel to the line of collimation, a horizontal line of sight can be set. This is achieved by

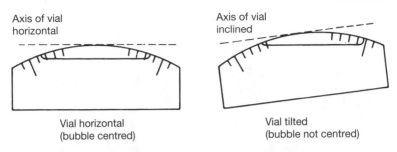

Figure 2.15 ● Spirit level tube (or vial).

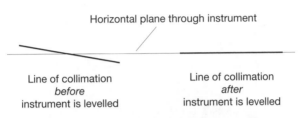

Figure 2.16 ● Principle of the tilting level.

adjusting the inclination of the telescope with the tilting screw until the bubble of the spirit level lies in the middle of its graduations.

In use, a tilting level is set up by attaching it to the tripod head and using the footscrews to centralise a circular bubble. As with the automatic level, this ensures that the instrument is almost level. Next, the telescope is turned until it is pointing in the direction in which the first staff reading is required and the tilting screw is rotated until the spirit level bubble is brought to the centre of its run. This ensures that the line of collimation is horizontal, as shown in Figure 2.16, but only in the direction in which the reading is being taken. When the telescope is rotated to other directions, the bubble will change its position for each direction of the telescope because the instrument is not exactly levelled. Consequently, the tilting screw must be reset before every reading is taken.

Digital levels

Shown in Figure 2.17, the digital level is similar in appearance to an automatic or tilting level and the same features can be identified including the telescope with focus and eyepiece, footscrews with a base plate and so on. In use, it is set up in the same way as an automatic level by attaching it to a tripod and centralising a circular bubble using the footscrews. A horizontal line of sight is then established by a compensator and readings could be taken in the same way as with an automatic level to a levelling staff, where all readings are taken and recorded manually.

However, this instrument has been designed to carry out all reading and data processing automatically via an on-board computer which is accessed through a

Figure 2.17 ● Digital levels (courtesy Topcon and Leica Geosystems).

display and keyboard. When levelling, a special bar-coded staff is sighted (see Figure 2.18), the focus is adjusted and a measuring key is pressed. There is no need to read the staff as the display will show the staff reading about two or three seconds after the measuring key has been pressed. When the bar-coded staff is sighted, it is interrogated by the level over a short span of about 500 mm using electronic image-processing techniques to produce a bar-coded image of the staff corresponding to the field of view of the telescope. The captured image is then compared by the on-board computer to the bar codes stored in the memory for the staff and when a match is found, this is the displayed staff reading. In addition to staff readings, it is also possible to display the horizontal distance to the staff with a precision of about 20–25 mm. All readings can be coded using the keyboard and as levelling proceeds, each staff reading and subsequently all calculations are stored in the level's internal memory.

In good conditions, a digital level has a range of about 100 m, but this can deteriorate if the staff is not brightly and evenly illuminated along its scanned section. The power supply for the digital level is standard AA or rechargeable batteries, which are capable of providing enough power for a complete day's levelling. If it is not possible to take electronic staff readings (because of poor lighting, obstructions such as foliage preventing a bar code from being imaged or loss of battery power), the reverse side of the bar-coded staff has a normal E type face and optical readings can be taken and entered manually into the instrument instead.

Figure 2.18 ● Bar-coded levelling staff (courtesy Leica Geosystems).

The digital level has many advantages over conventional levels since observations are taken quickly over longer distances without the need to read a staff or record anything by hand. This eliminates two of the worst sources of error from levelling – reading the staff incorrectly and writing the wrong value for a reading in the field book. As a digital level also calculates all the heights required, another source of error is removed from the levelling process – the possibility of making mistakes in calculations.

The data stored in a digital level can also be transferred to a removable memory card and then to a computer, where it can be processed further and permanently filed if required. To do this, the software provided by each manufacturer with the level can be used. As well as this, data can be transferred from the memory card directly into one of the many survey and design software packages now available. For on-site applications, design information can also be uploaded from a computer and memory card to the level for setting out purposes. All of the various options available for electronic data collection and processing are described in Section 5.4 for total stations.

Adjustment of the level

Whenever a level is set up, it is essential that the line of collimation, as viewed through the eyepiece, is horizontal. So far, the assumption has been made that once the circular bubble is centralised with the footscrews, the line of collimation is set exactly horizontal by the compensator and diaphragm (automatic and digital levels) or by centralising the bubble in the spirit level tube (tilting levels). However, because they are in constant use on site, most levels are not in perfect adjustment and if horizontal readings are not being taken when it has been set up properly, a *collimation error* is present in the level. Since most levels will have a collimation error, some method is required to check this to determine if the error is within accepted limits.

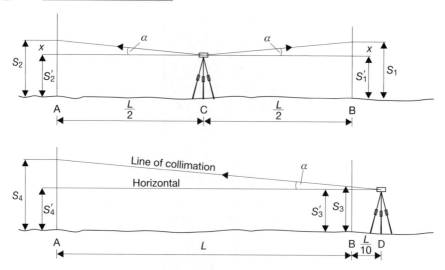

Figure 2.19 ● Two-peg test.

This is known as the *two-peg test* which should be carried out when using a new or different level for the first time and at regular intervals after this depending on how much the level has been used. Sometimes, the contract for a construction project will specify when the two-peg test should be carried out (say weekly) and will also specify a tolerance for the collimation error – because of this it is necessary to check all levels regularly when on site.

Referring to Figure 2.19, a two-peg test is carried out as follows.

● On fairly level ground, two points A and B are marked a distance of L m apart. In soft ground, two pegs are used and on hard surfaces, nails or paint can be used.

● The level is set up midway between the points at C and levelled. A levelling staff is placed at A and B in turn and staff readings S_1 (at B) and S_2 (at A) are taken.

● The two readings are

$$S_1 = (S_1' + x) \text{ and } S_2 = (S_2' + x)$$

where

S_1' and S_2' are the staff readings that would have been obtained if the line of collimation was horizontal

x is the error in each reading due to the collimation error, the effect of which is to tilt the line of sight by angle α.

● Although Figure 2.19 shows α and the line of collimation to be above the horizontal, it can be above or below. Since AC = CB, the error x in the readings S_1 and S_2 will be the same and the difference between readings S_1 and S_2 gives $S_1 - S_2 = (S_1' + x) - (S_2' + x) = S_1' - S_2'$, which gives the *true difference in height* between A and B. This demonstrates that if a collimation error is present in a level, the effect of this cancels out when height differences are computed provided readings are taken over equal sighting distances.

- The level is moved so that it is $\frac{L}{10}$ m from point B at D and readings S_3 and S_4 are taken (see Figure 2.19). The difference between readings S_3 and S_4 gives $S_3 - S_4 =$ the *apparent difference in height* between A and B.

- If the level is in perfect adjustment $(S_1 - S_2) = (S_3 - S_4)$. However, it usual that there is a difference between the true and apparent heights and since this has been measured over a distance of L m the collimation error for the level is given by

$$\text{collimation error } e = (S_1 - S_2) - (S_3 - S_4) \text{ per } L \text{ m} \tag{2.1}$$

- If the collimation error is found to be less than about ± 1 mm per 20 m (or some specified value) the level is assumed to be in adjustment.

- If the collimation error is found to be greater than about ± 1 mm per 20 m (or some specified value), the level has to be adjusted. To do this with the level still at point D, the horizontal reading S_4' that should be obtained at A is computed. The adjustment is carried out by a number of different methods. For automatic and digital levels, the diaphragm adjusting screws are loosened and the reticule is moved until reading S_4' is obtained. For some levels, the compensator has to be adjusted. A tilting level is adjusted by first turning the tilting screw to obtain S_4'. This causes the spirit level bubble to move from the centre of its run, so it is brought back to the centre by adjusting the vial. All of the mechanical adjustments described here for adjusting a level for collimation error are very difficult to do, especially on site, and if a level has an unacceptable collimation error, *it should be adjusted by a trained technician preferably under laboratory conditions*. This usually means returning it to the manufacturer.

Worked example 2.1: Two-peg test

Question

The readings obtained from a two-peg test carried out on an automatic level with a staff placed on two pegs A and B 50 m apart are:

- With the level midway between A and B
  Staff reading at A = 1.283 m   Staff reading at B = 0.860 m
- With the level positioned 5 m from peg B on line AB extended
  Staff reading at A = 1.612 m   Staff reading at B = 1.219 m

Calculate the collimation error of the level per 50 m sighting distance and the horizontal reading that should be observed on the staff at A with the level in position 5 m from B.

Solution

When solving problems of this type it is important that the numbering sequence shown in Figure 2.19 for staff readings is used otherwise an incorrect collimation error will be computed. In this case, the staff readings are identified as

$$S_1 = 0.860 \text{ m} \qquad S_2 = 1.283 \text{ m} \qquad S_3 = 1.219 \text{ m} \qquad S_4 = 1.612 \text{ m}$$

and from Equation (2.1):

$$\text{collimation error } e = (0.860 - 1.283) - (1.219 - 1.612) \text{ per } 50 \text{ m}$$
$$= (-0.423 - (-0.393)) = \textbf{-0.030 m per 50 m}$$

For the instrument in position 5 m from peg B, the horizontal reading that should have been obtained with the staff at A is

$$S_4' = S_4 - [\text{collimation error} \times \text{sighting distance DA}]$$

$$= 1.612 - \left[-\frac{0.030}{50}\right]55 = \textbf{1.645 m}$$

This is checked by computing S_3', where

$$S_3' = S_3 - [\text{collimation error} \times \text{sighting distance DB}]$$

$$= 1.219 - \left[-\frac{0.030}{50}\right]5 = 1.222 \text{ m}$$

and

$$S_3' - S_4' = 1.222 - 1.645 = -0.423 = S_1 - S_2 (\text{checks})$$

Reflective summary

With reference to levelling equipment, remember:

— Three different types of level are available – automatic, digital and tilting levels.

— The most widely used level on site is the automatic level but the digital level is gaining in popularity even though it is more expensive.

— All levels incorporate similar telescopes and to avoid errors occurring in staff readings, it is important that parallax is removed from a telescope before any readings are taken.

— Whatever type of level is used it must be checked for collimation error regularly.

2.3   Field procedure for levelling

After studying this section you should understand how height differences can be determined from staff readings and you should be familiar with the terms back sight, fore sight and intermediate sight, together with rise and fall in a levelling context. In addition, you should appreciate why it is important to start and finish all levelling at a bench mark.

How levelling is carried out

When a level has been correctly set up, the line or plane of collimation generated by the instrument coincides with or is very close to a horizontal plane. If the height of this plane is known, the heights of ground points can be found from it by reading a vertically held levelling staff.

In Figure 2.20, a level has been set up at point I_1 and readings R_1 and R_2 have been taken with the staff placed vertically in turn at ground points A and B. If the reduced level of A (RL_A) is known then, by adding staff reading R_1 to RL_A, the reduced level of the line of collimation at instrument position I_1 is obtained. This is known as the *height of the plane of collimation (HPC)* or the *collimation level*. This is given by

collimation level at $I_1 = RL_A + R_1$

In order to obtain the reduced level of point B (RL_B), staff reading R_2 must be subtracted from the collimation level to give

RL_B = collimation level − R_2 = ($RL_A + R_1$) − R_2 = $RL_A + (R_1 - R_2)$

The direction of levelling in this case is from A to B and R_1 is taken with the level facing in the opposite direction to this. For this reason it is known as a *back sight* (BS). Since reading R_2 is taken with the level facing in the direction from A to B, it is called a *fore sight* (FS). The height change between A and B, both in magnitude and sign, is given by the difference of the staff readings taken at A and B. Since R_1 is greater than R_2 in this case, ($R_1 - R_2$) is positive and the base of the staff has risen in moving from A to B. Because ($R_1 - R_2$) is positive it is known as a *rise*.

The level is now moved to a new position I_2 so that the reduced level of C can be found. Reading R_3 is first taken with the staff still at point B but with its face turned towards I_2. This will be the back sight at position I_2 and the fore sight R_4 is taken with the staff at C. At point B, both an FS and a BS have been recorded consecutively, each from a different instrument position and this is called a *change point* (CP).

From the staff readings taken at I_2, the reduced level of C (RL_C) is calculated from

$RL_C = RL_B + (R_3 - R_4)$

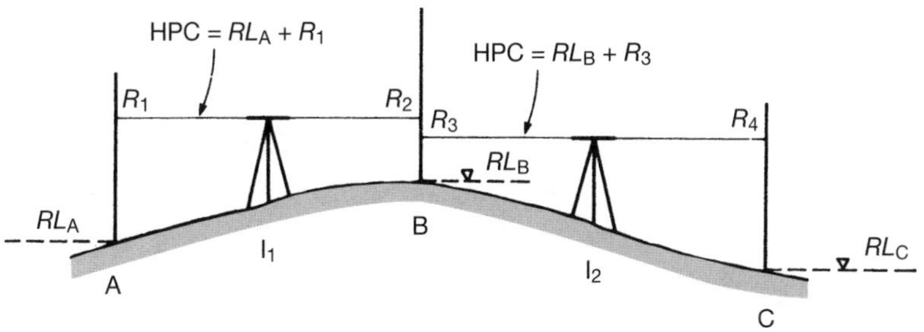

Figure 2.20 ● Principles of levelling.

The height difference between B and C is given both in magnitude and sign by $(R_3 - R_4)$. In this case, $(R_3 - R_4)$ is negative since the base of the staff has fallen from B to C. This time, the difference of the staff readings is known as a *fall*.

From the above, it can be seen that when calculating a rise or fall, this is always given by (back sight – fore sight). If this is positive, a rise is obtained and if negative, a fall is obtained.

In practice, a BS is the first reading taken after the instrument has been set up and is always to a bench mark or calculated reduced level. Conversely, a FS is the last reading taken at an instrument position. Any readings taken between the BS and FS from the same instrument position are known as *intermediate sights* (IS).

A more complicated levelling sequence is shown in cross-section and plan in Figure 2.21, in which an engineer has levelled between two TBMs to find the reduced levels of points A to E. The readings could have been taken with any type of level and the field procedure followed to determine the reduced levels is as follows.

- The level is set up at some convenient position I_1, and a BS of 2.191 m is taken to TBM 1, the foot of the staff being held on the TBM and the staff held vertically.

- The staff is then moved to points A and B in turn and readings are taken. These are intermediate sights of 2.505 m and 2.325 m respectively.

- A change point must be used in order to reach D owing to the nature of the ground. Therefore, a change point is chosen at C and the staff is moved to C and a FS of 1.496 m taken.

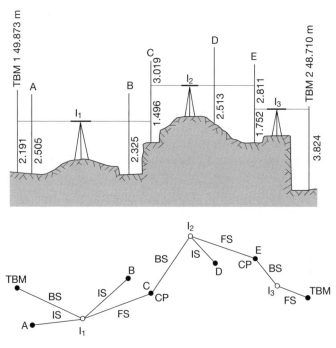

Figure 2.21 ● Line of levels.

- While the staff remains at C, the instrument is moved to another position, I_2. A BS of 3.019 m is taken from the new level position to the staff at change point C.

- The staff is moved to D and E in turn and readings of 2.513 m (IS) and 2.811 m (FS) are taken where E is another CP.

- Finally, the level is moved to I_3, a BS of 1.752 m taken to E and a FS of 3.824 m taken to TBM 2.

- The final staff position is at a TBM. This is most important as all levelling field-work must start and finish at a bench mark, otherwise it is not possible to detect errors in the levelling.

Reflective summary

With reference to field procedures for levelling, remember:

— Each reading must be identified as a back sight (BS), fore sight (FS) or intermediate sight (IS).

— It is essential that all levelling starts and finishes at a bench mark.

2.4 Calculating reduced levels

After studying this section you should know how to calculate heights (reduced levels) by the rise and fall method and the height of plane of collimation (HPC) method. You should be aware that it is very important to carry out arithmetic checks on these calculations.

This section includes the following topics:

- The rise and fall method
- The height of collimation method

The rise and fall method

In this section, the heights of points A to E described in the previous section are computed by the *rise and fall* method.

Table 2.1 shows all the readings for the levelling sequence shown in Figure 2.21 recorded in a rise and fall field book. Each row or line of the field book corresponds to a staff position and this is confirmed by the entries made in the *Remarks* column. The calculation of reduced levels proceeds in the following manner, in which the reduced level of points A to E are computed point-by-point starting at TBM 1.

Table 2.1 ● Rise and fall method.

BS	IS	FS	Rise	Fall	Initial RL	Adj	Adj RL	Remarks
2.191					49.873		49.873	TBM 1 49.873
	2.505			0.314	49.559	+0.002	49.561	A
	2.325		0.180		49.739	+0.002	49.741	B
3.019		1.496	0.829		50.568	+0.002	50.570	C (CP)
	2.513		0.506		51.074	+0.004	51.078	D
1.752		2.811		0.298	50.776	+0.004	50.780	E (CP)
		3.824		2.072	48.704	+0.006	48.710	TBM 2 48.710
Checks								
6.962		8.131	1.515	2.684	48.704			
8.131			2.684		49.873			
−1.169			−1.169		−1.169			

Note: The date, observer and booker (if not the observer), the survey title, level number, weather conditions and anything else relevant should be recorded as well as the staff readings.

● From the TBM 1 to A there is a small fall. A BS of 2.191 m has been recorded at TBM 1 and an IS of 2.505 m at A. So, for the fall from TBM 1 to A, the height difference is given by (2.192 − 2.505) = −0.314 m. The negative sign indicates a fall and this is entered in column for this on the line for point A. The fall is then subtracted from the RL of TBM 1 to obtain the initial reduced level of A as 49.873 − 0.314 = 49.559 m.

● The procedure is repeated and the height difference from A to B is given by (2.505 − 2.325) = +0.180 m. The positive sign indicates a rise and this is entered on the line for B. The RL of B is $(RL_A + 0.180)$ = 49.739 m.

● The rise from B to C up to the first CP is (2.235 − 1.496) = +0.829 m. To continue the level table, the height change from C to D is (3.019 − 2.513) = +0.506 m.

● The calculation is repeated until the initial reduced level of TBM 2 is calculated.

● When the *Initial* RL column of the table has been completed, a check on the arithmetic involved is possible and must always be applied. This check is

$$\Sigma \text{ BS} - \Sigma \text{ FS} = \Sigma \text{ RISES} - \Sigma \text{ FALLS} = \text{LAST } \textit{Initial} \text{ RL} - \text{FIRST RL}$$

It is normal to enter these summations at the foot of each relevant column in the levelling table (see Table 2.1). Obviously, agreement must be obtained for *all three parts of the check* and it is stressed that this only provides a check on all the *Initial* RL calculations and does not provide an indication of the accuracy of the readings.

The difference between the calculated and known values of TBM 2 is –0.006 m. This is known as the *misclosure* of the levelling and gives an indication of the accuracy of the levelling.

If the misclosure is greater than the *allowable misclosure* then the levelling must be repeated, but if the misclosure is less than the allowable value then it is distributed throughout the reduced levels. In this case, the allowable misclosure is ± 9 mm and the levelling is acceptable (see Section 2.5 for a full explanation of this).

The usual method of correction is to apply an equal, but cumulative, amount of the misclosure to each instrument position, the sign of the adjustment being opposite to that of the misclosure. Since there is a misclosure of –0.006 m in this example a total adjustment of +0.006 m must be distributed. As there are three instrument positions, +0.002 m is added to the reduced levels found from each instrument position. The distribution is shown in the *Adj* (adjustment) column, in which the following cumulative adjustments have been applied: Levels A, B and C + 0.002 m, levels D and E + (0.002 + 0.002) = +0.004 m and TBM 2 + (0.002 + 0.002 + 0.002) = +0.006 m. No adjustment is applied to TBM 1 since this level cannot be altered. The adjustments are applied to the *Initial* RL values to give the *Adj* (adjusted) RL values in Table 2.1.

The height of collimation method

Table 2.2 shows the field book for the reduction of the levelling of Figure 2.21 by the height of collimation method. This way of reducing levels is based on the HPC being calculated for each instrument position and proceeds as follows.

- If the BS reading taken to TBM 1 is added to the RL of this bench mark, then the HPC for the instrument position I_1 will be obtained. This will be 49.873 + 2.191 = 52.064 m and this is entered in the HPC column on the same line as the BS.

- To obtain the initial reduced levels of A, B and C the staff readings to those points are now subtracted from the HPC. The relevant calculations are

    RL of A = 52.064 – 2.505 = 49.559 m

    RL of B = 52.064 – 2.325 = 49.739 m

    RL of C = 52.064 – 1.496 = 50.568 m

- At point C, a change point, the instrument is moved to position I_2 and a new HPC is established. This collimation level is obtained by adding the BS at C to the RL found for C from I_1. For position I_2, the HPC is 50.568 + 3.019 = 53.587 m. The staff readings to D and E are now subtracted from this to obtain their reduced levels.

- The procedure continues until the *Initial* RL of TBM 2 is calculated and the misclosure found as before. With the *Initial* RL column in the table completed, the following check can be applied.

    Σ BS – Σ FS = LAST *Initial* RL – FIRST RL

Table 2.2 ● Height of collimation method.

BS	IS	FS	HPC	Initial RL	Adj	Adj RL	Remarks
2.191			52.064	49.873		49.873	TBM 1 49.873
	2.505			49.559	+0.002	49.561	A
	2.325			49.739	+0.002	49.741	B
3.019		1.496	53.587	50.568	+0.002	50.570	C (CP)
	2.513			51.074	+0.004	51.078	D
1.752		2.811	52.528	50.776	+0.004	50.780	E (CP)
		3.824		48.704	+0.006	48.710	TBM 2 48.710

Checks

6.962	7.343	8.131		300.420	

7.343 + 8.131 + 300.420 = 315.894

[52.064 × 3] + [53.587 × 2] + [52.528] = 315.894

6.962	8.131	48.704
8.131		49.873
−1.169		−1.169

Note: The date, observer and booker (if not the observer), the survey title, level number, weather conditions and anything else relevant should be recorded as well as the staff readings

However, this only verifies the reduced levels calculated using BS and FS readings. To check the reduced levels calculated from IS readings, a second check is used and is given by

Σ IS + Σ FS + Σ RLs except first = Σ [each HPC × number of applications]

Table 2.2 gives

Σ IS + Σ FS + Σ RLs except first = 7.343 + 8.131 + 300.420 = 315.894 m

The first HPC of 52.064 m has been used *three* times to calculate the reduced levels of A, B and C. The second of 53.587 m has been used *twice* to calculate the reduced levels of D and E and the last HPC of 52.528 m has been used *once* to close the levels onto TBM 2. This gives the second part of this check as

[52.064 × 3] + [53.587 × 2] + [52.528 × 1] = 315.894 m

● After applying the check, any acceptable misclosure is distributed as for the rise and fall method.

Reflective summary

With reference to calculating reduced levels, remember:

— There are two ways of calculating reduced levels – the rise and fall method and the height of plane of collimation (HPC) method.

— The arithmetic checks MUST be done for all levelling calculations.

— When establishing the heights of new TBMs and other important points, only BS and FS should be taken and the rise and fall method of calculation should be used.

— The HPC method of calculation can be much quicker when a lot of intermediate sights have been taken and it is a good method to use when mapping or setting out where many readings are often taken from a single instrument position.

— A disadvantage of the HPC method is that the check on reduced levels calculated from IS can be lengthy and there is a tendency for it to be omitted.

2.5 Precision of levelling

After studying this section you should be able to determine the allowable error for different types of levelling and you should realise how this helps to decide whether a line of levels is accepted or is rejected.

How good is my levelling?

In the previous section, a misclosure of –6 mm was obtained for the levelling by comparing the reduced level of the closing bench mark TBM 2 (48.710 m) with its *Initial* RL obtained from staff readings (48.704 m). By comparing the two reduced levels for TBM 2 in this way, an assessment of the quality or precision of the levelling can be made and it is usual to check that the misclosure obtained is better than some specified value called the *allowable misclosure*.

On construction sites and other engineering projects, levelling is usually carried out over short distances and it can include a lot of instrument positions. For this type of work, the *allowable misclosure* for levelling is given by

$$allowable\ misclosure = \pm m\sqrt{n}$$

where m is a constant and n is the number of instrument positions. A value often used for m is 5 mm.

When the misclosure obtained from staff readings is compared to the *allowable misclosure* and it is found that the misclosure is greater than the allowable value, the levelling is rejected and has to be repeated. If the misclosure is less than the allowable value, the misclosure is distributed between the instrument positions as described in the previous section. For the levelling given in Tables 2.1 and 2.2, the misclosure is −6 mm, the allowable misclosure is given by (with m = 5 mm) $\pm 5\sqrt{3} = \pm 9$ mm and the levelling has been accepted.

When assessing the precision of any levelling by this method, it may be possible for a site engineer to use a value of m based on site conditions. For example, if the reduced levels found are to be used to set out earthwork excavations, the value of m might be 30 mm but for setting out steel and concrete structures, the value of m might be 3 mm. Values of m may be specified as tolerances in contract documents or where they are not given, may simply be chosen by an engineer based on experience.

Specifications for levelling are also given in BS 5964: *Building setting out and measurement* and in the ICE Design and Practice Guide *The management of setting out in construction*. Typical accuracies expected for levelling are also listed in BS 5606: *Guide to accuracy in building* and it possible to evaluate the accuracy of anyone levelling by using BS 7334: *Measuring instruments for building construction*. For further details of these, see further reading and sources of information at the end of the chapter. The precision of levelling is also discussed in Chapter 9.

Reflective summary

With reference to the precision of levelling, remember:

— An allowable error should always be computed and a decision made to accept or reject levelling.

— The allowable error varies according to the number of times the level is set up and how the levels obtained are going to be used.

— If the levelling is accepted, the reduced levels obtained should be adjusted.

— If it is rejected, the levelling must be repeated – never try to correct or invent a reading to make a line of levels close.

2.6   Sources of error in levelling

After studying this section you should be aware of the various sources of error that can occur in levelling and you should appreciate how it is possible to minimise or even eliminate some of these by adopting sensible field procedures.

This section includes the following topics:

- Errors in the equipment
- Field or on-site errors
- The effects of curvature and refraction on levelling
- How to reduce the chance of errors occurring

Errors in the equipment

Collimation error

This can be a serious source of error in levelling if sight lengths from one instrument position are not equal, since the collimation error is proportional to the difference in these. So, in all types of levelling, sight lengths should be kept equal, particularly back sights and fore sights and before using any level it is advisable to carry out a two-peg test to ensure the collimation error is within acceptable limits.

Compensator not working

For an automatic or digital level, the compensator is checked by moving a footscrew slightly off level, by tapping the telescope gently or by pushing the compensator check lever to ensure that a reading remains constant. If any of these checks fail, the compensator is not working properly and the instrument must be returned to the manufacturer for repair.

Parallax

This effect must be eliminated before any staff readings are taken.

Defects of the staff

The base of the staff should be checked to see if it has become badly worn – if this is the case then the staff has a *zero error*. This does not affect height differences if the same staff is used for all the levelling, but introduces errors if two staffs are being used for the same series of levels.

When using a multi-section staff, it is important to ensure that it is properly extended by examining the graduations on either side of each section as it is extended. If any of the sections become loose, the staff should be returned for repair.

Tripod defects

The stability of tripods should be checked before any fieldwork commences by testing to see if the tripod head is secure, that the metal shoes at the base of each leg are not loose and that, once extended, the legs can be tightened sufficiently.

Field or on-site errors

Staff not vertical

Since the staff is used to measure a vertical difference between the ground and the plane of collimation, failure to hold the staff vertical will give incorrect readings. Since the staff is held vertical with the aid of a circular bubble, this should be checked at frequent intervals and adjusted if necessary.

Unstable ground

When the instrument is set up on soft ground and bituminous surfaces on hot days, an effect often overlooked is that the tripod legs may sink into the ground or rise slightly whilst readings are being taken. This alters the height of collimation and it is advisable to choose firm ground on which to set up the level and tripod, and to ensure that the tripod shoes are pushed well into the ground.

Similar effects can occur with the staff, and for this reason it is particularly important that change points should be at stable positions such as manhole covers, kerbstones, concrete surfaces, and so on. This ensures that the base of the staff remains at the same height in between a BS and FS.

For both the level and staff, the effect of soft or unstable ground is greatly reduced if readings are taken in quick succession.

Handling the instrument and tripod

As well as vertical displacement, the plane of collimation of a level may be altered for any set-up if the tripod is held or leant against. When levelling, avoid contact with the tripod and only use the level by light contact through the fingertips. If at any stage the tripod is disturbed, it will be necessary to relevel the instrument and to repeat all the readings taken from that instrument position.

Instrument not level

For automatic and digital levels this source of error is unusual, but for a tilting level in which the tilting screw has to be adjusted for each reading, this is a common mistake. The best procedure here is to ensure that the main bubble is centralised before and after a reading is taken.

Reading and booking errors

Many mistakes can be made during the booking of staff readings taken with an automatic or tilting level, and the general rule is that staff readings must be carefully entered into the levelling table or field book *immediately after reading*. As already noted, readings taken with a digital level are automatically stored by the instrument and there is no need for the operator to record anything by hand – this gives the digital level an advantage over automatic and tilting levels.

Weather conditions

In strong winds, a level can become unusable because the line of sight is always moving and it also very difficult to hold the staff steady. For these reasons, it is not possible to take reliable readings under these conditions which should be avoided when levelling.

In hot weather, the effects of refraction are serious and produce a shimmering effect near ground level that makes it very difficult to read the bottom metre of the staff.

The effects of curvature and refraction on levelling

In Section 2.1 it was shown that when a level is set up, it defines a horizontal line for measurement of height differences. In Figure 2.22 a level is shown set up at point A and it can be seen that the level and horizontal lines through the instrument diverge because level lines follow the curvature of the Earth which is defined as mean sea level. If not accounted for, this is a possible source of error in levelling since all readings taken at A are observed along the horizontal line instead of the level line.

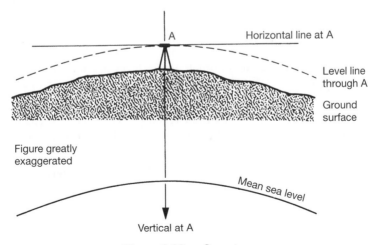

Figure 2.22 ● Curvature.

It is shown in Section 5.6 that the difference between a horizontal and level line is given by

$$c = 0.0785 \ D^2$$

where

c = curvature in *metres*

D = sighting distance in *kilometres*

For most levelling applications, sighting distances are relatively short and the correction for a length of sight of 100 m is less than 1 mm. Consequently, the difference between a horizontal line and a level line is small enough to be ignored.

The effect of atmospheric refraction on a line of sight is to bend it towards the Earth's surface. This has a value, under normal atmospheric conditions, about 1/7 that of curvature and is also ignored for most levelling work.

Whatever sight lengths are used, the effects of curvature and refraction will cancel if the sight lengths are equal. However, the effects of curvature and refraction cannot be ignored when measuring heights over long distances using total stations – this is discussed in Section 5.6.

How to reduce the chance of errors occurring

When levelling, the following procedures should be used if many of the sources of error are to be avoided.

- Levelling should always start and finish at bench marks so that misclosures can be detected. When only one bench mark is available, levelling lines must be run in loops, starting and finishing at the same bench mark.
- Where possible, all sights lengths should be below 50 m.
- The staff must be held vertically by suitable use of a circular bubble or by rocking the staff and noting the minimum reading.
- BS and FS lengths should be kept equal for each instrument position. For engineering applications, many IS readings may be taken from each setup. Under these circumstances it is important that the level has no more than a small collimation error.
- For automatic and tilting levels, staff readings should be booked immediately after they are observed and important readings, particularly at change points, should be checked. Use a digital level where possible as it takes staff readings automatically.
- The rise and fall method of reduction should be used when heighting reference or control points and the HPC method should be used when setting out.

2.7 Other levelling methods

After studying this section you should understand how it is possible to determine the heights of elevated points using inverted staff methods and you should be aware of the technique used to carry a line of levels over a wide gap avoiding any instrumental errors that may be present in the level.

This section include the following topics:

- Inverted staff
- Reciprocal levelling

Inverted staff

Occasionally, it may be necessary to determine the heights of points such as a ceiling or the soffit of a bridge, underpass or canopy. Usually, these points will be above the plane of collimation of the level. To obtain the reduced levels of these points, the staff is held upside down in an inverted position with its base on the elevated points. When booking an inverted staff reading, it is entered into the levelling table with a *minus sign*, the calculation proceeding in the normal way taking this sign into account.

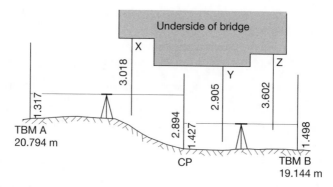

Figure 2.23 ● Inverted staff levelling.

Table 2.3 ● Inverted staff readings.

BS	IS	FS	Rise	Fall	Initial RL	Adj	Adj RL	Remarks
1.317					20.794		20.794	TBM A 20.794
	−3.018		4.335		25.129	−0.001	25.128	X
1.427		2.894		5.912	19.217	−0.001	19.216	CP
	−2.905		4.332		23.549	−0.002	23.547	Y
	−3.602		0.697		24.246	−0.002	24.244	Z
		1.498		5.100	19.146	−0.002	19.144	TBM B 19.144
Checks								
2.744		4.392	9.364	11.012	19.146			
4.392			11.012		20.794			
−1.648			−1.648		−1.648			

Note: The date, observer and booker (if not the observer), the survey title, level number, weather conditions and anything else relevant should be recorded as well as the staff readings

An example of a levelling line including inverted staff readings is shown in Figure 2.23 and the relevant calculations for this are in Table 2.3.

Each inverted reading is denoted by a minus sign and the rise or fall computed accordingly. For example, the rise from TBM A to point X is $1.317 - (-3.018) = 4.335$ m. Similarly, the fall from point Z to TBM B is $-3.602 - (1.498) = -5.100$ m.

An inverted staff position *must not be used as a change point* because it is often difficult to keep the staff vertical and to keep its base in the same position for more than one reading.

Reciprocal levelling

For all levelling, true differences in height between pairs of points are obtained by ensuring that their sight lengths are equal. This eliminates the effect of any collimation error and also the effects of curvature and refraction.

However, there are certain applications in engineering and site work when it may not be possible to take staff readings with equal sight lengths, as, for instance, when a line of levels has to be taken over a gap such as a river or ravine. In these cases, the technique of *reciprocal levelling* can be used.

Figure 2.24 shows two points A and B on opposite sides of a river. To obtain the true difference in height between A and B a level is placed at I_1 close to A on one side of the river and a levelling staff is held vertically at A and B. Staff readings of a_1 at A and b_1 at B are taken and the level is then moved to the other side of the river to position I_2 where readings a_2 and b_2 are taken.

The difference in height between A and B is obtained by treating each reading at A as a BS and each reading at B as a FS. This gives the height difference ΔH_{AB} as $(a_1 - b_1)$ or $(a_2 - b_2)$, but with a BS and FS taken over very unequal sight lengths from each instrument position. Since staff readings a_1 and b_2 are taken over short distances, it is assumed that the effect of any collimation error in the level and the effects of curvature and refraction on these readings are small and can be ignored. This will not be the case for staff readings b_1 and a_2, which have been taken over long distances. However, since these readings have been taken over the same distances with the same level, the combined effect ε of the collimation error and curvature and refraction will be the same in both readings. The corrected staff readings are $(b_1 + \varepsilon)$ and

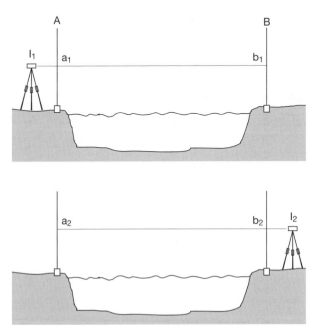

Figure 2.24 ● Reciprocal levelling.

$(a_2 + \varepsilon)$ and the true difference in height between A and B is given by the mean of the two observed differences from I_1 and I_2 as

$$\Delta H_{AB} = \tfrac{1}{2}[(a_1 - [b_1 + \varepsilon]) + ([a_2 + \varepsilon] - b_2)] = \tfrac{1}{2}[(a_1 - b_1) + (a_2 - b_2)]$$

in which the effects of the collimation error and curvature and refraction have cancelled.

For this type of work, several sets of readings are taken with the instrument being re-levelled in a slightly different position for each set. The average values for the staff readings are then used to compute ΔH_{AB}. When carrying out this procedure with one level, the two sets of observations must follow each other as soon as possible so that refraction effects are the same and are minimised. Where this is not possible, two levels have to be used simultaneously. As it is very unlikely that the two levels will have the same collimation error, the true height difference will not be obtained. This problem is overcome by interchanging the levels and repeating the whole procedure and then taking the mean value for the height difference from these. Another potential difficulty with this method of height measurement is its reliability as it becomes difficult to read a levelling staff as the sighting distance increases beyond about 100 m. It is recommended that, for distances of more than about 100 m, another method, such as trigonometrical heighting or GPS, should be used to transfer height across a wide gap.

Reflective summary

With reference to other levelling methods, remember:

— The inverted staff method is very useful on site for finding the heights of points up to about 4 or 5 m above the level (the length of a levelling staff) but because it can be difficult to hold the staff in an inverted position the accuracy is not as good as ordinary levelling.

— Reciprocal levelling can be used very effectively to transfer levels across an inaccessible gap but its accuracy is governed by how well a levelling staff can be read at long distances.

Exercises

2.1 Explain how parallax occurs and describe a procedure for removing it from a level.

2.2 Explain the difference between a level line and a horizontal line.

2.3 Why is it necessary to try and keep sight lengths as equal as possible when levelling?

2.4 What is the height datum in mainland Great Britain called and what is an OSBM?

2.5 What are the advantages of using a digital level compared to an automatic level?

2.6 Describe a test that can be carried out to determine the collimation error of a level.

2.7 Discuss the circumstances under which the rise and fall or HPC method would be used for reducing levels.

2.8 Some levelling is carried out on site where four instrument positions were used. What would be the allowable error for this levelling if it was to be used for setting out a steel structure or for earthworks?

2.9 State the sources of error that can occur in levelling and describe how these can be minimised.

2.10 Explain why it is not advisable to have a change point at an inverted staff position.

2.11 At what distance does Earth curvature have a value of 10 mm?

2.12 The observations listed below were recorded when testing the adjustment of an automatic level. Use them to calculate the collimation error in the level.

Level position	Staff position	Sighting distance	Staff reading
I_1	A	25 m	1.225 m
I_1	B	25 m	1.090 m
I_2	A	55 m	1.314 m
I_2	B	5 m	1.155 m

2.13 A digital level was checked for collimation error using a two-peg test and the following results were obtained.

With the level midway between two pegs B1 and B2 which are 40 m apart:
Staff reading at B1 = 1.476 m   Staff reading at B2 = 1.432 m

Level set up 10 m from B2 along the line B1–B2 extended:
Staff reading at B1 = 1.556 m   Staff reading at B2 = 1.472 m

Calculate the collimation error in the level and the readings that would have been obtained with the level in the second position close to B2 had it been in perfect adjustment.

2.14 The readings shown below were taken to find the heights of pegs A–D. Calculate adjusted reduced levels for the pegs.

BS	IS	FS	Remarks
1.603			TBM 40.825 m
	1.001		Peg A
1.761		1.367	CP
	1.297		Peg B
1.272		1.203	CP
	0.910		Peg C
1.979		2.291	CP
0.772		0.646	Peg D
		3.030	TBM 39.685 m

2.15 For the levelling shown below, calculate adjusted reduced levels for all points in the level table.

BS	IS	FS	Remarks
1.832			TBM 62.117 m
2.150		2.379	Change point
	1.912		A
	1.949		B
	2.630		C
1.165		1.539	D
2.381		2.212	Change point
	2.070		E
	2.930		F
	0.954		G
	2.425		H
		0.879	TBM 62.629 m

2.16 Select a suitable method and reduce the levels given below.

BS	IS	FS	Remarks
1.729			TBM 71.025 m
	0.832		Peg at chainage 200 m
	0.971		210 m
	1.002		220 m
	1.459		230 m
	1.031		240 m
	1.600		250 m
	1.621		260 m
	2.138		270 m
	2.076		280 m
		1.730	TBM 71.025 m

2.17 The extract given below is for levels taken between two TBMs. Calculate adjusted reduced levels for all entries in the book.

BS	IS	FS	Remarks
1.592			TBM 31.317 m
1.675		2.052	CP
1.354		1.704	CP
1.326		0.907	Peg A
	1.379		Peg B
	1.384		Peg C
	1.406		Peg D
0.940		1.315	Peg E
0.832		1.507	CP
	(2.938)		Soffit 1
	(2.833)		Soffit 2
	(2.717)		Soffit 3
		1.546	TBM 30.007 m

Note: The readings in brackets ( ) were taken with an inverted staff.

2.18 Reciprocal levelling involving a single level was used to transfer height across a river between two points R1 and R2 70 m apart. Using the following readings, calculate the height change from R1 to R2 and the collimation error in the level.

Level set up close to R1: Staff reading on R1 = 1.582 m
Staff reading on R2 = 0.792 m

Level set up close to R2: Staff reading on R1 = 2.112 m
Staff reading on R2 = 1.336 m

Further reading and sources of information

For assessing the accuracy of levelling, consult

BS 5606: 1990 *Guide to accuracy in building* (British Standards Institution [BSI], London). BSI web site http://www.bsi-global.com/.

BS 5964: 1996 (ISO 4463: 1995) *Building setting out and measurement* (British Standards Institution, London). BSI web site http://www.bsi-global.com/.

BS 7334-3: 1990 (ISO 8322-3:1989) *Measuring instruments for building construction. Methods for determining accuracy in use: optical levelling instruments.* (British Standards Institution, London). BSI web site http://www.bsi-global.com/.

Deutsches Institut für Normung DIN 18723-2:1990 *Field procedure for precision testing of surveying instruments; levels* (DIN e.V., Berlin). DIN web site http://www.din.de/.

ISO 17123–2: 2001 *Optics and optical instruments – Field procedure for testing geodetic and surveying instruments – Part 2: Levels* (International Organization for Standardization [ISO], Geneva). ISO web site http://www.iso.ch/.

For general guidance and assessing the accuracy of levelling on site refer to

ICE Design and Practice Guide (1997) *The management of setting out in construction.* Thomas Telford, London.

For the latest information on the equipment available, visit the following web sites:

http://www.leica.com/
http://www.nikon-trimble.com/
http://www.pentax.co.uk/
http://www.sokkia.com/
http://www.topcon.co.uk/
http://www.trimble.com/

chapter three

Angle measurement

 Aims

After studying this chapter you should be able to:

- Understand that theodolites and total stations measure horizontal and vertical angles
- Assess the accuracy of a theodolite and total station for site work
- Describe all the components of a theodolite and explain how these are used when measuring and setting out angles
- Outline the differences between electronic and optical theodolites
- Describe the field procedures that are used to set up and measure angles with a theodolite and total station
- Book and calculate horizontal and vertical angles from theodolite readings
- Recognise that the methods for setting up and measuring angles with a theodolite are subject to many sources of error and realise that these can be controlled provided the correct field procedures are used

This chapter contains the following sections:

3.1 Definition of horizontal and vertical angles

3.2 Accuracy of angle measurement

3.3 Components of an electronic theodolite

3.4 Optical theodolites

3.5 Measuring and setting out angles

3.6 Sources of error when measuring and setting out angles

Exercises

Further reading and sources of information

3.1 Definition of horizontal and vertical angles

After studying this section you should understand the exact definitions of horizontal and vertical angles and you should realise why it is necessary to centre and level a theodolite or total station when measuring these.

Angles, theodolites and total stations

The measurement of angles is one of the most important required in surveying and construction. Angles are usually measured using either a theodolite or a total station.

- *Horizontal angles* are used to determine bearings and directions in control surveys. They are used for locating detail when mapping and are essential for setting out all types of structure.

- *Vertical angles* are used when determining the heights of points by trigonometrical methods, and can be used to calculate slope corrections for horizontal distances.

- *Theodolites* are precision instruments and are either electronic (capable of displaying angle readings automatically) or optical, which need to be read manually. The Sokkia DT610 electronic theodolite is shown in Figure 3.1. Many different theodolites are available for measuring angles and they are often classified according to the smallest reading that can be taken with a particular instrument – this can vary from 1' to 0.1".

- *Total stations* are precision electronic instruments which are also capable of measuring angles, but the significant difference between these and theodolites is that a total station can also measure distances. Many of the features of theodolites are duplicated on total stations, as shown by the Leica TPS700 Series in Figure 3.1. The angle measuring system of a total station is equivalent to that of an electronic theodolite.

The electronic theodolite and total station are the predominant instruments for angle measurement on site and elsewhere, but optical theodolites are still in use. This chapter describes how these are used for angle measurement, but concentrates mostly on the electronic theodolite. Total stations are described in Chapter 5.

Figure 3.2 shows two points A and B and a theodolite or total station T set up on a tripod above a ground point G. Point A is higher than the instrument and is above the horizontal plane through T, whereas B is lower and below the horizontal plane. At T, the instrument is mounted a vertical distance h above G on its tripod.

The *horizontal angle* at T between A and B is not the angle in the sloping plane containing A, T and B, but the angle θ on the horizontal plane through T between the

Sokkia DT610 electronic theodolite   Leica TPS700 Series total station

Figure 3.1 ● Sokkia electronic theodolite compared to Leica total station (courtesy Sokkia Ltd and Leica Geosystems).

vertical planes containing the lines of sight TA and TB. The *vertical angles* to A and B from T are α_A (an *angle of elevation*) and α_B (an *angle of depression*).

Another angle often referred to is the *zenith angle*. This is defined as the angle in the vertical plane between the direction vertically above the instrument and the line of sight, for example z_A in Figure 3.2.

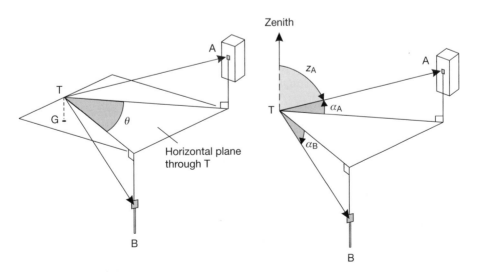

Figure 3.2 ● Horizontal, vertical and zenith angles.

In order to measure horizontal and vertical angles, a theodolite or total station must be *centred* over point G and must be *levelled* to bring the angle reading systems of the instrument into the horizontal and vertical planes. Although centring and levelling ensure that horizontal angles measured at T are the same as those that would have been measured if the instrument had been set on the ground at point G, the vertical angles from T are not the same as those from G, and the value of h, the *height of the instrument*, must be taken into account when height differences are being calculated.

Reflective summary

With reference to the definition of horizontal and vertical angles, remember:

— The only angles used in construction and surveying are horizontal and vertical angles, and it is important to understand what these are.

— Theodolites and total stations do not measure angles in sloping planes – they are designed to measure horizontal and vertical angles only.

3.2 Accuracy of angle measurement

After studying this section you should be able to specify the accuracy required for a theodolite in order to meet tolerances given for site work. You should also be aware that care is needed when assessing the angular accuracy of a theodolite using brochures and user manuals supplied by the manufacturers.

Angular specifications

In order to match the type of theodolite or total station to its intended application on site, some relationships between angle and sighting distance are given in Table 3.1. Using this as a guide it is evident that, if a 5 mm setting out tolerance was specified on site for work over sighting distances of up to 100 m, a 10" angular precision would be needed to meet this requirement.

When assessing the angular precision required for a theodolite or total station in this way care has to be taken, as the minimum reading of the instrument is not the same as its quoted accuracy. Many theodolites and total stations are tested to

Table 3.1 ● Angular and linear precision.

20"	is equivalent to	10 mm	at a sighting distance of 100 m
10"	is equivalent to	5 mm	at a sighting distance of 100 m
5"	is equivalent to	2.5 mm	at a sighting distance of 100 m
1"	is equivalent to	0.5 mm	at a sighting distance of 100 m

Note: Linear precision = angular precision (in radians) × sighting distance

Deutsches Institut Für Normung (DIN) Standard 18723 and have their angular accuracy quoted with reference to this. In addition to an instrumental accuracy, errors from setting up the instrument and sighting with it will make it difficult to achieve the theoretical values of angular precision given here, so an allowance may have to be made for this as well (this is discussed further in Section 9.3). When using a theodolite, some general guidance on the accuracy that can be expected is given in BS 5606 *Guide to accuracy in building* and a personal assessment of the accuracy that can be achieved can also be obtained by following the test procedures given in BS 7334 *Measuring instruments for building construction*. The ICE Design and Practice Guide *The management of setting out in construction* also gives details of good practice when using a theodolite. For further information on these, refer to the end of the chapter.

On many civil engineering projects, it is often assumed that a high specification theodolite or total station is required for all work – this is not necessary for most setting out, although such an instrument might be needed when establishing control or on construction projects demanding high-quality positioning.

At this point, it is worth noting that a full circle is 360° or 400ᵍ (400 gons or grads) and an angular reading system capable of resolving to 1" or a few mgon directly shows the degree of precision in the manufacture of theodolites and total stations. The gons angular unit is widely used in continental Europe instead of degrees, minutes and seconds for the measurement and setting out of angles. Angular units and conversions between these are described in the Appendix.

Reflective summary

With reference to accuracy of angle measurement, remember:

— When specifying the accuracy required for a theodolite (or total station) try to match this to the tolerances given on site – this may help reduce costs when purchasing or hiring instruments.

— The way in which the angular accuracy of a theodolite and total station is quoted varies according to each manufacturer – if this is important an accuracy specified to DIN 18723 is more reliable, but it is best to consult the manufacturer or supplier as to exactly what is given.

— To allow for centring and sighting errors when assessing accuracy requirements.

3.3 Electronic theodolites

After studying this section you should be familiar with all of the various parts of a theodolite and how each of these helps the operator to set up and use a theodolite. You should understand what single and dual-axis compensators are and be aware that a theodolite fitted with one of these is capable of producing better results.

This section includes the following topics:

- Mechanical parts of the theodolite
- Electronic components
- Single and dual-axis compensation
- Specialised theodolites

Mechanical parts of the theodolite

The features of a typical electronic (or digital) theodolite are shown in Figure 3.3. As shown in Chapter 5, all of the features described in this section are similar to those found on the angle-measuring systems of total stations.

The *tribrach* forms the base of the instrument. In order to be able to attach the theodolite to a tripod, most tripods have a clamping screw, which locates into a threaded centre on the base of the tribrach. The tribrach also consists of three *threaded levelling footscrews* and a *circular bubble*. Some instruments have the facility for detaching the upper part of the theodolite from the tribrach, as shown in Figure 3.4. A special target or other piece of equipment can then be centred in exactly the same position occupied by the theodolite. This ensures that angular and linear measurements are carried out between the same positions without having to set up tripods more than once – this is known as *forced centring* and helps reduce centring errors (centring errors are discussed in Section 3.6).

The two main features of the upper part of the instrument are the *telescope*, which is supported on the *standards*. The imaginary axis about which the telescope is rotated vertically (or tilted) between the standards is called the *tilting axis*. The whole instrument and telescope can also be turned horizontally about another imaginary axis called the *vertical axis*. Movement about the tilting and vertical axes enables the instrument to be pointed along any direction or line of sight for the measurement of angles.

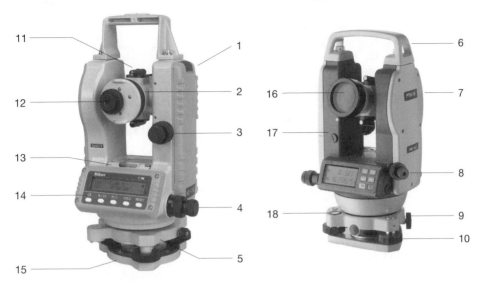

Figure 3.3 ● Nikon NE100 and Pentax PTH-10 electronic theodolites: 1 battery compartment; 2 telescope focus; 3 vertical clamp and tangent screw; 4 horizontal clamp and tangent screw; 5 tribrach; 6 carrying handle; 7 vertical circle; 8 optical plummet; 9 tribrach clamp; 10 footscrew; 11 collimator (sight); 12 eyepiece; 13 plate level; 14 keyboard and display; 15 trivet; 16 objective; 17 standard; 18 circular bubble (courtesy Survey Supplies and Pentax UK Ltd).

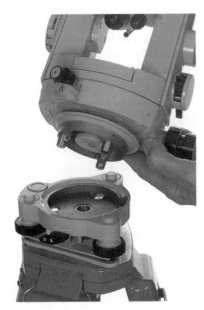

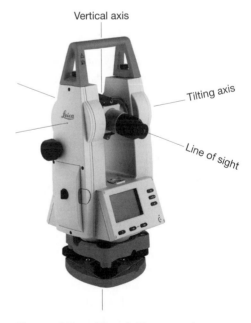

Figure 3.4 ● Detaching theodolite and tribrach.

Figure 3.5 ● Theodolite axes shown on Leica T105 electronic theodolite (courtesy Leica Geosystems).

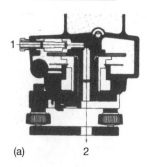
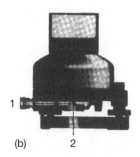

(a)   2   (b)   2

Figure 3.6 ● Sections through theodolite showing (a) optical plummet mounted on upper part of instrument (b) optical plummet on tribrach: 1 eyepiece 2 line of sight along vertical axis.

The arrangement of the axes of the theodolite is shown in Figure 3.5. When the instrument is levelled, the vertical axis is made to coincide with the vertical at the point where the instrument is set up, and the tilting axis is assumed to be horizontal. This is achieved by using the levelling footscrews and *plate level*, which is identical to the spirit level tube of an optical level (for a description of a spirit level, see Section 2.2).

Centring the theodolite involves setting the vertical axis directly above a ground point, and this can be done using an *optical plummet*. This consists of a small eyepiece which is built into the upper part of the instrument or into the tribrach (see Figure 3.6), the line of sight of which is deviated by 90° so that a point corresponding to the vertical axis can be viewed on the ground underneath the instrument. Some instruments now incorporate a *laser plummet* which produces a visible red laser beam along the vertical axis.

Rotation of the instrument about the vertical axis is controlled by a *coaxial horizontal clamp* and *tangent screw*, and the rotation of the telescope about the tilting axis by a *coaxial vertical clamp* and *tangent screw*. When sighting a point, one of the *collimators* located on the top of the telescope is used to obtain an approximate pointing, both clamps are locked and the two tangent screws are then used to obtain an exact pointing. It is usual that the tangent screws will not operate until their corresponding clamps are locked. Some instruments have endless horizontal and vertical drives in which no clamps are required and where the instrument is simply rotated to the approximate pointing and the tangent screws used to locate a target exactly.

For the measurement of angles, the theodolite has two circles. Once the theodolite has been levelled using the plate level and footscrews, the *horizontal circle* is positioned in the horizontal plane for the measurement of horizontal angles and the *vertical circle* is automatically positioned perpendicular to this in a vertical plane for the measurement of vertical angles. Theodolites are usually classified according to the minimum circle reading that can be obtained, which can vary from 20" to 0.1" in the electronic versions.

The construction of the main telescope is similar to that used in optical levels, as described in Section 2.2, but the *focusing screw* of the telescope is fitted concentrically with the barrel of the telescope instead of on the side. The eyepiece is focused in the same way as the eyepiece of the level in order to remove parallax and to give a sharp view of the cross hairs. Additionally, the diaphragm can be illuminated for night or tunnel work. When the main telescope is rotated in altitude about the tilting axis from one direction to face in the opposite direction it has been *transited*. The side of the main telescope, viewed from the eyepiece, containing the vertical circle is called the *face*, which is labelled as face left (or face I) and face right (or face II) – transiting the telescope changes the face.

Built into the standard containing the vertical circle is a device called an *automatic vertical index*. This is similar to the compensator in an automatic level (see Section 2.2) and its function, after the instrument has been levelled, is to set the vertical circle such that it normally reads 0° at the zenith with 90° or 270° defining the horizontal.

Electronic components

Both the horizontal and vertical circle measuring systems fitted into electronic theodolites use glass circles with binary codes etched on them. Within the theodolite, light is passed through the encoded circles and the light pattern emerging through the circles is detected by photodiodes. The measurement system most commonly used in electronic theodolites is known as an *incremental encoder*, where the varying intensity of light passing through the horizontal or vertical circle as it is rotated during a measurement varies in proportion to the angle through which the theodolite has been rotated. This varying light intensity is converted into digital signals by the photodiodes, and these in turn are passed to a microprocessor, which converts the signal into an angular output.

The microprocessor is accessed through a keyboard and Liquid Crystal Display (LCD) as shown for the Nikon NE203 in Figure 3.7. On some instruments, the display and keyboard are mounted on both sides (faces) of the theodolite, but to save costs a few instruments have them on one side only. The various keys perform the following functions

☼   Provides illumination for the LCD panel and telescope cross hairs (diaphragm).

%/VA   The vertical circle of some electronic theodolites can be made to read the vertical angle or percentage of slope instead of the zenith angle (see Figure 3.8). This is assessed by pressing the %/VA key.

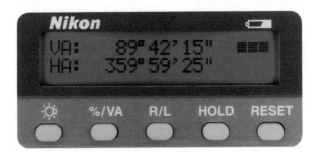

Figure 3.7 ● Control panel of Nikon NE203 (courtesy Survey Supplies).

R/L   Conventionally, the horizontal circle of a theodolite is graduated clockwise when viewed from above. This means the readings increase when the telescope is rotated clockwise. By pressing the R/L key, this direction can be reversed to make readings increase when the telescope is rotated anti-clockwise. This is a very useful feature for setting out left-handed road curves (see Chapters 12 and 13) and in other types of setting out (see Chapter 11).

HOLD   Once pressed, this holds the current horizontal angle reading on the display until it is pressed a second time. This is used when setting the horizontal circle to a particular value.

RESET   This enables a reading of 00°00'00" to be set on the horizontal circle in the direction in which the theodolite is pointing. On some theodolites, this function is called *ZERO SET* or *0 SET*.

In addition to the basic keyboard operations, a menu system can be accessed, and it is possible to change the angular output from 360° degrees-minutes-seconds to 360° decimal or to 400 gons. Electronic theodolites often use AA type batteries or similar as their power source, from which it is possible to get about 20 hours use. Also on the display is a three-level bar graph showing the remaining battery level. In order to save power, most instruments have a power-saving function which automatically turns off the instrument after 10 or 30 minutes.

Although it is possible to record readings manually in field books when using an electronic theodolite, since angle information is generated in a digital format this can be transmitted by some theodolites to a suitable storage device for recording and processing by a computer. Data is transferred by connecting a storage device to a data port on the side of the theodolite or to a memory card. Electronic storage devices and methods of data capture are described in more detail in Section 5.4.

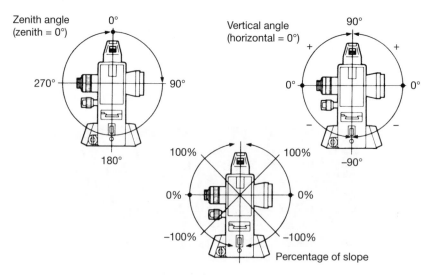

Figure 3.8 ● Vertical angle measurement modes (courtesy of Sokkia Ltd).

Single and dual-axis compensation

Many theodolites and total stations use single or dual-axis compensation. Both instruments have to be levelled manually using the footscrews, and the procedure for doing this with a plate level or electronic bubble is described in Section 3.5. However, no matter how carefully either instrument is levelled, it is unusual for the vertical axis to coincide exactly with the vertical, and this tilt, even though it is small, can give rise to errors in displayed horizontal and vertical angles. Electronic theodolites and total stations correct for the effects of vertical axis tilt using a liquid or pendulum type compensator.

The effect of vertical axis tilt in the direction in which the telescope is pointing is shown in Figure 3.9. As can be seen, this causes an error in vertical angles and this is compensated automatically by some electronic theodolites, which apply a correction to the vertical angle. This is known as *single-axis compensation* and most devices for this usually have a working range of about ±3'.

Dual-axis compensators measure the effect of an inclined vertical axis not only in the direction in which the telescope is pointing (single-axis compensation) but also in the direction of the tilting axis. The effect of an inclined tilting axis is to produce errors in horizontal angles (see Figure 3.10) and it can be shown that the error is proportional to the tangent of the vertical angle of the telescope pointing. Consequently, for steep sightings and for precise work, this error could be significant. A dual-axis compensator will correct not only for vertical axis tilt in the direction of the telescope, but will also measure tilting axis dislevelment and correct horizontal circle readings automatically for this error.

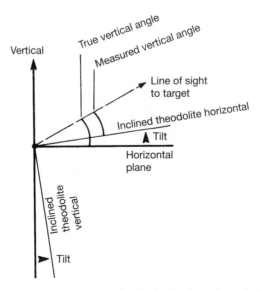

Figure 3.9 ● Effect of vertical axis tilt along line of sight.

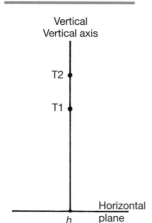 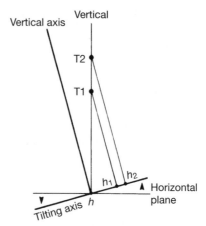

(a) Vertical axis not tilted:
horizontal reading *h* recorded
for targets T1 and T2

(b) Vertical axis tilted:
horizontal reading h_1 and
h_2 recorded for targets
T1 and T2 instead of *h*

Figure 3.10 ● Effect of vertical axis tilt along tilting axis.

Specialised theodolites

Industrial theodolites are electronic theodolites that are used for precise measurement and setting out on site. They have a very high angular accuracy specification of 0.5″, incorporate features such as motorised drives and, when operated on-line with computer and special applications software, can process spatial data in real time. The Leica TM5100A industrial theodolite is shown in Figure 3.11.

Laser theodolites are conventional theodolites that have a visible laser built into them for alignment purposes. Sokkia's LDT50 electronic laser theodolite has a coaxial beam that projects along the line of sight of the instrument – further details of this instrument and its applications are given in Section 11.5.

Figure 3.11 ● Leica TM5100A industrial theodolite (courtesy Leica Geosystems).

Reflective summary

With reference to electronic theodolites, remember:

— The instrument is much more complicated than a level and before attempting any measurements with a theodolite it is worthwhile becoming familiar with all the controls and especially the electronic functions – this may save time and reduce the chance of errors occurring when on site.

— Most instruments are fitted with a single or dual-axis compensator – the advantages of and the difference between these must be fully understood.

— A total station is very similar in appearance to an electronic theodolite and is used in much the same way – the difference between them is the ability of the total station to measure distances as well as angles.

— Because the instrument is electronic, it is necessary to ensure that replacement batteries are always available when on site. A lot of time can be wasted if a battery runs out or is not charged sufficiently while work is being carried out at a remote location or a long way from the site office.

3.4   Optical theodolites

After studying this section you should be aware of the differences between the reading systems of electronic and optical theodolites.

How to read an optical theodolite

These instruments also measure horizontal and vertical angles and have features similar to electronic theodolites. However, the reading systems are very different and rely on manual operation and recording. When taking a reading with an optical theodolite, light is directed into the instrument and is passed through the horizontal and vertical circles, both of which are made of glass and have angular graduations rather than being binary coded. The images of the circles are then viewed through a special *reading telescope* situated next to the main telescope. There are three types of reading system in use.

An *optical scale reading system* is shown in Figure 3.12 for the Wild T16 theodolite. The readings shown are obtained directly to 1' from two fixed scales shown

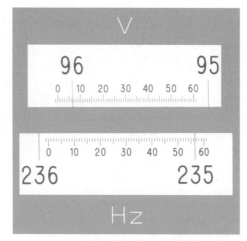

Vertical circle reading 96°06.5'
Horizontal circle reading 235°56.4'

Figure 3.12 ● Wild T16 optical scale reading system (courtesy Leica Geosystems).

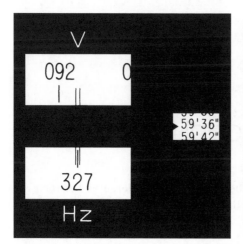

Horizontal circle reading 327°59'36"

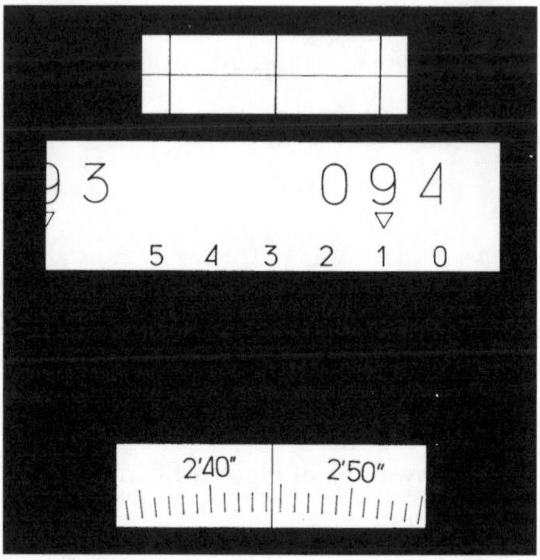

Horizontal or vertical circle reading 94°12'44"

Figure 3.13 ● Wild T1 single reading optical micrometer system (courtesy Leica Geosystems).

Figure 3.14 ● Wild T2 double reading optical micrometer system (courtesy Leica Geosystems).

superimposed on portions of the horizontal and vertical circles corresponding to the pointing of the instrument and telescope.

The *single reading optical micrometer system* of the Wild T1 theodolite is shown in Figure 3.13. In this case, a reading is made up of two parts: a circle reading added to a micrometer reading. In order to take the horizontal reading shown, a *micrometer screw* situated on one of the standards has been adjusted such that the 327° mark lies in between the two fixed lines on the Hz frame. The amount of rotation required to achieve this setting is recorded on the micrometer scale as 59'36". If the vertical circle is to be read at the same telescope setting, it is necessary to reset the microm- eter and read the vertical angle separately. The reading interval of the Wild T1 is 6".

The *double reading optical micrometer system* of the Wild T2 is shown in Figure 3.14. In this case, the micrometer screw is turned until the lines seen in the upper frame of the reading telescope are all coincident. The reading is obtained by adding a circle reading of 94°10' to a micrometer reading of 2'44". These instruments usually do not show the horizontal and vertical scales together and a change-over switch is provided to transfer from one scale to the other. The reading interval of the Wild T2 is 1".

Reflective summary

With reference to optical theodolites, remember:

— Although still in use, this type of theodolite has been replaced by the electronic theodolite and total station for nearly all work on site.

— Optical theodolites are not as easy to read as electronic theodolites and great care is needed when measuring angles with these, as it is easy to make a mistake when reading the scales and to forget to set a micrometer.

3.5   Measuring and setting out angles

After studying this section you should be familiar with the methods used to centre and level a theodolite and total station. You should also know how to use a theodo- lite and total station to measure and set out angles.

This section includes the following topics:

● Setting up a theodolite

● Measuring angles

● Booking and calculating angles

● Setting out angles

Setting up a theodolite

The process of setting up a theodolite is carried out in three stages: centring the theodolite, levelling the theodolite and elimination of parallax.

It is possible to centre a theodolite using a number of different methods and everyone has their own preferred method. Any technique used to centre a theodolite is perfectly acceptable as long as it is quick, accurate and not likely to damage the theodolite. Remember that the object of centring involves setting the vertical axis of the theodolite directly over a point of some sort. The procedure described below is recommended, where it is assumed that the theodolite is to be centred over a nail in the top of a peg which has been driven into the ground – this is a typical point or reference mark used in construction and setting out.

- Leaving the instrument in its case, the tripod is first set up over the peg. The legs of the tripod are placed an equal distance from the peg and are extended to suit the height of the observer. The tripod head should be made as level as possible by eye.

- Standing back a few paces from the tripod, the centre of the tripod head is checked to see if it is vertically above the peg – this should be done by eye from two directions at right angles. If the tripod is not centred, each leg is moved a distance equal to the amount the tripod is judged to be off-centre and in the same direction in which it is not centred. It is important to keep the tripod head level when changing its position. When the tripod has been centred in this way, the tripod legs are pushed firmly into the ground. If one foot goes in more than the others, making the tripod head go off level, this can be allowed for by loosening the clamp of the tripod leg affected, adjusting the length and then re-clamping.

- The theodolite is carefully taken out of its case, its exact position being noted to help in replacement, and it is securely attached to the tripod head. Whenever carrying a theodolite, always hold it by the standards and *not* the telescope. Never let go of the theodolite until it is firmly screwed onto the tripod.

- The ground mark under the theodolite is now observed through the optical plummet. The focus and reference mark (either cross hairs or a circle) are adjusted such that, in this case, the nail in the peg and the plummet's reference mark are seen together in clear focus. By adjusting the three footscrews, the image of the nail seen through the plummet is moved until it coincides with the reference mark. If the instrument is fitted with a laser plummet, the footscrews are adjusted so that the beam is centred on the ground mark (nail).

- The circular bubble on the tribrach or upper part of the theodolite is now centred by adjusting the length of individual tripod legs, as required. Take care to ensure that the tripod feet do not move at all during this process.

At this stage, the theodolite is almost centred and is almost level. To level the instrument exactly, the plate level has to be used. The procedure for this is as follows and is the only one recommended for a three-footscrew instrument.

- The theodolite is rotated until the plate level axis is parallel to the line through any two footscrews, as shown in Figure 3.15(a). These two footscrews are turned until the plate level bubble is brought to the centre of its run. The levelling footscrews should be turned in opposite directions simultaneously, remembering that the bubble will move in a direction corresponding to the movement of the left thumb.

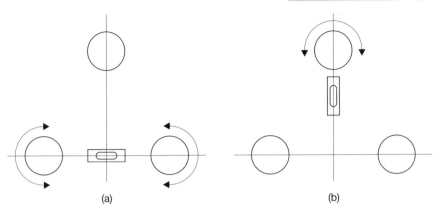

Figure 3.15 ● Levelling a theodolite.

● The instrument is turned through 90° (see Figure 3.15(b)) and the bubble centred again, but using the third footscrew only.

● This process is repeated until the plate level bubble is central in both positions.

● The instrument is now turned until the plate level is in a position 180° from the first. If the plate level bubble is still in the centre of its run, the theodolite is level and no further adjustment is needed. If the bubble is not central, it has an error equal to half the amount the bubble has run off centre. If, for example, the bubble moves off centre by two plate level divisions to the left, the error in the bubble is one division to the left.

● If the plate level has an error, the theodolite is returned to its initial position and the bubble is moved off-centre an amount equal to the error in the bubble. This is done using the two footscrews in line with the axis of the plate level. Using the example already quoted, the bubble would be placed one division to the left.

● The instrument is then turned through 90° and the plate level bubble is again moved one division to the left but using the third footscrew.

● The instrument is now slowly rotated through 360°, and the plate level bubble should remain in the same position throughout (one division to the left in this case). The theodolite has now been levelled and the vertical axis coincides with the vertical through the instrument.

On examining this procedure, it can be seen that the plate level bubble is not necessarily in the centre of its run when the theodolite is level. However, as long as the plate level is set up so that the bubble remains in the same position, the theodolite is level and can be used perfectly satisfactorily until the plate level is adjusted (see Section 3.6).

It may be possible to level a theodolite electronically after the rough levelling has been carried out. In this case, the position of an electronic bubble is shown on the display and the three levelling footscrews are used to centre it, as shown in Figure 3.16. All instruments capable of electronic levelling will give a warning if they are not levelled properly and will not function until they have been re-levelled.

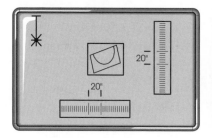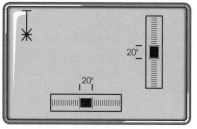

Figure 3.16 ● Electronic levelling (courtesy Leica Geosystems).

At this stage, the theodolite is levelled but will not be centred exactly. Centring is carried out after the theodolite has been levelled as follows:

● The clamping screw on the tripod is undone and the theodolite is moved by sliding it on the tripod head until it appears centred by looking through the optical plummet or by centring a laser plummet.

● The levelling should be checked, as centring in this way can change the position of the plate level bubble. If it has changed, it should be re-levelled and the centring checked afterwards.

When using the optical or laser plummet for centring, it is essential that the theodolite is properly levelled before this is done. If the theodolite is not levelled, the axis of the plummet will not be vertical and even though it may appear to be centred, the theodolite will be miscentred.

When the theodolite has been levelled and centred, parallax is eliminated by accurately focusing the cross hairs of the telescope against a light background and focusing the instrument on a distant target.

At this stage the theodolite is ready for reading angles or for setting out. If any point is occupied for a long time, it is necessary to check the levelling and centring at frequent intervals, especially when working on soft ground or in hot sunshine.

All of the procedures given in this section for setting up a theodolite are also used when centring and levelling a total station. They are also used for setting up a tripod-mounted GPS antenna over a control point.

Measuring angles

This section assumes that a theodolite has been levelled and centred over survey point W and the horizontal and vertical angles to three distant points X, Y and Z are to be measured as shown in Figure 3.17.

In order to be able to measure the directions to X, Y and Z, it may be necessary to establish *targets* at these points. All targets, whatever type of survey work they are being used for, should be centred exactly, whether this is over a ground marker, a nail in a peg or some other survey point, and it should be easy to take accurate sightings to them.

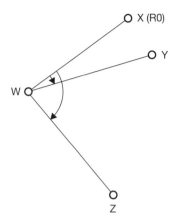

Figure 3.17 ● Measurement of angles at point W.

Figure 3.18 ● Survey target and prism (courtesy York Survey Supply Centre).

The targets used for the majority of observations on site are specially manufactured for this purpose and are often combined with a reflecting prism to enable distances to be measured to a given point at the same time as angles. A typical arrangement is shown in Figure 3.18 in which the target is fixed to a tribrach and is tripod mounted. When the tribrach has been levelled and centred (the procedure is the same as for a theodolite), the points for measuring horizontal and vertical angles are set in their correct positions. This type of target is used when taking measurements with a total station and further examples are given in Section 5.2.

As an alternative to a tripod mounted target, a *detail pole* (see Figure 3.19) can be held on a control or setting out point during a survey. However, if care is not taken centring errors will be introduced into angular observations if the pole is not held vertically – this is shown in Figure 3.20, where it can be seen that sightings taken lower on the pole will result in smaller centring errors. To avoid this error, most detail poles are fitted with a circular bubble vial which helps keep it vertical when hand held. The effect of miscentring a target or theodolite is discussed further in Section 3.6. As well as being centred and vertical, the width of the pole should be proportional to the length of sight and, ideally, should be about the same size as a cross hair when viewed through the telescope. Clearly, it is a waste of time trying to observe a direction to a detail pole when the line of sight is short, since accurate bisection is difficult. Table 3.1 can be of use here again, but for assessing the required width of a target in this case. For example, a 20 mm wide detail pole is equivalent to an angle of 80" over a distance of 50 m – measuring angles to a precision of 5" or 10" with this would not be possible.

Another target available for angle and distance measurements is the handheld *mini-prism* assembly shown in Figure 3.21. If required, this can be used on its own or it can be attached to a detail pole.

On construction sites, the survey point to be sighted is usually a nail in the top of a wooden peg – the nail should be observed directly if possible, but holding a pencil on

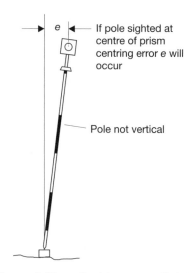

Figure 3.19 ● Detail pole (courtesy York Survey Supply Centre).

Figure 3.20 ● Centring errors if detail pole not held vertical.

Figure 3.21 ● Handheld mini-prism assembly (courtesy York Survey Supply Centre).

it can help. For some measurements, a tripod can be set up and a plumb bob can be suspended from it directly over the survey point. The plumb line can then be observed, but it is better to attach a string target to it.

Assuming suitable targets are in place at X, Y and Z in Figure 3.17, the observation procedure at W starts with the horizontal angles. For this, one of the points has to be chosen as the *reference object* (RO). Point X is chosen in this case and all the horizontal angles are referred to this point, in which the horizontal angles XWY and XWZ are to be measured. Table 3.2 shows some example readings for both the horizontal and vertical angles observed at point W.

All observations require the operator to bisect each target exactly. To do this, one of the sights on the telescope (or *collimators*) is first used to point the telescope approximately at a target and the horizontal and vertical clamps are tightened. For horizontal angles, it is necessary that the vertical hair of the cross hairs is made

Table 3.2 ● Horizontal and vertical angle bookings with calculations.

Angles at *POINT W*

Horizontal readings

POINT	FACE LEFT	FACE RIGHT	MEAN	ANGLE
X (RO)	00 03 50	180 04 30	00 04 10	
Y	17 22 10	197 23 10	17 22 40	17 18 30
Z	83 58 50	264 00 00	83 59 25	83 55 15
X (RO)	45 12 30	225 13 30	45 13 00	
Y	62 31 10	242 32 10	62 31 40	17 18 40
Z	129 07 30	309 08 40	129 08 05	83 55 05

FINAL HORIZONTAL ANGLES

$$X\hat{W}Y = 17\ 18\ 35 \qquad X\hat{W}Z = 83\ 55\ 10$$

Vertical readings

POINT with target height	FACE LEFT	FACE RIGHT	REDUCED FACE LEFT	REDUCED FACE RIGHT
X 1.16 m	88 10 30	271 51 20	+01 49 30	+01 51 20
Y 1.52 m	89 34 50	270 27 30	+00 25 10	+00 27 30
Z 1.47 m	92 48 20	267 13 40	–02 48 20	–02 46 20

FINAL VERTICAL ANGLES

$$X = +01\ 50\ 25 \qquad Y = +00\ 26\ 20 \qquad Z = -02\ 47\ 20$$

Note: The date, observer and booker (if not the observer), the survey title, theodolite number, weather conditions, a sketch similar to Figure 3.17 and anything else relevant should be recorded as well as the angle readings

coincident with the target when viewed through the telescope. To do this, the target is sighted by adjusting the horizontal tangent screw such that the vertical hair bisects the target exactly. This is very important – any small difference between the target and hair will cause an error in the displayed angle.

At the start of the measuring procedure, a reading is set on the theodolite in the *face left* (FL) position along the direction to the RO. When using an electronic theodolite, a reading of exactly 00°00'00" can be set by pressing the *0 SET* key when the theodolite is pointing at the RO, which is point X in this case. If some other reading is required to an RO, the following procedure is used. The theodolite is rotated until a horizontal reading close to the one required is displayed, the horizontal circle is clamped and the tangent screw is used to obtain the desired reading. Following this, the *HOLD* key is pressed and the RO sighted in the usual manner: the required reading should now be set along the direction to the RO. Finally, the *HOLD* key is pressed again and readings will now change as the theodolite is rotated.

Having set the required direction to X as 00°03'50" on FL in this case, the procedure for measuring angles continues as follows.

● Points Y and Z are sighted in turn and the horizontal circle readings are recorded as 17°22'10" and 83°58'50". The telescope is transited so that the theodolite is now in the *face right* (FR) position and horizontal circle readings are recorded at Z, Y and X in the reverse order.

● At this stage, *one round* of angles has been completed. The theodolite is changed to face left and the *zero* changed by setting the horizontal circle to read something different from the reading set for the first round when sighting X, the RO. It is not necessary to set an exact reading on the second round, but it is important to realise that as well as changing the degrees setting to the RO, the setting of the minutes and seconds should also be different from that of the first round. In this case, the face left reading to X at the start of the second round is 45°12'30" compared with 00°03'50" for the first round.

● A second round of angles is taken. At least two rounds of angles should be taken at each survey point in order to detect errors when the angles are computed, since each round is assumed to be independently observed. Both rounds must be computed and compared *before* the instrument and tripod are moved to another point in case it is necessary to re-observe.

The procedure for measuring zenith or vertical angles is similar to that for horizontal angles but in this case, the vertical circle is orientated by the automatic vertical index. Vertical angles should be read after the horizontal angles to avoid confusion when booking and they can be observed in any order. It is usual to take all face left readings first and it is normal that only one round of angles is taken. For vertical angles, the vertical tangent screw is adjusted such that the horizontal hair bisects the target exactly. Finally, and when required, the height of each target should also be measured.

Having completed all angular observations, the levelling and centring are finally checked in case any unnoticed movement has occurred whilst observing the angles. If these are satisfactory, the theodolite is carefully removed from the tripod head and put back in its case. Before removing the theodolite from the tripod head, the three footscrews should be set central in their runs. If it is found that the instrument is not level or is not centred for some reason, these should be corrected and another round of angles taken and compared with those already measured.

If other survey points are to be occupied, the theodolite must *never* be left on the tripod when moving between stations, since this can distort the axes and, if the operator trips or falls, the instrument may be severely damaged.

All of the procedures given here for measuring angles with a theodolite can also be used with total stations.

Booking and calculating angles

Table 3.2 shows the horizontal and vertical angle booking and calculation for points X, Y and Z observed from station W. Many different formats exist for recording and calculating angles and only one method is shown here.

For the *horizontal angle readings*, all the points observed for both rounds are entered in the POINT, FACE LEFT and FACE RIGHT columns as appropriate. The mean horizontal circle readings are obtained by averaging each pair of face left and face right readings. To simplify these calculations, the degrees of the face left readings are carried through and only the minutes and seconds values are averaged and written in the MEAN column. The mean horizontal circle readings are then reduced to the RO in the ANGLE column to give the horizontal angles. The FINAL HORIZONTAL ANGLES are obtained by taking the average of the two rounds.

From the *vertical circle readings* obtained, it can be seen that the vertical circle is set with 0° at the zenith. Although zenith angles can be used in survey calculations, when angles are reduced by hand in this way, it is usual to reduce the vertical FL and FR zenith readings to their equivalent vertical angles. These are given by

FL vertical angle = 90° − FL zenith reading
FR vertical angle = FR zenith reading − 270°

On the angle booking form, these are computed using

REDUCED FACE LEFT = 90° − FACE LEFT
REDUCED FACE RIGHT = FACE RIGHT − 270°

Each FINAL VERTICAL ANGLE is obtained by averaging the reduced FL and FR values.

Zenith and vertical angles are often measured to enable height differences to be calculated. The procedures and calculations required for this are given in Section 5.6.

In addition to the procedures given above for the calculation of horizontal and vertical angles, the following should also be adopted as good practice when recording and processing angular readings.

● If a single figure occurs in any reading, for example, a 2 or a 4, this should be recorded as 02 or 04. If a mistake is made, the number should always be rewritten, for example, if an 8 is written and should be 9, this should be recorded as 8̶ 9.

● *Never* copy out observations from one field sheet or field book to another.

● As readings are entered into the form, they should be checked for consistency in *horizontal collimation* in the horizontal angles and *vertical collimation* in the vertical angles. Referring to Table 3.2, the checks are as follows. For horizontal angles, the difference between each FL and FR reading is first checked to see if they differ by 180°, for example 0° and 180° at X, 17° and 197° at Y and so on. Secondly, the difference (FL − FR) is computed for each sighting considering minutes and seconds only. This gives the following results for the first round:

Station	(FL – FR)
X	–00'40"
Y	–01'00"
Z	–01'10"

Assuming a 10" theodolite was used to record the two rounds, this shows the readings to be satisfactory since (FL – FR), for a 10" theodolite, should agree to within about 30" for each point observed. If, for example, the difference for station Z was – 11'10" then an *operator error* of 10' is immediately apparent. In such a case, the readings for station Z would be checked. For a 1" theodolite (FL – FR) should agree within a few seconds, depending on the length of sight and the type of target used. A similar process is applied to the vertical circle readings to check for consistency in vertical collimation. In this case FL + FR *should* = 360° and, for station X, FL + FR = 88°10'30" + 271°51'20" = 360°01'50". For stations Y and Z, 360°02'20" and 360°02'00" are obtained. All three values agree very closely, which shows the readings to be consistent and therefore acceptable.

The method of reading angles may be thought to be somewhat lengthy and repetitious, but it is necessary to use this so that instrumental errors are eliminated (see Section 3.6).

The procedure described above can be applied to any type of theodolite and is used when angles are recorded and processed by hand on a booking form. However, if an electronic theodolite or total station is used for measuring angles, it is not usually necessary for the operator to write anything on a booking form and all readings will either be stored directly in some sort of data collector or internally within the instrument. At the press of a key, the horizontal and vertical angles can be calculated and then stored as well, but this might be done at a later stage by downloading the data into a computer installed with appropriate software. For setting out work, horizontal angles can also be viewed directly on the instrument's display by sighting the RO and pressing the *0 SET* key to read 00°00'00". Any subsequent pointing will then be an angle, but this will be measured on one face only and unless the instrument is properly calibrated, may be liable to instrumental errors – this is discussed further in Section 3.6.

Setting out angles

Although it is possible to measure angles with a theodolite using the procedures described in the previous section, it is usual to do this with a total station and measure distances at the same time. However, some work on site only requires angular measurements and can be carried out with a theodolite, although a total station could be used in angular mode.

Setting out is a term used to describe that part of engineering surveying where the points marking new structures are put in place ready for construction. Many different methods are used for this, and because it is such an important topic it is covered separately in Chapter 11. However, when appropriate, some parts of setting out are described in other chapters, as is the case here.

For some types of construction work, it is often necessary to *prolong a straight line* when setting out. To extend the line AB shown in Figure 3.22, the theodolite is set

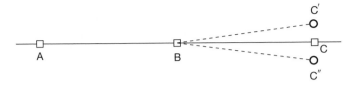

Figure 3.22 ● Prolonging a straight line.

up at B and point A is sighted on face left. Keeping the horizontal clamp locked, the telescope is transited to face right, and because the instrument is likely to have a horizontal collimation error, point C′ is established on the line of sight instead of C, the exact continuation of AB. The horizontal clamp is now released, the instrument is rotated and A sighted once again. When the telescope is transited back to face left again, point C″ can be established next to C′ – point C will then be located midway between C′ and C″. If this procedure is not adopted when prolonging a straight line, a serious error can occur when setting out point C, especially over long distances. Another source of error when prolonging a straight line can be caused by not levelling the theodolite properly, but errors arising from this source are usually small and can be ignored. However, care must be taken to level the theodolite properly when steep sightings are taken to extend a line to the upper floor of a building or into a deep trench. *If the theodolite has a dual axis compensator*, the following procedure can be used to extend a line on both faces and remove any collimation errors plus those caused by not levelling the theodolite exactly. On face left, point A is sighted as usual, but instead of transiting the telescope, it is rotated through exactly 180° and point C′ is established (as in Figure 3.22). Changing to face right, A is sighted again, the telescope rotated through exactly 180° and point C″ is established. Point C is once again midway between C′ and C″.

Another operation frequently carried out on a construction site is to *set out intermediate points* between two fixed points. When it is possible to occupy and observe between the two fixed points (say A and B), a theodolite is set up at A, B is sighted on either face and all intervening points can then established directly on the line of sight of the telescope. A good example of this on site occurs when establishing column or other positions on a structural grid. In this case, two theodolites are set up to define two lines of sight at right angles to each other and column positions are fixed at the intersection of these two lines (this is described further in Section 11.4). This is the best way of setting out points in straight lines – this should not be done with a total station or GPS methods using coordinates. For this type of work, it is not necessary to set out the points on both faces from A provided the telescope is positioned at about the same elevation to that when B was sighted. When it is necessary to tilt the telescope to very large vertical angles compared to that when B was sighted, the intervening points should be set out on both faces and the mean positions used.

On some projects, it is sometimes necessary to set up the theodolite on the line between two fixed points. Sometimes called *lining in*, this is usually done when control points A and B cannot be occupied or when they are not intervisible. The procedure for this is as follows. The theodolite is set up at an estimated point C, as close as possible to the line, from which both A and B can be observed. Point A is intersected, the telescope is transited or rotated through 180° and a sighting taken to B where the position of the line of sight is noted. The distance that the theodolite must be moved in order to intersect B is estimated, the theodolite is repositioned and the procedure repeated until point B is intersected exactly after transiting or rotating the telescope. For best results, this procedure should be done on both faces of the instrument which will result in two positions being established at C – the point required on the line AB will be midway between these.

The most common procedure for which theodolites can be used on site is to *set out angles*. Since this is done at the same time as distances, it is usual that total stations will be used for most of this work. However, theodolites are sometimes used on their own to set out or establish right angles which is one of the most frequently performed setting out tasks involving angles. To do this, the theodolite is set up at one end of a line and sighted onto the other, and a reading of exactly 00°00'00" obtained by pressing the *0 SET* key. The instrument is then rotated clockwise until a reading of 90°00'00" is obtained or anticlockwise until a reading of 270°00'00" is obtained to define the right angle to the line along which the instrument was referenced. Alternatively, the *R/L* key can be pressed to give a reading of 90°00'00" when the theodolite is rotated anticlockwise to define the right angle. Once again, it is necessary to set out any points by this method on both faces of the theodolite to remove any systematic instrumental errors, and the theodolite must be levelled carefully for steep sightings.

Reflective summary

With reference to measuring and setting out angles, remember:

— When levelling and centring a theodolite or total station try to follow known procedures rather than take short cuts as these can often lead to mistakes which waste time. If, at any point in the setting up procedure it goes wrong, it is best to put the instrument back in its case, lift the tripod out of the ground and start again.

— For all work on site and elsewhere, good levelling and centring of a theodolite or total station are *essential* and parallax must be removed before starting observations.

— For best results when measuring angles and setting out, proper targets should be used to reduce the chance of errors occurring. Although detail poles are often used on site as targets, using these is not recommended because they are difficult to keep vertical and they are too wide to take accurate sightings to.

— As with centring and levelling, always use the proper procedures when measuring angles – if measurements do not include two independent rounds taken on both faces of the theodolite, the results obtained are subject to gross and systematic errors.

— Again for best results, all setting out carried out with a theodolite should be done on both faces unless the instrument is calibrated or adjusted within acceptable limits.

— When measuring horizontal angles, at least two independent rounds should be taken so that mistakes can be detected. If necessary, additional rounds should be taken until satisfactory results are obtained.

3.6 Sources of error when measuring and setting out angles

After studying this section you should be aware of many of the sources of error that can occur when using a theodolite and you should know how to implement field procedures to eliminate or reduce the effects of these errors.

This section includes the following topics:

● Errors in the equipment
● Field or on site errors
● Observing on face left and face right

Errors in the equipment

Figure 3.5 shows the arrangements of the axes of a theodolite when it is in perfect adjustment, in which

● the axis of the plate level vial *should be* perpendicular to the vertical axis
● the line of sight (collimation) *should be* perpendicular to the tilting axis
● the tilting axis *should be* perpendicular to the vertical axis

This configuration is rarely achieved in practice and any variation from these conditions will cause errors in observed angles. The effects of the theodolite axes not being in perfect adjustment and of other instrumental errors are discussed in the following sections. As with most of the other sections in this chapter, the information given here also applies to total stations. For optical theodolites, the instrumental errors described in this section require mechanical adjustments to the theodolite, but electronic theodolites and total stations are calibrated electronically.

Plate level not in adjustment

The purpose of levelling a theodolite or total station is to make its vertical axis coincide with the vertical through the instrument. If the plate level is not in adjustment, it is possible that when the instrument appears to be level and the plate level bubble is centred, the vertical axis may be tilted. If the instrument is not level, it is not possible to remove any errors caused by this when observing and setting out angles on both faces. It is quite easy to detect whether the plate level vial is out of adjustment in the procedure given in the previous section for levelling a theodolite. In this, the plate level bubble is centred in two positions at right angles and the instrument is then rotated through 180°. If the bubble moves off centre in this third position, the plate level is out of adjustment. To correct this at the time of setting up, the footscrews are adjusted so that half the error is removed and the bubble stays in the same, but off centre, position for a complete rotation of the instrument. The vertical axis of the theodolite will now coincide with the vertical.

Although it is still possible to measure angles even if the plate level is not in adjustment, it can be difficult to level the instrument properly when the error becomes large. At this point, the plate level must be adjusted and the manufacturer's handbook should be consulted for guidance on how to do this.

If the theodolite is levelled electronically, it will usually be fitted with a dual-axis compensator and it can calculate corrections for any errors caused by the instrument not being levelled properly (vertical axis tilt) and will apply these to displayed horizontal and vertical angles. If a single-axis compensator is fitted, *corrections are only applied to vertical angles*. However, the compensator itself may be out of adjustment. To correct for this, an on-board electronic calibration can be carried out in which the compensator index errors are measured and then automatically applied to all readings taken using the theodolite. Further details of this are given in Section 5.5, where electronic calibration for total stations is discussed. If the compensator index errors exceed a specified limit, the theodolite should be returned to the manufacturer for adjustment.

It can be shown that the error in horizontal angles caused by the theodolite not being level is proportional to the tangent of the vertical angle of the line of sight. Consequently, it is important to ensure that the theodolite is carefully levelled for any steep sightings, such as those taken to tall buildings and into deep excavations when on site.

Horizontal collimation error

This error occurs when the line of sight is not perpendicular to the tilting axis and is detected by taking face left and face right horizontal circle readings to the same point – if these do not differ by *exactly* 180°, the theodolite has a horizontal collimation error. In Table 3.2, the theodolite used to take the readings has a horizontal (FL – FR) difference of about 1'00", which shows the presence of a horizontal collimation error.

The error is removed by taking the average of face left and face right readings to any given point and by taking the mid-point when setting out angles on both faces. It can also be removed in an electronic calibration – see Section 5.5 for a description of this. For best results it is also necessary to keep the telescope at similar elevations when observing.

Tilting axis not horizontal

If the tilting axis of the theodolite is not perpendicular to the vertical axis, it will not be horizontal when the theodolite has been levelled. Since the telescope rotates about the tilting axis it will not move in a vertical plane, which will give rise to errors in measured horizontal angles.

As with the horizontal collimation error, this error is also removed by taking the average of face left and face right readings, by setting out on two faces or by carrying out an electronic calibration on the instrument.

Vertical collimation or vertical circle index error

When a theodolite is levelled, it is assumed that the automatic vertical circle index normally sets the vertical circle to read 0° at the zenith such that 90° is horizontal on face left and 270° is horizontal on face right. To detect this error, the same point is sighted on face left and face right and vertical circle readings taken – when added these should be *exactly* 360° or a vertical collimation error is present in the theodolite. This error is shown in Table 3.2, where the theodolite used has a (FL + FR) consistency of about 2'00".

The vertical collimation error is cancelled by taking the mean of face left and face right readings after reducing them. Since this error is caused by the automatic vertical index being out of adjustment, the theodolite should be returned to the manufacturer or supplier when adjustment becomes necessary. Alternatively, an electronic calibration can be carried out.

Plummet

The plummet is an important part of the theodolite and accurate results cannot be obtained for horizontal angles if it is out if adjustment. The line of collimation of a plummet must coincide with the vertical axis of the theodolite, and to check if this is in adjustment the following tests are carried out.

For an *optical plummet*:

● *If the plummet is mounted on the upper part of the instrument and can be rotated about the vertical axis (as in Figure 3.6(a))* secure a piece of paper on the ground below the instrument and make a mark where the plummet intersects it. Rotate the theodolite through 180° and make a second mark – if the marks coincide, the plummet is in adjustment. If not, the correct position of the plummet axis is given by a point midway between the two marks.

- *If the plummet is mounted on the tribrach and cannot be rotated without disturbing the levelling (as in Figure 3.6(b))* set the theodolite on its side on a bench with its base facing a wall and mark the point on the wall intersected by the plummet. Rotate the tribrach through 180° and again mark the wall. If both marks coincide, the plummet is in adjustment. If not, the correct position is the midway point as before.

In both cases, it is difficult to adjust the plummet precisely under site conditions. If the plummet needs adjusting, it is best to return the theodolite to the manufacturer or supplier.

Laser plummets are checked by setting up the theodolite on its tripod and levelling the instrument. The laser plummet is switched on and the centre of the red spot is marked on the ground. The instrument is rotated slowly through 360° and the position of the spot is observed. If the centre of the spot makes a circular movement of more than 1–2 mm instead of remaining stationary, the plummet needs adjusting. For most instruments, this is carried out by the manufacturer.

Tribrachs and tripods

The clamping mechanism of *tribrachs* can become worn with time and it is possible that the theodolite will move on the tripod head after the clamp has been tightened. If this occurs regularly, the tribrach should be repaired or replaced. Some instruments and tribrachs have circular bubbles fitted to them. Although these need not be precisely adjusted, they are used to help centre and level the instrument and adjustment may be necessary from time to time. To do this, the theodolite is levelled using the plate level or electronic bubble – the circular bubble should now be centred in its circle. If it is not, adjustment is carried out on site according to the manufacturer's instructions.

A problem that can sometimes occur with *tripods* is that the joints between the various parts become loose. Where adjusting screws are provided, these should be kept tightened as instructed by the manufacturer. The shoes at the base of each leg should also be inspected regularly to check that they have not become loose.

Importance of observing procedure with a theodolite

The methods given in the previous sections for reading and setting out angles may be thought to be lengthy and sometimes repetitious, but it is necessary to use these so that some instrumental errors are eliminated.

- By taking the mean of face left and face right readings for horizontal and vertical angles or by setting out on both faces of the theodolite, the effects of systematic instrumental errors such as *horizontal collimation, vertical collimation* and *tilting axis dislevelment* are all eliminated.

- Observing on both faces also removes any errors caused if the diaphragm is inclined provided the same positions are used on each cross hair for observing.

- The effect of an *inclined vertical axis* (plate level not set correctly) is not eliminated by observing on both faces, but any error arising from this is negligible if the theodolite is carefully levelled. Since this error is proportional to the tangent of the vertical angle of the sighting, care should be taken when recording angles to points at significantly different elevations, as is often the case on construction sites. However, when using an electronic theodolite with a dual-axis compensator, the effect of improper levelling is corrected provided the theodolite is levelled such that the tilt sensor is within its working range.

Quick procedure for checking a theodolite and total station

On site, it is good practice to test a theodolite and total station regularly to ensure that it is free of any serious error or malfunction before it is used to measure or set out angles.

For *optical theodolites,* the following checks should be carried out on any instrument that has not been used before or one that not been used for some time. The checks can be carried out anywhere convenient and it is not necessary to centre the theodolite over a point.

- Carefully level the theodolite in the usual way noting how much the plate level bubble is off centre when the instrument is level.

- Measure horizontal and vertical angles on both faces of the theodolite using any clearly defined point as a target and check for horizontal and vertical collimation errors.

- Check the plummet.

If any of these checks show that the instrument needs to be adjusted, it is better not to do this on site and the instrument should be returned to the manufacturer or supplier. If the amount by which the theodolite is out of adjustment is not serious, the instrument can still be used for measuring and setting out angles provided the proper field procedures are used to eliminate or reduce the effects of any errors. However, every attempt should be made to change the theodolite as soon as this is possible.

For *electronic theodolites* and *total stations,* an electronic calibration should be done regularly. This can be carried out on site using the procedures described in Section 5.5. When any electronic calibration parameter is too big, the instrument will usually tell the operator and under these circumstances the instrument will have to be returned to the manufacturer for a service and re-calibration. In addition to this, the mechanical parts of the instrument, such as the plummet and tribrach, should also be checked.

Field or on site errors

Instrument not level

Failure to level a theodolite properly will cause the vertical axis to be tilted. If the plate level is in good adjustment, the main bubble is set in the middle of its run and in the same, but not central, position if it is out of adjustment. If either of these

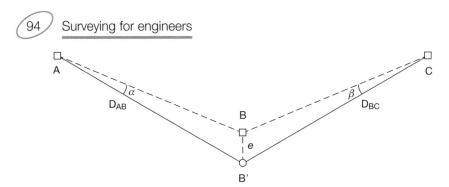

Figure 3.23 ● Miscentring.

conditions are not met and the instrument has been poorly levelled, errors will occur in measured angles that are not eliminated by observing on face left and face right. Although instruments fitted with dual-axis compensators can correct for the effects of a tilted vertical axis, it is still good practice to take some care when levelling theodolites that have a compensator. If a theodolite is found to be off level whilst measuring or setting out an angle, it is best to re-level the instrument and repeat the measurements.

Miscentring

If a theodolite is not centred exactly over a point, incorrect horizontal angles will be measured. In Figure 3.23, horizontal angle ABC is to be measured but, owing to miscentring, the theodolite is set up over B' instead of B and horizontal angle AB'C is measured instead. The error in angle ABC will be $(\alpha + \beta)$, where

$$\alpha = \frac{e}{D_{AB}} \sin\left(\frac{A\hat{B}C}{2}\right) \text{ radians} \qquad \beta = \frac{e}{D_{BC}} \sin\left(\frac{A\hat{B}C}{2}\right) \text{ radians}$$

As an example, if ABC = 120°, $D_{AB} = D_{BC}$ = 50 m and the theodolite is miscentred by 5 mm, the error in the measured angle will be 36". This is a significant error, and for smaller values of D, both α and β will increase and *great care must be taken with centring when sighting over the short distances that are often used on site and in engineering surveying*.

The same errors can occur if a tripod-mounted target is not centred properly and when a detail pole is either miscentred or is not held vertical.

Provided it is in adjustment, it should be possible to centre a theodolite and targets using plummets to within 1 or 2 mm, whereas a plumb line may only be accurate to 5 mm. To avoid centring errors completely, the theodolite should be mounted on a pillar instead of a tripod (see Section 6.3 and Figure 8.9), but it is only necessary to use these when high precision is required.

Not using theodolite properly

It is essential that the telescope of a theodolite is adjusted correctly. At the start of observations, parallax must be removed and as work proceeds, the focus has to be changed for each target so that accurate intersection is possible. When sighting a target or survey point, the tangent screws have to be adjusted to intersect it exactly or errors will be present in measured angles and when setting out. As well as proper intersection of targets, care should be taken not to lean on the tripod at any time as this may cause the instrument to go off level.

Weather and ground conditions

If a tripod is set up on soft ground or tarmac on a hot day, it could gradually settle and the operator may not realise, whilst measurements are being taken, that the theodolite is no longer level. When it is necessary to set a theodolite up on unstable ground, measurements should be taken as quickly as possible and the plate level should be checked frequently to ensure that the theodolite remains level. If the theodolite is fitted with a compensator, it will display a warning when dislevelment exceeds the range of the tilt sensor. It is also possible for a theodolite to become dislevelled when it has been set up in hot sunshine and to prevent this happening, it may be necessary to shade the instrument in extreme conditions. Again in hot weather, refraction can cause a heat shimmer close to the ground and observations may become impossible in such conditions. In addition, lateral refraction in tunnels can cause large errors in horizontal angles unless special observing procedures are adopted and frequent checks are made.

Temperature differences between the theodolite and its environment may effect the compensator and give rise to measurement errors. As a rule of thumb, the time it takes a compensator to adjust to atmospheric conditions in which the instrument is placed is 2 minutes per °C. For example, if the site office is heated to 20 °C and the theodolite is taken outside from this on a cold day at 5 °C, the operator should wait 30 minutes for the compensator to adjust to the lower temperature before using the instrument. Readings and measurements taken on both faces will reduce the effect of any errors from this source.

The effects of high winds when using a theodolite are to cause the tripod and instrument to vibrate so that accurate centring and accurate bisection of targets is either difficult or not possible. It is best to avoid these conditions when observing.

Observing on face left and face right

It has already been stated that taking observations or setting out using both faces of the theodolite produces the best results for angular work as it removes the effects of a number of systematic instrumental errors in the observation procedure. To save time, it is often tempting to observe on one face only – if this is to be done and the results are to be reliable, a properly calibrated instrument is required.

This is usually not possible with optical theodolites because they rely on a set of mechanical adjustments that need to be done regularly by a manufacturer or supplier.

However, since electronic theodolites are software calibrated, it is possible for the user to determine calibration values which can then be applied to raw angles to correct for any maladjustment of the theodolite. The procedures for doing this are given in the user's manual supplied with each instrument and the calibrations that can be carried out will vary according to the make and model of each theodolite (see Section 5.5). User calibrations must be done when the instrument is suspected of being out of adjustment, for example after periods of long storage or rough transport and in extreme temperature differences.

Even if a theodolite is in perfect adjustment and single face readings are considered, dual face readings are always more accurate than single face readings because of the increase in accuracy due to statistical error propagation (this is discussed further in Section 9.3). The angular accuracy quoted by most manufacturers for their theodolites applies to dual face measurements.

In summary, measurements on face left and face right are recommended when:

● The adjustment of the theodolite is uncertain or when calibration values do not seem to be correct and cannot be determined for some reason.

● Where the best possible accuracy is required for measuring and setting out angles.

● Large differences of temperature occur during measurements.

Reflective summary

With reference to sources of error when measuring and setting out angles, remember:

— Always carry out a check before using a new theodolite or when it has not been used for some time (*see* Quick procedure for checking a theodolite and total station; p. 93). If it is not in good adjustment, carry out an electronic calibration or exchange it for another instrument.

— Always measure and set out with a theodolite and total station using both faces unless it is known to be in good adjustment.

— All *instrumental errors* can be avoided by adjusting or calibrating the theodolite regularly and by using all the appropriate field procedures that help to eliminate and reduce these.

— All *field errors* are caused by incorrectly setting up and not using the theodolite properly – the worst sources of error here are not levelling or centring the theodolite with sufficient care.

— As with all survey fieldwork, working under extreme weather conditions is always difficult and should be avoided where possible when using theodolites. One of the important environmental effects to be aware of is that exposing the instrument to large changes of temperature can cause serious errors to occur in angular measurements.

Exercises

3.1 With the aid of diagrams, explain how a horizontal angle differs from a vertical angle. What is a zenith angle?

3.2 Explain why it is necessary to centre a theodolite and describe a procedure for doing this.

3.3 Describe the procedure for levelling a theodolite.

3.4 List, with a brief explanation of each, the functions that are found on the keyboard and displayed by most electronic theodolites.

3.5 Explain what single and dual-axis compensators are and describe the different functions they perform.

3.6 What is an industrial theodolite?

3.7 What is the difference between an optical and electronic theodolite?

3.8 When measuring angles with a theodolite it is always advisable to take face left and face right readings and to take two rounds. Discuss the reasons for this.

3.9 Why is a detail pole not recommended as a target for angle measurement?

3.10 Describe a procedure that can be used to prolong a straight line using a theodolite that does not have a compensator. What alternative method should be used if the theodolite was fitted with a dual-axis compensator?

3.11 What are the angular configurations that should exist between the vertical axis, tilting axis and line of sight in a theodolite. Draw a diagram to show these relationships.

3.12 You have just completed some setting out with a theodolite and notice that it has not been levelled properly. How would this affect your work if the sightings had been taken over level ground in comparison to any sightings taken to elevated points?

3.13 As you move from one part of a construction site to another, the theodolite you have been using for setting out is badly knocked. Describe the tests you would carry out on the instrument to find out if it is still in adjustment or not.

3.14 Give reasons why it is advisable to measure and set out with a theodolite on both faces.

3.15 You are required to work to a linear tolerance of 2.5 mm on a construction site where the longest sighting distance expected is 50 m. What theodolite is required to meet this tolerance?

3.16 What angles does a 5 mm width target subtend at 10, 20 and 50 m?

3.17 The horizontal angle between two points is approximately 85°. If this is measured by a theodolite that is miscentred by 5 mm, what will be the centring error in the measured angle if both sighting distances are 30 m? If the sighting distances are increased to 65 m, what does the centring error reduce to?

3.18 The horizontal circle readings shown below were taken using a 5" reading theodolite correctly set up and levelled at point T. Book the readings in a suitable format and calculate values for the horizontal angles.

Point	Face left reading	Face right reading
A	00°17'35"	180°17'15"
B	38°22'20"	218°22'00"
C	69°30'10"	249°29'40"
D	137°09'55"	317°09'40"
A	45°39'10"	225°38'55"
B	83°43'20"	263°43'00"
C	114°52'00"	294°51'50"
D	182°31'30"	02°31'10"

3.19 Some zenith angles are measured with a theodolite and the readings below were taken:

Point	Face left reading	Face right reading
D1	87°23'38"	272°38'26"
D2	91°48'09"	268°13'49"
D3	95°19'52"	264°42'10"
D4	89°48'17"	270°13'40"

Calculate each of the vertical angles and the value of the vertical collimation error.

3.20 Make a copy of the theodolite booking form given in Section 3.5, enter the readings below and calculate values for the horizontal and vertical angles observed.

Point sighted and face	Horizontal circle reading	Vertical circle reading
W FL	02°17'16"	87°06'26"
X FL	137°52'20"	93°16'31"
Y FL	209°22'37"	91°17'49"
Z FL	312°14'50"	88°32'44"
Z FR	132°15'01"	271°26'51"
Y FR	29°22'48"	268°41'41"
X FR	317°52'33"	266°43'09"
W FR	182°17'29"	272°53'14"

Further reading and sources of information

For assessing the accuracy of a theodolite and angle measurement, consult

BS 5606: 1990 *Guide to accuracy in building* (British Standards Institution [BSI], London). BSI web site http://www.bsi-global.com/.

BS 5964: 1996 (ISO 4463: 1995) *Building setting out and measurement* (British Standards Institution, London). BSI web site http://www.bsi-global.com/.

BS 7334-4: 1992 (ISO 8322-4: 1991) *Measuring instruments for building construction. Methods for determining accuracy in use of theodolites.* (British Standards Institution, London). BSI web site http://www.bsi-global.com/.

ISO 17123–3: 2001 *Optics and optical instruments – Field procedure for testing geodetic and surveying instruments – Part 3: Theodolites* (International Organisation for Standardisation [ISO], Geneva). ISO web site http://www.iso.ch/.

Deutsches Institut für Normung DIN 18723-3: 1990 *Field procedure for precision testing of surveying instruments; theodolites* (DIN e.V., Berlin). DIN web site http://www.din.de/.

For general guidance and assessing the accuracy of angle measurement on site refer to

ICE Design and Practice Guide (1997) *The management of setting out in construction.* Thomas Telford, London.

For the latest information on the equipment available, visit the following web sites

http://www.leica.com/
http://www.nikon-trimble.com/
http://www.pentax.co.uk/
http://www.sokkia.com/
http://www.topcon.co.uk/

chapter four

Distance measurement: taping

 Aims

After studying this chapter you should be able to:

- Understand the difference between slope, horizontal and vertical distances and why all of these are used in engineering surveying

- Describe on-site procedures that are used to measure distances with tapes

- Recognise the mistakes that occur frequently when taping and how to avoid these

- Calculate and apply a range of corrections to taped measurements in order to remove systematic errors from readings

- Discuss the precision that can be obtained when taping

This chapter contains the following sections:

4.1 Measurements and methods

After studying this section you should be aware that distances are defined as slope, horizontal and vertical and that all of these are used in construction and civil engineering.

Distance measurement

Alongside height and angle measurement, distance measurement is another essential component of any survey. In construction work, distances are measured every day when setting out all types of structure. They are needed for plotting the position of detail when mapping and they provide scale in control surveys.

Two methods can be used to measure, record and set out distances. This chapter deals with direct measurement by taping, whereas electronic methods used by total stations and distance meters are covered in the next chapter.

In engineering surveying, three types of distance are used: slope distance, horizontal distance and vertical distance (or height difference).

With reference to Figure 4.1:

$$\text{Slope distance} = AB = L$$
$$\text{Horizontal distance} = AB' = A'B = D$$
$$\text{Vertical distance} = AA' = BB' = V = \Delta H$$

Horizontal and vertical distances are used in mapping, control surveys and engineering design work. Slope distances and vertical distances are used when setting out on construction sites.

Slope distances are usually measured by laying the tape on the surface of the ground or structure, as shown in Figure 4.2(a). However, when measuring over very steep surfaces or undulating ground, the tape may be held horizontally, as in Figure

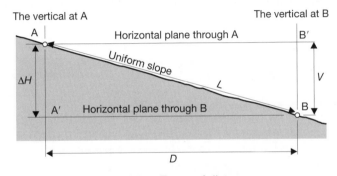

Figure 4.1 ● Types of distance.

(a)

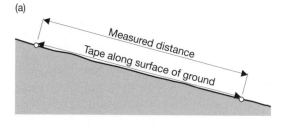

Measured distance
Tape along surface of ground

(b)

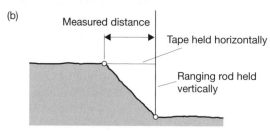

Measured distance

Tape held horizontally

Ranging rod held vertically

(c)

Steel tape in catenary

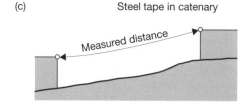

Measured distance

Figure 4.2 ● Tape measurement methods.

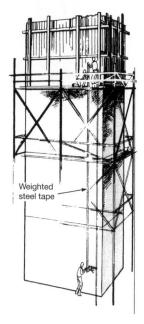

Weighted steel tape

Figure 4.3 ● Transfer of height in a multi-storey building with a tape.

4.2(b), this technique being known as *stepping*. Occasionally, it may be necessary to suspend a tape between two points, as in Figure 4.2(c).

Vertical distances (or height differences) are obtained by allowing a tape to hang freely with a weight attached to its zero end. A common application of this is in the transference of height from floor to floor in a multi-storey building by measuring up or down vertical columns, as shown in Figure 4.3. This is usually done in lift wells or service ducts such that an unrestricted line is available for taping and where the tape can be held vertical.

Reflective summary

With reference to measurements and methods, remember:

— If a *horizontal distance* is required, it is usual that a slope distance is measured with the tape positioned along the ground – this is then converted to its horizontal equivalent.

— For engineering and site work, the horizontal distance is always known and this has to be converted to a *slope distance* for setting out purposes.

— The measurement of *vertical distance* can be a convenient and accurate method of transferring height for dimensional control in tall buildings, but this requires some skill by site operatives if it is to be done properly.

4.2 Equipment and fieldwork for taping

After studying this section you should know what the difference is between steel and fibreglass tapes and you should be able to describe the procedures for measuring short and long distances with either of these. You should be aware of the mistakes that can occur when taping and the procedures that can be used to reduce these to a minimum.

This section contains the following topics:

- Tapes used in surveying
- Measuring a distance
- Common mistakes when taping

Tapes used in surveying

Survey tapes are available in various lengths up to 100 m, but 30 m is the most common length. They are either encased in steel or plastic boxes with a recessed winding lever (case tapes) or are mounted on open aluminium or plastic frames with a folding winding lever (frame tapes), as shown in Figure 4.4. Closed case tapes prevent an excess of dirt getting onto the blade and are cleaned by wiping with a cloth or glove as the tape is rewound after use in muddy or gritty conditions. Open frames make cleaning easier by simply rinsing the tape in a bucket of water or by hosing, but the tape must be allowed to dry after cleaning. Most tapes incorporate a small loop or grip at the end of the tape and various styles are possible for this, as shown in Figure 4.5. Usually, the claw end loop is fitted as standard to most tapes in the UK. Because the position of the zero point varies, it is essential to check where this is before using any tape.

When a *steel tape* is manufactured, the steel band and its printed graduations are protected by covering them with coats of polyester or nylon which give a tape its characteristic colour (usually yellow or white). Various methods are used for

Figure 4.4 ● Measuring tapes (courtesy Fisco Tools Ltd).

graduating these tapes, as shown in Figure 4.6. All steel tapes are manufactured so that they measure their nominal length at a specific temperature and under a certain pull. These standard conditions, 20 °C and 50 N, are printed somewhere on the first metre of the tape. The effects of variations from the standard conditions are discussed in Section 4.3. With care, it is possible to take measurements with a steel tape with an accuracy of better than 1 in 10,000 (a precision of 3 mm for a 30 m measured distance).

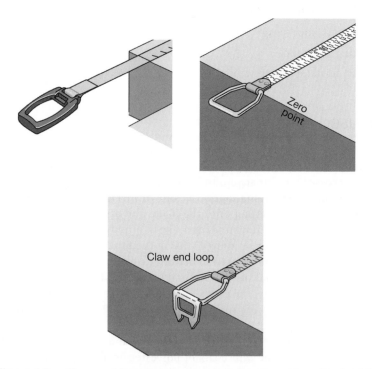

Figure 4.5 ● Tape end loops and zero points (courtesy Fisco Tools Ltd).

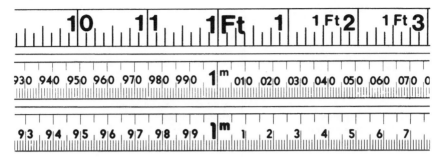

Figure 4.6 ● Steel tape graduations (courtesy Fisco Tools Ltd).

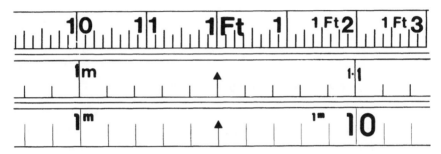

Figure 4.7 ● Fibreglass tape graduations (courtesy Fisco Tools Ltd).

In addition to steel tapes, *fibreglass tapes* are available in a variety of lengths with typical graduations being shown in Figure 4.7. These are made from fibreglass strands embedded in PVC. Compared with steel tapes they are lighter, more flexible and less likely to break, but they tend to stretch much more when pulled. However, their advantages are that they are rust-free and rot-proof, and are electrically non-conductive when dry.

Because they stretch a lot, this type of tape should only be used in mapping, sectioning and setting out where precisions in the order of 1 in 1000 (a precision of 30 mm for a 30 m measured distance) are acceptable for linear measurements. The difference in precision between the two types of tape is evident in their graduations – steel tapes are always graduated in millimetres whereas most fibreglass tapes are only graduated in centimetres.

Measuring a distance

Distance measurement using tapes involves determining the straight-line distance between two points.

When the length to be measured is less than that of the tape, measurements are carried out by unwinding and positioning the tape along the straight line between the points. The zero of the tape (or some convenient graduation) is held against one

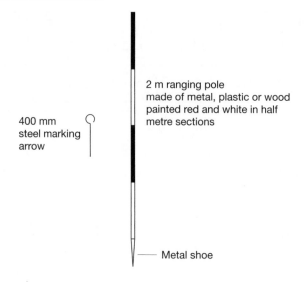

400 mm
steel marking
arrow

2 m ranging pole
made of metal, plastic or wood
painted red and white in half
metre sections

Metal shoe

Figure 4.8 ● Ranging pole and marking arrow.

point, the tape is straightened and pulled taut, and the distance is read directly on the tape at the other point. This will be the normal procedure on construction sites, where short distances tend to be measured with a tape instead of a total station because it is a quicker and more convenient method of distance measurement.

When it is necessary to measure a distance that exceeds that of the tape, some form of alignment of the tape is necessary. This is known as *ranging* and is achieved using *ranging poles* or *rods*, *survey* or *marking arrows* (see Figure 4.8) and two people to measure known as the *leader* and the *follower*. The procedure for measuring line AB is as follows.

● If starting at A, a ranging pole is pushed into the ground as vertical as possible at point B or it is held in a support tripod if this is on a hard surface or hard ground.

● The leader, carrying another ranging pole, unwinds the tape and walks towards point B, stopping just short of a tape length, at which point the ranging pole is held vertically.

● The follower steps a few paces behind point A and lines up the ranging pole held by the leader with point A and with the pole at B. This is known as *ranging by eye* and should be done by the follower sighting as low as possible on the poles.

● The tape is now straightened. The zero point or some convenient graduation is set against A by the follower and it is laid against the pole at B by the leader. With the tape in this position, it is pulled taut and the tape length marked by placing an arrow in the ground next to the full length of the tape or some convenient graduation. On hard ground, the arrow can be laid on the ground and fixed in position by some means (using a weight for example) or the point can be marked with paint, chalk or crayon.

- For the next tape length, the leader and the follower move ahead simultaneously with the tape unwound, the procedure being repeated but with the follower now at the first survey arrow or mark. Before leaving point A, the follower erects a ranging pole at A, as this will be sighted on the return measurement from B to A, which should always be taken as a check for gross errors.

- As measurement proceeds, the follower picks up each arrow and, on completion, the number of arrows held by the follower indicates the number of whole tape lengths measured. This number of tape lengths plus the section at the end less than a tape length gives the total length of the line.

Common mistakes when taping

The error that occurs most often in taping is to misread the tape. To help detect reading errors, long distances should be checked by repeating the measurement in the reverse direction. A good way of checking shorter distances is to again take a second measurement but to use different parts of the tape – in this case the difference of each pair of readings should give the same result.

Because the end loop on most tapes can be difficult to use as a zero, it is common practice to use another point on the tape as a 'zero' and subtract this from the reading obtained when measuring a distance. The false zero used is usually 100 mm or 1 m, being chosen for convenience. In practice, great care is needed when doing this not to forget that a different zero is being used or to confuse 100 mm with 1 m – to avoid mistakes it is good practice for the person holding the tape at the end loop to call out what the actual zero is just before each reading is taken. Alternatively, an initial (check) reading can be taken using the end loop as zero to give an approximate result which can then be followed by a more precise reading, but remembering the whole number of metres.

Because of the difficulties of using a tape with a zero in the end loop, the manufacturer sometimes positions the zero further along the band. Whatever type of tape is being used, it is necessary to check where the zero is by a simple inspection of the tape graduations *but before work commences.*

The procedure for ranging by eye may seem to be inaccurate when precisions of a few millimetres are expected for taped measurements. In fact, extreme care is not necessary and it can be shown that if the tape is misaligned by about 0.15 m at the centre of a 30 m span, this produces an error of about 1 mm per 30 m of measured distance. Whilst every care should be made to align the tape properly, there is often a tendency to overdo this at the expense of other corrections.

In use, although steel tapes can produce better results for distance measurements, they can be more difficult to use on construction sites. One of the worst problems is their tendency to permanently kink or even break if bent severely when trodden on or if they are run over by a vehicle of some sort. Never fully unwind a steel tape unless absolutely necessary and take care with its use on a busy site. If a tape has any permanent kinks in it or has been repaired, it should not be used.

Many of the measurements that are taken with steel and fibreglass tapes can now be measured electronically using a handheld laser distance meter. These have a range up to 200 m with an accuracy of 3–5 mm and require only one person to use them. They are described in Section 5.3.

Reflective summary

With reference to equipment and fieldwork for taping, remember:

— Although it is possible to measure a distance longer than a tape by ranging, this is seldom worthwhile on site and it is probably better to use a total station for distances exceeding a 50 m tape.

— The two most common sources of error in taping are misreading the tape and incorrectly identifying where the zero is. Always check distances with a repeat measurement, preferably using different parts of the tape, and if the end loop is not being used, always call out the value of the graduation being used as a 'zero', especially when setting out.

— If steel tapes are not used properly, they can be broken quite easily on site, whereas fibreglass tapes are lighter, highly flexible and less likely to be damaged.

— For engineering work, the important difference between steel and fibreglass tapes is their accuracy – because fibreglass tapes tend to stretch a lot their accuracy is limited to about 30 mm per 30 m, but steel tapes are capable of measuring to better than 3 mm per 30 m.

4.3 Systematic errors in taping

After studying this section you should understand that taping is subject to a series of systematic errors that must be accounted for in order to improve the precision of a measured distance. You should also be aware of the techniques involved and corrections that have to be applied to remove each of these errors.

This section contains the following topics:

● Slope measurements and slope corrections

● Standardisation

● Tension

● Temperature variations

● Sag (catenary)

● Combined formula

Slope measurements and slope corrections

The following corrections can be applied to taped distances in order to improve their precision: slope, standardisation, tension, temperature and sag. These are all classified as systematic errors (see Section 9.1) and, however carefully a distance has been measured, they cannot be removed unless a special field technique is employed to do this or a mathematical correction is computed and then applied to the recorded tape reading.

A distinction is made at this point between *observing and measuring* a distance that is required for, say, a control survey and *setting out a distance* for construction purposes. This section assumes that a measured distance is to be corrected – the procedures to follow when applying corrections to distances that are to be set out are described in a worked example in Section 4.5.

The method of ranging described in Section 4.2 can be carried out for any line, either sloping or level. Since all surveying calculations, plans and setting-out designs are based or drawn in the horizontal plane, any sloping length measured must be reduced to the horizontal before being used for calculations or plotting. This can be achieved by calculating a *slope correction* for the measured length or by *measuring the horizontal equivalent of the slope* directly in the field.

Consider Figure 4.9(a), which shows a sloping line AB. To record the horizontal distance D between A and B, the *method of stepping* may be employed in which a series of horizontal measurements is taken. To measure D_1, the tape zero or a whole metre graduation is held at A and the tape then held horizontally and on line towards B against a previously lined-in ranging pole. The horizontality of the tape should, if

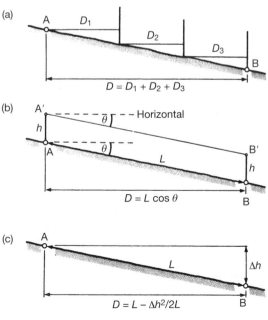

Figure 4.9 ● Slope measurements.

possible, be checked by a third person viewing it from one side some distance away. At another convenient tape graduation (preferably a whole metre mark again) the horizontal distance is transferred to ground level using a plumb line (a string line with a weight attached). The tape is now moved forward to measure D_2 and D_3 in a similar manner. It is recommended that the maximum length of an unsupported tape should be 10 m and that this should be considerably shorter on steep slopes, since the maximum height through which a distance is transferred should be 1.5 m. To do stepping accurately requires considerable skill and experience, and it is often better to choose another method of measuring the distance if possible.

As an alternative to stepping, the slope angle of the ground, θ, can be measured using a theodolite which is set up (say) at A and the slope angle measured along A'B' parallel with the ground (see Figure 4.9(b)). In this case, h will be the height of the theodolite above ground level.

The horizontal distance D can be calculated from the measured slope distance L as $D = L\cos\theta$, as shown in Figure 4.9(b). Alternatively, a correction can be computed from

$$\text{slope correction} = -L(1 - \cos\theta) \tag{4.1}$$

When converting an observed slope distance to its horizontal equivalent, this correction is always negative and is applied to the measured length L.

On comparing these methods of obtaining a horizontal distance, we can see that stepping is more useful when the ground between points is very irregular, whereas the slope angle relies on measurements taken on uniform slopes.

A third method is available if the height difference between the two points is known (for example, by levelling) and the slope between them is again uniform. In Figure 4.9(c), if Δh is the height difference between A and B, then

$$\text{slope correction} = -\frac{\Delta h^2}{2L} - \frac{\Delta h^4}{8L^3}$$

For slopes less than 10% the last term in this expression can be ignored and

$$\text{slope correction} = -\frac{\Delta h^2}{2L} \tag{4.2}$$

Standardisation

Under given conditions a tape has a certain nominal length. However, with a lot of use, a tape tends to stretch, and this effect can produce serious errors in length measurement. Therefore standardisation of tapes should be carried out frequently against a reference tape or baseline. If using a reference tape, standardisation should be done on a smooth, flat surface such as a road or footpath. The reference tape should not be used for any fieldwork and should be checked by the manufacturer as often as possible. From standardisation measurements a correction is computed as follows:

$$\text{standardisation correction} = \frac{L(l' - l)}{l} \tag{4.3}$$

where

L = recorded length of a line
l = nominal length of field tape (say 30 m)
l' = standardised length of field tape (say 30.011 m)

The sign of the correction depends on the values of l and l'. However, since l' is usually greater than l, this correction is usually positive and shows that when a tape is stretched and is too long it reads too short.

On a construction site, a baseline for standardising tapes is often used and should consist of two fixed points, located such that they are unlikely to be disturbed. These points could be nails in pegs, but marks set into concrete blocks are preferable. The length of the field tape is compared to the length of the baseline and the standardisation correction is given by

$$\text{standardisation correction} = \frac{L(l_B - l_F)}{l_B} \tag{4.4}$$

where

l_B = length of baseline
l_F = length of field tape along baseline

For a tape that is too long, l_F will be less than l_B.

Proper application of this correction is advisable as any error in standardisation gives the same proportional error in corrected distances. Care is also needed on site when using many different tapes – these should all be calibrated regularly on the same baseline.

As an alternative to the methods already given here, some companies offer a calibration service for tapes, and instead of performing this on site they can be standardised by the manufacturer or supplier, who should be contacted if this service is required.

Tension

This correction is only considered for steel tapes.

Steel, in common with many metals, is elastic, and the tape length varies with applied tension. If this is ignored, the effects of a varying tension on the precision of taping can be serious.

Every steel tape is manufactured and calibrated with a standard tension of 50 N applied, or it may be standardised on site at a different tension. So, instead of merely pulling the tape taut, an improvement in precision is obtained if the tape is pulled at its standard (or at least a known) tension. This can be achieved using a *spring balance* specially made for use in ground taping. When measuring, the spring balance is attached to the loop end of the tape and is pulled until its sliding index indicates that the correct tension is applied, as shown in Figure 4.10. This tension is then maintained while measurements are taken.

Figure 4.10 ● Spring balance for tensioning (courtesy York Survey Supplies).

Should a tape be subjected to a pull other than the standardising value, it can be shown that a correction to an observed length is given by

$$\text{tension correction} = \frac{L(T_F - T_S)}{AE} \tag{4.5}$$

where

T_F = tension applied to the tape (N)
T_S = standard tension (N)
A = cross-sectional area of the tape (mm^2)
E = modulus of elasticity for the tape material (N mm^{-2})
    (for steel tapes, typically 200,000 N mm^{-2})

The sign of the correction depends on the magnitudes of T_F and T_S, but as with the standardisation correction, if the tape is pulled at a tension greater than its standardised value it will stretch, giving rise to a positive correction.

Temperature variations

As with tension, this correction is only considered for steel tapes.

In addition to the effects of standardisation and tension, steel tapes contract and expand with temperature variations and are calibrated at a standard temperature of 20 °C by the manufacturer. Alternatively, a tape could be standardised on a site baseline at a temperature that is not 20 °C – the temperature at which this is done then becomes the on-site standardising temperature.

In order to improve precision, the temperature of the tape has to be recorded, since it will seldom be used at 20 °C or the on-site standardising temperature, and special surveying thermometers are used for this purpose. When using the tape along the ground, measurement of the air temperature can give a different reading from that obtained close to the ground, so it is normal to place the thermometer alongside the tape at ground level. For this reason, the thermometers are usually metal-cased for protection. When in use they should be left in position until a steady reading is obtained, since the metal casing can take some time to reach a constant temperature. It is also necessary to have the tape in position for a period before readings are taken to allow it also to reach the ambient temperature. It is bad practice to measure a distance on site in winter with a tape that has just been removed from a heated office.

The temperature correction is given by

$$\text{temperature correction} = \alpha L(t_f - t_s) \qquad (4.6)$$

where

α = the coefficient of expansion of the tape material
(for example 0.0000112 per °C for steel)
t_f = mean field temperature (°C)
t_s = temperature of standardisation

The sign of this correction is given by the magnitudes of t_F and t_S.

Sag (catenary)

As with tension and temperature, this correction is only applied to steel tapes.

On a construction site, this correction is only considered when measuring distances less than a tape length between elevated points on structures when the tape may be suspended for ease of measurement. If this is done, the tape will sag under its own weight in the shape of a catenary curve, as shown in Figure 4.11, and will give an incorrect measurement if this is ignored.

Since the tape reading required is the chord between the end points of the distance, a sag correction must be applied to the catenary length measured. This correction is given by

$$\text{sag correction} = -\frac{w^2 L^3 \cos^2 \theta}{24 T_F^2} = -\frac{W^2 L \cos^2 \theta}{24 T_F^2} \qquad (4.7)$$

where

θ = the angle of slope between tape supports
w = the weight of the tape per metre length (N m^{-1})
W = the total weight of the tape (N)
T_F = the tension applied to the tape (N)

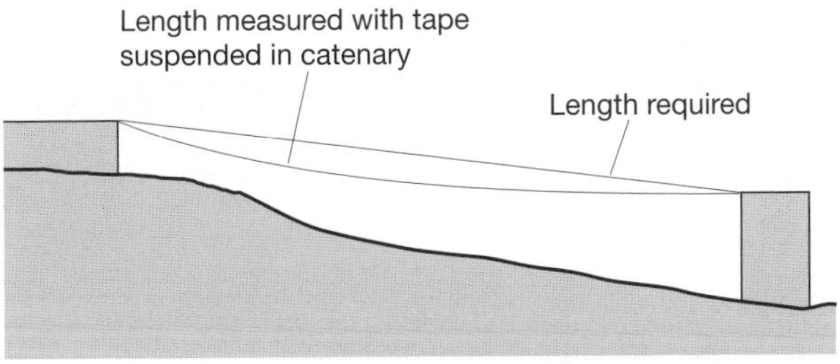

Length measured with tape
suspended in catenary

Length required

Figure 4.11 ● Measurement in catenary.

When converting a measurement to its equivalent chord length, this correction is always negative.

Combined formula

The corrections discussed in the preceding sections are usually calculated separately and then used in the following equations:

For converting slope distances L to horizontal distances D:

$$D = L - \text{slope} \pm \text{standardisation} \pm \text{tension} \pm \text{temperature} - \text{sag} \qquad (4.8)$$

For vertical measurements V:

$$V = V_M \pm \text{standardisation} \pm \text{tension} \pm \text{temperature} \qquad (4.9)$$

where

V_M = measured vertical distance

Both of these can be used when measuring and setting out distances, as shown in the worked examples in Section 4.5.

Reflective summary

With reference to taping corrections, remember:

— Slope, standardisation, tension, temperature and sag are all systematic errors and must not be confused with mistakes in reading, misaligning or zeroing the tape or any other errors of a personal nature.

— No matter how carefully a distance has been measured, the effects of slope, standardisation, tension, temperature and sag will still be present unless they are removed by some means.

— The usual method of accounting for slope, standardisation, tension, temperature and sag is to use a special field procedure or to calculate and apply a mathematical correction to the measured distance.

— For all measurements with any tape, it is *always* necessary to apply a slope correction to a distance whether this is being measured or set out.

4.4 Precision and applications of taping

After studying this section you should appreciate that the precision of taping can vary according to which systematic errors are accounted for in the measurement process.

How good should my taping be?

The general rules for the precision of taping for construction work can be summarised as follows.

For a precision of the order of 1 in 1000 (that is ± 30 mm per 30 m), a fibreglass tape can be used, but it is necessary to apply slope and standardisation corrections. This would be suitable for the location of earthworks, pipelines and soft detail.

To obtain a precision better than this, steel tapes must be used.

For a maximum precision of 1 in 5000 with a steel tape (that is ± 6 mm per 30 m), measurements can be taken over most ground surfaces if only standardisation and slope corrections are applied. This order of precision would be suitable for drainage works and for locating hard detail.

If the tape is tensioned correctly and temperature variations are taken into account, the precision of steel taping can be increased to the order of 1 in 10,000 (that is ± 3 mm per 30 m). On specially prepared surfaces or over spans less than a tape length, the precision may be improved further. This is the precision to which most sites would attempt to measure distances, and covers general setting out of roads, buildings, bridges and most other structures. Any distances measured for control surveys should also meet this specification.

To obtain the best precision of 1 in 20,000 (that is ± 1.5 mm per 30 m), it is necessary to measure with the tape on a very smooth surface or with the tape suspended. If suspended, sag corrections must be applied to the measurement in addition to all the other corrections. A precision of 1 in 20,000 or better is often required on engineering projects requiring high-quality setting out and control, such as dams, long tunnels and bridges, nuclear power stations, fabrication of offshore structures, and tall buildings.

All of the systematic errors, corrections and precisions given for steel taping are summarised in Table 4.1.

Specifications for distance measurement in control surveys and setting out are given in BS 5606, BS 5964 and in the ICE Design and Practice Guide – these are listed in further reading at the end of the chapter.

Table 4.1 ● Taping precisions.

Systematic error	Correction formula	Procedure required to achieve stated precision	
		1:5000	**1:10,000**
Slope Equations (4.1) and (4.2)	$\dfrac{\Delta h^2}{2L}$ or $L(1 - \cos\theta)$	Slope correction *always* applied and is usually the largest correction	
Standardisation Equations (4.3) and (4.4)	$\dfrac{L(l' - l)}{l}$ or $\dfrac{L(l_B - l_F)}{l_B}$	Standardise tape and apply correction On construction sites, it is better to establish a baseline rather than rely on having a reference tape	
Tension Equation (4.5)	$\dfrac{L(T_F - T_S)}{AE}$	Negligible effect if tape pulled 'sensibly'	Best to apply standard tension where possible or apply correction
Temperature Equation (4.6)	$\alpha L(t_f - t_s)$	Only important in hot or cold weather	Measure temperature and apply correction
Sag Equation (4.7)	$\dfrac{w^2 L^3 \cos^2\theta}{24 T_F^2}$ or $\dfrac{W^2 L \cos^2\theta}{24 T_F^2}$	No correction necessary	Apply correction for suspended measurements

Reflective summary

With reference to the precision and applications of taping, remember:

— The precision of taping varies according to the type of tape used and the systematic errors that have been accounted for.

— On construction sites, the effects of tension on a taped distance are often ignored – this can have a serious effect on the precision obtained if the tape is pulled too hard or is too slack.

— As with tension, the effect of temperature is also ignored on site for most taped measurements – this can also have a serious effect on the precision obtained in hot or cold weather or at temperatures very different from those at which the tape was standardised.

— If the precision of taping is to be improved to allow for tension and temperature, extra equipment such as spring balances and thermometers is required, and more information about the tape has to be found

- all of this complicates the field procedure and involves more calculations.

— Because it is difficult to monitor the tension and temperature of a 100 m tape, their use on site is not recommended where a high precision is required.

— The precision of a measured distance should be carefully matched with its application. For example, setting out the positions of steel columns with a fibreglass tape is not recommended and using a steel tape with a full set of corrections applied to each measurement is not required when setting out the alignment of a pipeline.

4.5   Steel taping worked examples

After studying this section you should be able to calculate all the necessary corrections associated with systematic errors in taping. You should also understand the difference between problems in taping that involve the measurement of horizontal distances, the setting out of slope distances for construction work and vertical measurements.

This section contains the following topics:

- Measuring a horizontal distance with a steel tape
- Setting out a slope distance with a steel tape
- Measuring a vertical distance with a steel tape

Note: In each of the following problems, Young's modulus (E) for steel is assumed to be 200 kN mm^{-2} and the coefficient of thermal expansion (α) for steel is assumed to be 0.0000112 per °C.

Worked example 4.1: Measuring a horizontal distance with a steel tape

Question

A steel tape of nominal length 30 m was used to measure the distance between two points A and B on a structure. The following measurements were recorded with the tape suspended between A and B:

Line	Length measured	Slope angle	Mean temperature	Tension applied
AB	29.872 m	3°40'	5 °C	120 N

The standardised length of the tape against a reference tape is 30.014 m at 20 °C and 50 N tension. The tape weighs 0.17 N m^{-1} and has a cross-sectional area of 2 mm^2.

Calculate the horizontal length of AB.

Solution

A series of corrections is computed as follows

From equation (4.1):

slope correction $= -L(1 - \cos \theta) = -29.872 \, (1 - \cos 3°40') = $ **−0.0611 m**

From equation (4.3):

standardisation correction $= \dfrac{L(l' - l)}{l} = \dfrac{29.872(30.014 - 30)}{30} = $ **+0.0139 m**

From equation (4.5):

tension correction $= \dfrac{L(T_F - T_S)}{AE} = \dfrac{29.872(120 - 50)}{2 \times 200,000} = $ **+0.0052 m**

From equation (4.6):

temperature correction $= \alpha L(t_f - t_s) = 0.0000112 \times 29.872(5 - 20) = $ **−0.0050 m**

From equation (4.7):

sag correction $= -\dfrac{w^2 L^3 \cos^2 \theta}{24T_F^2} = -\dfrac{0.17^2 \times 29.872^3 \times \cos^2 3°40'}{24 \times 120^2} = $ **−0.0022 m**

Using equation (4.8):

Horizontal length AB $= 29.872 - 0.0611 + 0.0139 + 0.0052 - 0.0050 - 0.0022$

$= 29.8228$

$= $ **29.823 m** (to the nearest mm)

Worked example 4.2: Setting out a slope distance with a steel tape

Question

On a construction site, a point R is to be set out from a point S using a 50 m steel tape. The horizontal length of SR is designed as 35.000 m.

During the setting out the steel tape is laid on the ground and pulled at a tension of 70 N, the mean temperature being 4 °C. The reduced levels of S and R are 22.75 m and 24.86 m.

The tape was standardised on site as 40.983 m at 50 N tension and 12 °C on a baseline of length 41.005 m and it has a cross-sectional area of 2.4 mm^2.

Calculate the length that should be set out on the tape along the direction SR to establish the exact position of point R.

Solution

In this case, the horizontal distance D is known and the slope distance L must be calculated.

The slope, standardisation, tension and temperature corrections all apply, but the sag correction does not since the tape is laid along the ground.

Although L is not known, for the purposes of calculating the corrections it is sufficiently accurate to use D instead of L in each formula.

Substituting D for L in equation (4.2) gives

$$\text{slope correction} = -\frac{\Delta h^2}{2D} = -\frac{(24.86 - 22.75)^2}{2 \times 35.000} = \mathbf{-0.0636\ m}$$

Substituting D for L in equation (4.4) gives

$$\text{standardisation correction} = \frac{D(l_B - l_F)}{l_B} = \frac{35.000(41.005 - 40.983)}{41.005} = \mathbf{+0.0188\ m}$$

Substituting D for L in equation (4.5) gives

$$\text{tension correction} = \frac{D(T_F - T_S)}{AE} = \frac{35.000(70 - 50)}{2.4 \times 200,000} = \mathbf{+0.0015\ m}$$

Substituting D for L in equation (4.6) gives

$$\text{temperature correction} = \alpha L(t_f - t_s) = 0.0000112 \times 35.000(4 - 12) = \mathbf{-0.0031\ m}$$

The slope length SR is obtained using equation (4.8), where

$$D_{SR} = L_{SR} - \text{slope} \pm \text{standardisation} \pm \text{tension} \pm \text{temperature}$$

From which

$$35.000 = L_{SR} - 0.0636 + 0.0188 + 0.0015 - 0.0031$$

or

$$L_{SR} = 35.000 + 0.0636 - 0.0188 - 0.0015 + 0.0031$$

As can be seen, the sign of each correction is reversed for *setting out* problems compared to *measurement* problems.

This gives

$$L_{SR} = 35.0464 = \mathbf{35.046\ m} \text{ (to the nearest mm)}$$

Worked example 4.3: Measuring a vertical distance with a steel tape

Question

A steel tape of nominal length 30 m was used to transfer a level from a reference line near the base of a vertical concrete column to a reference line near its top.

A 100 N weight was attached to its zero end and the tape was hung down the side of the column such that its 100 mm mark was against the bottom reference line. A reading of 14.762 m was obtained at the top reference line.

The tape used had a cross-sectional area of 1.9 mm² and was standardised on the flat as 30.007 m at a tension of 50 N and a temperature of 20 °C. During the measurement the mean temperature of the tape was 29 °C.

Calculate the vertical distance between the two reference lines.

Solution

For this measurement, only the standardisation, tension and temperature corrections apply as follows

$$\text{measured length} = 14.762 - 0.100 = \textbf{14.662 m}$$

$$\text{standardisation correction} = \frac{14.662(30.007 - 30)}{30} = \textbf{+0.0034 m}$$

$$\text{tension correction} = \frac{14.662(100 - 50)}{1.9 \times 200{,}000} = \textbf{+0.0019 m}$$

This calculation does not account for the weight of the tape which adds to the tension – this has been ignored, as it will have a negligible effect on the correction and vertical distance.

$$\text{temperature correction} = 0.0000112 \times 14.662(29 - 20) = \textbf{+0.0015 m}$$

The vertical distance is given by equation (4.9) as

$$V = L \pm \text{standardisation} \pm \text{tension} \pm \text{temperature}$$
$$= 14.662 + 0.0034 + 0.0019 + 0.0015 = 14.6688$$
$$= \textbf{14.669 m}$$

Reflective summary

With reference to steel taping worked examples, remember:

— When calculating all of the individual corrections for systematic errors, it is better to do this to 0.1 mm and to round the final answer to 1 mm.

— For setting out work, calculate all the corrections as if the distance had been measured – then change the sign of each correction and apply these to the known horizontal distance to give the required setting out distance.

— Take care with the magnitude for each correction, as they are all small (except for the slope correction on steep slopes) and should only be a few millimetres – if answers much bigger than this are obtained a mistake has been made and the calculation must be checked.

Exercises

Note: Where appropriate, assume that Young's modulus (E) for steel is 200 kN mm^{-2} and the coefficient of thermal expansion of steel (α) is 0.0000112 per °C.

4.1 Discuss the circumstances under which you might choose to use a fibreglass tape instead of a steel tape for distance measurements on a construction site.

4.2 List, with their formulae, the five corrections that apply to taping.

4.3 The horizontal distance between two points approximately 65 m apart is to be measured with a steel tape. Describe the field procedure for this measurement if an accuracy of 1 in 10,000 is required.

4.4 What are the sources of error that occur most in taping and how can they be avoided?

4.5 Describe three ways of determining a horizontal distance on sloping ground.

4.6 A steel tape has the following properties:

Nominal length: 50 m
Standardised length: 50.010 m at 50 N tension and 20 °C
Cross-sectional area: 2.5 mm^2
Weight: 0.15 N m^{-1}

For a recorded length of 40.000 m with this tape, calculate the following corrections:

Slope, where the height differences between the ends of the line are 1 m and 2 m
Standardisation
Tension if 35 N and 75 N are applied to the tape for the measurement
Temperature if this is 10 °C and 25 °C during the measurement
Sag with tensions of 50 and 100 N applied to the tape (assume the slope angle is zero)

4.7 A steel tape of nominal length 30 m was used to check the distance between two offset pegs A1 and A2 on a construction site. The following results were obtained with the tape suspended between A1 and A2.

Length recorded on tape	Heights		Temperature	Tension applied
	A1	A2		
23.512 m	21.50 m	23.50 m	28 °C	100 N

When compared to a 25.000 m baseline, the tape read 24.994 m with 50 N tension applied to it at 15 °C. The cross-sectional area of the tape is 2.0 mm^2 and it weighs 4.5 N.

Calculate the horizontal length A1 to A2.

4.8 The distance between two points A and B is recorded as 27.554 m but without applying any corrections to the measurement.

Later, this was checked and found to be 27.567 m but at a recorded tension of 50 N and temperature of 20 °C. The difference in level between A and B was known to be 0.15 m.

If the 30 m tape used had been standardised as having a length of 30.007 m at a tension of 50 N and temperature of 10 °C and has dimensions of 13 × 0.2 mm, determine by how much the original measurement was in error.

4.9 A steel tape of nominal length 30 m was used to transfer a level from point C1 near the base of a reinforced concrete column to another point C2 near its top.

Just before the measurement was done, the tape was standardised against a 30 m reference tape as 30.015 m at a tension of 50 N. From data published by the manufacturer, the cross-sectional area of the tape was found to be 1.7 mm².

On site, a 100 N weight was attached to the tape and it was hung down the side of the column in a vertical position with its 1 m mark held against C1. A reading of 20.839 m was obtained at C2.

If the reduced level of C1 at the bottom of the column is 12.365 m, calculate the reduced level of C2 at the top.

4.10 A line XY is to be set out from point X using a 30 m steel tape. The horizontal length of XY given on drawings is 20.000 m.

Records show that the tape was standardised on the 30.000 m site baseline as 29.995 m with 50 N tension applied to it at 18 °C.

At the time of setting out, the tape is laid on the ground and pulled at a tension of 50 N and the temperature is 26 °C. A theodolite was used to measure the slope angle as 04°12'. If the 1 m graduation is held at X, calculate the length that should be set out on the tape along the direction XY to establish the exact position of Y.

4.11 To check the setting out of the corners of a 20 × 25 m rectangular building, the diagonals were carefully measured with a 50 m steel tape and the following results were obtained with the tape pulled at 50 N tension and laid on the ground.

Diagonal B1–B3: Tape readings 1.137 and 33.157 m

Diagonal B2–B4: Tape readings 0.927 and 32.952 m

Using a level, the following staff readings were taken at each corner

B1: 2.392 m   B2: 1.683 m   B3: 1.679 m   B4: 1.201 m

Immediately following this, the tape was standardised on the 48.125 m site baseline as 48.124 m at 50 N tension.

Calculate the errors in the diagonals assuming the ground slopes uniformly along these.

Further reading and sources of information

For assessing the accuracy of taping and distance measurement, consult

BS 5606: 1990 *Guide to accuracy in building* (British Standards Institution [BSI], London). BSI web site http://www.bsi-global.com/.

BS 5964:1996 (ISO 4463: 1995) *Building setting out and measurement* (British Standards Institution, London). BSI web site http://www.bsi-global.com/.

BS 7334-2: 1990 (ISO 8322-2:1989) *Measuring instruments for building construction. Methods for determining accuracy in use: measuring tapes.* (British Standards Institution, London). BSI web site http://www.bsi-global.com/.

For general guidance and assessing the accuracy of taping on site refer to

ICE Design and Practice Guide (1997)*The management of setting out in construction.* Thomas Telford, London.

Total stations

 Aims

After studying this chapter you should be able to:

- Understand how total stations measure distances electronically

- Identify and describe all the features of total stations that are used for angle and distance measurement

- Distinguish between the different categories of total station and assess the best applications for each of these on site

- Discuss the various methods by which survey data can be stored and transferred between total stations and computers

- Evaluate the effect of instrumental and other errors on angle and distance measurements taken with a total station and how these can be minimised

- Understand how a total station can be used for measuring heights

This chapter contains the following sections:

5.1 Integrated total stations

After studying this section you should be aware of what an integrated total station does, how it is used on site and the accuracies it can achieve.

What is a total station?

In the previous chapter, the various types of distance that can be measured were described, together with taping methods. Although taping and theodolites are used regularly on site, *total stations* are also used extensively in surveying, civil engineering and construction because they can measure both distances and angles simultaneously, with relative ease and to a high degree of precision. This chapter serves as an introduction to the total station and its applications in surveying.

A typical total station is shown in Figure 5.1. The appearance of this is very similar to an electronic theodolite, but the difference is that it is combined with a distance measurement component which is fitted around the telescope. Because the instrument combines both angle and distance measurement in the same unit, it is known as an *integrated total station* which can measure horizontal and vertical angles as well as slope distances. Using the vertical angle, the total station can calculate the horizontal and vertical distance components of the measured slope distance and display these. As well as these basic functions, total stations are capable of performing a number of different survey tasks and associated calculations and they can store relatively large amounts of data. As with the electronic theodolite, all the functions of a total station are controlled by its microprocessor (or computer) which is accessed through a keyboard and display.

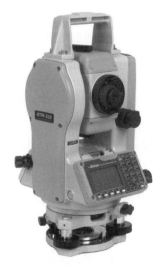

Figure 5.1 ● Nikon DTM-332 integrated total station (courtesy Survey Supplies).

Figure 5.2 ● Measuring with a total station (courtesy Topcon).

Figure 5.3 ● Robotic total station in use (courtesy Trimble Navigation Ltd).

To use a total station, it is set over one end of the line to be measured and some form of reflector is positioned at the other end such that the line of sight between the instrument and the reflector is unobstructed. This is shown in Figure 5.2, where the reflector is a prism attached to a detail pole. The telescope is aligned and pointed at the prism, the measuring sequence is initiated and a signal is transmitted from the instrument towards the reflector, where part of it is returned to the instrument. This is processed, in a few seconds, to give the slope distance together with the horizontal and vertical angles. A total station can also be used in reflectorless mode, in which the telescope is aimed at the point to be measured but without using a reflector. Some instruments have motorised drives and can use automatic target recognition to search and lock onto a prism – this process is fully automated and does not require an operator. Taking this a stage further, some total stations can be controlled from the detail pole, enabling surveys to be carried out by one person, as shown in Figure 5.3.

A total station is centred and levelled in the same way as a theodolite and the procedures described in Chapter 3 for measuring angles and setting out with a theodolite can all be done with a total station. However, an integrated total station can perform many more measurement and setting out tasks than an electronic theodolite, as will be described in later sections of this chapter.

Most total stations have a distance measuring range of up to a few kilometres when using a single prism, a range of at least 100 m in reflectorless mode and an accuracy of 2–3 mm at short ranges, which will decrease to about 4–5 mm at 1 km. Most of them have angular accuracies varying from 1" to 10".

Although angles and distances can be measured and used separately, the most common applications for total stations occur when these are combined to define position in control surveys, mapping and setting out. All of these are discussed in subsequent chapters.

As well as the total station, site surveying is increasingly being carried out using GPS equipment. Some predictions have been made that this trend will continue, and in the long run GPS methods will replace all others. Although the use of GPS is increasing, total stations are one of the predominant instruments used on site for surveying and will be for some time. Eventually, the two will find applications that complement rather than compete with each other.

Reflective summary

With reference to integrated total stations, remember:

— The advantage of the total station compared with other survey equipment is that it can measure angles, distances and heights simultaneously.

— Although an integrated total station can be a highly sophisticated precision instrument, its basic function is simply to measure angles and distances. However, it is also a computer that is capable of storing and processing survey data in many different ways, which can be useful for site surveying.

— Even though they are relatively easy to operate, total stations must be used properly and must be checked regularly in order to ensure that the high degree of accuracy they are capable of is achieved.

— In common with other survey equipment, the development of total stations continues and new models are introduced at regular intervals by each manufacturer.

5.2 Electromagnetic (or electronic) distance measurement

After studying this section you should be able to explain how total stations measure distances using the phase shift and pulsed laser methods. You should be familiar with the different types of reflector than can be used for distance measurement and understand that total stations can also measure in reflectorless mode. Since lasers are now used in nearly all total stations, you should be aware of safety issues when using these.

This section includes the following topics:

- Distance measurement
- Reflectors and reflectorless measurements
- Laser safety and total stations

Distance measurement

When a distance is measured with a total station, an electromagnetic wave or pulse is used for the measurement – this is propagated through the atmosphere from instrument to reflector or target and back during a measurement. Distances are measured by one of two methods: the *phase shift* method, which uses continuous electromagnetic waves, or the *pulsed laser* method, in which pulses of laser radiation are used.

Phase shift method

This technique uses continuous electromagnetic waves for distance measurement. Although these are extremely complex in nature, electromagnetic waves can be represented in their simplest form as periodic sinusoidal waves, as shown in Figure 5.4. The wave completes a *cycle* when moving between identical points on the wave and the number of times in one second the wave completes a cycle is called the *frequency* of the wave. The frequency is represented by f hertz, 1 hertz (Hz) being 1 cycle per second. The *wavelength* is the distance which separates two identical points on the wave or is that length travelled in one cycle by the wave and is denoted by λ metres.

The *speed of propagation v* of an electromagnetic wave depends on the medium through which it is travelling. In a vacuum or free space, the speed of propagation is called the *speed of light* and is given the symbol c. This is known at the present time as

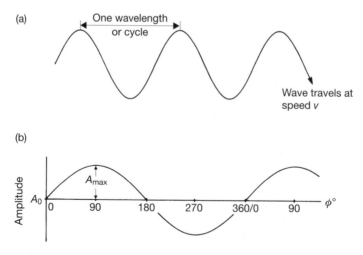

Figure 5.4 ● Sinusoidal wave motion: (a) as a function of distance or time (b) as a function of phase angle ϕ.

$c = 299{,}792{,}458 \text{ m s}^{-1}$

Interestingly, the definition of the metre is that it is the distance travelled by light in a vacuum during a time interval of $1/299{,}792{,}458$ of a second, the inverse of the speed of light c.

An exact knowledge of c, which is called a universal constant, is essential to electromagnetic distance measurement. The speed of electromagnetic radiation when propagated through the atmosphere will vary from the free space value according to the temperature, humidity and pressure at the time of measurement. However, using well-established formulae, v can be calculated for any set of atmospheric conditions (this is discussed further in Section 5.5 in *Atmospheric effects*; p. 168).

All of the above properties of electromagnetic waves are related by

$$\lambda = \frac{v}{f}$$

A further term associated with periodic waves is the *phase* of the wave. As far as distance measurement is concerned, this is a convenient method of identifying fractions of a wavelength or cycle. A relationship that expresses the instantaneous amplitude of a sinusoidal wave is (see Figure 5.4(b))

$$A = A_{max} \sin \phi + A_0$$

where A_{max} is the maximum amplitude developed by the wave, A_0 is the reference amplitude and ϕ is the phase angle. Angular degrees are often used as units for a phase angle up to a maximum of $360°$ for a complete cycle and it is important to note that a phase angle between 0 and $360°$ can apply to the same position on *any* cycle or wavelength.

The phase shift method determines distance by measuring the difference in phase angle between transmitted and reflected signals. This phase difference is usually expressed as a fraction of a cycle, which can be converted into distance when the frequency and velocity of the wave are known.

The methods involved in measuring a distance by the phase shift method are as follows.

In Figure 5.5(a), a total station has been set up at A and a reflector at B so that distance AB = D can be measured.

Figure 5.5(b) shows the same configuration as in Figure 5.5(a), but only the details of the electromagnetic wave path have been shown. The wave is continuously transmitted from A towards B, is instantly reflected at B (without change of phase angle) and received back at A. For clarity, the same sequence is shown in Figure 5.5(c) but the return wave has been opened out. Points A and A′ are the same since the transmitter and receiver would be side by side in the instrument at A.

From Figure 5.5(c) it is apparent that the distance covered by the wave in travelling from A to A′ is given by

$$2D = n\lambda_m + \Delta\lambda_m$$

where D is the distance between A and B, λ_m the wavelength of the measuring wave, n the whole number of wavelengths travelled by the wave and $\Delta\lambda_m$ the fraction of a

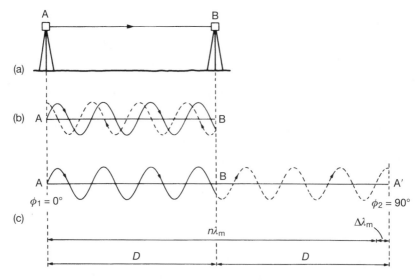

Figure 5.5 ● Phase shift method.

wavelength travelled by the wave. Since the double distance is measured, the effective measuring wavelength is $\lambda_m/2$.

The distance D is made up of two separate parts which are determined by two processes.

● The *phase shift* or $\Delta\lambda_m$ measurement is carried out by measuring phase angles. An electronic phase meter or detector built into the total station at A measures the phase of the electromagnetic wave as it is transmitted. Let this be ϕ_1 degrees. Assume the same detector also measures the phase of the wave as it returns from the reflector at A' (ϕ_2°). These two can be compared to give a measure of $\Delta\lambda_m$ using the relationship

$$\Delta\lambda_m = \frac{\text{phase difference in degrees}}{360} \times \lambda_m = \frac{(\phi_2 - \phi_1)°}{360} \times \lambda_m$$

● Since the phase value ϕ_2 can apply to any incoming wavelength at A', the phase shift can only provide a means of determining by how much the wave travels in excess of a whole number of wavelengths. Therefore, some method of determining $n\lambda_m$, the other part of the unknown distance, is required. This is referred to as *resolving the ambiguity* of the phase shift and can be carried out by one of two methods. Either the measuring wavelength is increased in multiples of 10 until a coarse measurement of D is eventually made, or the distance can be found by measuring the line with different but related wavelengths to form simultaneous equations of the form $2D = n\lambda_m + \Delta\lambda_m$. These are solved to give a value for D.

Whatever techniques are used by a total station to carry out phase shift measurements and to resolve ambiguities, they are fully automated and the instrument will measure and display a distance at the press of a key – no calculations or any further action are required.

Although they might appear to be very different, the same methods are used by GPS equipment to measure distances and to determine position. In this case, L-Band signals are transmitted by the satellites with a wavelength of about 0.2 m, from which a $\Delta\lambda_m$ is obtained using phase shift methods. Resolving the ambiguity to give n λ_m is also necessary, but is a much more difficult process than with a total station because of the long distances to the satellites. The processes by which GPS determines distance are described in Chapter 7.

Modulation

Total stations use the wavelength λ_m of an electromagnetic wave as the basic unit for measuring a distance by the phase shift method. The value chosen for λ_m depends to a great extent on the desired accuracy of the instrument and on *phase resolution*, the smallest fraction of a cycle that the instrument is capable of resolving. By combining a distance resolution of 1 mm with 1/10,000 digital phase resolution, a typical value chosen for λ_m is 10 m (1 mm × 10,000), which corresponds to a frequency for the measuring wave of 30 MHz. This and similar frequencies used for distance measurement by total stations are in the VHF part of the electromagnetic spectrum and, although it is possible to generate and transmit a VHF signal fairly easily, problems occur when these are to be propagated through the atmosphere. To transmit this order of frequency over any distance without significant attenuation of the signal would require either a very large transmitter or a small but very inefficient transmitter that would require considerable power to drive it. Both of these alternatives are unacceptable for portable surveying equipment. A solution to these problems might be to decrease the value of λ_m and therefore increase the frequency of measurement and accuracy of length measurement. This could be done until a suitable compromise is reached for both transmission and measurement. Unfortunately, the phase measurement process tends to become unstable at high frequencies, and use of a very short measuring wavelength would result in difficulties with resolving the ambiguity of measurement.

In order to be able to use a typical measuring wavelength of 10 m and combine this with efficient propagation, the process of *modulation* is used, in which the measuring wave is mixed with a *carrier wave* of much higher frequency. The carrier wave is chosen to be a type of radiation that can be transmitted through the atmosphere without serious attenuation over long distances. The type of modulation used in the majority of total stations is *amplitude* or *intensity modulation* (Figure 5.6), in which the measuring wave is used to vary the amplitude of the carrier wave. During a distance measurement, although it is the carrier wave that is transmitted, the phase shift is carried out as if the measuring wave was transmitted directly. The carrier waves used in most instruments are either infrared or visible red lasers and this is due to the carrier source, which is a semiconductor diode. These can be very easily amplitude modulated at the high frequencies required for distance measurement and provide a simple and inexpensive method of producing a modulated carrier wave, as shown in Figure 5.7.

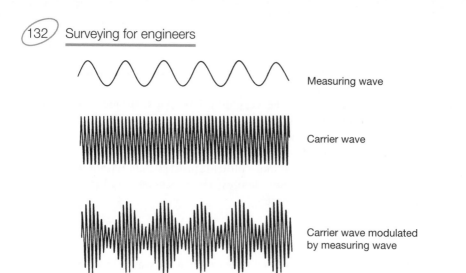

Figure 5.6 ● Amplitude or intensity modulation.

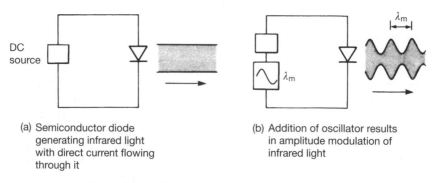

(a) Semiconductor diode generating infrared light with direct current flowing through it

(b) Addition of oscillator results in amplitude modulation of infrared light

Figure 5.7 ● Modulation of semiconductor diode.

Phase measuring system

The schematic diagram of Figure 5.8 shows the essential parts of a phase shift distance-measuring system of an integrated total station.

The sinusoidal modulation signal or measuring wave is derived from a crystal-controlled oscillator, the frequency value of which is typically 10–100 MHz. It is necessary that the frequency of the measuring wave be held at a constant value within a few parts per million (ppm) of the nominal frequency, as this determines the accuracy for distances when scaled by the velocity of the wave (remember $\lambda = v/f$).

The intensity-modulated carrier is transmitted from the total station towards a reflector or target at the remote end of the line to be measured. Since the carrier is an infrared or visible laser, optical components are used to focus and transmit the carrier along the line of sight of the telescope as a highly collimated beam with a low angular divergence. This helps to increase the measuring range of the instrument.

The receiving optics are usually mounted coaxially with the transmitting optics and they occupy as large an area as possible so as to collect sufficient return signal for

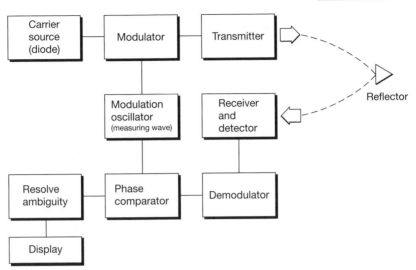

Figure 5.8 ● Phase shift distance measuring system.

measurement purposes. Upon re-entering the instrument, the modulated carrier is detected and demodulation takes place (the separation of the measuring and carrier waves).

From the demodulator, the return signal is fed into the phase comparator. A reference signal, also derived from the modulation oscillator, is fed into the phase comparator and these two signals are processed to produce a $\Delta\phi$ or $\Delta\lambda$ value for the relevant line. In addition to this, further measurements are taken to resolve the ambiguity of measurement.

This process has been described as if one wave was transmitted from instrument to reflector. However, a standing wave is established between the total station and reflector and the phase shift is sampled many times during a measurement to improve precision. In addition to this, several patented improvements have been made to the optical and electronic components of the system described to enhance the accuracy and the range of the method.

Pulsed laser distance measurement

In many total stations, distances are obtained by measuring the time taken for a pulse of laser radiation to travel from the instrument to a prism (or target) and back. As in the phase shift method, the pulses are derived from an infrared or visible laser diode and they are transmitted through the telescope towards the remote end of the distance being measured, where they are reflected and return to the instrument. Since the velocity v of the pulses can be accurately determined, the distance D can be obtained using $2D = vt$, where t is the time taken for a single pulse to travel from instrument–target–instrument. This is also known as the *timed-pulse* or *time-of-flight* measurement technique, in which the *transit time t* is measured using electronic

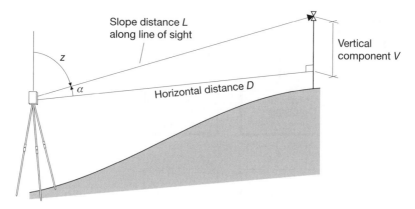

Figure 5.9 ● Slope and horizontal distances with total station.

signal-processing techniques. Although only a single pulse is necessary to obtain a distance, the accuracy obtained would be poor. To improve this, a large number of pulses (typically 20,000 every second) are analysed during each measurement to give a more accurate distance.

The pulsed laser method is a much simpler approach to distance measurement than the phase shift method, which was originally developed about 50 years ago at a time when electronic distance measurement was first invented. Although methods based on transit times were considered at that time, it was not possible to measure these with sufficient accuracy for distance measurement under site conditions. Consequently, the phase shift method was adopted, but it is now being superseded by the pulsed laser method due to advances in technology.

Slope and horizontal distances

Both the phase shift and pulsed laser methods will measure a slope distance L from the total station along the line of sight to a reflector or target. For most surveys, the horizontal distance D is required as well as the vertical component V of the slope distance. With reference to Figure 5.9, these are calculated by the total station using

Horizontal distance $D = L \cos \alpha = L \sin z$

Vertical component $V = L \sin \alpha = L \cos z$

where α is the vertical angle and z is the zenith angle.

As far as the user is concerned, these calculations are seldom done because the total station will either display D and V automatically or will display L first and then D and V after pressing keys.

How accuracy of distance measurement is specified

All total stations have a linear accuracy quoted in the form

$\pm (a \text{ mm} + b \text{ ppm})$

The constant *a* is independent of the length being measured and is made up of internal sources within the instrument that are normally beyond the control of the user. It is an estimate of the individual errors caused by such phenomena as unwanted phase shifts in electronic components, errors in phase and transit time measurements and differences between the mechanical and electrical centres of the instrument.

The systematic error *b* is proportional to the distance being measured, where 1 ppm (part per million) is equivalent to an additional error of 1 mm for every kilometre measured. It depends on the atmospheric conditions at the time of measurement and on any frequency drift in the oscillator. At short distances, this part of the distance error is small in comparison to instrumental and centring errors and it can be ignored for most survey work. However, at longer ranges, atmospheric conditions can be the worst source of error for electronic distance measurements, and since these are proportional to distance, extra care should be taken in the recording of meteorological conditions when these are used to calculate corrections that are applied to long lines.

Typical specifications for a total station vary from ± (2 mm + 2 ppm) to ± (5 mm + 5 ppm). Taking the ± (2 mm + 2 ppm) specification as an example, at 100 m the error in distance measurement will be ± 2 mm but at 1.5 km, the error will be ± (2 mm + [2 mm/km × 1.5 km]) = ± 5 mm.

Reflectors and reflectorless measurements

When taking electronic distance measurements, a return or reflected signal is required for measurement purposes. This can be obtained using a special reflector or directly from an uncooperative target (a target that is not a reflector).

Reflectors used in distance measurement

Since the waves or pulses transmitted by a total station are either visible or infrared (which behaves like light but is invisible), a plane mirror could be used to reflect them. Unfortunately, this would require very accurate alignment of the mirror, because the transmitted wave or pulses have a narrow spread. To overcome this problem, special reflecting prisms are used instead. These are constructed from glass cubes or blocks and will always return a wave or pulse along a path exactly parallel to the incident path, but over a range of angles of incidence of about 20° to the normal of the front face of the prism, as shown in Figure 5.10. As a result, the alignment of the prism is not critical and it is quickly set when on site.

A wide range of reflecting prisms are available to suit short range measurements (small prisms) and long range measurements (large and multiple prisms). Single and triple *prism sets* (a combination of a prism with an optical target) for tripod mounting are shown in Figure 5.11, together with a handheld or pole-mounted *mini prism* (a mini prism is also shown in Figure 3.21). These would normally be used for control surveys, but for setting out and mapping, a pole-mounted reflector (or *detail pole*) similar to that shown in Figures 5.2 and 3.19 can be used. When surveying with a

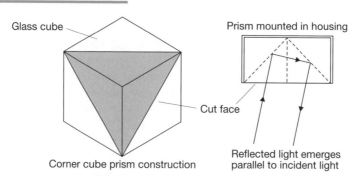

Glass cube

Prism mounted in housing

Cut face

Corner cube prism construction

Reflected light emerges parallel to incident light

Figure 5.10 ● Corner cube reflecting prism (retroreflector).

Single and triple prism sets

Hand-held mini prism   Mini prism for pole mounting

Figure 5.11 ● Single and triple prism sets with tribrachs for tripod mounting (top); hand-held and pole-mounted mini prisms (bottom) (courtesy Sokkia Ltd and Leica Geosystems).

motorised total station, a 360° prism is used (see Figure 5.22) which allows measurements to be taken from any direction.

Associated with all reflecting prisms is a *prism constant* – this is the distance between the effective centre of the prism and its plumbing point. Owing to the refractive properties of glass, which slows down the carrier wave or pulse when it passes

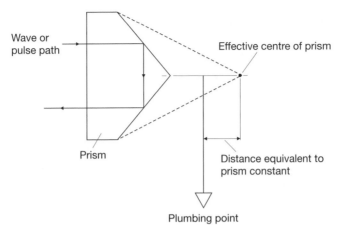

Figure 5.12 ● Prism constant.

through a prism, its effective centre is normally well behind the physical centre, as shown in Figure 5.12. A prism constant is typically −30 or −40 mm and this value is keyed into a total station as a correction that is applied automatically to each distance measured. If ignored, or applied incorrectly, this is a systematic error present in all measured distances and it is not eliminated by applying any field procedure. It is therefore very important that the correct prism constant is identified for the prism in use with a total station and that it is applied to all measurements.

As an alternative to reflecting prisms and at short distances, special plane reflecting targets or foil, as shown in Figure 5.13, can be used. These are made in various sizes, formats and colours, are self-adhesive and are available in sheet or roll form. This type of target is capable of reflecting a signal over a range of intersection angles and will work with most total stations – they are ideal for monitoring where targets have to be left in the same positions for long periods of time. When using plane reflective targets, the measuring range that can be obtained depends on their size, but a 25–30 mm square target will generally be sufficient to measure up to 100 m. The angle of

Figure 5.13 ● Reflecting targets (courtesy Leica Geosystems).

intersection of the target can affect the measuring range and accuracy considerably, and, where possible, these targets should be sighted such that the line of sight of the telescope is as close as possible to the normal to the target surface.

Reflectorless distance measurement

For many applications in construction and surveying, it often difficult or inconvenient to place a reflector at one end of the distance to be measured. In *reflectorless mode*, a total station can measure distances without using a reflector. To take a measurement, the telescope cross hairs are simply aimed at the point to be measured in order to record the distance. There is no need to locate a prism or special reflector at the point. Measurements not requiring a reflector can have considerable advantages over distance measurement with reflectors, especially when measurements have to be taken to difficult targets that are inaccessible or dangerous to reach. Because there is no need for a reflector, other advantages are that measurements can be taken by one person and in a reduced time as there is no need to wait whilst the reflector is moved from point-to-point. Because target location can be a problem with reflectorless measurements, those systems using a visible red laser (instead of infrared lasers) have the advantage that this can be used to locate the measurement point exactly instead of using the telescope cross hairs.

The following are examples of applications for reflectorless distance measurements:

- Monitoring of deformation in bridges, cooling towers and other large structures
- Monitoring of waste disposal sites, slurry pits and other hazardous areas
- Measuring volumes in quarries and open cast mines
- Recording dimensions on elevated objects such as roofs, ceilings and high walls for the surveys of buildings
- Tunnel profiling
- Mapping building and other façades, cliffs and rock faces without the need to erect scaffolding for access
- Setting out in deep trenches and foundations where safety is an issue
- Surveys of roads and railways avoiding traffic disruption
- Property surveys where there is no land access

Comparison of techniques

At present, a variety of total stations are available that can take distance measurements to a prism and in reflectorless mode using the phase shift or pulsed laser methods.

Phase shift measurements taken to a prism tend to be the most accurate, and those systems using a laser carrier wave will have measuring ranges of up to about 3–5 km to a single prism. In reflectorless mode, the phase shift method can measure distances up to about 100 m.

Those instruments using pulsed laser technology can measure longer distances than the phase shift method, and in reflectorless mode the range can be several hundreds of metres. Although the accuracy obtainable with a pulsed laser system is slightly less than with the phase shift method, the difference is small enough to be ignored for most work on site.

Although atmospheric conditions and visibility can affect these, the ranges quoted by each manufacturer for distance measurements taken to a prism are well defined because the reflecting surface is always the same (a glass prism). However, when specifying the range for reflectorless measurements, the situation is quite different, because the maximum distance that can be obtained will depend on the reflectivity of the object surveyed. For example, reflectorless measurements taken to white surfaces are much more efficient than those taken to dark surfaces; smooth surfaces are also better reflectors, as are dry surfaces compared to wet ones.

In order to be able to compare the ranges produced by different instruments in reflectorless mode, a standard surface known as the *Kodak Grey Card* is used. This is a card which is white on one side and reflects exactly 90% of the light that is incident on it and grey on the other, reflecting exactly 18% of the light that is incident upon it. Most manufacturers use the *Kodak Grey Card 18% Reflective Standard* for specifying the ranges that can be obtained in reflectorless mode with their total stations. The range to the 18% reflective Kodak Grey Standard is the most reliable indicator of the range that will be obtained on site because most everyday objects have an average reflectivity of about 18%. If a range is quoted to the 90% reflective standard, this only shows the distance that the total station can achieve when measuring to a very highly reflective target under ideal conditions.

The laser source used in all total stations diverges (or spreads out) as it travels from the instrument to a reflector or target and the size of the laser at the point of measurement is called its *footprint*. For reflectorless measurements, a large laser footprint at long range can be both a disadvantage an advantage: a disadvantage as it reduces signal strength and the chance of a successful measurement, and an advantage as it enables small objects – especially power lines and antennas – to be detected and measured. On the other hand, a small laser footprint has considerable advantages when structures such as buildings and tunnels are to be surveyed, especially at close range, as the narrow beam precisely marks a target. For all reflectorless measurements, the size of a laser footprint is greatly affected when observations are taken at very oblique angles to the line of sight of the telescope. In situations where distances are to be measured under difficult conditions (poor reflectivity, at oblique angles or to small objects), those total stations using pulsed laser methods are more likely to be able to take satisfactory measurements. However, phase shift systems tend to have a smaller divergence than a pulsed system and can locate objects more precisely.

Another comparison that can be made between different instruments is their time of measurement. Generally, the pulsed laser technique is faster in the reflectorless mode because the phase shift method has a measurement time that increases as a function of the distance being measured. For applications where reflectorless measurements are taken and where a short measuring time is critical, pulsed laser systems should be used.

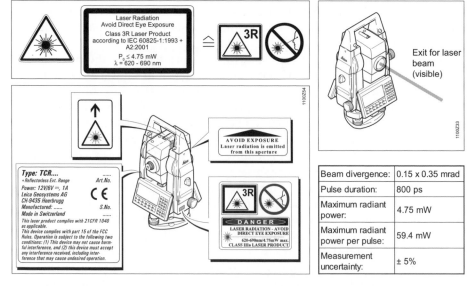

Figure 5.14 ● Laser safety labels for total stations (courtesy Leica Geosystems).

Laser safety and total stations

Although laser safety on site is discussed in detail in Section 11.5, a summary is given here that is relevant to total stations.

The main hazard associated with lasers in surveying and construction is damage to the eyes, and to help assess this lasers used in total stations are classified as Class 1, Class 2 or Class 3R lasers.

● *Class 1* laser products are safe under all reasonably foreseeable conditions and are not harmful to the eyes even when viewed directly through optical instruments. All infrared lasers used in distance measurement fall into this laser classification because they are invisible.

● *Class 2* lasers emit visible laser radiation and all those working on site in the vicinity of the total station must be careful not to stare directly into the beam with the naked eye or with any optical instrument such as binoculars, a level or theodolite. In addition, the instrument must not be pointed at anyone. Many total stations use Class 2 lasers for reflectorless measurements and most laser plummets are Class 2.

● *Class 3R* lasers are also used by some total stations in reflectorless mode. The use of these high-powered lasers extends the range of the instrument but increases the safety risk, and looking directly into the emitted beam is dangerous to the eyes. A number of rules apply to the use of Class 3R laser products – these include not aiming the instrument at reflective areas such as windows or metal surfaces which could emit unwanted reflections, ensuring that it is not possible for anyone on site to look directly into the beam with or without an optical instrument and the possible erection of warning signs in the area in which the total station is to be used. Consequently, it may only be practical for Class 3R total stations to be used in protected construction environments, and not in open and public areas.

All total stations must have labels attached to them giving details of the lasers used and the points at which they are emitted from the instrument – examples of some of these are shown in Figure 5.14. If the labels are missing and it is not possible to determine the classification of the laser, the total station should not be used.

Reflective summary

With reference to electromagnetic (or electronic) distance measurement, remember:

— Total stations can measure distances by the phase shift or pulsed laser methods. From a practical point of view and for site work, there is not much difference between these and both will give acceptable results.

— The phase shift method measures distance as a phase difference with ambiguity resolution, whereas the pulsed method measures transit times. These are not direct methods of measuring distances like taping because the phase difference or transit time has to be converted into a distance.

— Distances measured with reflectors give the best accuracy and range, but reflectorless distance measurements have become a standard feature of total stations – these can be taken using both the phase shift and pulsed methods.

— Although reflectorless measurements have some advantages over those requiring reflectors, the distance over which they can be taken varies according to the reflectivity of the target the total station is aimed at – this can be unpredictable.

— The maximum distance that a total station can measure in reflectorless mode should be specified, by the manufacturer, using the Kodak Grey Card 18% Reflective Standard – those that have a different specification may appear to have longer ranges but may not perform so well on site.

— Nearly all total stations now use lasers for distance measurements and for centring – everyone using these must be aware of the hazards associated with the use of lasers both to personnel on site and to the general public.

5.3   Instrumentation

After studying this section you should be familiar with the various categories of total station and the applications for which each is intended. You should also be aware of all the components and features of a total station that are used for measuring angles and distances in site surveying.

This section includes the following topics:

- Features of total stations
- Applications software
- Motorised total stations
- Handheld laser distance meters

Features of total stations

All total stations are capable of measuring angles and distances simultaneously and combine an electronic theodolite with a distance-measuring system and microprocessor. The microprocessor not only controls both the angle and distance measuring systems but is also used as a computer that can calculate slope corrections, vertical components, rect-angular coordinates and other information as well as setting out data. Most total stations can also process and store observations directly using an internal memory.

Three broad categories of total station are available. Instruments intended for use in *building and construction* have a shorter measuring range and lower angular specifi-cation than others, but they are made to be more robust and will resist water and dust penetration to a higher degree. It has been shown that 95% of all site distance measurements are under 500 m and that a 10" angular accuracy is adequate for most setting out procedures. Examples of instruments in this category are Sokkia's Series 10 and the Topcon GPT 3005 shown in Figure 5.15. The next category of total station covers those intended for *surveying applications*. These will have better angle and distance specifications, more functions, and better data storage and processing capa-bilities. Examples of instruments in this category are Sokkia's 030R Series and Trimble's 3600 DR Total Stations, shown in Figure 5.16. The remaining category covers *motorised total stations*, which tend to have the best specifications but are the most expensive – these are shown in Figure 5.20.

Although many different total stations are available with differing technical speci-fications, they tend to be made to a similar format. Those features common to the majority of instruments are described in this section.

Angle measurement

All of the components of the electronic theodolite described in Chapter 3 form part of a total station and these can all be seen on the total stations shown in Figures 5.15 and 5.16.

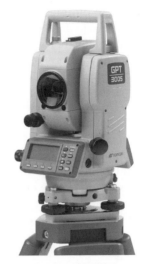

Sokkia Series 10   Topcon GPT3005

Figure 5.15 ● Total stations for building and construction (courtesy Sokkia Ltd and Topcon).

Sokkia 030R   Trimble 3600DR

Figure 5.16 ● Total stations for survey applications (courtesy Sokkia Ltd and Trimble Navigation Ltd).

The axis configuration is identical and comprises the vertical axis, the tilting axis and line of sight (or collimation), and these should all be mutually perpendicular. Other parts include the tribrach with levelling footscrews, the keyboard with display and the telescope which is mounted on the standards and which rotates around the tilting axis. Levelling is carried out in the same way as for a theodolite by adjusting the footscrews to centralise a plate level or electronic bubble. The telescope can be transited and used in the face left (or face I) and face right (or face II) positions. Horizontal rotation of the total station about the vertical axis is controlled by a horizontal clamp and tangent screw and rotation of the telescope about the tilting axis is controlled by a vertical clamp and tangent screw. These can be replaced by endless friction drives that do not require a clamp, and some total stations incorporate dual-speed drives – coarse for rapid target location and fine for exact target location. All total stations will have a horizontal and vertical circle for measurement of angles, which are measured in digital form and displayed as degrees–minutes–seconds or gons. The angular accuracy of a total station varies from instrument to instrument but this will be in the range 1–10" for most of the instruments that are likely to be used in construction surveying. The majority of total stations will have dual-axis compensation and all have either an optical or laser plummet.

The total station can be used to measure angles in the same way as an electronic theodolite using the procedures described in Section 3.5. This gives details for measurements taken on both faces. Some instruments will have software installed for measuring rounds of angles on both faces, but it is usual to measure on one face only with a total station. This is acceptable provided the total station is properly calibrated (as described in Section 5.5).

Distance measurement

All total stations will measure a slope distance which the onboard computer uses, together with the zenith (vertical) angle recorded by the theodolite along the line of sight (line of distance measurement) to calculate the horizontal distance. In addition, the height difference between the tilting axis and prism centre is also calculated and displayed.

In Section 5.2, the phase shift and pulsed laser methods by which a total station measures distances have been described. Three types of measurement are possible when using either of these.

For distances taken to a prism or reflecting foil, the most accurate is *precise* (*fine* or *standard*) *measurement*. For phase shift systems, a typical specification for this is a measurement time of about 1–2 s, an accuracy of (2 mm + 2 ppm) and a range of 3–5 km to a single prism. For pulsed laser systems, the range that can be measured to a prism is slightly longer, but the measurement times can be up to 5 s with an accuracy of (3 mm + 3 ppm). Although all manufacturers quote ranges of several kilometres to a single prism, there are practical difficulties with this and ranges beyond several hundreds of metres are seldom measured on site.

For those construction projects where long distances are required to be measured, GPS methods are used in preference to total stations. There is no standard distance at which the change from one to the other occurs, as this will depend on a number of factors, including the accuracy required and site topography. However, it would be unusual to measure distances well over 500 m with a total station.

Rapid (*coarse* or *fast*) *measurement* reduces the measurement time to a prism to between 0.5 and 1 s for both phase shift and pulsed systems, but the accuracy for both may degrade slightly.

These two measurement techniques would normally be used for control surveys and mapping.

Tracking measurements are taken extensively when setting out or for machine control, since readings are updated very quickly and vary in response to movements of the prism, which is usually pole-mounted. In this mode, the distance measurement is repeated automatically at intervals of less than 0.5 s.

For reflectorless measurements taken with a phase shift system, the range that can be obtained is about 100 m, with a similar accuracy to that obtained when using a prism or foil. However, this can vary according to the reflectivity of the target and the angle of incidence of the measurement. The measurement time will also vary and is typically 3 s + 1 s for every 10 m measured over 30 m. Reflectorless measurements with pulsed systems have a range of 200–300 m and their measurement time is of the order of 5 s irrespective of distance. The accuracy is similar to measurements taken to a prism.

The specifications given here are representative only and do not apply to any particular total station – for further information, the brochure for each instrument or web site provided by each manufacturer should be consulted. The web sites of the principal manufacturers are listed at the end of the chapter.

Keyboard and display

A total station is activated through its control panel, which consists of a keyboard and multiple line liquid crystal display (LCD). Total station displays are moisture-proof and can be illuminated; some incorporate contrast controls to accommodate different viewing angles. A number of instruments have two control panels (one on each face) which makes them easier to use. The keyboard enables the user to select and implement different measurement modes, enables instrument parameters to be changed and allows special software functions to be accessed. Some keyboards incorporate multi-function keys that carry out specific tasks, whereas others use keys to activate and display menu systems which enable the total station to be used as a computer might be.

In addition to controlling the total station, the keyboard is often used to code data generated by the instrument. Angles and distances are usually recorded electronically by a total station in digital form as raw data (slope distance, vertical angle and horizontal angle). For mapping, if a code is entered from the keyboard to define the feature being observed, the data can be processed much more quickly when it is downloaded into and processed by an office-based computer and plotter. On numerical keyboards, codes are represented by numbers only, whereas on alphanumeric keyboards, codes can be represented by numbers and/or letters, which gives greater versatility and scope. Many total stations now have large graphic screens that make it possible for data to be edited on site. Feature codes and their application to large-scale surveys and mapping are discussed in Section 10.9.

Some examples of keyboards and displays are given in Figure 5.17.

On some total stations, the keyboard and display can be detached from the instrument and interchanged with other total stations and with GPS receivers, in what is known as *integrated surveying*. This enables data to be shared between different instruments and systems using a single interface. These combined keyboards and

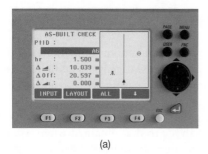
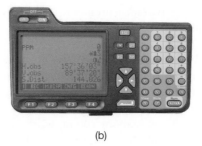

(a)   (b)

Figure 5.17 ● Examples of total station keyboards and displays: (a) Leica TCR410C with basic functions required on site (b) Sokkia 030R with greater functionality required for mapping and other surveys (courtesy Leica Geosystems and Sokkia Ltd).

displays not only control the instrument, but are also data storage devices (these are described in Section 5.4).

Power supply

Three types of rechargeable battery are used in surveying instruments: these are Nickel Metal hydride (NiMh), Nickel Cadmium (NiCad) and Lithium-ion batteries. The NiMh battery is the most popular because it is compatible with standard camcorder batteries and has a better capacity than the NiCad battery. However, the NiCad battery has been available for many years, is still in widespread use and has more charging cycles than the NiMh battery. For most total stations, both types will give up to about 10 hours use for angle measurements, about 7–8 hours for angle and distance measurements taken to a reflector and about half this for reflectorless distance measurements and in motorised total stations. Lithium-ion batteries, which have the advantage of being easy to charge and maintain, have an operating time of about half that of NiMh and NiCad batteries. As an alternative to rechargeable batteries, some instruments will accept AA size alkaline batteries. Most total stations are capable of giving a battery power indication and some have an auto power save feature which switches the instrument off or into some standby mode after it has not been used for a specific time.

Useful accessories

An *optical guidance system* for setting out has been produced by a number of manufacturers. One or two light-emitting diodes, situated just above or below the telescope objective, emit a visible light pattern which enables a detail pole to be set directly on the line of sight and at the correct distance without the need for hand signals from the total station. In one system, the device consists of two visible beams of red light, one steady and one flashing: if the prism is to the left of centre of the line of sight, a steady red light is seen and if the prism is to the right, a flashing red light is seen – as shown in Figure 5.18. During setting out, the frequency of the flashing changes to indicate whether the prism needs to move forward or back in order to set out the

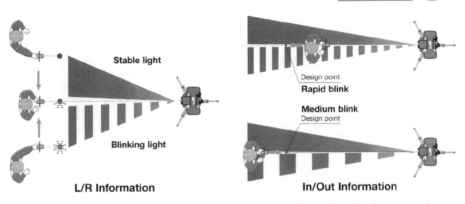

Figure 5.18 ⬤ Optical guidance system for setting out with total station (courtesy Survey Supplies).

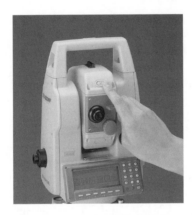

Figure 5.19 ⬤ Autofocus button for Topcon GTS-603AF (courtesy Topcon).

correct distance to the design point. In other systems, different coloured lights are used to indicate setting out to the left or right and the rate at which these flash helps position the prism along the line of sight.

In some total stations, an *autofocus* facility is available. To use this, the telescope is aimed at a prism or target, an *AF* button is pressed (see Figure 5.19) and the focus is set automatically – this considerably reduces the time taken to locate points compared with manual focusing. However, if needed, manual focusing can be carried on all total stations fitted with an autofocus system.

Applications software

The microprocessor built into a total station is the equivalent of a small computer and the primary function for this is controlling the measurement of angles and distances – to do this, the microprocessor issues on-screen instructions which guide the operator through each measuring sequence. However, the microprocessor is also pre-programmed to enable the operator to perform a number of other survey tasks and to carry out coordinate and other calculations. It can process and store survey data and it is also used to help the operator carry out calibration checks on the instrument.

All of these functions are activated using a menu structure, the complexity of which varies from instrument to instrument. The applications software available on many total stations includes the following:

- Slope corrections and reduced levels
- Horizontal circle orientation
- Coordinate measurement
- Traverse measurements
- Resection (or free stationing)
- Missing Line Measurement (MLM)
- Remote Elevation Measurement (REM)
- Areas
- Setting out

Further details of these programs are given in subsequent chapters.

Motorised total stations

The latest generation of total stations have many of the features described in the previous sections but are also fitted with servo-motors which control their horizontal and vertical movement. These are known as *motorised total stations* and examples of these are the Leica TPS1200 and Trimble S6 series shown in Figure 5.20.

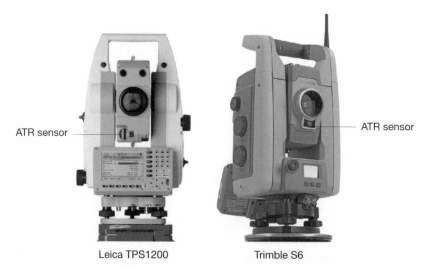

Leica TPS1200   Trimble S6

Figure 5.20 ● Motorised total stations (courtesy Leica Geosystems and Trimble Navigation Ltd).

When using these instruments, the operator does not have to look through the telescope to align with a prism or a target because the servo-motors – after a key press to initiate a measurement – take over pointing the instrument. This has some advantages over a manually pointed system, since a motorised total station can aim and point much more quickly and with a better precision. For setting out, a point number is entered and the instrument instantly computes setting out data and automatically positions itself on the calculated bearing. If an elevation is stored, the total station will also align itself along the correct vertical angle.

A servo-assisted pointing is made possible in these total stations by using *Automatic Target Recognition (ATR)*, which requires the total station to be fitted with an ATR sensor. At the start of a survey, the search routine used to find a prism or reflective target is initiated by the operator. The ATR sensors of the Leica and Trimble total stations shown in Figure 5.20 transmit a vertical laser fan whilst the instrument is rotated horizontally through an angle up to 360° or through a user-defined window. As soon as the prism is located within the fan, the instrument stops rotating and switches to a vertical scan with narrow beam to complete the search. In the method used by the Topcon GPT-8003A series (see Figure 5.21), four fixed sensors mounted on each side of a special carrying handle transmit a laser beam to detect the prism. In each case, part of the sensor beam is reflected back to the total station and is processed by a digital camera mounted inside the ATR sensor. For the final part of the search and lock-on, the exact position of the centre of the sensor beam is determined and then horizontal and vertical offsets from this to the telescope cross hairs are computed. These are then used to control the servo-motors which will rotate the instrument to minimise the offsets and precisely locate the prism. Once locked on, the total station will follow the movement of the prism and measurements are taken without the need for any manual pointing of the telescope. However, if the line of sight is obscured or the prism moves too quickly, a loss of prism lock might occur.

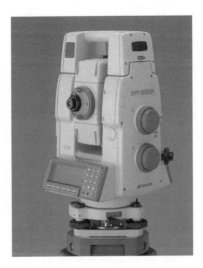

Figure 5.21 ● Topcon GPT-8003A motorised total station (courtesy Topcon).

Robotic total stations

Although all motorised total stations can be used as conventional instruments might be, their full potential is realised when they are remote controlled and used as *robotic total stations*. By providing remote control of the total station from the prism, these are surveying systems that permit single-user operation for either mapping or setting out. To do this, the instrument works together with a special detail pole, as shown in Figure 5.22. This has a 360° prism fitted to it as well as a small computer (called a remote control unit or simply *controller*) and radio or optical communication between the prism and total station.

Although a survey is carried out with a robotic total station from the detail pole, it is possible to access and use all the functions of the total station as if the operator were standing at the instrument. As soon as the total station and controller are switched on, a signal is transmitted by the operator at the detail pole to instruct the total station to search for and lock onto the prism. As measurements are taken, the instrument automatically follows the movements of the prism and if contact is lost, this is re-established by the search routine. When collecting data for topographical surveys, the operator places the detail pole at a point of interest and, by pressing keys on the controller, the angles and slope distance are measured to the prism from the total station. A feature code is allocated and all of this data is stored in the controller. All field data is eventually processed by detaching the controller from the reflector pole and taking it to the site or survey office or by using email and a cellular modem on site to connect it to a computer for file transfer. When setting out, the procedure is reversed and coordinates are downloaded from a computer file into the controller. On site, the controller is attached to the detail pole, and the total station is set up, orientated and locked onto the prism. The detail pole is held at the approximate position of the point to be set out and its number or identity code is keyed into the controller in order to generate the relevant setting out data. The controller will, after an angle and distance measurement has been taken, use this data to display the amounts by which the reflector pole has to be moved so that the exact position of the point can be set out.

Handheld laser distance meters

As well as taping and total stations, distances can also be measured using devices such as the Leica DISTO, a handheld distance meter, which is shown in use in Figure 5.23. Although these are not total stations, they are included here as they use electronic methods for distance measurement.

To use one of these, all the operator has to do is hold the DISTO at one end of the feature to be measured, point it at the other and press a button – the distance is then displayed automatically. As it is a handheld device, distances can be measured in any direction by simply pointing the DISTO as required. The method used for the measurement of distance is a mixture of the phase shift and pulsed laser methods described in Section 5.2. In this case, a laser diode in the DISTO emits pulses which are transmitted towards a target and reflected (without using a prism) back to

Leica TPS1200 pole-mounted 360° prism and remote control unit (uses radio communication to total station)

Topcon RC-2II optical data link, 360° prism and FC-1000 Field Controller

Figure 5.22 ● Robotic total stations (courtesy Leica Geosystems and Topcon).

Figure 5.23 ● Leica DISTO hand-held laser distance meter in use (courtesy Leica Geosystems).

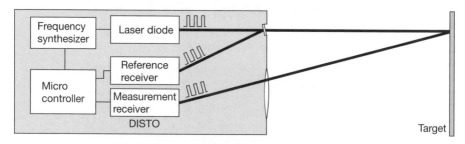

Figure 5.24 ● Measurement principle of DISTO (courtesy Leica Geosysytems).

the DISTO. Due to the difference between an internal reference path and the external measurement path, the light pulses reflected from the target and received by the DISTO experience a phase shift compared with the light pulses received through the internal reference path (see Figure 5.24). As shown in Section 5.2, the phase difference between these two paths is proportional to the distance between the instrument and the target. This is measured by the DISTO and a coarse frequency measurement is used to resolve the ambiguity in distance.

A close-up of the DISTO Classic[5]a is shown in Figure 5.25. This has a quoted accuracy of ±3 mm, a range of 200 m and a measuring time of up to 4 s. These figures depend on the surface characteristics of the target and may degrade under difficult measuring conditions. The DISTO Classic can calculate areas and volumes, it has a tracking facility and can be used as a pocket calculator with seven functions. It also has a soft-touch keypad with illuminated display, and a built-in bubble is provided for horizontal and vertical measurements.

Similar devices to the DISTO are also available from other equipment manufacturers, including the Hilti PD 30 and HD150 from Trimble.

Handheld laser distance meters can be used to measure distances and heights in almost any application on site. They are particularly useful where it is not possible to use

Figure 5.25 ● Leica DISTO Classic[5]a
(courtesy Leica Geosystems).

a tape when measuring across wide gaps, through obstacles and to elevated or inaccessible points, and they are becoming the standard method for height transfer in tall buildings and other large structures (this is described in Section 4.1 for taping). Another advantage is that only one person is needed to take distance measurements. However, because these devices use a Class 2 laser, it is necessary to ensure that no one can stare into the beam and that it is not accidentally pointed at anyone during a measurement. For these reasons, some care is required when using a handheld laser distance meter on site and in public spaces – amongst other safety precautions that might be taken, aiming the beam above or below possible eye levels is essential.

Reflective summary

With reference to instrumentation, remember:

— There are a lot of integrated total stations available to suit almost any application on site. Because of this, some care is needed when choosing a system that balances the accuracy and features required against costs.

— The technical specifications published for total stations show them to be high-quality precision instruments for measuring angles and distances. However, they can also perform a variety of other survey tasks such as traversing and calculation of coordinates.

5.4 Electronic data recording and processing

After studying this section you should be aware of all the various methods used with total stations and other survey equipment to store and manage survey data.

This section includes the following topics:

- Data collectors
- Field computers
- Internal memories and memory cards

Data collectors

One of the earliest and most successful applications of computers in surveying was the creation of software and hardware for the automatic drawing of maps and plans. These improvements, however, caused problems with data transfer because at the time computers were first used in surveying, it was only possible to record observations by hand in field books. This meant that all the data collected on site had to be taken to an office and entered manually, via a keyboard, into a computer so that the required plan could be compiled and eventually plotted. This is a relatively slow process, prone to error. These problems were overcome when total stations were introduced because they can generate computer-compatible angle and distance readings. Today, electronic data recording and transfer are carried out for all survey and site work.

A number of different devices for recording data electronically have evolved. These include *data collectors* (a form of handheld computer), which can be programmed to ask the operator to record data from an instrument, and *field computers*, which are laptop and tablet computers adapted to survey data collection. It is also possible to use *memory cards*, which take the form of plug-in cards onto which data is encoded by a total station, and data can also be stored using *onboard* or *internal memory*.

Typical examples of data collectors (also referred to as *handhelds* or *controllers*) used for data storage and processing with total stations are the Topcon FC-1000 Field Controller already shown in Figure 5.22, the SDR8100 data collection platform from Sokkia, Trimble's TSCe Controller and the Allegro CE, all of which are shown in Figure 5.26. When interfaced with a total station, these data collectors take control of the measurement process, and as each survey point is sighted, for example, in a detail survey, the total station is instructed to transmit angle and distance readings taken at each feature directly to them. For these surveys, data storage and entry are carried out through a series of step-by-step instructions displayed by the data collector and the type of feature code entered for each point depends on the software to be used to edit and plot a survey. The use of total stations and data collectors in detail surveying and mapping is described in Chapter 10.

Because the software and menu structures installed in each data collector vary between manufacturers, it is not always possible to use a total station and data collection system without some sort of training. However, as soon as an engineer or surveyor has gained sufficient experience, the high degree of sophistication that can be achieved with data collectors can be used very effectively for site surveying and data management.

As well as performing the basic function of data storage, data collectors and controllers have some resident programs installed to make it possible for them to

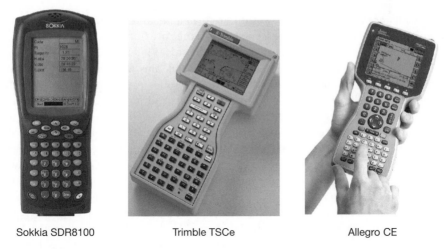

Sokkia SDR8100   Trimble TSCe   Allegro CE

Figure 5.26 ● Data collectors (courtesy Sokkia Ltd, Trimble Navigation Ltd and Thales).

collect and process data in a variety of ways. For example, observations normally stored as angles and distances (raw data) can be converted to three-dimensional coordinates prior to transfer to a computer. Further programs can range from the editing of recorded data into a number of different formats to the on site calculation of areas and volumes. Coordinate and other information can also be uploaded to data collectors and controllers from a computer or other source for construction surveys. In this case, stored data is used with data measured by the total station to display setting out information in real time.

Referring to the SDR8100, the term SDR is an abbreviation for Sokkia Data Recorder, but these devices are sometimes referred to as electronic field books. When used as an electronic field book, this data collector is not connected to a total station or other survey instrument and readings are entered manually. This enables data to be recorded in a computer format ready for further processing, even though the instrument being used is either not electronic or cannot be interfaced directly with a computer or data collector. In common with other handhelds, the SDR8100 uses a Microsoft Windows CE (CE = *Consumer Electronics*) operating system to run a variety of Sokkia applications programs and it can be interfaced with Sokkia total stations and GPS receivers. After completion of a survey or at intermediate stages, data must be transferred from the SDR8100 to a computer or backup storage device using its RS232 serial communication port.

Built with field survey in mind, the SDR8100 is *ruggedised*, a term used to describe survey equipment that is waterproof and can withstand exposure to dust, vibration, dropping and a wide range of working temperatures. The *Ingress Protection* (IP) rating provides a scale against which this can be measured. An IP54 rating means that the unit (data collector or survey instrument) is protected against dust on a $\frac{5}{6}$ scale and water on a $\frac{4}{8}$ scale. A high rating of IP67 means that the unit is completely protected against dust and to immersion in water. Drop testing provides another scale of ruggedness and is measured 1.2 m above a surface – a typical drop from someone's hand. Units are dropped 26 times at different heights and at different angles onto plywood-covered concrete and pass the IP test if they are fully functional after this.

The Trimble TSCe Controller is also installed with a Windows CE operating system and has features similar to the Sokkia SDR8100. The data collection software provided by Trimble for use with the TSCe Controller can be accessed using the alphanumeric keyboard or full-colour touchscreen and it is ruggedised to operate under most conditions expected on site. For data transfer, it can be connected directly to a computer, to any Trimble total station or to a GPS receiver, and communication is provided by RS232 serial or USB ports. Data can be taken from different instruments and the TSCe will automatically combine the different measurements into a single dataset. In addition, data can be stored on 512 MB compact flash cards (sometimes referred to as PC cards). Flash memory is particularly useful for portable computing because it is a non-volatile memory that is retained even if there is a total power loss in the data collector. If fitted with the Trimble BlueCap module, the TSCe can utilise *Bluetooth wireless communication* with Trimble GPS receivers and with other data collectors. Bluetooth wireless technology is used to link portable electronic devices – such as PCs, mobile phones and handheld devices – with each other. In surveying, Bluetooth enables cable-free connections to be made between data collectors, survey equipment and mobile phones so that data can be transmitted between different equipment and also via a mobile phone using email and the Internet. Bluetooth technology is described in more detail in Section 7.7.

The SDR8100 and TSCe are produced by Sokkia and Trimble predominantly for use with their total stations. However, many different types of data collector are available from a variety of manufacturers and although these can be used for data collection with almost any total station, it is necessary to purchase and install a software package in them for doing this from one of several companies specialising in survey software. These are known as *third party data collection systems* because they are made independently of the companies that manufacture survey instruments. An example of this type of data collector is the Allegro CE from Juniper Systems, which is shown in Figure 5.26.

Figure 5.27 ● Compaq iPAQ pocket computer (courtesy Survey Supplies).

Figure 5.28 ● Trimble Recon Controller (courtesy Survey Supplies).

Survey data can also be collected using other third party handheld devices such as the Compaq iPAQ pocket computer (see Figure 5.27). Compared to the SDR8100 and TSCe, this has a larger graphical display without keypad but a 'thumb type' keypad can be attached if required. The iPAQ is fully waterproof and it can be used in a sealed environmental case in harsh conditions. Mobile communications are possible if it is fitted with a wireless Ethernet PC card for connection to the Internet and email; otherwise serial/USB connections or a 64 MB compact flash memory card are used for data transfer.

Another device featuring a large display is the Trimble Recon Controller shown in Figure 5.28. This is similar to the iPAQ but is fully ruggedised for site use. The Recon has a Windows CE operating system, a full colour touchscreen and data storage provided through two waterproof compact flash slots.

Trimble also markets a device called the ACU controller (ACU = *Attachable Control Unit*) which clips onto a total station in place of a conventional keyboard and display (see Figure 5.29). When connected, the ACU controller is used to access all the functions of the instrument, but is also a detachable data collector that can be used with other total stations and GPS receivers. When a survey is complete, the ACU controller can be taken to the office and data is transferred to a host computer using RS232, USB and other ports. Alternatively, the ACU is Internet-capable and, with an external wireless modem, files can be transmitted and received by the ACU controller when on site. Sharing data between ACUs is also possible using built-in Bluetooth technology. The ACU controller offers Windows CE data recording with Trimble's Survey Controller Software, a TFT colour touch screen with full alphanumeric keyboard, and an interactive graphic display for real-time data editing. The memory capacity is 128 MB of flash memory.

The ACU controller is an example of integrated or modular surveying in which data can be shared between different equipment using a single interface, as shown in Figure 5.30, in which the ACU is being exchanged between equipment whilst on site. Because this is becoming more popular, survey equipment manufacturers have now started to develop integrated systems so that data generated by different instruments can be combined.

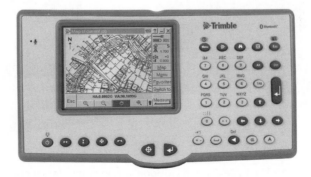

Figure 5.29 ● Trimble ACU controller (courtesy Trimble Navigation Ltd).

Figure 5.30 ● Exchange of ACU Controller from GPS to total station (courtesy Trimble Navigation Ltd).

Field computers

These are fully functional laptop and tablet computers made suitable for outdoor use. Compared to a dedicated data collector, they can offer a more flexible approach to data collection and processing since they can be programmed for many forms of data entry from any instrument to suit the individual requirements of any user.

Many different IT companies produce field computers, and examples in this category of data collector are the Itronix Husky fex21 and Itronix GoBook notebook laptops, as shown in Figure 5.31. The fex21 is a ruggedised handheld PC with a Windows CE operating system whereas the GoBook series are ruggedised laptop computers using the Windows XP operating system. Both have wireless communications and can be used as mobile phones for voice calls, as fax machines and they can be connected to the Internet or local and wide area networks. In effect, they are field

Husky fex21   GoBook laptop

Figure 5.31 ● Itronix Husky fex21 and GoBook laptop computers (courtesy Positioning Resources Ltd and Itronix (UK) Ltd).

Figure 5.32 ● Tablet PCs (courtesy Leica Geosystems and Itronix (UK) Ltd).

computers that transmit and receive data in real time. However, data can also be stored and the fex21 has 32 MB of compact flash memory available, compared with the GoBooks which are fitted with 20 or 40 GB hard drives. The GoBook laptops all feature Intel Pentium processors with RAM of up to 1024 MB. When used in surveying, the fex21 and GoBooks are third party computers and have to be installed with specialist land survey software and appropriate interfaces before they can be used for data collection and processing.

Other devices gaining popularity in surveying for data collection are tablet PCs, shown in Figure 5.32. In common with many of the computers and data collectors already described, these are third party devices that can also have Windows CE operating systems, compact flash memory for data storage and built-in communications for real-time data collection and transmission.

Internal memories and memory cards

The method of storing and processing information using data collectors and field computers is thought by some surveyors and engineers to be inconvenient since it involves using an extra piece of equipment. Internal memories and memory cards overcome this problem and offer an alternative approach to data processing and storage.

Internal memories are perhaps the most convenient method for storing survey data and nearly all total stations have an on board storage capacity of up to about 5000–10,000 coded survey points, but this can increase to 30,000 points in some of the more expensive instruments. For data transfer, the total station can be connected directly to a computer using a cable and communications port, as shown in Figure 5.33.

Alternatively, data can be stored on *memory cards*, which are plugged into the total station or data collector as shown in Figure 5.34. Data is then exchanged through a PC card drive on a computer (rather like a small DVD or floppy disk drive) or separate

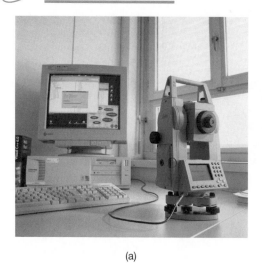

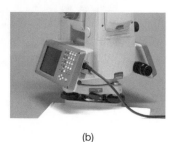

(a)   (b)

Figure 5.33 ● Data transfer from total station to computer: (a) with RS232 port; (b) using USB connection (courtesy Leica Geosystems and Topcon).

Figure 5.34 ● Memory cards in total station and data collector (courtesy Sokkia Ltd and Topcon).

PC card reader connected to the computer as in Figure 5.35. By reversing this process, setting out and other data can also be uploaded to the total station.

All memory cards (including flash cards) should conform to PCMCIA (Personal Computer Memory Card International Association) standards.

Figure 5.35 ● Data transfer for memory cards (courtesy Survey Supplies).

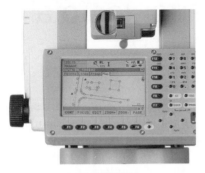

Leica TPS1200   Topcon GTS-720

Figure 5.36 ● Leica TPS1200 and Topcon GTS-720 Series total stations with large graphics screens (courtesy Leica Geosystems and Topcon).

Total stations such as Leica's TPS1200 and Topcon's GTS-720 Series have large colour graphics screens (see Figure 5.36) and Windows CE operating systems as well as sophisticated software on board. With displays like these, it is much easier to check and edit measured data on site before storage.

Reflective summary

With reference to electronic data recording and processing, remember:

— Electronic data storage is a much better method for recording survey data because mistakes in reading and recording by hand are eliminated.

— As the amount of data generated by electronic survey equipment is increasing, the capacity of data storage systems is also increasing and these, like total stations, are developing and changing rapidly.

— The trends in data storage and transfer are towards internal memory, wireless communication, better presentation of data on site for edit, view and review plus the use of the Internet and mobile phones.

— Although total station data collectors have large graphics screens onto which data can be plotted and edited, it is always a good idea to supplement any survey with field sketches and notes.

5.5 Sources of error for total stations

After studying this section you should be aware of the sources of error that can occur when using a total station. You should also know that a total station should be calibrated regularly and that proper field procedures must be implemented to eliminate or reduce the effect of these errors. In addition, you should be aware of some of the common causes of error when setting up and using total stations and how good practice and handling can help prevent these.

This section includes the following topics:

● Errors in the equipment
● Atmospheric effects
● Using the total station properly to avoid errors and other problems

Errors in the equipment

Calibration of total stations

To maintain the high level of accuracy offered by modern total stations, there is now much more emphasis on monitoring instrumental errors, and, with this in mind, some construction projects require total stations and other survey equipment to be checked regularly according to procedures given in quality manuals. Some instrumental errors are eliminated by observing on two faces of the total station and averaging, but because one face measurements are the preferred method on site, it is important to determine the magnitude of instrumental errors and correct for them. For total stations, instrumental errors are measured and corrected using electronic calibration procedures that are carried out at any time and can be applied to the instrument on site. These are preferred to the mechanical adjustments that used to be done in laboratories by trained technicians. Because this is carried out on site by the user rather than in the laboratory, each calibration process must be completed carefully, following the manufacturer's instructions, because any single face measurements taken after this will be affected by the accuracy of the various calibration parameters determined.

When a total station is manufactured or serviced, a full calibration is carried out and a certificate should be issued to confirm this has been done in accordance with an appropriate standard, such as that required by ISO9001 or DIN 18723. At this time, all instrumental errors are set to zero.

Since calibration parameters can change because of mechanical shock, temperature changes and rough handling of what is a high-precision instrument, an electronic calibration should be carried out on a total station as follows:

- Before using the instrument for the first time
- After long storage periods
- After rough or long transportation
- After long periods of work
- Following big changes of temperature
- Regularly for precision surveys

Before each calibration, it is essential to allow the total station enough time to reach the ambient temperature, that it is set up on solid ground and that it is not in direct sunshine. As soon as electronic calibration values have been determined, the instrument's software will automatically apply the corrections obtained to all measurements until another calibration is carried out. For quality assurance purposes, some total stations will record calibration parameters in a protocol file before and after each instrument calibration together with the date, time and the units set on the instrument.

Figure 3.5 shows the arrangements of the axes of a theodolite when it is in perfect axial adjustment – these are the same for a total station. This arrangement is seldom achieved in practice and gives rise to instrumental errors in total stations which are the same as those for a theodolite, some of which have already been described in Section 3.6. The degree to which a total station can be calibrated depends on the make and model, but the most sophisticated can measure and correct for horizontal and vertical collimation, tilting axis error, compensator index error and ATR calibration, as described in the following sections.

Horizontal collimation (or line-of-sight error)

This axial error is caused when the line of sight is not perpendicular to the tilting axis. It affects all horizontal circle readings and increases with steep sightings, but is eliminated by observing on two faces. For single face measurements, an on-board calibration function in the total station is used to determine c (see Figure 5.37), the deviation between the actual line of sight and a line perpendicular to the tilting axis. A correction is then applied automatically for this to all horizontal circle readings. If c exceeds a specified limit, the total station should be returned to the manufacturer for adjustment.

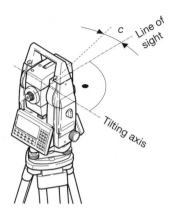

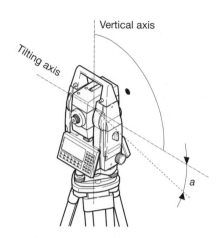

Figure 5.37 ● Horizontal collimation error (courtesy Leica Geosystems).

Figure 5.38 ● Tilting axis error (courtesy Leica Geosystems).

Tilting axis error

This axial error occurs when the tilting axis of the total station is not perpendicular to its vertical axis. This has no effect on sightings taken when the telescope is horizontal, but introduces errors into horizontal circle readings when the telescope is tilted, especially for steep sightings. As with the horizontal collimation error, this error is eliminated by two face measurements, or the tilting axis error a (see Figure 5.38) is measured in a calibration procedure and a correction applied for this to all horizontal circle readings. As before, if a is too big, the instrument should be returned to the manufacturer for adjustment.

Compensator index error

As explained in Section 3.6, any errors caused by not levelling a theodolite or total station carefully (sometimes called the vertical axis error) cannot be eliminated by observing on two faces. If the instrument is fitted with a compensator and this is switched on, it will measure any residual tilts of the instrument and will apply corrections to horizontal and vertical angles for these. However, all compensators have a longitudinal error l in the direction of the line of sight of the total station and a transverse error t perpendicular to this (see Figure 5.39). These are known as *zero point errors* and although they are eliminated when averaging face left and face right readings, for single face readings values for l and t must be determined using the on-board calibration function so that measured angles can be corrected.

Vertical collimation (or vertical index error)

A vertical index error exists in a total station if the 0°–180° line of the vertical circle does not coincide with the vertical axis. This zero point error is present in all vertical circle readings, and like the horizontal collimation error, it is eliminated by taking two face readings or by determining i, the index error shown in Figure 5.40, in

Vertical axis

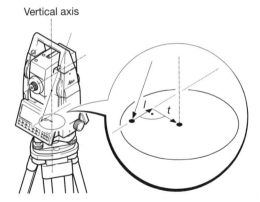

Figure 5.39 ● Compensator index error (courtesy Leica Geosystems).

another calibration procedure. In this case, corrections for *i* are applied directly to vertical angles.

ATR collimation error

An ATR collimation error is caused in robotic total stations when there is an angular divergence between the line of sight and the ATR camera axis. In other words, the centre of the ATR sensor beam defines a different position from the centre of the cross hairs. To determine this error, the centre of a prism is sighted accurately and the horizontal and vertical circle readings to it are noted. The ATR system is switched on and the position this gives is compared with the manual pointing at the centre of the prism. If these are different, the total station has an ATR collimation error with horizontal and vertical components shown in Figure 5.41 – these are determined in an ATR calibration routine. It is essential that these corrections are measured

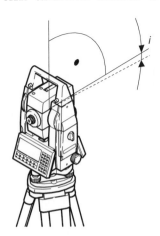

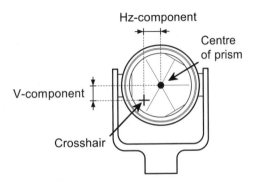

Figure 5.40 ● Vertical index error (courtesy Leica Geosystems).

Figure 5.41 ● ATR calibration errors (courtesy Leica Geosystems).

regularly and applied to surveys where a mixture of ATR readings and those taken manually to prisms have been taken. Because of atmospheric shimmer, the image of the centre of the prism will move a small amount during calibration and the ATR camera will change its computed position to account for this. These effects are filtered out by taking several sets of readings during a calibration until the total station indicates an accepted set of corrections have been computed.

Collimation of laser pointer

Although distances measured in reflectorless mode by a total station can be advantageous on site, problems can occur with target location. These are overcome in some total stations by using a visible red laser pointer – this is sometimes combined with the measuring beam. For these instruments, the laser pointer or beam must be collimated with the line of sight, otherwise less accurate measurements might be taken because the laser beam will not be reflected from the same point at which the cross hairs are pointing. To carry out a check on the collimation of the laser and line of sight a special target plate can be used (see Figure 5.42). Using the telescope cross hairs, the total station is aligned with the centre of the target plate and the laser pointer is switched on. If the red laser spot illuminates the cross at the centre of the target plate, the instrument is in adjustment. If not, a mechanical adjustment is carried out on site to align the laser spot onto the cross. If no target plate is available, any well-defined target can be used for the check and adjustment.

Instrumental distance errors

The performance of the distance-measuring components of total stations, like their angular components, can also be affected by constant use on site and with age. For

Figure 5.42 ● Target plate for collimation of laser pointer (courtesy Leica Geosystems).

Figure 5.43 ● Determining the zero error for a total station.

these reasons, these should also be tested regularly and calibrated. As for axial and zero point errors, distance calibrations may also be a contractual requirement for some site work if quality assurance has to be guaranteed.

Zero error (or index error) occurs if there are differences in the mechanical, electrical and optical centres of the total station and reflectors, and includes the prism constant. It is of constant magnitude and is not dependent on the length of line measured. A check should be carried out periodically and the simplest method for determining a zero error involves taking distance measurements along a three-point baseline, as shown in Figure 5.43. All three distances are measured with the total station and l_{12}, l_{23} and l_{13} are obtained. If each measured distance has the same zero error z:

$$d_1 = l_{12} + z \qquad d_2 = l_{23} + z \qquad (d_1 + d_2) = l_{13} + z$$

where d_1 is the correct distance from 1 to 2 and d_2 is the correct distance from 2 to 3.

Since the measurement of $(d_1 + d_2)$ should equal the sum of the two separate components d_1 and d_2 we have

$$l_{13} + z = (l_{12} + z) + (l_{23} + z)$$

or

$$z = l_{13} - (l_{12} + l_{23})$$

This is sometimes called the *three peg test* and it should be repeated several times and an average value computed for z. The advantage of this method for determining the zero error is that it is based on unknown baseline lengths and the three points are usually only put in place for the test. The value of a zero error obtained from this calibration procedure applies to a total station combined with reflector and can be used when the prism constant is not known to correct for this. However, if the reflector is changed, the zero constant changes and a new calibration is required.

A *scale error (or frequency drift)* is caused by variations in the frequency of the modulation oscillator in a phase shift system or the frequency of the synthesizer in a pulsed distance measuring system. The error is proportional to the distance being measured and determination of scale error is best carried out by taking a series of distance measurements along a baseline of known lengths.

Cyclic error (or instrument non-linearity) is caused by unwanted interference (called cross-talk) between signals generated and processed inside the total station. It can be investigated by measuring known distances spread over the measuring wavelength of the instrument. If a calibration curve of (observed – measured) distances is plotted against distance and a periodic wave is obtained, the total station has a cyclic error.

In Section 5.2, the accuracy of distance measurement for a total station was specified as ± (a mm + b ppm). The constant a is a function of zero and cyclic errors and b is a function of scale error.

For most construction projects, establishing a complicated multi-point baseline for calibration purposes is impractical and all that might be used for periodically checking the distances displayed by total stations (and for standardising tapes) is a simple two-point baseline. A comparison is made of the distance measured by the instrument with the baseline value – if an unacceptable difference is found between these, the instrument must be returned to the manufacturer for servicing and calibration. The zero error might also be determined on site, but it is very unlikely that any attempt will be made to determine scale or cyclic errors as these are small and can usually be ignored for the short distances measured on site. However, for precise work, more care is required when calibrating the distance function. Under these circumstances, a total station can be returned to the manufacturer where a commercial calibration service is available. Most manufacturers have a short baseline (typically 20 m) with accurately known intermediate points for calibration purposes and they also perform electronic tuning of a total station using frequency measurements. To be of real value, however, a calibration should be related to a national standard. At present, the only accredited source for this in the UK is the National Physical Laboratory (http://www.npl.co.uk/). For further details of the calibration of total stations, check the RICS web site (http://www.rics.org/) for publication of guidelines on this. It is also possible to assess the accuracy of distance measurement with a total station by following the test procedures given in BS 7334, ISO 17123 and DIN 18723 (see further reading and sources of information at the end of the chapter).

Auxiliary equipment

The plummet of a total station is checked and adjusted using the procedures given for this in Section 3.6; guidelines for the care of tripods and tribrachs are also given in that section. Detail poles should be inspected regularly and their circular bubbles adjusted to ensure that they are held vertically when centred over points that are to be measured. A cracked or chipped prism should never be used for distance measurement.

Atmospheric effects

All electromagnetic waves and pulses, when travelling in a vacuum, travel at the so-called speed of light, a universal constant. However, when travelling in the atmosphere, their speed v is reduced from the free-space value c owing to the retarding action of the atmosphere. Consequently, v will be a variable depending on atmospheric conditions, and for phase shift systems the modulation wavelength will vary for all measurements since $\lambda_m = v/f$. The significance of this is that the measuring unit λ_m is not constant and the distance recorded by an instrument will include a systematic error. The same effect occurs in pulsed systems where the speed of the pulses is affected. This is analogous to steel taping, where variations in temperature cause the tape to contract and expand from some reference value.

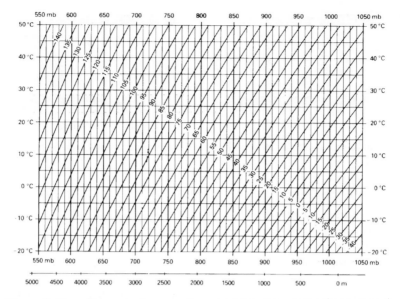

Figure 5.44 ● Atmospheric correction chart (courtesy Leica Geosystems).

To remove any errors caused by atmospheric effects, a correction is entered into a total station according to the air pressure, air temperature and relative humidity prevailing at the time of measurement. A correction, usually in ppm, is deduced from atmospheric conditions using a chart supplied with the instrument, an example of which is shown in Figure 5.44. Usually, the relative humidity is not measured, and on this chart, a value of 60% has been assumed. This introduces a maximum error of 2 ppm into the distance and for most work it is usually sufficient to take an atmospheric correction from the chart using temperature and pressure only and to enter this into the instrument. As far as the accuracy of distance measurement is concerned, atmospheric effects are a factor in the constant *b* in the specifications and are dependent on the length of line being measured. Note that on the chart in Figure 5.44, only small corrections close to 0 ppm are required for temperature and pressures that might be normally encountered and because short distances are measured on site, instrumental and centring errors are much more important than atmospheric effects. These are important, however, in extreme weather and are always critical on long lines.

Using the total station properly to avoid errors and other problems

The discussion of errors given in Section 3.6 for theodolites is also relevant here, but some additional material for total stations is given.

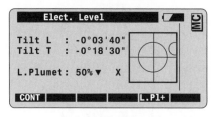 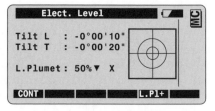

Instrument not level   Instrument level

Figure 5.45 ● Tilt errors displayed by total station (courtesy Leica Geosystems).

Levelling and centring

The usual requirement at the start of any work with a total station is to level and centre it over a point – this applies to tripod-mounted reflectors as well. The procedure for levelling a total station is the same as for a theodolite and is achieved by adjusting the three footscrews and observing a plate level or electronic bubble on the instrument's display. Most total stations allow the user to view the amount by which the instrument is off level at any time, as shown in Figure 5.45, and all will issue a warning and stop working when the instrument is not levelled within specified limits.

Once levelled, centring of the total station is carried out using either an optical or laser plummet.

The ability to level and centre total stations and tripod-mounted reflectors is taken for granted on site and everyone is expected to be able to do this quickly and efficiently. However, levelling and centring are important and care must be taken to ensure they are done properly, no matter how long it takes to achieve this. If a total station or reflector is set up for a long time at the same point, it is necessary to check the levelling and centring at frequent intervals – this is especially important if they have been set up on soft ground or work is being carried out in hot sunshine.

Prism constant

When measuring a distance to a prism, great care must be taken to ensure that the correct prism constant is entered into the total station otherwise all measurements will be subject to a systematic error. Not all prism constants are the same, and these vary from prism to prism and between different manufacturers. *It is also necessary to enter a constant even when measuring to reflecting foil and in reflectorless mode.* As already noted, a zero error can be determined for a total station which can be applied to measurements to correct for the prism constant.

Angle and distance measurements

As with any optical instrument, parallax must be removed before taking readings through the telescope – this also applies to the most sophisticated total stations.

If angles are to be measured, the procedures given in Section 3.5 for a theodolite are recommended, but there may be alternative procedures programmed into the total station or its data collector for doing this. Whatever method is used, it must account for any instrumental errors present in the total station.

As with angles, the procedures used for measuring distances will vary from instrument-to-instrument and it is necessary to follow the instructions for this given by each manufacturer. Important distances should be checked by measuring them in both directions with the instrument and the reflector interchanged.

Reflectorless measurements

Care is needed with reflectorless measurements to check that the laser beam is not reflected by any surfaces close to the line of sight instead of the intended target, as the instrument might record the wrong distance without the operator's knowledge. Sometimes, sightings are taken under difficult measuring conditions involving acute angles of incidence to poorly reflecting targets – it may not be possible to obtain sufficient return signal for measurement in these circumstances, and the accuracy in distance measurement might also degrade significantly. When a reflectorless distance measurement is triggered, the total station will measure a distance to the object which is in the beam path at that moment – if something is obstructing the measurement path, an incorrect distance will be recorded, again without the operator's knowledge. This may happen on site when someone moves across the beam or when plant and machinery temporarily interrupt the beam. A false reading can also occur in heavy rain or fog. It is always advisable to check important work in advance where possible to see whether reflectorless measurements can be taken before these are used on complicated setting out routines.

Temperature effects

It is essential that a total station is allowed to acclimatise to environmental conditions before taking any measurements with it. Although temperature differences can result in changes to any of the characteristics of the instrument, the compensator is most affected and the time suggested for temperature adjustment is 2 minutes per 1 °C.

Transporting and storing a total station

Having finished observations, if the total station is to be used at another point it should be taken off the tripod and carried separately before it is moved, as shown in Figure 5.46. As with a theodolite, a total station should never be carried or moved fixed to the tripod, as this could damage it in some way.

Upon completion of work, the total station should always be put in its case, again to protect it against any accidental damage. If, however, the instrument has been used outside in rain or in damp and humid conditions, it must not be left in a closed case for any length of time as this could permanently damage the instrument. If the total station is wet, it must be dried and then placed in its case, which should be left open for an extended period (for example, overnight). If the case itself gets wet, this must also be left open until dried.

Figure 5.46 ● Correct way to transport total station (courtesy Leica Geosystems).

One final note regarding storing and transporting equipment – it should not be left inside a vehicle in summer in direct sunlight as it might be damaged by exceeding its operating temperature.

Data recording and transfer

When data for a survey is recorded by hand, great care has to taken not to misread any angles and distances displayed by the total station and not to write these incorrectly in a field book. These problems do not occur with electronic data collection, but because many software packages can be used, it is advisable to be completely familiar with the instrument and data collection system before going on site with it – this avoids wasting time and making mistakes.

Once recorded, electronic data is very vulnerable, and data held in a total station, data collector or card should be transferred to another source at frequent intervals – at least once a day. Never erase the original data from any source until a backup copy has been created and read successfully.

Batteries

Total stations and other electronic survey equipment are supplied with batteries of different types, all of which have differing operating times (capacities) under different measuring conditions. It is good practice, no matter what assurance a manufacturer may give about the capacity of a battery, to have fully charged spares available at all times with the instrument.

When batteries are charged, the charging temperature has a significant effect on them as charging at high temperatures can cause an irreversible loss of capacity. For optimal charging, most batteries should be charged in an ambient temperature of 10–20 °C. If a total station is used at low operating temperatures, the operating time of the battery can be significantly reduced, and continuous working in high temperatures can shorten the life of a battery. All of this suggests that more batteries are needed when carrying out surveys in extreme climates.

Both NiCad and NiMh batteries, but particularly NiCad batteries, are subject to the *memory effect* if the same charging and discharging conditions are continually applied to them, and this can seriously affect their capacity in the long term. Whenever the memory effect occurs, a battery can be refreshed by fully charging and discharging it several times. Compared to the NiCad and NiMh batteries, lithium-ion batteries can be charged at any time by any amount without reducing their capacity.

Sometimes it is necessary to store batteries for extended periods. NiCad batteries can be stored in any charged state for unlimited periods, but NiMh batteries must be stored fully charged and must be recharged every six months whilst in storage.

Reflective summary

With reference to sources of error for total stations, remember:

— There are many different types of error that can occur, but these can be minimised by regular maintenance and proper handling and use of the total station. Always follow recommended procedures when measuring – avoid taking short cuts.

— An electronic calibration of the total station should be carried out regularly – be careful to follow the manufacturer's instructions for doing this.

— Calibration of the total station on a baseline should also be carried out frequently. If the instrument is not reading accurately it must be returned to the manufacturer for servicing.

— Like the electronic theodolite, using a total station in adverse weather can be difficult and this should be avoided where possible. If work must proceed, care must be taken to apply atmospheric corrections correctly, to ensure that the instrument is given enough time to adapt to any changes of temperature and that its tripod or other platform is securely fixed.

— If the best accuracy is required from a total station, even from a high-precision instrument, the following measures should be taken:

   • Use a sunshade to protect instrument and tripod
   • Let the instrument reach the ambient temperature (allow 2 minutes per °C)
   • Calibrate the instrument regularly
   • Measure on face left (I) and face right (II)
   • Measure the temperature and pressure and apply atmospheric corrections
   • Carry out measurements in the morning or under an overcast sky

5.6 Measuring heights (reduced levels) with total stations

After studying this section you should understand how a total station computes heights from raw data and you should be able to carry out the calculation of heights using trigonometrical methods. You should also be aware of the accuracy of total station heights and why these are not as good as those obtained by levelling.

Trigonometrical heighting

The raw data measured by a total station are horizontal angle, zenith (or vertical) angle and slope distance, all of which are usually converted into the three-dimensional coordinates of the position of the reflector. This section gives details of how the height of the point sighted is obtained – the procedures for obtaining the horizontal position are given in Chapter 6, which covers traversing and coordinate calculations.

To illustrate how a total station is used to determine heights, consider Figure 5.47, which shows two points A and B. A total station is levelled and centred at A and a reflector is set up at B. The slope distance L and zenith angle z (or vertical angle α) between A and B are measured together with the height hi of the total station above A (height of instrument) and the height hr of the reflector above B (height of reflector). If the height of A is known (H_A), the height of B (H_B) is given by

$$H_B = H_A + hi + V_{AB} - hr \qquad (5.1)$$

The vertical component V_{AB} is obtained from

$$V_{AB} = L \cos z = L \sin \alpha$$

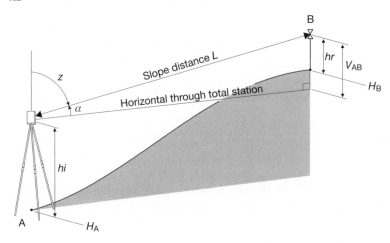

Figure 5.47 ● Measurement of height with total station.

and this will be positive if the telescope is tilted above the horizontal through the total station (z less than 90° or positive vertical angle as shown in Figure 5.47) and will be negative if it is below the horizontal (z greater than 90° or negative vertical angle).

This is known as *trigonometrical heighting* and is the basis for all height measurement with a total station. It gives heights of sufficient accuracy for most detail surveys and for some setting out up to a distance of about a hundred metres. It is also used in other applications programs by a total station when heights are required for three-dimensional traverses, when performing a resection with elevations, and for remote height measurements.

For all trigonometrical heighting, it is important that the reflector is sighted accurately and that zenith (or vertical) angles are measured carefully, especially for traverses, resections and other control surveys. It is also important that the correct instrument and reflector heights are measured and entered into the instrument. When taking readings to a detail pole, the pole is often extended so that it is the same length as the height of the instrument – this simplifies any manual calculations of height since hi will equal hr to give $H_B = H_A + V_{AB}$. However, there is a tendency for telescopic detail poles to shorten gradually during a long setup and their lengths should be checked frequently. When a lot of points are to be measured from a single instrument position, it is better to use a pole fully extended or, better still, to use a fixed length pole – these are available in various lengths.

As well as being able to measure ground heights at a reflector, a total station can also be used in reflectorless mode to determine the heights of any point targeted including inaccessible ones where it is not possible to locate a prism. In this case, there is no height of reflector to measure and $H_B = H_A + hi + V_{AB}$.

Most total stations also have a remote elevation program for measuring height differences. To use this, the total station is set up anywhere convenient and a detail pole is held at the point where height information is required. With the reflector height entered into the instrument, the slope distance is measured. The remote elevation program will now calculate and then display the height difference from ground level to any point along the vertical through the detail pole as the telescope is tilted – there is no need to sight a prism or target to do this. Using remote elevation, the underside of the bridge B in Figure 5.48 could be sighted directly above the prism

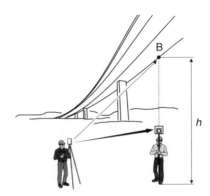

Figure 5.48 ● Remote elevation measurement (courtesy Leica Geosystems).

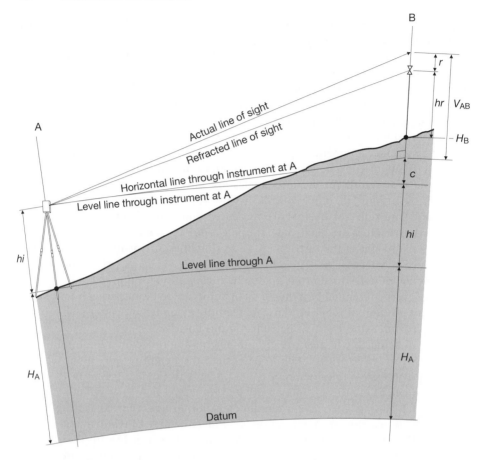

Figure 5.49 ● Measuring heights over long distances with total station.

and its height h above ground level recorded. This facility is often used for measuring clearances between the ground and overhead objects such as bridges and overhead cables.

If the height difference between points some distance apart is to be determined, although levelling will give the best result, it can be a very time-consuming process, especially in hilly terrain. Under these circumstances, GPS might be used to obtain any unknown heights, but if this is not available, a total station can be used. When trigonometrical heighting is carried out over long distances, Earth curvature and atmospheric refraction have to be accounted for. Figure 5.49 is similar to Figure 5.47 but shows the measurement of the height difference between two points A and B with curvature and refraction introduced. Working upwards along the vertical through B from the datum, the height of B is given by

$$H_B = H_A + hi + c + V_{AB} - r - hr$$

or

$$H_B = H_A + hi + V_{AB} - hr + (c - r) \qquad\qquad (5.2)$$

As can be seen, the only difference between this and the calculation of height for short sighting distances (equation 5.1) is the combined correction for curvature and refraction $(c - r)$. In deriving equation (5.2), it has been assumed that the two verticals are parallel and that the horizontal surface through the total station at A is perpendicular to the vertical through B. For very long lines this would not be true, but for the distances measured during most engineering and site surveys, any errors arising from this can be ignored.

A value for the combined correction for curvature and refraction $(c - r)$ can be obtained as follows.

The curvature correction c is given by

$$c = \frac{D^2}{2R}$$

where D is the horizontal distance between A and B and R is the radius of the Earth.
Taking R to be 6370 km:

$$c \text{ (in metres)} = 0.0785 \, D^2 \text{ (D in km)}$$

The effect of refraction in the atmosphere is often assumed to bend the line of sight of the total station towards the Earth so that it reduces the effect of curvature by a ratio of about $\frac{1}{7}$ and the combined correction for curvature and refraction is sometimes quoted as

$$(c - r) \text{ (in metres)} = \tfrac{6}{7} \, 0.0785 \, D^2 = 0.0673 \, D^2 \text{ (D in km)}$$

However, the refraction correction should be derived from a parameter known as the *coefficient of refraction k* using

$$r = k \frac{D^2}{R}$$

and the combined curvature and refraction correction from

$$(c - r) = (1 - 2k) \frac{D^2}{2R}$$

or

$$(c - r) \text{ (in metres)} = 0.0785 \, (1 - 2k) \, D^2 \text{ (D in km)}$$

Assuming refraction to be $\frac{1}{7}$ of curvature gives a value for k of 0.07 and a value similar to this for the coefficient of refraction is commonly used in trigonometrical heighting and in some total stations for correcting measured height differences.

The effect of refraction on the measurement of heights is usually quite small under normal atmospheric and ground conditions. However, the value of k is dependent on atmospheric conditions and it can sometimes vary considerably from a value of 0.07, especially at dawn or dusk and in an unpredictable manner close to the ground. If this variation is not accounted for in field procedures when measuring height differences over long distances with a total station, errors can occur. Figures 5.47 and 5.49 show *single-ended trigonometrical heighting* in which the total station is set up at A and measurements are taken from A to the reflector at B. This would be the normal method for the measurement of heights with a total station and is routinely used up to distances of about 100 m, where the effects of curvature and refraction are usually small and are ignored. However, when sightings are taken over longer distances, curvature and refraction corrections should be computed using an accurate value for *k*, which can be very difficult to determine. In *reciprocal trigonometrical heighting*, the total station takes measurements from A to B and then it is taken to B, where a second set of measurements are taken from B to A. If these two measurements are averaged, the effects of curvature and refraction will cancel out provided the value of *k* is the same for each measurement. Therefore, in reciprocal measurements, the height difference between A and B is obtained by computing them as two single-ended measurements and then taking the mean value. The two sets of measurements should be taken as quickly as possible to reduce the effects of any change that may occur in the value of *k* in between them. In *simultaneous reciprocal trigonometrical heighting*, measurements are taken at each end of a line with two instruments at exactly the same time in order to remove the effects of curvature and refraction.

For all methods of trigonometrical heighting, the accurate measurement of zenith (or vertical) angles is critical. In fact, under normal atmospheric conditions, angular errors are the most serious source of error in measured height differences – see Section 9.3 for an analysis of this.

Since measurements are carried out along the line of sight when using an integrated total station, the telescope is tilted and pointed at the centre of the reflector when measuring – great care must be taken to do this accurately. The distance over which reciprocal and simultaneous reciprocal measurements can be carried out may be several hundreds of metres, but both methods are limited by how well the reflector can be sighted – this becomes more difficult at long ranges. To overcome these problems and maintain precision, zenith (and vertical) angles should be measured on both faces and several sets of these taken and averaged. The accuracy of sightings can also be improved using prism sets fitted with target plates (see Figure 5.11).

Another way to extend the range of a total station in trigonometrical heighting is to use it in a similar manner to a level. To do this, it is set up halfway between two points A and B where the height difference is required and single-ended measurements are taken to each point in the normal way.

For the sighting taken to A equation (5.2) gives
$$H_A = H_i + hi + V_A + (c - r)_A - hr_A$$

For the sighting taken to B equation (5.2) gives
$$H_B = H_i + hi + V_B + (c - r)_B - hr_B$$

where

H_A and H_B are the heights of A and B
H_i is the height of the point at which the total station is set up
hi is the height of the instrument at the setup

V_A and V_B are the vertical components to A and B

$(c - r)_A$ and $(c - r)_B$ are the curvature and refraction corrections to A and B

hr_A and hr_B are the reflector heights at A and B

The difference between these gives

$$(H_A - H_B) = (V_A - V_B) - (hr_A - hr_B)$$

Because H_i has been eliminated, it is not necessary to know the height of the point at which the total station has been set up – this means that it can be set up in any position. It is also not necessary to measure the height hi of the instrument. The effects of curvature and refraction also cancel out provided the two sighting distances are approximately equal.

This technique has the advantage of extending the range of a total station for trigonometrical heighting, but has the disadvantage that two zenith angles have to be measured.

Worked example 5.1: Reciprocal trigonometrical heighting

Question

The height difference between two points A and B is measured with a total station and the following results are obtained.

With the total station at A	With the total station at B
Slope distance = 508.118 m	Slope distance = 508.125 m
Zenith angle = 88°19'44"	Zenith angle = 91°41'45"
Height of instrument at A = 1.617 m	Height of instrument at B = 1.652 m
Height of reflector at B = 1.515 m	Height of reflector at A = 1.572 m

If the height of point A is 57.225 m, calculate the height of point B.

Solution

The height of B is obtained by calculating two single-ended observations, ignoring the curvature and refraction correction.

With the total station at A, equation (5.1) gives

$$H_B = H_A + hi_A + V_{AB} - hr_B$$
$$= 57.225 + 1.617 + 508.118 \cos 88°19'44" - 1.515$$
$$= 72.145$$

With the total station at B, equation (5.1) gives

$$H_A = H_B + hi_B + V_{BA} - hr_A$$
$$57.225 = H_B + 1.652 + 508.125 \cos 91°41'45" - 1.572$$

and

$$H_B = 72.182 \text{ m}$$

Note that two different results have been obtained because the effects of curvature and refraction have been ignored. The height of B is given by the mean value as

$$H_B = \frac{72.145 + 72.182}{2} = 72.164 \text{ m}$$

Reflective summary

With reference to measuring heights (reduced levels) with total stations, remember:

— Total stations do not measure heights directly and use a slope distance and zenith (vertical) angle to compute these.

— The advantage of measuring heights with a total station is that sightings can be taken over longer distances than are possible with levelling. However, although the accuracy of heights from total stations is satisfactory for most survey work on site, for best results levelling should always be used.

— If height differences are to be measured over long distances, the effects of curvature and refraction must be accounted for either by correction or by field procedures.

— When taking readings for height differences over long distances, great care must be taken with each sighting because the worst source of error occurs when measuring the zenith (vertical) angle.

Exercises

5.1 Explain the difference between the phase shift and pulsed laser methods for measuring distances.

5.2 What is an integrated total station and what is integrated surveying?

5.3 Explain what a third party data collection system is.

5.4 Draw diagrams to show the instrumental errors that can be measured by a total station in an electronic calibration.

5.5 What are the best applications for handheld laser distance meters on construction sites?

5.6 What are the differences between precise, rapid and tracking distance measurement modes?

5.7 Explain how ATR enables a motorised total station to lock onto a prism. How does a robotic total station differ from a motorised one?

5.8 What are the advantages and disadvantages of reflectorless distance measurements?

5.9 Describe the methods by which data can be transferred from a total station to a computer.

5.10 What are zero, scale and cyclic errors?

5.11 How do variations in atmospheric conditions affect distance measurements taken with a total station?

5.12 Explain how a total station measures height differences. What are the problems with these measurements when they are taken over long distances?

5.13 Discuss the precautions that should be taken on site when using a total station that has a Class 3R laser.

5.14 Discuss the advantages and disadvantages of data collectors and internal memory for data storage with total stations.

5.15 Explain what a prism constant is and why a value for this must be entered into a total station when measuring distances.

5.16 List the field procedures that should be taken to ensure that the best possible accuracy is obtained from a total station for both angle and distance measurements.

5.17 What is the memory effect in batteries and how can it be removed?

5.18 The distance specification for a total station is quoted as ± (3 mm + 3 ppm). According to this, what is the accuracy in distance measurement for this instrument at 50 m and 2 km?

5.19 The following measurements were taken to determine the zero error of a total station: (all points are on the same line) AB = 13.100 m, BC = 25.017 m and AC = 38.162 m. Calculate a value for the zero error.

5.20 Assuming the coefficient of atmospheric refraction to be 0.10, calculate values for the combined curvature and refraction correction at 100 m intervals for distances up to 500 m.

5.21 The height difference between two points A and B is measured with a total station and the following results are obtained.

With the total station at A	*With the total station at B*
Slope distance = 256.347 m	Slope distance = 256.352 m
Zenith angle = 92°25'51"	Zenith angle = 87°37'42"
Height of instrument at A = 1.550 m	Height of instrument at B = 1.625 m
Height of reflector at B = 1.405 m	Height of reflector at A = 1.530 m

If the height of point A is 19.072 m, calculate the height of point B.

Further reading and sources of information

For assessing the accuracy of a total station, consult

BS 7334-8: 1992 (ISO 8322-8: 1992) *Measuring instruments for building construction. Methods for determining accuracy in use of electronic distance measuring instruments up to 150 m.* (British Standards Institution [BSI], London). BSI web site http://www.bsi-global.com/.

Deutsches Institut für Normung DIN 18723-6:1990 *Field procedure for precision testing of surveying instruments; electro-optical distance measuring instruments for short ranges* (DIN e.V., Berlin). DIN web site http://www.din.de/.

ISO 17123–4: 2001 *Optics and optical instruments – Field procedure for testing geodetic and surveying instruments – Part 4: Electro-optical distance meters (EDM instruments)* (International Organization for Standardization [ISO], Geneva). ISO web site http://www.iso.ch/.

For general guidance on using a total station on site refer to

ICE Design and Practice Guide (1997) *The management of setting out in construction.* Thomas Telford, London.

For the latest information on the equipment available, visit the following web sites:

http://www.leica.com/
http://www.nikon-trimble.com/
http://www.pentax.co.uk/
http://www.sokkia.com/
http://www.topcon.co.uk/
http://www.trimble.com/

The following journals include annual reviews of total stations and include frequent articles on them: *Civil Engineering Surveyor, Engineering Surveying Showcase, Geomatics World* and *GIM International* (see Chapter 1 for further details).

Traversing and coordinate calculations

 Aims

After studying this chapter you should be able to:

- State what control surveys are and why these are an essential part of surveying
- Discuss the advantages of defining position for most surveying projects in the form of plane rectangular coordinates
- Carry out coordinate calculations using *rectangular* → *polar* and *polar* → *rectangular* conversions
- Derive transformation parameters for converting coordinates from one plane rectangular grid to another
- Explain what a traverse is and describe all of the fieldwork involved in traversing
- Perform all of the necessary calculations that are required to obtain traverse coordinates from measured angles and distances
- Describe how total stations are used for traversing
- Calculate the coordinates of points fixed by intersection and resection
- Discuss how networks are used in control surveys

This chapter contains the following sections:

6.1   Control surveys

6.2   Rectangular and polar coordinates

6.3   Coordinate transformations

6.4   Planning and fieldwork required for traversing

6.5   Traverse calculations

6.1 Control surveys

After studying this section you should know what control surveys are and be aware that they can include both horizontal and vertical components. You should appreciate that different methods are used for control surveys depending on the size of the site.

The need for survey control

All measurements taken for engineering surveys are based on a network of horizontal and vertical reference points called *control points*. These networks are used on site in the preparation of maps and plans, they are required for dimensional control (setting out) and are essential in deformation monitoring. Because all survey work needs control points, at the start of any engineering or construction work a *control survey* must be carried out in which the positions of all the control points to be used are established.

For the majority of engineering work, the positions of *horizontal control* points are specified as plane rectangular coordinates (equivalent to X and Y coordinates used in mathematics). This is normal practice for construction sites as survey work is greatly simplified and fewer mistakes are made when using rectangular coordinates for setting out and other dimensional work. One of the techniques used to determine the positions of control points for small construction sites is *traversing*, which can be extended using *intersection* and *resection*. All of these methods involve the measurement of angles and distances which are used to calculate the coordinates of the control points – all of the field procedures and calculations required for this are described in this chapter.

Alongside horizontal control, all sites will have some form of *vertical control*. The method used to provide vertical control on most small construction sites is levelling, which is used to establish a series of benchmarks around a site. However, trigonometrical heighting with a total station can also be used for this.

As the size of a site increases, a *network* is often observed to determine the coordinates of control points. These also involve the measurement of angles and distances. Plane surveying can still be assumed for these surveys, but when the size of the site exceeds 10–15 km the curvature of the Earth becomes too large to be ignored and coordinates based on a plane or flat grid are no longer valid. In this case, coordinates based on a special map projection must be used. Ultimately, control surveys can be carried out over large areas and cover whole countries. Throughout Great Britain, Ordnance Survey National Grid coordinates and heights based on the Ordnance Datum at Newlyn can be used for site work. Whatever coordinate systems are used, the control for construction sites that cover large areas is usually surveyed using GPS methods.

Reflective summary

With reference to control surveys, remember:

— For all engineering work, a control survey is carried out to fix the positions of reference points required for mapping, setting out and other dimensional work.

— For most construction sites, a control survey is made up of *horizontal control* in which position is defined as plane rectangular coordinates and *vertical control* in which position is defined as heights. In this case, levels, theodolites, tapes and total stations are used to provide the data from which the positions of control points are calculated.

— On large construction and engineering projects, position must be defined taking into account the curvature of the Earth. In this case, horizontal and vertical control is usually surveyed using GPS methods.

6.2 Rectangular and polar coordinates

After studying this section you should know that the positions of control and design points on site are defined in the form of plane rectangular coordinates. You should also know what polar coordinates are and how they define position. You should be able to calculate rectangular coordinates from bearings and distances and vice versa.

This section includes the following topics:

- Plane rectangular coordinates
- Calculation of rectangular coordinates
- Polar coordinates
- Coordinate measurements with total stations

Plane rectangular coordinates

As noted in the previous section, the horizontal positions of control points for most applications in engineering surveying are defined using plane rectangular coordinates. Because these are used extensively for dimensional control on site, it is essential for everyone using instruments that can process and display coordinate information to understand how these are defined and to be able to perform calculations with coordinates.

The coordinate system adopted for most survey purposes is a rectangular system using two axes at right angles to one another called the north axis and the east axis. The scale along both axes is always the same and, with reference to Figure 6.1, any point P has coordinates which are known as the *easting* E_P and *northing* N_P quoted in the order E_P, N_P unless otherwise stated. Any number of points can be defined on a rectangular system and the relative positions of each control point in a network are expressed in the differences between coordinates.

For all types of survey and engineering work, the origin of a local coordinate system is chosen so that only positive eastings and northings are used. If, at the planning stage of survey, it appears that negative eastings and northings might occur, the origin should be moved such that all coordinates will be positive. Changing the origin of a survey after construction work has commenced is not recommended as this could cause mistakes to occur when changing from one coordinate system to

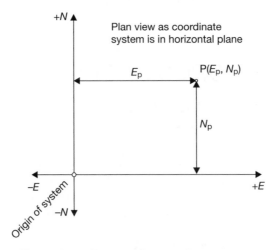

Figure 6.1 ● Rectangular coordinate system.

another. Extreme caution must be exercised when working with negative coordinates if they should occur.

The north axis of a rectangular coordinate system has to be orientated to a specified direction – a number of options for defining a north direction are available as follows.

The definition of *true north* and details of how this might be realised on site are given in Chapter 8. The accurate determination of this is only carried out for special construction projects and true north is not normally used for the majority of engineering surveys. If required for information purposes, an approximate value can be obtained by scaling from an Ordnance Survey map of the area.

Magnetic north is determined by a freely suspended magnetic needle and when required for survey work, is best measured using a prismatic compass. When taking compass bearings, it is advisable to be aware of any metallic objects close to the compass that might cause incorrect readings to be obtained. Another source of error is *magnetic declination*, which is the difference between true north and magnetic north. This is a function of time and is mostly caused by *secular variation*, the behaviour of which can only be predicted from observations of previously recorded declinations at known locations. The *daily* or *diurnal variation* in magnetic north also causes it to swing through a range of values according to the time of day, and *irregular variations* can cause magnetic north to become unstable. Allowing for all of these, it is not possible to determine the direction of magnetic north to better than ±0.5° and it is only used to give a general indication of north when no reliable maps are available and when an arbitrary north is chosen for a survey.

Grid north is based on the Ordnance Survey National Grid, which is discussed in detail in Chapter 8. To adopt this north direction, GPS is used to give the National Grid coordinates of control points.

Arbitrary north is the north direction most commonly used on site to define bearings and coordinates – the coordinate grid for a construction project could therefore be aligned using any convenient direction to represent north even though this is not a true or magnetic north. For example, a coordinate grid may be aligned by using existing control points or it could be aligned along some feature to suit site conditions (such as the centreline of a long bridge or along a building line).

For those sites where traversing is used to provide control, it is usual that an independent or local coordinate grid is adopted in which an arbitrary north is used. This chapter deals with the fieldwork and calculations required for this type of control survey.

Calculation of rectangular coordinates

On a coordinate grid, the direction of a line between two points is known as its bearing. The *whole-circle bearing* of a line is measured in a clockwise direction in the range 0 to 360° from a specified reference or north direction. Examples of whole-circle bearings are given in Figure 6.2.

Figure 6.3 shows the plan position of two points A and B on a plane rectangular grid. If the coordinates of A (E_A, N_A) are known, the coordinates of B (E_B, N_B) are obtained from A as follows

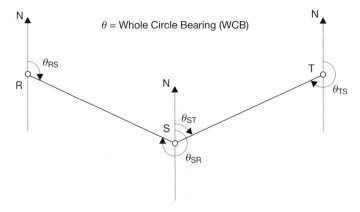

Figure 6.2 ● Whole-circle bearings.

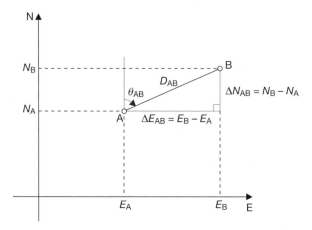

Figure 6.3 ● Calculation of rectangular coordinates

$$E_B = E_A + \Delta E_{AB} = E_A + D_{AB} \sin \theta_{AB} \qquad (6.1a)$$

$$N_B = N_A + \Delta N_{AB} = N_A + D_{AB} \cos \theta_{AB} \qquad (6.1b)$$

where

ΔE_{AB} = the *eastings difference* from A to B
ΔN_{AB} = the *northings difference* from A to B
D_{AB} = the horizontal distance from A to B
θ_{AB} = the whole-circle bearing from A to B

The calculation of easting and northing differences from a bearing and distance is known as a *polar to rectangular coordinate conversion*. The coordinates of B are obtained by computing the eastings and northings differences first and then applying these to the coordinates of A. If a calculator is used, values of ΔE and ΔN can be obtained directly from the equations. Alternatively, the *polar → rectangular* function found on most calculators can be used to calculate ΔE and ΔN. Since the method for doing this varies with the make and model of each calculator, the handbook supplied with the calculator should be consulted for details of how to perform these calculations. There are also several software packages available that can carry out coordinate calculations and they can also be carried out by the majority of total stations.

Worked example 6.1: Polar to rectangular conversions

Question

The coordinates of point A are 311.617 mE, 447.245 mN. Calculate the coordinates of point B where $D_{AB} = 57.916$ m and $\theta_{AB} = 37°11'20"$ and point C where $D_{AC} = 85.071$ m and $\theta_{AC} = 205°33'55"$.

Solution

With reference to Figure 6.3 and equations (6.1a) and (6.1b)

$$E_B = E_A + D_{AB} \sin \theta_{AB} \qquad\qquad N_B = N_A + D_{AB} \cos \theta_{AB}$$
$$= 311.617 + 57.916 \sin 37°11'20" \qquad = 447.245 + 57.916 \cos 37°11'20"$$
$$= 311.617 + 35.007 \qquad\qquad = 447.245 + 46.139$$
$$= \mathbf{346.624\ m} \qquad\qquad\qquad = \mathbf{493.384\ m}$$

Similarly

$$E_C = E_A + D_{AC} \sin \theta_{AC} \qquad\qquad N_C = N_A + D_{AC} \cos \theta_{AC}$$
$$= 311.617 + 85.071 \sin 205°33'55" \qquad = 447.245 + 85.071 \cos 205°33'55"$$
$$= 311.617 - 36.711 \qquad\qquad = 447.245 - 76.742$$
$$= \mathbf{274.906\ m} \qquad\qquad\qquad = \mathbf{370.503\ m}$$

Polar coordinates

Another coordinate system used in surveying is the polar coordinate system shown in Figure 6.4. When using this, the position of point B is located with reference to point A by polar coordinates D and θ where D is the horizontal distance between A and B and θ the whole-circle bearing of the line A to B. Although polar coordinates could be used to define absolute position using an origin as eastings and northings do on a rectangular coordinate system, they are only used in surveying for defining the relative position of one point with respect to another.

In Worked example 6.1, it was shown that the coordinates of one end of a line joining two points

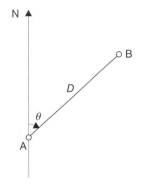

Figure 6.4 Polar coordinates.

can be computed from the other using the whole-circle bearing and horizontal distance.

For the reverse case where the coordinates of both points at each end of a line are known, it is possible to compute the horizontal distance and whole-circle bearing of the line between the points. This is known as an *inverse calculation* or *rectangular to polar coordinate conversion* and is frequently used in engineering surveying when setting out with coordinates. Horizontal distances and whole-circle bearings can be calculated by a number of methods – these include determining the *quadrant* of the line and applying formulae with a calculator, using the *rectangular → polar function* found on most calculators or by using a commercial software package. Many total stations will also perform these calculations directly from coordinates.

Worked example 6.2: Rectangular to polar conversion

Question

The coordinates of two points A and B are known as $E_A = 469.721$ m, $N_A = 338.466$ m and $E_B = 501.035$ m, $N_B = 310.617$ m. Calculate the horizontal distance D_{AB} and whole-circle bearing θ_{AB} of line AB.

Solution

The procedure for calculating the distance and bearing using *quadrants* is as follows.

The distance between A and B is given by $D_{AB} = \sqrt{\Delta E_{AB}^2 + \Delta N_{AB}^2}$.
Since $\Delta E_{AB} = (E_B - E_A)$ and $\Delta N_{AB} = (N_B - N_A)$ it follows that

$$D_{AB} = \sqrt{(E_B - E_A)^2 + (N_B - N_A)^2}$$
$$= \sqrt{(501.035 - 469.721)^2 + (310.617 - 338.466)^2}$$
$$= \sqrt{31.314^2 + (-27.849)^2}$$
$$= \mathbf{41.906 \ m}$$

A sketch showing the relative positions of the two points is drawn to show which quadrant the line is in. Doing this correctly is important as the most common source of error in this type of calculation is the wrong identification of quadrant. For whole-circle bearings, the four quadrants are shown in Figure 6.5 together with the set of rules for calculating whole-circle bearings in each quadrant. For this problem, Figure 6.6 shows the sketch for points A and B and the whole-circle bearing AB to be in quadrant II. Applying the rule for quadrant II gives

$$\theta_{AB} = \tan^{-1}\left[\frac{\Delta E_{AB}}{\Delta N_{AB}}\right] + 180° = \tan^{-1}\left[\frac{31.314}{-27.849}\right] + 180°$$
$$= \tan^{-1}[-1.124421] + 180° = -48°21'07" + 180°$$
$$= \mathbf{131°38'53"}$$

When using the rectangular → polar function on a calculator, values of D and θ are obtained directly. *However, the coordinate values must be entered into the calculator in the*

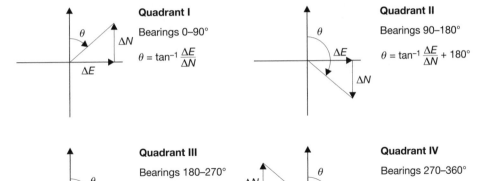

Quadrant I
Bearings 0–90°
$\theta = \tan^{-1} \dfrac{\Delta E}{\Delta N}$

Quadrant II
Bearings 90–180°
$\theta = \tan^{-1} \dfrac{\Delta E}{\Delta N} + 180°$

Quadrant III
Bearings 180–270°
$\theta = \tan^{-1} \dfrac{\Delta E}{\Delta N} + 180°$

Quadrant IV
Bearings 270–360°
$\theta = \tan^{-1} \dfrac{\Delta E}{\Delta N} + 360°$

Figure 6.5 ● Quadrants.

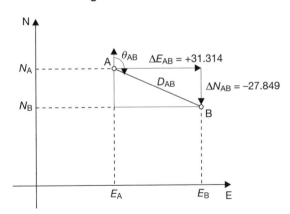

θ_{AB} $\Delta E_{AB} = +31.314$
D_{AB} $\Delta N_{AB} = -27.849$

Figure 6.6 ● Rectangular to polar conversion.

correct sequence, otherwise the wrong bearing will be obtained. The same care with data entry must be taken when using computer software or a total station.

Coordinate measurements with total stations

The horizontal circle of a total station can be set to read whole-circle bearings directly using the following procedure. First, the coordinates of the instrument station are entered into the total station, followed by the coordinates of a reference point, as shown in Figure 6.7. Second, the reference point is sighted and an orientation program in the total station is activated to carry out a rectangular → polar conversion and calculate the whole-circle bearing *A* from the instrument station to the reference station and to set the horizontal circle to display this

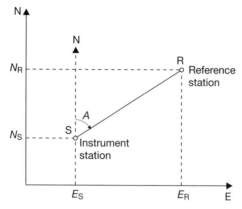

Figure 6.7 ● Horizontal circle orientation: E_S, N_S and E_R, N_R are entered into total station which calculates A and orientates to this.

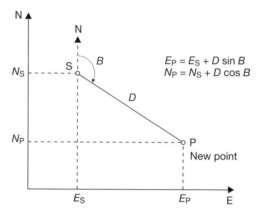

$$E_P = E_S + D \sin B$$
$$N_P = N_S + D \cos B$$

Figure 6.8 ● Coordinate measurement: point to P and measure D – total station calculates E_P, N_P from instrument station.

bearing. This is known as *horizontal circle orientation*, and whatever direction the total station is pointed in after this, it will display the whole-circle bearing along that direction.

Having orientated the horizontal circle of a total station, the coordinates of other points can be determined fairly easily. In Figure 6.8, a new point P is sighted and the total station will display the whole-circle bearing B to this. The horizontal distance D is measured and, using the *coordinate measurement* program, the instrument will now carry out a polar → rectangular conversion to calculate and display the coordinates of the new point. This can be extended to three dimensions if the height of the instrument station S and appropriate instrument and reflector heights are also entered into the total station, which will then display the height of the new point. This is calculated using trigonometrical methods described in Section 5.6.

Reflective summary

With reference to rectangular and polar coordinates, remember:

— Horizontal position is nearly always defined in site surveying as eastings and northings (in the order *E, N*) on a plane (or flat) rectangular grid.

— Most rectangular grids are based on an arbitrary north direction as this is usually the most convenient and simple way of defining coordinates for a construction project.

— A lot of setting out on construction sites is done using rectangular coordinates. Everyone doing this must be able to carry out the required calculations even when setting out information has already been pre-computed as it may be wrong, need checking or be computed for the wrong control points.

— Polar coordinates involve bearings and distances and are used in surveying to define the relative position of one point with respect to another.

— Coordinate calculations either involve converting measured bearings and distances into rectangular coordinates (a polar → rectangular conversion) or, when setting out, converting coordinates into bearings and distances (a rectangular → polar conversion). These can all be carried out using a calculator, a data collector or computer.

— Total stations are used extensively for setting out with coordinates. To do this, they have a number of special applications programs installed for calculating and displaying coordinates.

6.3   Coordinate transformations

After studying this section you should understand what coordinate transformations are and how to calculate transformation parameters for these.

The use of coordinate transformations on site

A problem that frequently occurs in engineering surveying is that of transforming coordinates of points from one reference system into their equivalent positions on another. Without proper management, coordinate transformations can cause many misunderstandings and errors to arise on site and they should be avoided where possible. If they are to be used, it is good practice to have them done as early as possible in a project and any transformation parameters that are to be used on site must be specified in contract documents.

For sites where GPS is to be used, extreme care is required when defining GPS coordinates in relation to a site datum. This can involve the use of fairly complicated coordinate transformations. In the simplest case, however, a site may only require conversions between eastings and northings on two different plane rectangular coordinate grids. This requires a two-dimensional coordinate transformation, and the calculations required for this are given below.

Two-dimensional coordinate transformations

Suppose points on an old $e–n$ coordinate system are to be transformed into those on a new $E–N$ coordinate system. To do this, the $e–n$ coordinate system has to be rotated through angle θ so that its axes are parallel to the $E–N$ coordinate system and its origin has to be shifted (or translated) through E_0 and N_0, as shown in Figure 6.9. There will also be a change of scale s between the two coordinate systems. This gives rise to *four transformation parameters*, which are given by

$$a = s \cos \theta \qquad b = s \sin \theta$$

$$c = E_0 \qquad d = N_0$$

These parameters are determined by knowing the coordinates of at least two control points in both coordinate systems. If these are point A with coordinates (e_A, n_A), (E_A, N_A) and point B with coordinates (e_B, n_B), (E_B, N_B) it can be shown that

$$E_A = ae_A - bn_A + c \tag{6.2a}$$

$$N_A = be_A + an_A + d \tag{6.2b}$$

and

$$E_B = ae_B - bn_B + c \tag{6.3a}$$

$$N_B = be_B + an_B + d \tag{6.3b}$$

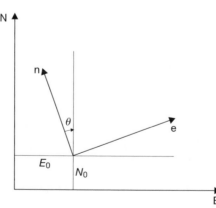

Figure 6.9 ● Two-dimensional coordinate transformation.

These equations are solved by entering the coordinates of A and B into them from both coordinate systems. The parameters obtained can then be used for transforming coordinates between the two coordinate systems.

Worked example 6.3: Four parameter coordinate transformation

Question

On a construction site, the coordinates of the control points used for the site survey are to be transformed into coordinates on a new setting out grid. The coordinates of control points A and B in the outdated e–n and revised E–N coordinate systems and those of point C in the e–n system are as follows, where all units are in metres.

	e	n	E	N
A	250.000	450.000	198.463	569.836
B	337.367	522.240	268.100	659.294
C	309.500	475.250		

Calculate values for the four transformation parameters between the two coordinate systems and the coordinates of point C on the new E–N system.

Solution

The known coordinates of control points A and B are entered into the transformation equations as:

For point A from equations (6.2a) and (6.2b):

$$198.463 = 250.000a - 450.000b + c$$

$$569.836 = 450.000a + 250.000b + d$$

For point B from equations (6.3a) and (6.3b):

$$268.100 = 337.367a - 522.240b + c$$

$$659.294 = 522.240a + 337.367b + d$$

Solution of these four equations gives

$$a = 0.9762540 \qquad b = 0.2167135$$

$$c = 51.921 \text{ m} \qquad d = 76.343 \text{ m}$$

The coordinates of point C are obtained as follows

$$E_C = ae_C - bn_C + c = 0.9762540(309.500) - 0.2167135(475.250) + 51.291$$
$$= 251.079 \text{ m}$$
$$N_C = be_C + an_C + d = 0.2167135(309.500) + 0.9762540(475.250) + 76.343$$
$$= 607.381 \text{ m}$$

This solution for the transformation parameters relies on two control points only, and for a better solution with a check, more control points must be introduced whose positions are known in both coordinate systems.

Transformation equations can be solved manually, but as the number of points used in the transformation increases, it is better to use a computer to do this.

Reflective summary

With reference to coordinate transformations, remember:

— Using well-established methods, it is always possible to convert coordinates from one plane rectangular grid to another provided the positions of at least two points are known on both grids.

— There are many different types of transformation possible in surveying varying from those with four parameters for converting coordinates in two dimensions between plane rectangular grids to those requiring seven (or more) parameters when converting three-dimensional GPS coordinates to Ordnance Survey or local coordinates.

— Where possible, it is always best to work with one coordinate system on site. As soon as another is introduced, for whatever reason, there is a good chance that mistakes will be made by using the wrong system or when converting coordinates between them.

6.4 Planning and fieldwork required for traversing

After studying this section you should understand what traverses are. You should also appreciate that proper planning is essential for these and you should be aware of the methods used to measure angles and distances for traverses.

This section includes the following topics:

● Types of traverse
● Reconnaissance
● Station marking
● Angle and distance measurements

Types of traverse

A traverse is a means of providing horizontal control in which the rectangular coordinates of a series of control points located around a site are determined from a combination of angle and distance measurements. Control points in traversing are often called *traverse stations*.

In Figure 6.10, a traverse has been run from existing control point A to unknown traverse stations 1–2–3 and then to existing control point X. This traverse opens along AB, fixes the positions of 1, 2 and 3 by measuring the angles and distances shown and then closes along XY. For this reason this traverse is called a *link* traverse (or sometimes a *connecting* or *closed-route* traverse). The origin and north direction of a link traverse are set by the coordinates of the opening control points A and B.

In Figure 6.11(a), a traverse starts at point A and returns to A via traverse stations 1, 2 and 3. In this case the traverse is called a *polygon* traverse (or sometimes a *loop* or *closed-ring* traverse) since it closes back on itself to form a polygon. The origin for this traverse is defined by point A which can be an existing control point or a traverse station with an assumed position – an arbitrary bearing is assigned to one of the lines to define the orientation and north direction. Figure 6.11(b) shows another polygon

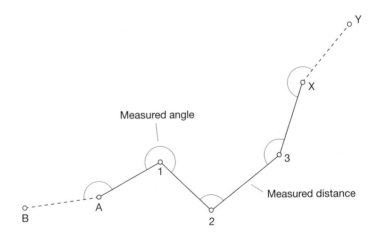

Figure 6.10 ● Link traverse.

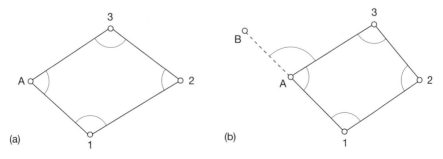

Figure 6.11 ● Polygon traverses.

traverse based on two existing control points in which the origin and orientation are defined by the coordinates of A and B.

Both the polygon and link traverses described here are known as *closed traverses*, since there is an external check on the observations and fieldwork as they both start and finish at known or assumed points.

Reconnaissance

This is one of the most important parts of any control survey and must *always* be done before any angles or distances are measured. The main aim of the reconnaissance is to locate suitable positions for the traverse stations and it cannot be over-emphasised that a poorly executed reconnaissance can result in difficulties at later stages on site, leading to wasted time and increased costs.

To start a reconnaissance, information relevant to the survey area should be gathered, especially that relating to any previous surveys. Such information may include existing paper or digital maps, aerial or orthographic photographs, and any site surveys already prepared for the project. Using this, a diagram can be drawn showing the proposed locations of the traverse stations. Following this, it is *essential that the site is visited* at which time the final positions for the stations are chosen. For small sites and where no previous information is available, the site visit becomes the reconnaissance.

When locating traverse stations, an attempt should be made to keep the number of these to a minimum, and short traverse lines should be avoided to minimise the effect of any centring errors (the effect of these are discussed in Section 3.6). If the traverse is to be used for mapping purposes and measurements are to be taken with a total station, a polygon traverse is usually sited around the area at points of maximum visibility. It should be possible to observe cross checks or lines across the area to assist in the location of any angular errors. On a construction site, traverse stations are put in place for the best accuracy and ease of setting out, but the effect of the construction must be taken into account, as this may block lines of sight as work proceeds. Traverses for roadworks and pipelines generally require a link traverse, since these sites tend to be long and narrow. In this case, the shape of the road or pipeline dictates the shape of the traverse.

Although these would normally be measured with a total station, if distance measurements are to be carried out using tapes, the ground conditions between stations should be suitable for this purpose. Steep slopes or badly broken ground along the traverse lines should be avoided, and it is better if there are as few changes of slope as possible. Roads and paths that have been surfaced are usually good for ground measurements.

Stations should be located such that they are clearly intervisible, preferably at ground level, so that it is possible to see adjacent stations and as many others as possible – this makes it easier to measure the angles and improves their accuracy. However, owing to the effects of lateral refraction and shimmer, traverse lines of sight should be kept well above ground level (greater than 1 m) for most of their length. This avoids any possible angular errors due to observations passing close to ground level (grazing rays), the effects of which are serious in hot weather.

Stations should be placed in firm, level ground so that the total station and tripod are supported adequately when observing at the stations. Very often, stations are used for a site survey and at a later stage for setting out. Since some time may elapse between the site survey and the start of the construction, the choice of firm ground in order to prevent the stations moving in any way becomes even more important. It is sometimes necessary to install semi-permanent stations.

When the positions of stations have been chosen, a sketch of the traverse should be prepared, *approximately to scale*, to help in the planning and checking of fieldwork. On this, the stations are given reference letters or numbers.

Based on the specification for the traverse, the final part of the planning for the traverse is to choose the instrument(s) to be used for the survey. This is discussed further in Section 6.5.

Station marking

When the reconnaissance has been completed, the stations have to be marked for the duration, or longer, of the survey. Station markers must be permanent and not easily disturbed, and it should be easy to set up and centre an instrument over them. The construction and type of station depend on the requirements of the survey.

For general purpose traverses, 50 mm square *wooden pegs* can be used, which are hammered into soft ground until only the top 50 mm of the peg is showing above the ground. If it is not possible to drive the whole length of the peg into the ground the excess should be sawn off to avoid it being knocked. A nail should be tapped into the top of the peg to define the exact position of the station. Figure 6.12(a) shows such a station from which several months use is possible. To increase the duration of this type of station, it can be encased in concrete, as shown in Figure 6.12(b).

Ground markers provide a more permanent method of station marking in soft ground. As shown in Figure 6.13, several different versions of these are available. They consist of a ribbed steel bar of length up to 1 m, which can be driven into the

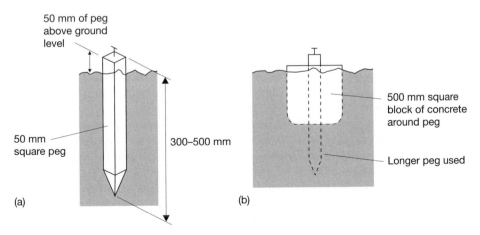

Figure 6.12 ● Wooden peg: (a) typical dimensions; (b) longer duration in concrete.

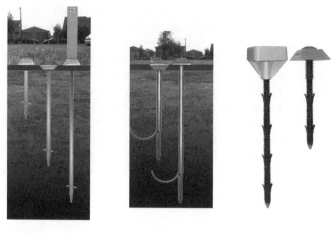

Ground marker designs

Installation

Washer and stud to mark station

Figure 6.13 ● Ground markers (courtesy York Survey Supply Centre).

ground with a lump hammer. The bar is fitted with retention lugs or an anchor to provide a stable survey station that is resistant to accidental displacement and vandalism. The bar is also attached to a head of some sort into which is located a coloured washer that is secured to the top of the marker with a domed stud. The stud acts as the plumbing point for centring instruments and as a high point for levelling purposes.

Stations in hard surfaces can be marked with a variety of *survey nails*, some of which are shown in Figure 6.14. These include the P-K masonry nail, which is 6 mm in diameter and is available in lengths up to 64 mm. To help identify a station, they can be used with a steel or plastic marking washer. Survey point and Hilti-type nails can also be used, and for placing points in very hard concrete the Pin-Mark nail is available. To use a Pin-Mark nail, it is first inserted into the base of its setting tool, the tool is positioned and the nail is driven into concrete by hitting the top of the setting tool with a lump hammer. When finished, the setting tool is removed from the Pin-

P–K masonry nail and installation

Pin–Mark nail and installation

Figure 6.14 ● Survey nails (courtesy York Survey Supply Centre).

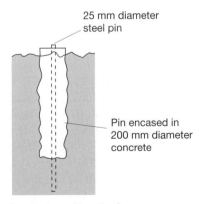

25 mm diameter steel pin

Pin encased in 200 mm diameter concrete

Depth of concrete and length of pin vary according to ground conditions

Figure 6.15 ● Steel pin in concrete.

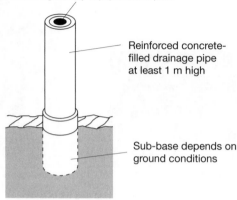

Plate for attaching survey equipment to pillar

Reinforced concrete-
filled drainage pipe
at least 1 m high

Sub-base depends on
ground conditions

Methods for sub-base
in rock – use reinforcing bars fixed to rock
in granular soil – use additional concrete-filled pipes
in clay – use concrete pile driven to refusal

Figure 6.16 ● Observation pillar.

Mark nail and it is left proud or the head is snapped off. All of these options for marking survey points in hard surfaces are permanent.

As the need for longer term stability or better precision arises for a control survey, the complexity of station marks increases. The *steel pin* set in concrete shown in Figure 6.15 and the *observation pillar* shown in Figure 6.16 are designs for long-term survey stations. The Ordnance Survey triangulation pillar shown in Figure 8.9 is another example. Observation pillars are the most expensive type of survey station and they are only used when high accuracy and a very long-term station are required.

All stations in a traverse should have a *station description* prepared that provides all the information shown in Figure 6.17. On this, measurements are taken from the station to nearby permanent features to enable it to be relocated.

Angle and distance measurements

After the traverse stations have been put in place, the next stage in the field procedure is to use a theodolite or total station to measure the included angles between the lines. This requires two operations: setting the instrument over each station and measuring an angle. In most cases it will be necessary to have targets in place, since the station marks may not be directly visible. Suitable types of target are described in Sections 3.5 and 5.2.

The *measurement of traverse angles* requires that the theodolite, total station and targets be located in succession at each station. If this operation is not carried out accurately centring errors are introduced, the effect of which depends on the length

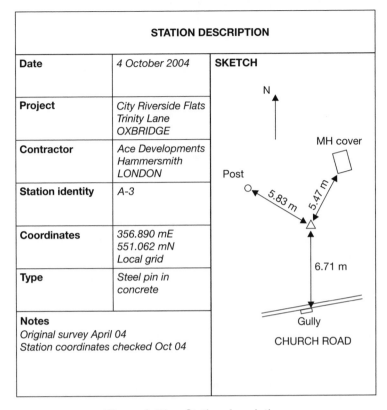

STATION DESCRIPTION		
Date	*4 October 2004*	**SKETCH**
Project	*City Riverside Flats* *Trinity Lane* *OXBRIDGE*	
Contractor	*Ace Developments* *Hammersmith* *LONDON*	
Station identity	*A-3*	
Coordinates	*356.890 mE* *551.062 mN* *Local grid*	
Type	*Steel pin in concrete*	
Notes *Original survey April 04* *Station coordinates checked Oct 04*		

Figure 6.17 ● Station description.

of an individual traverse line. If a target displacement of 5 mm occurs on a 100 m traverse line, the resulting angular error can be as large as 10". Although this should be avoided, the same displacement on a 25 m traverse line will produce an unacceptable angular error of 40". As centring errors are carried through a traverse, all bearings calculates after the error has occurred will be incorrect and if the theodolite or total station is also not centred properly, another source of error arises.

As can be seen, care with centring is vital, especially when traverse lines are short. In engineering surveying short lines are often unavoidable – for example, in surveys in mines and tunnels and on congested sites. One way of reducing the effects of centring errors in such cases is to use three or more tripods during a survey and to use theodolites and total stations that can be detached from their tribrachs and interchanged with a target or prism set (see Figure 3.4). This is known as *three-tripod traversing* and operates as follows, with reference to Figure 6.18.

When angle ABC is measured
● At A a tripod is set up and a tribrach attached to the tripod head. A target or prism set is placed into the tribrach and clamped in position. The target or tribrach will

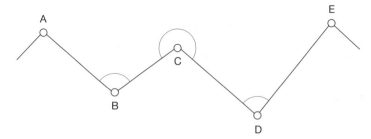

Figure 6.18 ● Three-tripod traversing.

have a tube or circular bubble attached so that the target can be set vertical by levelling using the tribrach footscrews. In order to be able to centre the target, the tribrach has a plummet.

● At B the total station (or theodolite) is set up in the normal manner.

● At C a tripod and target is set up as at A.

   This enables the horizontal angle at B to be observed and, if a total station is being used, enables distances BA and BC to be measured.

When angle BCD is measured

● At A the tripod and target are moved to D, where the target is centred and set vertical.

● At B the total station or theodolite is unclamped, removed from its tribrach and interchanged with the target at C. Hence, at B and C, the tripods and tribrachs remain undisturbed and there is no need for recentring.

With the equipment set in this position, the horizontal angle at C and distances CB and CD are measured. A check can be made on the horizontal distances BC and CB at this stage.

When angle CDE is measured

● At B the tripod and target are moved to E.

● The total station (theodolite) and target at C and D are interchanged, the tribrachs (and centring) remaining undisturbed.

   The process is repeated for the whole traverse. If four tripods (or more) are used this speeds up the fieldwork considerably, as tripods can be moved and positioned whilst angles and distances are being measured.

   When three-tripod traversing, care is still needed to avoid centring errors when setting up each tripod, but the difference between this and conventional traversing is that a centring error is confined to the station at which it occurs and errors do not accumulate through the traverse.

AT STATION C

Angles

POINT	FACE LEFT	FACE RIGHT	MEAN	ANGLE
B	00 07 22	180 07 18	00 07 20	
D	192 23 38	12 23 44	192 23 41	192 16 21
B	87 32 35	267 32 26	87 32 31	
D	279 49 10	99 49 04	279 49 07	192 16 36

FINAL ANGLE = 192 16 29

Distances

CB = 47.580 m   CD = 65.254 m

Diagram

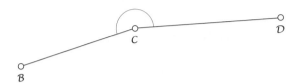

Figure 6.19 ● Recording traverse data.

The method given in Section 3.5 for measuring, booking and calculating angles should be used whenever possible. In the case where no standard booking forms are available, angles and distances can be entered in a field or note book as shown in Figure 6.19, in which two complete rounds of angles have been observed and the zero changed between rounds. When using an electronic theodolite or total station, all readings are stored in a data collector or internal memory, and these are either processed on site or transferred to a host computer – traversing with a total station is described in Section 6.5.

Errors that can occur in angular measurements are discussed in detail in Section 3.6, but the various sources of error that may occur when measuring traverse angles are summarised here. These errors are:

● Inaccurate centring of the theodolite, total station or target
● Non-verticality of targets
● Inaccurate bisection of targets
● Parallax not eliminated
● Lateral refraction, wind and atmospheric effects
● Theodolite or total station not level or not in adjustment
● Incorrect use of the theodolite or total station
● For manually recorded surveys, mistakes in reading and booking

Measurement of the lengths of traverse lines is done using steel tapes or a total station. If a total station is being used, this has the considerable advantage that both angular and distance measurements are combined at each traverse station. The various sources of error that may occur when using steel tapes and total stations are covered in Chapters 4 and 5, but particular attention should be paid to the following:

- Mistakes and systematic errors in taping
- Inaccurate centring of a total station or prism affecting distances in this case
- Total station not level or not in adjustment
- Incorrect use of the total station
- For manually recorded surveys, mistakes in reading and booking

When all the traverse fieldwork has been completed, a single sheet or file containing the mean angles observed and mean horizontal (corrected) distances measured should be prepared. It is preferable to show all the data on a sketch of the traverse, as this helps in the subsequent calculations and can minimise the chance of a mistake occurring. Such a sketch is known as a *traverse abstract*.

A typical abstract of field data is shown in Figure 6.20, the angles and distances being entered on to a traverse diagram – this will be referred to in the following section, which deals with traverse calculations.

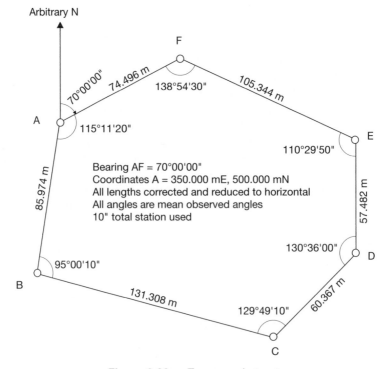

Figure 6.20 ● Traverse abstract.

Reflective summary

With reference to the planning and fieldwork required for traversing, remember:

— For small construction sites, horizontal survey control is usually provided by traversing.

— The first stage in the traverse is to carry out a reconnaissance which locates the positions of the stations – it is very important that this is done properly to ensure that adequate control is available on site and that it is in the right place. *A site visit is an essential part of a reconnaissance.*

— Other aspects of planning at this stage include making a decision as to the type of traverse to be used – polygon or link. For a polygon traverse, the origin for coordinates has to be chosen and an orientation defined. For a link traverse, existing control must be available.

— There are many different types of control point available for site work, ranging from a simple peg to an observation pillar – some care is needed when choosing these. Remember to take into account the length of the project, ground conditions, the accuracy required and whether vandalism is an issue.

— Traversing requires angles and distances to be measured – this can be done separately by theodolite and tape or simultaneously using a total station. With reference to the specification given for the accuracy of the traverse, suitable equipment has to be selected and correct field procedures adopted.

— When the fieldwork is completed, it is good practice to prepare an abstract of the traverse to show all of the information required for calculating the coordinates of the traverse stations from the measured angles and distances.

6.5 Traverse calculations

After studying this section you should have a full understanding of how to calculate the coordinates of traverse stations from measured angles and distances.

This section includes the following topics:

- Calculation of whole circle bearings
- Calculation of coordinate differences
- Calculation of coordinates

Calculation of whole-circle bearings

As shown in Section 6.1, to calculate the coordinates of a control point, distances and bearings must be known. For a traverse, this procedure starts with the calculation of all the bearings for the traverse; then coordinate differences are determined using these with the measured distances, and finally the station coordinates are calculated.

The first part of a traverse calculation is to check that the observed angles sum to their required value.

The observed angles of a *polygon traverse* can be either the *internal* or *external* angles, and angular misclosures are found by comparing the sum of the observed angles with one of the following theoretical values

$$\text{sum of } internal \text{ angles} = (2n - 4) \times 90°$$
$$\text{or sum of } external \text{ angles} = (2n + 4) \times 90°$$

where n is the number of angles measured.

When the bearings in a *link traverse* are calculated, an *initial back bearing* can usually be determined from known points at the start of the traverse and, to check the observed angles, a *final forward bearing* is computed from known points at the end of the traverse. The angular misclosure in a link traverse is found using

$$\text{sum of angles} = (\text{final forward bearing} - \text{initial back bearing}) + (m \times 180°)$$

In this equation, m is an integer, the value of which depends on the shape of the traverse. In most cases, m will be $(n-1)$, n or $(n+1)$, where n is the number of angles measured between the initial back bearing and final forward bearing. Worked example 6.5 for a link traverse given later in this section shows how m is determined.

For both types of traverse, care must be taken to ensure that the correct angles have been abstracted and summed. These are the internal *or* external angles in a polygon traverse and the angles on the same side of a link traverse.

Owing to the effects of occasional miscentring, slight misreading and small bisection errors, a small misclosure will usually result when the summation check is made. For most site traverses, the allowable misclosure E in the measured angles is given by (see Section 9.3: *Propagation of angular errors in traversing*; p. 371)

$$E'' = \pm KS\sqrt{n}$$

where

K is a multiplication factor of 1 to 3 depending on weather conditions and the number of rounds taken

S is the smallest reading interval on the theodolite or total station in seconds, for example 20", 5", 1"

n is the number of angles measured.

Because a 10" total station was used, the allowable misclosure for the traverse shown in the abstract of Figure 6.20 varies from (with $K = 1$ and $n = 6$) $\pm1\times10 \times \sqrt{6} = 24$" to (with $K = 3$) $\pm3\times10 \times \sqrt{6} = 73$".

When the actual misclosure is known and this is compared to the allowable value, two cases will arise.

- *If the misclosure is acceptable* (less than the allowable) it is divided equally between the observed angles. An equal distribution is the only acceptable method, since each angle is measured in the same way and there is an equal chance of the misclosure having occurred in any of the angles. No attempt should be made to distribute the misclosure in proportion to the size of an angle.

- *If the misclosure is not acceptable* (greater than the allowable) the angles must be re-measured if no gross error can be located in the angle bookings or summation. It may be possible to isolate a gross error in a small section of the traverse if check lines have been observed across it.

The determination of the misclosure and adjustment of the angles of the polygon traverse ABCDEFA given in the abstract of Figure 6.20 are shown in Table 6.1.

Worked example 6.5 at the end of this section shows how the angles in a link traverse are adjusted.

The next stage in the traverse calculation is to determine the forward bearings of all the traverse lines.

Table 6.1 Adjustment of angles for polygon traverse.

Station	Observed angle	Adjustment	Adjusted angle
A	115°11'20"	−10"	115°11'10"
B	95 00 10	−10	95 00 00
C	129 49 10	−10	129 49 00
D	130 36 00	−10	130 35 50
E	110 29 50	−10	110 29 40
F	138 54 30	−10	138 54 20
Sums	720 01 00	−01'00"	720 00 00

Required sum = [(2 × 6) − 4] × 90° = 720°
Misclosure = +01'00" (acceptable for 10" total station with $K= 3$)
Adjustment per angle = −01'00"/6 = −10"

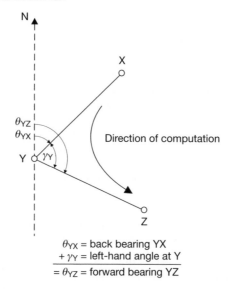

θ_{YX} = back bearing YX
+ γ_Y = left-hand angle at Y
= θ_{YZ} = forward bearing YZ

Figure 6.21 ● Calculation of whole-circle bearings.

Consider Figure 6.21, which shows two lines of a traverse on each side of a traverse station Y. The decision has been made to calculate the traverse in the direction ... X → Y → Z ... which defines the bearings between the stations as follows.

Bearings XY and YZ are *forward bearings* since they are in the same direction in which the traverse is to be calculated.

Bearings YX and ZY are *back bearings* since they are opposite to the direction in which the traverse is to be calculated.

Bearings XY and YX differ by ±180°, as do those of YZ and ZY. Therefore the forward bearing of a line differs from the back bearing by ±180°.

For the direction of computation shown in Figure 6.21, γ_Y is known as the *left-hand angle* at Y since it lies to the left at station Y relative to the direction X → Y → Z.

If γ_Y is added to the back bearing YX it can be seen from Figure 6.21 that the resulting angle will be the forward bearing YZ. Therefore

forward bearing YZ = back bearing YX + γ_Y

and in general for any traverse station

forward bearing = back bearing + left-hand angle

For polygon traverses when working in an *anticlockwise* direction around the traverse, the left-hand angles will be the *internal* angles of the traverse and when working in a *clockwise* direction, the left-hand angles will be the *external* angles.

Some of the forward whole-circle bearings of the lines of the traverse ABCDEFA shown in Figure 6.20 will now be computed using the adjusted left-hand angles of Table 6.1. Figures 6.22 and 6.23 show the relevant sections of this traverse for these calculations.

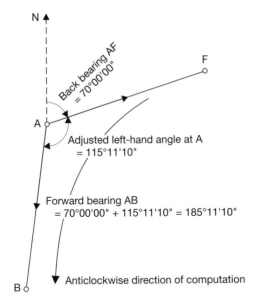

Figure 6.22 ● Calculation of forward bearing AB.

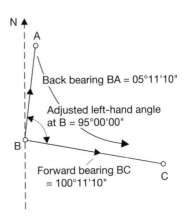

Figure 6.23 ● Calculation of forward bearing BC

The calculation of whole-circle bearings *must* start with a known bearing or an assumed arbitrary bearing. For the traverse here, the bearing AF is given as 70°00'00". In this case, because the angles have been measured and recorded as internal angles, the traverse is calculated in an anticlockwise direction and bearing AF is a back bearing. As a result, at *station A* in Figure 6.22:

forward bearing AB = back bearing AF + adjusted left-hand angle at A

Note the letter A here in forward bearing <u>A</u>B, back bearing <u>A</u>F and station <u>A</u> – if the station letter or reference is not the first one in each case, a wrong bearing is being used. To continue with the calculation of bearing AB:

forward bearing AB = 70°00'00" + 115°11'10" = 185°11'10"

At *station B* in Figure 6.23:

forward bearing BC = back bearing BA + adjusted left-hand angle at B

To convert the forward bearing AB into the back bearing BA use

back bearing BA = forward bearing AB ± 180°
= 185°11'10" ± 180°
= 365°11'10" or 05°11'10"
= 05°11'10" (to keep the bearing in the range 0 to 360°)

This gives

forward bearing BC = 05°11'10" + 95°00'00" = 100°11'10"

The bearings of all the other lines can be computed in a similar manner.

If at some stage the result for a forward bearing is greater than 360°, then 360° must be subtracted from it to give a bearing in the range 0 to 360°. For example, the forward bearing CD is given by

$$\text{forward bearing CD} = \text{back bearing CB} + \text{adjusted left-hand angle at C}$$
$$= 280°11'10" + 129°49'00"$$
$$= 410°00'10"$$

Since this is greater than 360°:

$$\text{forward bearing CD} = 410°00'10" - 360° = 50°00'10"$$

Assuming the observed angles have been adjusted correctly, the bearing calculation *must* finish by recalculating the start bearing, which in this case is the back bearing AF. The computed version of this bearing *must* be in agreement with the start value – if any difference occurs, an *arithmetic mistake* has been made and the calculation of all of the whole-circle bearings must be checked before proceeding to the next stage in the traverse calculation.

When calculating a traverse, a systematic approach is required to minimise the chance of any mistakes occurring. To achieve this, each stage of the calculation should be tabulated and it is essential to show that a series of checks has been applied successfully. There are many formats that can be adopted for this, but Table 6.2 is recommended as a traverse form, although any design will do provided it meets the criteria of a proper tabulation of each stage in the calculation and a clear display of all the various checks.

Table 6.3 shows the complete calculation on the traverse form for all the whole-circle bearings for traverse ABCDEFA and gives details of the angular misclosure for the traverse. Note that the check has been applied and is correct as forward bearing FA = 250°00'00" which is equivalent to back bearing AF = 70°00'00".

Calculation of coordinate differences

The next stage in the traverse computation is the determination of the coordinate differences of the traverse lines.

The information available at this point will be the whole-circle bearings and horizontal lengths of all the lines.

Traverse ABCDEFA is again used in the following examples together with the whole-circle bearings given in Table 6.3. With reference to Section 6.1, consider line AB in Figure 6.24 in which the eastings difference is given by

$$\Delta E_{AB} = D_{AB} \sin \theta_{AB}$$
$$= 85.874 \sin 185°11'10" = 85.874(-0.0903491)$$
$$= -7.762 \text{ m}$$

Similarly, the northings difference is

Table 6.2 Traverse form.

Line	Back bearing	Whole-circle bearing θ	Horizontal distance D	Coordinate differences								Coordinates		Station
Station	Adjusted LHA			Calculated		Adjustments		Adjusted				E	N	
Line	Forward bearing			ΔE	ΔN	δE	δN	ΔE	ΔN					

Table 6.3 ● Bearings entered on traverse form.

Line / Station / Line	Back bearing — Adjusted LHA / Forward bearing (° ′ ″)			Whole-circle bearing θ (° ′ ″)			Horizontal distance D	Calculated ΔE	Calculated ΔN	Adjustments δE	Adjustments δN	Adjusted ΔE	Adjusted ΔN	Coordinates E	Coordinates N	Station
AF	70	00	00													
A	115	11	10													
AB	185	11	10	185	11	10										
BA	05	11	10													
B	95	00	00													
BC	100	11	10	100	11	10										
CB	280	11	10													
C	129	49	00													
CD	410	00	10	50	00	10										
DC	230	00	10													
D	130	35	50													
DE	360	36	00	00	36	00										
ED	180	36	00													
E	110	29	40													
EF	291	05	40	291	05	40										
FE	111	05	40													
F	138	54	20													
FA	250	00	00	250	00	00										

Sum of left-hand angles = 720°01′00″
Required sum = $(2 \times 6 - 4) \times 90° = 720°$
Misclosure = +60″
Adjustments are –10″ to each angle

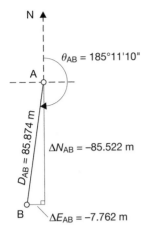

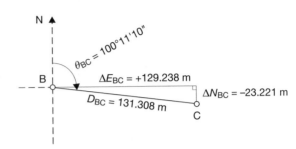

Figure 6.24 ● Calculation of coordinate differences along AB.

Figure 6.25 ● Calculation of coordinate differences along BC.

$$\Delta N_{AB} = D_{AB} \cos \theta_{AB}$$
$$= 85.874 \cos 185°11'10" = 85.874(-0.995906)$$
$$= -85.522 \text{ m}$$

For line BC shown in Figure 6.25, the coordinate differences are given by

$$\Delta E_{BC} = D_{BC} \sin \theta_{BC}$$
$$= 131.308 \sin 100°11'10" = 131.308(+0.984239) = +129.238 \text{ m}$$

$$\Delta N_{BC} = D_{BC} \cos \theta_{BC}$$
$$= 131.308 \cos 100°11'10" = 131.308(-0.176846) = -23.221 \text{ m}$$

Table 6.4 shows all the calculations for the coordinate differences added to the traverse form.

When all of the coordinate differences ΔE and ΔN have been computed, further checks can be applied to the traverse.

For polygon traverses these are

$$\Sigma \Delta E \ should = 0 \text{ and } \Sigma \Delta N \ should = 0$$

since the traverse starts and finishes at the same point.

For link traverses, the checks are

$$\Sigma \Delta E \ should = E_X - E_A \text{ and } \Sigma \Delta N \ should = N_X - N_A$$

where station A is the starting point and station X the final point of the traverse. Since stations A and X are of known position, the values of $E_X - E_A$ and $N_X - N_A$ can be calculated.

In both cases, owing to on site errors in measuring the distances, there will normally be a misclosure on returning to the starting point in a polygon traverse or on arrival at the final known station in a link traverse.

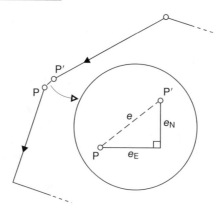

Figure 6.26 ● Traverse misclosure.

In order to assess the accuracy of the traverse, the ΔE and ΔN values are summed and the misclosures in easting e_E and in northing e_N are computed by comparing the summations with those expected. These misclosures form a measure of the linear misclosure of the traverse and can be used to determine its accuracy. Consider Figure 6.26, which shows the starting point P of a polygon traverse. Because of measurement errors, the traverse ends at P′ instead of P. The *linear misclosure e* is given by

$$e = \sqrt{e_E^2 + e_N^2}$$

To obtain a measure of the accuracy of the traverse, the value of e is compared with the total length of the traverse lines ΣD to give the *fractional linear misclosure*, where

$$fractional\ linear\ misclosure = 1\ \text{in}\ \frac{\Sigma D}{e}$$

The fractional linear misclosure for traverse ADCDEFA can be obtained from Table 6.4, remembering that the traverse is a polygon. From Table 6.4

$$\Sigma\Delta E = -7.762 + 129.238 + 46.246 + 0.602 - 98.285 - 70.003 = +0.036\ \text{m}$$

or

$$e_E = +\ 0.036\ \text{m since } \Sigma\Delta E \text{ should be zero}$$

$$\Sigma\Delta N = -85.522 - 23.221 + 38.801 + 57.479 + 37.914 - 25.479 = -0.028\ \text{m}$$

or

$$e_N = -\ 0.028\ \text{m since } \Sigma\Delta N \text{ should be zero}$$

Therefore

$$Linear\ misclosure = e = \sqrt{(0.036)^2 + (-0.028)^2} = 0.046\ \text{m}$$

From Table 6.4

$$\Sigma D = 514.871\ \text{m}$$

Table 6.4 ● Coordinate differences entered on traverse form.

Line / Station / Line	Back bearing: Adjusted LHA / Forward bearing			Whole-circle bearing θ			Horizontal distance D	Coordinate differences – Calculated ΔE	Calculated ΔN	Adjustments δE	Adjustments δN	Adjusted ΔE	Adjusted ΔN	Coordinates E	Coordinates N	Station
AF	70	00	00													
A	115	11	10													
AB	185	11	10	185	11	10	85.874	−7.762	−85.522							
BA	05	11	10													
B	95	00	00													
BC	100	11	10	100	11	10	131.308	+129.238	−23.221							
CB	280	11	10													
C	129	49	00													
CD	410	00	10	50	00	10	60.367	+46.246	+38.801							
DC	230	00	10													
D	130	35	50													
DE	360	36	00	00	36	00	57.482	+0.602	+57.479							
ED	180	36	00													
E	110	29	40													
EF	291	05	40	291	05	40	105.344	−98.285	+37.914							
FE	111	05	40													
F	138	54	20													
FA	250	00	00	250	00	00	74.496	−70.003	−25.479							

$$\Sigma = 514.871$$

$$e_E = +0.036 \qquad e_N = -0.028$$

Sum of left-hand angles = 720°01'00"

Required sum = (2 × 6 − 4) × 90° = 720°

Misclosure = +60"

Adjustments are −10" to each angle

Linear misclosure = $[(+0.036)^2 + (-0.028)^2]^{1/2} = 0.046$

Fractional linear misclosure = 1 in 514.871 ÷ 0.046 = 1 in 11,200

and

$$\text{Fractional linear misclosure} = 1 \text{ in } \frac{514.871}{0.046} = 1 \text{ in } 11{,}200$$

This calculation is also shown at the bottom of Table 6.4.

The fractional linear misclosure is *always* computed for a traverse *before* adjustments to the coordinate differences and the station coordinates are calculated – this is compared to its specification and a decision made to accept or reject the traverse.

If, on comparison, the fractional linear misclosure is better than the specified value, the traverse fieldwork is satisfactory and the misclosures, e_E and e_N, are distributed throughout the traverse.

If, on comparison, the fractional linear misclosure is worse than that specified, there may be an error in one or more of the measured traverse lines or compensating errors in two or more of the observed traverse angles. The calculation should, however, be thoroughly checked before re-measuring any distances or angles.

The accuracy required for a traverse depends on the survey work that it is to be subsequently used for on site and three types of traverse can be classified, as shown in Table 6.5. The most common type of traverse for general engineering work and

Table 6.5 Traverse specifications.

Type	Typical accuracy	Purpose	Angular measurement	Distance measurement
Geodetic or precise	1 in 50,000 or better	• Provision of very accurate reference points for engineering surveys • Primary control surveys	Use 0.1" industrial theodolite	Accuracy better than 1 mm required – use specialised electronic distance measuring equipment or industrial total station
General	1 in 5000 to 1 in 50,000	• General engineering work including site surveys and setting out of structures and roadworks • Adding secondary control to primary control networks	Electronic theodolite or total station with accuracy between 1" and 10" required	Use total station or steel tapes to match accuracy requirement
Low accuracy	1 in 500 to 1 in 5000	• Control for drainage schemes and earthworks • Small-scale mapping	Electronic or optical theodolite of 20" or lower accuracy required	Total station, steel or synthetic tapes can be used

Note: See also BS5964 (ISO4463): 1996 *Building setting out and measurement* and the ICE Design and Practice Guide *The management of setting out in construction* for further information on traverse specifications.

site surveys would be of typical accuracy 1 in 10,000 – this chapter is concerned mainly with an expected accuracy range of about 1 in 5000 to 1 in 20,000.

In order to achieve a stated accuracy or specification, the correct equipment must be used and an important factor when selecting traversing equipment is that the various instruments should produce roughly the same order of precision. For example, it is pointless using a 1" theodolite to measure traverse angles if the distances are being measured with a synthetic tape. Conversely, using a total station with a high-order specification of 1" or 2" for angle measurement and ±(2 mm + 2 ppm) for distances is inappropriate for low accuracy traversing. Table 6.5 gives a general indication of the grouping of suitable equipment.

Returning to the example being developed here, a fractional linear misclosure of 1 in 11,200 has been obtained and the traverse is accepted.

Further methods for assessing the accuracy of traverses are given in Section 9.4.

The next stage in the traverse calculation is to adjust the coordinate differences such that they sum to zero. Many methods of adjusting the linear misclosure of a traverse are possible, but for everyday engineering traverses of accuracy up to 1 to 20,000 one of two methods is normally used.

Bowditch method

The values of the adjustment found by this method are directly proportional to the length of the individual traverse lines. The method is best suited to taped traverse lines.

Adjustment to ΔE (or ΔN) for a traverse line

$$= \delta E \text{ (or } \delta N) = -e_{\mathrm{E}} \text{ (or } -e_{\mathrm{N}}) \times \frac{\text{length of traverse line}}{\text{total length of the traverse}}$$

For this method, the adjustment of the first two sets of ΔE and ΔN values in Table 6.4 is as follows (the misclosures have already been determined as $e_{\mathrm{E}} = 0.036$ m, $e_{\mathrm{N}} = -0.028$ m and the total length of the traverse is 514.871 m)

For line AB

$$\delta E_{\mathrm{AB}} = -0.036 \times \frac{85.874}{514.871} = -0.006 \text{ m} \qquad \delta N_{\mathrm{AB}} = +0.028 \times \frac{85.874}{514.871} = + 0.005 \text{ m}$$

For line BC

$$\delta E_{\mathrm{BC}} = -0.036 \times \frac{131.308}{514.871} = -0.009 \text{ m} \qquad \delta N_{\mathrm{BC}} = +0.028 \times \frac{131.308}{514.871} = +0.007 \text{ m}$$

Equal adjustment

For traverses measured by total station, the likely error in each distance will be independent of the distance measured for normal work where traverse lines rarely exceed 100–200 m. For these traverses, the error in each measured distance will be of the same order of magnitude and an equal distribution of the misclosure is acceptable. In such cases

$$\delta E \text{ (or } \delta N) \text{ for each line} = \frac{-e_E (\text{or } -e_N)}{n}$$

where n is the number of traverse lines.

An equal adjustment to the same example gives the following for all the traverse lines:

$$\delta E = \frac{-0.036}{6} = -0.006 \text{ m} \qquad \delta N = \frac{+0.028}{6} = +0.005 \text{ m}$$

When all the adjustments have been calculated, they can be checked using

$$\Sigma \delta E \text{ should} = -e_E \text{ and } \Sigma \delta N \text{ should} = -e_N$$

If these checks are successful, adjusted values of ΔE and ΔN are obtained by applying the corrections to the calculated values. Some care must be taken when doing this somewhat simple arithmetic as this stage of the calculation is where, from experience, mistakes are often made, especially when the coordinate differences are in metres and the corrections are recorded in mm. Table 6.6 shows the Bowditch adjustments and the adjusted coordinate differences for traverse ABCDEFA. Note that the adjustments and adjusted coordinate differences are also shown to be checked (the adjusted coordinate differences should summate to zero for a polygon traverse).

Note: All methods of adjustment to a traverse will alter the original bearings and distances entered on the traverse form by small amounts. If the bearings and distances between the traverse stations are required for any setting out or other work, they must be recalculated either from the adjusted coordinate differences or the adjusted coordinates using polar to rectangular conversions. *Do not use* the original bearings and distances entered on the traverse from.

Calculation of coordinates

To calculate the coordinates of the stations for *polygon* traverses, the coordinates of the starting point can either be allocated to give positive coordinates for a site or to align it with a structure or they may already be known if an existing station is used to start the traverse.

To calculate the coordinates of the stations for *link* traverses, the coordinates of the start and end points will be known and the coordinates of the new stations will be fixed relative to these.

For both types of traverse, the coordinates of each station are obtained by adding or subtracting the adjusted coordinate differences as necessary, working around the traverse. This continues until the coordinates of the start point for a polygon are recalculated or the coordinates of the end point in a link traverse are calculated – at this point, the final check is made on the traverse calculation.

For a *polygon* traverse, the final and start coordinates should be the same.

Table 6.6 Bowditch adjustments and adjusted coordinate differences entered on traverse form.

Line / Station / Line	Back bearing / Adjusted LHA / Forward bearing			Whole-circle bearing θ			Horizontal distance D	Calculated ΔE	Calculated ΔN	Adjustment δE	Adjustment δN	Adjusted ΔE	Adjusted ΔN	Coord E	Coord N	Station
AF	70	00	00													
A	115	11	10													
AB	185	11	10	185	11	10	85.874	-7.762	-85.522	-0.006	+0.005	-7.768	-85.517			
BA	05	11	10													
B	95	00	00													
BC	100	11	10	100	11	10	131.308	+129.238	-23.221	-0.009	+0.007	+129.229	-23.214			
CB	280	11	10													
C	129	49	00													
CD	410	00	10	50	00	10	60.367	+46.246	+38.801	-0.004	+0.003	+46.242	+38.804			
DC	230	00	10													
D	130	35	50													
DE	360	36	00	00	36	00	57.482	+0.602	+57.479	-0.004	+0.003	+0.598	+57.482			
ED	180	36	00													
E	110	29	40													
EF	291	05	40	291	05	40	105.344	-98.285	+37.914	-0.008	+0.006	-98.293	+37.920			
FE	111	05	40													
F	138	54	20													
FA	250	00	00	250	00	00	74.496	-70.003	-25.479	-0.005	+0.004	-70.008	-25.475			

Σ = 514.871

eE = +0.036 eN = -0.028

Σ = -0.036 Σ = +0.028

Σ = 0.000 Σ = 0.000

Sum of left-hand angles = 720°01′00″
Required sum = (2 × 6 − 4) × 90° = 720°
Misclosure = +60″
Adjustments are −10″ to each angle

Adjustments by Bowditch method
Linear misclosure = [(+0.036)² + (−0.028)²]^{1/2} = 0.046
Fractional linear misclosure = 1 in 514.871 ÷ 0.046 = 1 in 11,200

For a *link* traverse, the final coordinates should be the same as those of the end point.

If this check shows a difference in these and assuming the coordinate differences have been adjusted correctly, an arithmetic mistake has been made in the calculation of the final coordinates – the best procedure in this situation is to recalculate the coordinates.

Returning to traverse ABCDEFA, the adjusted coordinate differences have been determined so far. Using the start coordinates given for A of 350.000 mE, 500.000 mN and Bowditch adjustments, the coordinates of station B are given by

$$E_B = E_A \pm \Delta E_{AB} = 350.000 - 7.768 = 342.232 \text{ m}$$

$$N_B = N_A \pm \Delta N_{AB} = 500.000 - 85.517 = 414.483 \text{ m}$$

The coordinates of station C are calculated from station B as

$$E_C = E_B \pm \Delta E_{BC} = 342.232 + 129.229 = 471.461 \text{ m}$$

$$N_C = N_B \pm \Delta N_{BC} = 414.483 - 23.214 = 391.269 \text{ m}$$

This calculation is repeated until station A is re-coordinated as a check – see Table 6.7 for the full calculation on the traverse form.

Worked example 6.4: Polygon traverse

Question

The traverse diagram of Figure 6.27 is an abstract for a polygon traverse A1234A which starts at existing control point A (642.515 mE, 483.980 mN) and is orientated to existing control point B (548.005 mE, 594.279 mN).

Calculate the adjusted coordinates of stations 1–4 and the fractional linear misclosure for the traverse.

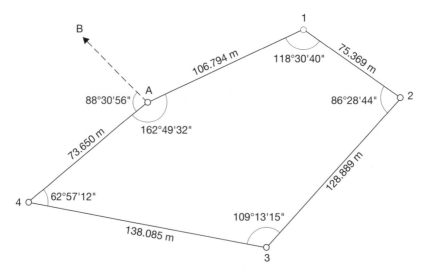

Figure 6.27 ● Diagram for Worked example 6.4: Polygon traverse.

Table 6.7 ● Coordinates of stations entered on traverse form.

Line / Station	Back bearing / Adjusted LHA / Forward bearing (° ' ")	Whole-circle bearing θ (° ' ")	Horizontal distance D	Calculated ΔE	Calculated ΔN	Adjustments δE	Adjustments δN	Adjusted ΔE	Adjusted ΔN	E	N	Station
										350.000	500.000	A
AF	70 00 00											
A	115 11 10											
AB	185 11 10	185 11 10	85.874	−7.762	−85.522	−0.006	+0.005	−7.768	−85.517	342.232	414.483	B
BA	05 11 10											
B	95 00 00											
BC	100 11 10	100 11 10	131.308	+129.238	−23.221	−0.009	+0.007	+129.229	−23.214	471.461	391.269	C
CB	280 11 10											
C	129 49 00											
CD	410 00 10	50 00 10	60.367	+46.246	+38.801	−0.004	+0.003	+46.242	+38.804	517.703	430.073	D
DC	230 00 10											
D	130 35 50											
DE	360 36 00	00 36 00	57.482	+0.602	+57.479	−0.004	+0.003	+0.598	+57.482	518.301	487.555	E
ED	180 36 00											
E	110 29 40											
EF	291 05 40	291 05 40	105.344	−98.285	+37.914	−0.008	+0.006	−98.293	+37.920	420.008	525.475	F
FE	111 05 40											
F	138 54 20											
FA	250 00 00	250 00 00	74.496	−70.003	−25.479	−0.005	+0.004	−70.008	−25.475	350.000	500.000	A
A	250 00 00											
			Σ = 514.871	Σ = +0.036	Σ = −0.028	Σ = −0.036	Σ = +0.028	Σ = 0.000	Σ = 0.000			

$e_E = +0.036$ $e_N = -0.028$

Sum of left-hand angles = 720°01'00"
Required sum = (2 × 6 − 4) × 90° = 720°
Misclosure = +60"
Adjustments are −10" to each angle

Adjustments by Bowditch method
Linear misclosure = $[(+0.036)^2 + (-0.028)^2]^{1/2} = 0.046$
Fractional linear misclosure = 1 in 514.871 + 0.046 = 1 in 11,200

Solution

The complete solution is given on the traverse form shown in Table 6.8. On the form the following should be noted:

- All the various checks have been applied to the traverse calculation as follows
  — whole-circle bearings: *bearing A–4 is recomputed at the bottom of the form*
  — the δ adjustments: $\Sigma\delta E = -e_E$ $\Sigma\delta N = -e_N$
  — the adjusted coordinate differences: $\Sigma\delta E = 0$ $\Sigma\delta N = 0$
  — station coordinates: *coordinate calculation closes back to A*
- The sum of the observed angles is 539°59'23" giving an angular misclosure of –37", since the angles should sum to 540°. The adjustments are applied as equally as possible to the nearest second and because 37 ÷ 5 is not an integer, small variations are possible in the adjustment. These are not important and in this case, the misclosure is allocated to the observed angles as A = +7", 4 = +7", 3 = +7", 2 = +8" and 1 = +8".

- Since *internal* angles are given, the traverse must be computed *anticlockwise*. So, even though it is numbered 1–2–3–4, it *must* be computed in the direction 4–3–2–1 starting at existing control point A.

- To obtain an opening whole-circle bearing on the traverse, an additional calculation is required. To determine this, the whole-circle bearing from A to B is first computed from the given coordinates following the method given in Section 6.1 for a rectangular → polar conversion. Since AB is in quadrant IV:

$$\theta_{AB} = \tan^{-1}\left[\frac{\Delta E_{AB}}{\Delta N_{AB}}\right] + 360° = \tan^{-1}\left[\frac{548.005 - 642.515}{594.279 - 483.980}\right] + 360°$$

$$= \tan^{-1}\left[\frac{-94.510}{110.299}\right] + 360° = \tan^{-1}[-0.856853] + 360°$$

$$= -40°35'30" + 360° = 319°24'30"$$

This is connected to the traverse through the observed angle $B\hat{A}4 = 88°30'56"$, and by inspection of Figure 6.27 the bearing from A to 4 is given by

$$\theta_{A4} = \theta_{AB} - 88°30'56" = 230°53'34"$$

This is the first bearing entered at the top of the traverse form and is a forward bearing.

- The fractional linear misclosure of 1 in 16,300 for the traverse would be acceptable for most engineering work.

- An *equal adjustment* has been used to correct the coordinate differences. Equal adjustments to the coordinate differences can vary by small amounts when the misclosure ÷ the number of stations is not an integer. As these only vary by a millimetre in magnitude, their differences are not important.

Table 6.8 ● Traverse form for Worked example 6.4: Polygon traverse.

Line (Station / Line)	Back bearing — Adjusted LHA / Forward bearing	Whole-circle bearing θ	Horizontal distance D	Calculated ΔE	Calculated ΔN	Adjustments δE	Adjustments δN	Adjusted ΔE	Adjusted ΔN	Coordinates E	Coordinates N	Station
										642.515	483.980	A
A–4	230 53 34	230 53 34	73.650	−57.150	−46.456	+0.002	+0.006	−57.148	−46.450			
4–A	50 53 34									585.367	437.530	4
4	62 57 19											
4–3	113 50 53	113 50 53	138.085	+126.295	−55.829	+0.002	+0.006	+126.297	−55.823			
3–4	293 50 53									711.664	381.707	3
3	109 13 22											
3–2	403 04 15	43 04 15	128.889	+88.019	+94.155	+0.002	+0.006	+88.021	+94.161			
2–3	223 04 15									799.685	475.868	2
2	86 28 52											
2–1	309 33 07	309 33 07	75.369	−58.113	+47.993	+0.003	+0.006	−58.110	+47.999			
1–2	129 33 07									741.575	523.867	1
1	118 30 48											
1–A	248 03 55	248 03 55	106.794	−99.063	−39.893	+0.003	+0.006	−99.060	−39.887			
A–1	68 03 55									642.515	483.980	A
A	162 49 39											
A–4	230 53 34		Σ = 522.787	eE = −0.012	eN = −0.030	Σ = +0.012	Σ = +0.030	Σ = 0.000	Σ = 0.000			

Sum of left-hand angles = 539°59′23″
Required sum = (2 × 5 − 4) × 90° = 540°
Misclosure = −37″
Adjustments are 3 × 7″ and 2 × 8″

Adjustments by equal method
Linear misclosure = [(−0.012)² + (−0.030)²]^{1/2} = 0.032
Fractional linear misclosure = 1 in 522.787 + 0.032 = 1 in 16,300

Worked example 6.5: Link traverse

Question

A link traverse was run between stations A and X as shown in the traverse diagram of Figure 6.28. The coordinates of the existing control stations at the ends of the traverse are

	mE	mN
A	375.369	543.008
B	264.507	604.938
X	601.624	404.041
Y	698.076	384.945

Calculate the adjusted coordinates of stations 1–4 and the fractional linear misclosure for the traverse.

Solution

The complete solution is given on the traverse form shown in Table 6.9. On the form the following should be noted:

- All the usual checks have been applied to the calculation.
- The solution follows the direction A to X as this will give the observed angles as left-hand angles.
- When link traversing, the *initial back bearing* at the start of the traverse and the *final forward bearing* at the end of the traverse may either be given directly or implied by the coordinates of existing control stations. In this case, coordinates are given and it is necessary to compute these bearings using rectangular to polar conversions.
  - *Initial back bearing AB* is in quadrant IV and is given by

$$\theta_{AB} = \tan^{-1}\left[\frac{\Delta E_{AB}}{\Delta N_{AB}}\right] + 360° = 299°11'19"$$

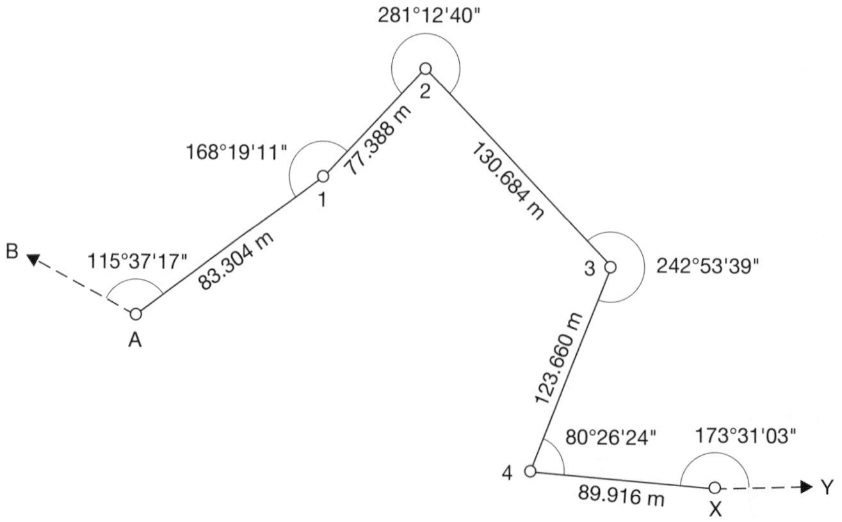

Figure 6.28 ● Traverse diagram for Worked example 6.5: Link traverse.

Table 6.9 — Traverse form for Worked example 6.5: Link traverse.

Line / Station / Line	Back bearing / Adjusted LHA / Forward bearing (° ′ ″)	Whole-circle bearing θ (° ′ ″)	Horizontal distance D	Calculated ΔE	Calculated ΔN	δE	δN	Adjusted ΔE	Adjusted ΔN	E	N	Station
A–B	299 11 19											
A	115 37 21									375.369	543.008	A
A–1	414 48 40	54 48 40	83.304	+68.081	+48.006	+0.003	−0.002	+68.084	+48.004			
1–A	234 48 40											
1	168 19 15									443.453	591.012	1
1–2	403 07 55	43 07 55	77.388	+52.909	+56.476	+0.003	−0.002	+52.912	+56.474			
2–1	223 07 55											
2	281 12 44									496.365	647.486	2
2–3	504 20 39	144 20 39	130.684	+76.178	−106.185	+0.004	−0.002	+76.182	−106.187			
3–2	324 20 39											
3	242 53 43									572.547	541.299	3
3–4	567 14 22	207 14 22	123.660	−56.600	−109.946	+0.004	−0.002	−56.596	−109.948			
4–3	27 14 22											
4	80 26 28									515.951	431.351	4
4–X	107 40 50	107 40 50	89.916	+85.669	−27.308	+0.004	−0.002	+85.673	−27.310			
X–4	287 40 50											
X	173 31 06									601.624	404.041	Y
X–Y	461 11 56		Σ = 504.952	ΣΔE = 226.237	ΣΔN = −138.957	Σ = +0.018	Σ = −0.010	Σ = 226.255	Σ = −138.967			

Sum of left-hand angles = 1062°00'14"

Required sum = (101°11'56" − 299°11'19") + (7 × 180°)

= −197°59'23" + 1260° = 1062°00'37"

Misclosure = −23"   Adjustments are 5 × 4" and 1 × 3"

$e_E = \Sigma\Delta E - (E_X - E_A) = 226.237 - (601.624 - 375.369) = 226.237 - 226.255 = -0.018$

$e_N = \Sigma\Delta N - (N_X - N_A) = -138.957 - (404.041 - 543.008) = -138.957 - (-138.967) = +0.010$

Linear misclosure = $[(-0.018)^2 + (+0.010)^2]^{1/2} = 0.021$

Fractional linear misclosure = 1 in 504.952 ÷ 0.021 = 1 in 24,000

— *Final forward bearing* XY is in quadrant II and is given by

$$\theta_{XY} = \tan^{-1}\left[\frac{\Delta E_{XY}}{\Delta N_{XY}}\right] + 180° = 101°11'56"$$

● The sum of the observed angles is 1062°00'14". The required sum is given by

(final forward bearing – initial back bearing) + m × 180° = (101°11'56" – 299°11'19") + m × 180° = –197°59'23" + m × 180°

— The value of m will be $(n - 1)$, n or $(n + 1)$, where n is the number of left-hand angles (n = 6 for this traverse) and it is obtained by assuming that no gross error was made when measuring the left-hand angles and that their actual sum is approximately correct. In this case, the value of m needed to give a required sum close to 1062°00'14" is m = 7 = $(n + 1)$ and the required sum of the left-hand angles is

–197°59'23" + 7 × 180° = –197°59'23" + 1260° = 1062°00'37"

— This gives an angular misclosure of 1062°00'14" – 1062°00'37" = –23". The adjustments are applied as equally as possible to the nearest second as A = +4", 1 = +4", 2 = +4", 3 = +4", 4 = +4" and X = +3".

● The misclosures in eastings and northings are given by

$e_E = \Sigma\Delta E - (E_X - E_A)$ = 226.237 – (601.624 – 375.369) = –0.018

$e_N = \Sigma\Delta N - (N_X - N_A)$ = –138.957 – (404.041 – 543.008) = +0.010

— These are distributed using an equal adjustment.

● The fractional linear misclosure is 1 in 24,000 and is accepted.

Reflective summary

With reference to traverse calculations, remember:

— The various stages in a traverse calculation are
  • check and adjust the observed angles
  • calculate all the forward bearings
  • calculate the coordinate differences
  • determine the fractional linear misclosure – reject or accept traverse
  • calculate the final coordinates

— To avoid errors, it is recommended that all of these are tabulated in some form – this will also make it easier to apply all of the arithmetic checks to the traverse calculation.

— The procedures described in this section are for the manual calculation of a traverse. Following the step-by-step instructions given aids an understanding between bearings, distances and coordinates – for site surveying it is important to develop this. After some experience has been gained calculating coordinates, it may be more convenient to use one of the many commercial software packages available for this. Another alternative is to use a total station with a traverse program installed.

6.6 Traversing with total stations

After studying this section you should be aware that some total stations can be used in traverse mode to measure, compute and store a set of on-site traverse coordinates with a check and adjustment. You should also know that programs used for determining the coordinates of control points with a total station should not be used unless the results obtained can be checked.

How to use a total station for traversing

The main advantage of using a total station for traversing is that angles and distances are measured simultaneously at each station. This can reduce the time taken for a survey, and because they can carry out coordinate calculations, another advantage of the total station is that it is possible to compute traverse coordinates at each station as angles and distances are measured.

The procedure for doing this with a total station for the traverse shown in Figure 6.29 is as follows. The total station is set up at existing control point A and a backsight is taken to existing control point B. The coordinates of A and B are entered into the instrument and the horizontal circle is orientated to display the whole-circle bearing

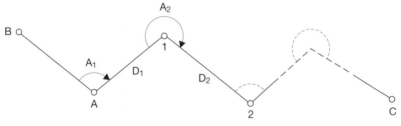

Figure 6.29 ● Traverse measurements with a total station.

from station A to the backsight B. Following this, the total station is rotated through angle A_1 and station 1 is sighted on the traverse – the instrument will now display the whole-circle bearing along line A1. The horizontal distance D_1 is measured and the coordinates of station 1 are calculated by the total station. The instrument is moved and set up at station 1, point A is sighted as a backsight and the horizontal circle is orientated along this direction. By rotating through angle A_2 to sight station 2 and by measuring distance D_2, the coordinates of station 2 are obtained. The instrument is now moved to station 2 and the process is repeated to give the coordinates of the next station. To end the traverse, measurements are taken from the last traverse station to a closing point C whose coordinates are known. As soon as this is done, the total station will display the misclosures in easting e_E and in northing e_N for the traverse, and if these are accepted it will adjust the coordinates of the traverse stations. Because an on-site calculation and check of the traverse can be carried out, any errors in the traverse can be corrected immediately before leaving the site. All of the coordinates for the traverse are stored by the total station using its internal memory, data collector or field computer.

It is also possible to measure *three-dimensional traverses* with a total station. In this case, the heights of the first control point A and those of the closing point C are entered into the total station together with each instrument and reflector height as individual measurements are taken. Heights are then calculated for each station in the traverse and these are checked at point C. As with the eastings and northings, if the misclosure in the heights is accepted, they are adjusted and then stored by the total station.

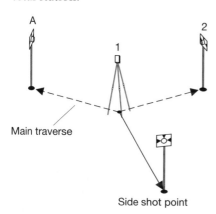

Figure 6.30 Side shot (courtesy Leica Geosystems).

Some traverse programs allow the user to fix the positions of additional points along the traverse called *side shots*. To coordinate these, the side shot point is sighted in between the backsight and next station (see Figure 6.30) and the distance is measured to it. The instrument will calculate the coordinates of the point and store these. As side shots are not part of the traverse, there is no check on the coordinates obtained and they should not be used to determine the positions of control points.

Another feature found on some total stations is *radial traversing*. This is simply an extension of the coordinate measuring program and is carried out by setting up the total station at an existing control station and orientating onto another. Further points are then coordinated by measuring a distance to them and using polar to rectangular conversions to compute their coordinates. Although this offers the ability to coordinate a series of control points much more quickly than by conventional traversing, the drawback of radial traversing is that it is only possible to check the results obtained by measuring the distances between the points surveyed. As this can

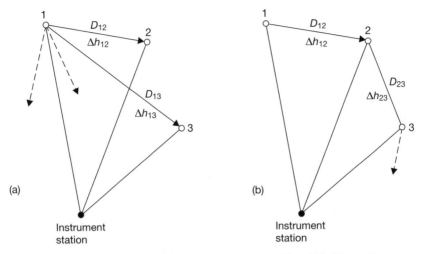

Figure 6.31 ● Missing Line Measurement (MLM): (a) radial; (b) continuous.

involve as much work as ordinary traversing, the need for checking cancels the advantage of a single station set up.

The *Missing Line Measurement (MLM)* software option installed in many total stations allows a user, from a single instrument position, to determine the horizontal distance and height difference between a start point and a series of subsequently selected points. In Figure 6.31, points 1 and 2 are sighted and the distances and circle readings to them recorded at the instrument station. The MLM program then computes the horizontal distance D_{12} and height difference Δh_{12} between these two points. If the distance and circle reading to a third point are included in the sequence, the total station can display D_{13} and Δh_{13} (*radial* MLM) or it can display D_{23} and Δh_{23} (*continuous* MLM). Any number of points can be added to the sequence.

As can be seen, a total station can perform many of the coordinate calculations required on site, and as far as engineering surveying is concerned this has created a situation where the emphasis is turning away from hand calculations and associated checking procedures to *good site practice and field checking procedures*.

There is no doubt that most total stations are very sophisticated instruments. Although this has obvious advantages in control surveys, it can have some disadvantages as well. As an example, consider a traverse to be a recognised surveying procedure for obtaining the coordinates of control points that can be checked with a total station by closing onto a known point. As noted above, by using radial traversing a total station could produce the same set of coordinates for a series of control points much more quickly than by conventional traversing, since the observations could be taken from a single instrument position. In this case, the time taken to arrive at the end result when using a total station is reduced by such a large factor compared with an established field procedure that the temptations are often too great and mistakes could occur if the checks are not done properly. Consequently, although there are a number of different methods used by total stations for obtaining coordinates, they are very risky if used in the wrong circumstances, especially when fixing the positions of control points. While every opportunity should be made to make full use of a total station, *any field procedure involving a total station that does not include an independent check on fieldwork must be treated as incomplete.*

Reflective summary

With reference to traversing with total stations, remember:

— If used properly, a total station is capable of measuring and computing a traverse much quicker than by other methods. However, it is critical that a check is applied by closing the traverse – if the total station does not allow the user to do this it should not be used for traversing.

— When using a total station to fix the coordinates of a control point, always be aware that without a check on any of the measurements taken, the coordinates or other data obtained are not reliable.

— There are still good reasons why checks must be applied to *all* work even when the most up-to-date total station is being used.

6.7   Intersection and resection

After studying this section you should be aware that it is possible to locate the positions of additional control points from an existing network by performing an intersection or resection. In addition, you should also be able to calculate coordinates for intersections and resections.

This section includes the following topics:

● Intersection
● Resection

Intersection

This is a method of locating a control point without occupying it. On construction sites, prominent marks around a site, such as tall buildings, church spires and other clearly defined features may be useful as control points during construction. It is obviously not possible to set up an instrument at these but it is possible to obtain their coordinates by using intersection. Since they are usually in elevated positions they can be seen when the lines of sight to other control points at ground level become obscured as construction proceeds.

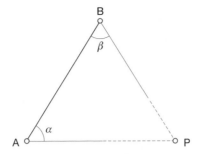

Figure 6.32 ● Intersection.

In Figure 6.32, A and B are points in a control network with known coordinates (E_A, N_A) and (E_B, N_B). To coordinate unknown point P which lies at the intersection of the lines from A and B, a total station or theodolite is set up at A and B and the horizontal angles α and β are observed. The coordinates of P can be calculated by a number of different methods, two of which are given here.

Intersection by solution of triangle

In triangle ABP of Figure 6.32, the length and bearing of the baseline AB are obtained from their coordinates and are given by rectangular → polar conversions as

$$D_{AB} = \sqrt{\Delta E_{AB}^2 + \Delta N_{AB}^2} \qquad \theta_{AB} = \tan^{-1}\left[\frac{\Delta E_{AB}}{\Delta N_{AB}}\right]$$

where $\Delta E_{AB} = E_B - E_A$ and $\Delta N_{AB} = N_B - N_A$.

The sine rule gives

$$D_{BP} = \frac{\sin \alpha}{\sin(\alpha + \beta)} D_{AB} \qquad D_{AP} = \frac{\sin \beta}{\sin(\alpha + \beta)} D_{AB}$$

The whole-circle bearings in the triangle are given by

$$\theta_{AP} = \theta_{AB} + \alpha \qquad \theta_{BP} = \theta_{BA} - \beta$$

These distances and bearings are used to calculate the coordinates of P along AP using polar → rectangular conversions as

$$E_P = E_A + D_{AP} \sin \theta_{AP} \qquad N_P = N_A + D_{AP} \cos \theta_{AP}$$

The calculations are checked along BP using

$$E_P = E_B + D_{BP} \sin \theta_{BP} \qquad N_P = N_B + D_{BP} \cos \theta_{BP}$$

Intersection using the observed angles

If we adopt the *clockwise lettering sequence* used in Figure 6.32, the coordinates of P can be obtained directly from

$$E_P = \frac{(N_B - N_A) + E_A \cot \beta + E_B \cot \alpha}{\cot \alpha + \cot \beta}$$

$$N_P = \frac{(E_A - E_B) + N_A \cot \beta + N_B \cot \alpha}{\cot \alpha + \cot \beta}$$

A disadvantage of this method compared with solving the triangle is that there is no check on the calculations.

Intersection from two baselines

When solving intersections using the formulae given above, it is not possible to check the fieldwork because a unique solution is obtained for the position of point P. The method that should be used to detect errors in the observed angles and hence errors in the coordinates of P is to observe the intersection from at least two baselines and to determine the coordinates of the intersected point by solving two or more separate triangles. If the differences between the sets of coordinates obtained are small, it is assumed that no gross errors have occurred and the mean coordinates are taken as the final values.

For all intersections, it must be realised that the accuracy of the coordinates obtained depends to some extent on the shape of the intersection triangle and in particular, the intersection angle between the lines of sight. To avoid any serious errors from this, only *well-conditioned* intersection triangles should be used in which the intersection angle is not less than 25°.

Worked example 6.6: Intersection from two baselines

Question

The coordinates of three control points S, A and L are as follows

	mE	mN
S	1309.652	1170.503
A	1395.454	1078.806
L	1268.855	1028.419

Calculate the coordinates of point B which has been located by observing the following clockwise angles

$A\hat{S}B$ 122°21'43"
$B\hat{A}S$ 29°34'50"
$L\hat{A}B$ 39°01'16"
$B\hat{L}A$ 105°20'36"

Solution

Using the data provided, the layout of the stations and the observed angles are shown in Figure 6.33. A diagram similar to this must be drawn for these calculations so that all the control points and angles can be properly identified. For triangle SAB, the *clockwise sequence* SAB is equivalent to ABP in Figure 6.32 and the coordinates of B are given by the angles method as

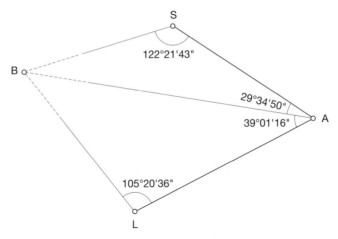

Figure 6.33 ● Intersection from two baselines.

$$E_B = \frac{N_A - N_S + E_S \cot B\hat{A}S + E_A \cot A\hat{S}B}{\cot A\hat{S}B + \cot B\hat{A}S}$$
$$= \frac{1078.806 - 1170.503 + 1309.652 \cot 29°34'50" + 1395.454 \cot 122°21'43"}{\cot 122°21'43" + \cot 29°34'50"}$$
$$= 1180.161 \, m$$

$$N_B = \frac{E_S - E_A + N_S \cot B\hat{A}S + N_A \cot A\hat{S}B}{\cot A\hat{S}B + \cot B\hat{A}S}$$
$$= \frac{1309.652 - 1395.454 + 1170.503 \cot 29°34'50" + 1078.806 \cot 122°21'43"}{\cot 122°21'43" + \cot 29°34'50"}$$
$$= 1145.951 \, m$$

For triangle ALB, the *clockwise sequence* ALB is equivalent to ABP in Figure 6.32 and the coordinates of B are given by the angles method as

$$E_B = \frac{N_L - N_A + E_A \cot B\hat{L}A + E_L \cot L\hat{A}B}{\cot L\hat{A}B + \cot B\hat{L}A}$$
$$= \frac{1028.419 - 1078.806 + 1395.454 \cot 105°20'36" + 1268.855 \cot 39°01'16"}{\cot 39°01'16" + \cot 105°20'36"}$$
$$= 1180.146 \, m$$

$$N_B = \frac{E_A - E_L + N_A \cot B\hat{L}A + N_L \cot L\hat{A}B}{\cot L\hat{A}B + \cot B\hat{L}A}$$
$$= \frac{1395.454 - 1268.855 + 1078.806 \cot 105°20'36" + 1028.419 \cot 39°01'16"}{\cot 39°01'16" + \cot 105°20'36"}$$
$$= 1145.942 \, m$$

Since the two sets of results for E_B and N_B differ by 0.015 m and 0.009 m respectively, no gross error has occurred in the observations and the final coordinates are the mean values from the two sets, hence

$$E_B = 1180.154 \text{ m} \qquad N_B = 1145.947 \text{ m}$$

Resection

This is a method of locating a point by taking observations from it to other known control points in a network. Two types of resection are possible – *angular resections* in which horizontal angles are measured and *distance resections* in which horizontal distances are measured. Both of these are particularly useful for coordinating temporary control points on site which are called *free station points* because they can be established anywhere convenient.

Angular resections

These are used to coordinate a point by taking observations from it to existing control points – an advantage of the method is that a resection can be done without occupying any of the control points to which observations are taken. A good example of this occurs on high-rise buildings when ground control is no longer available or cannot be transferred through to the upper floors. In this case, intersected points can be fixed around the site prior to construction and used later as control for resections as the building is constructed.

A point can be coordinated in an angular resection by observing angles from it to at least three existing control points in a *three-point resection*, as shown in Figure 6.34. In

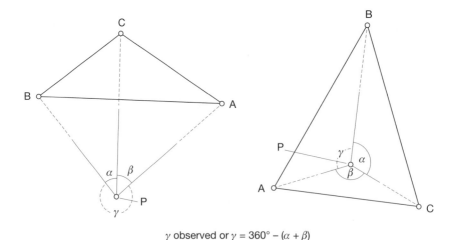

γ observed or $\gamma = 360° - (\alpha + \beta)$

Figure 6.34 ● Angular resection.

this, horizontal angles α and β subtended at the resection point P by control points A, B and C are observed.

If the triangles ABC of Figure 6.34 are *lettered clockwise*, the coordinates of P are given by

$$E_P = \frac{k_1 E_A + k_2 E_B + k_3 E_C}{k_1 + k_2 + k_3} \qquad N_P = \frac{k_1 N_A + k_2 N_B + k_3 N_C}{k_1 + k_2 + k_3}$$

where

$$\frac{1}{k_1} = \cot \hat{A} - \cot \alpha \qquad \alpha = \text{clockwise angle between directions PB and PC}$$

$$\frac{1}{k_2} = \cot \hat{B} - \cot \beta \qquad \beta = \text{clockwise angle between directions PC and PA}$$

$$\frac{1}{k_3} = \cot \hat{C} - \cot \gamma \qquad \gamma = \text{clockwise angle between directions PA and PB}$$

$\hat{A}$, $\hat{B}$ and $\hat{C}$ are the angles within the control triangle ABC.

It must be noted that if points A, B, C and P all lie on the circumference of the same circle, the resection will be indeterminate – this will occur when $\hat{A} = \alpha$, $\hat{B} = \beta$ and $\hat{C} = \alpha + \beta$. As well as this, if the control points A, B and C are co-linear, the resection cannot be solved by this method. This will occur when the angles A, B and C within the control triangle are close to either 0 or 180°. As with intersections, well-conditioned triangles should also be used for angular resections so that any errors due to poor geometry are avoided.

Worked example 6.7: Three-point resection

Question

At a resection point P, the following horizontal angles were observed to three control points L, M and N:

$$L\hat{P}N = 112°15'03'' \qquad N\hat{P}M = 126°42'41''$$
$$M\hat{P}L = 121°02'16''$$

The coordinates of L, M and N are

	mE	mN
L	571.895	684.528
M	613.076	439.187
N	780.004	644.132

Calculate the coordinates of point P.

Solution

Using the coordinates and angles provided, Figure 6.35 is sketched to show that P is

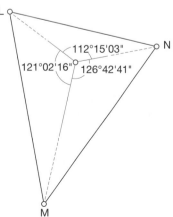

Figure 6.35 ⬤ Three-point resection.

within the control triangle which is *lettered clockwise* as MLN. Taking the lettering sequence MLN to be equivalent to ABC in Figure 6.34, the resection angles are

$$\alpha = L\hat{P}N = 112°15'03"$$

$$\beta = N\hat{P}M = 126°42'41"$$

$$\gamma = M\hat{P}L = 121°02'16"$$

Using rectangular to polar conversions, the bearings in triangle MLN are

$$\theta_{ML} = 350°28'18" \quad \theta_{LN} = 100°59'06" \quad \theta_{NM} = 219°09'46"$$
$$\theta_{LM} = 170°28'18" \quad \theta_{NL} = 280°59'06" \quad \theta_{MN} = 39°09'46"$$

The angles in triangle MLN are obtained from these bearings as

$$\hat{M}(=\hat{A}) = \theta_{MN} - \theta_{ML} = 39°09'46"-350°28'18" = -311°18'32" = 48°41'28"$$

$$\hat{L}(=\hat{B}) = \theta_{LM} - \theta_{LN} = 170°28'18"-100°59'06" = 69°29'12"$$

$$\hat{N}(=\hat{C}) = \theta_{NL} - \theta_{NM} = 280°59'06"-219°09'46" = 61°49'20"$$

Check: $\hat{M} + \hat{L} + \hat{N} = 180°00'00"$

The calculations for the coordinates of P are tabulated below.

	Angle	k	E	kE	N	kN
$\hat{M}(\hat{A})$	48°41'28"		613.076		439.187	
α	112°15'03"	$k_1 = 0.7764432$		476.0187		341.0038
$\hat{L}(\hat{B})$	69°29'12"		571.895		684.528	
β	126°42'41"	$k_2 = 0.8929877$		510.6952		611.2751
$\hat{N}(\hat{C})$	61°49'20"		780.004		644.132	
γ	121°02'16"	$k_3 = 0.8791561$		685.7453		566.2926
		$\Sigma k =$ 2.5485870		$\Sigma kE =$ 1672.4592		$\Sigma kN =$ 1518.5715
P			656.230		595.848	

In a three-point resection, any errors that occur in the observed angles will not be detected, and to provide a check more angles have to be measured. For most work, it is sufficient to take observations to four control points and to use the additional angle for checking purposes. This is shown in the following worked example.

Worked example 6.8: Four-point resection

Question
With reference to Worked example 6.7, suppose a fourth point R is observed where $E_R = 828.172$ m, $N_R = 556.268$ m and the horizontal angle $N\hat{P}R = 34°16'30"$. Use this information to check the coordinates of P for possible mistakes in the fieldwork.

Solution

At the start of a four-point resection calculation, three out of the
four control points that give the
best geometry are used to calculate the resection coordinates. In
the four-point resection worked
example given here, control
points N and R would not be used
together in the three-point resection because the angle between
them is too small. In this case, the
decision has been made to use
point R for checking purposes
only.

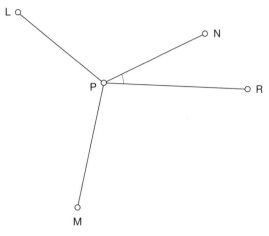

Figure 6.36 ● Four-point resection.

Figure 6.36 shows control
point R in relation to the three-
point resection computed in
Worked example 6.7. Taking the coordinates of P obtained from this and those given
for points N and R, the bearings along PN and PR are (by rectangular → polar
conversions)

$$\theta_{PN} = 68°41'22" \qquad \theta_{PR} = 102°57'48"$$

This gives

$$N\hat{P}R = \theta_{PR} - \theta_{PN} = 102°57'48" - 68°41'22" = 34°16'26"$$

Since the observed value of $N\hat{P}R = 34°16'30"$, the two values differ by only 4" and
this checks the observations and coordinates of P.

Distance resections

These are carried out on site using total stations. Shown in Figure 6.37, a total station
is set up at unknown point P and the horizontal distances D_{PA} and D_{PB} to two existing
control points A and B are measured together with the horizontal angle α subtended
by the control points. Unlike an angular resection, it must be possible to occupy the
control points in this case and place reflectors at A and B.

To determine the coordinates of the total station at P, the distance and bearing
along AB are first calculated from their coordinates using rectangular → polar
conversions. Following this, all three angles in the triangle ABP are calculated using
the distances and the cosine rule – these can be checked by ensuring they sum to
180° and by comparing the measured value of α with its calculated value. It is now
possible to compute the whole-circle bearings along AP and BP and these are used,
with the measured distances, to calculate the coordinates of P as an intersection by
solution of triangle.

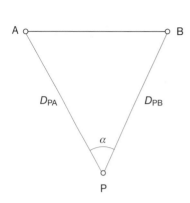

Figure 6.37 ● Distance resection.

Figure 6.38 ● Free stationing (resection): E_1, N_1 and E_2, N_2 are entered into total station and D_1, D_2 with A are measured. Instrument then calculates and displays E_S and N_S.

To check the fieldwork and calculations, a second resection can be observed and calculated using different control points. On site, it is normal practice to include a third station and a resection is carried out with this and control point A or B.

The majority of total stations have an applications program installed for performing distance resections – this program is usually called *free stationing*. Shown in Figure 6.38, all the operator is required to do when using these is to enter the coordinates of two control points and measure the distances to them – the instrument will then calculate and display the coordinates of the instrument position. As with other coordinate programs, if the heights of the control points and the reflectors are also entered into the total station together with the instrument height, the height of the instrument position can be calculated and displayed.

When using a total station for a distance resection, it is essential to check the coordinates obtained by performing a second resection.

Reflective summary

With reference to intersection and resection, remember:

— The fieldwork for an intersection is fairly straightforward and involves measuring horizontal angles only from known control points to fix the position of a new point.

— An advantage of intersection is that the point sighted (whose position is being determined) does not have to be occupied. This enables the coordinates of clearly defined landmarks surrounding a site to be determined – these can be very useful for setting out.

— Although it involves more fieldwork, it is good practice to check the fieldwork and calculations for an intersection by observing and computing it using two or more baselines.

— Like an intersection, the fieldwork for an angular resection involves measuring horizontal angles, but in this case to known control points from the unknown point in order to fix its position. Resection can be very useful for establishing *free station points* on congested sites.

— It is essential that at least four control points are observed in an angular resection so that it is possible to check the fieldwork and calculations.

— A total station is used for taking the measurements required for a distance resection – it also calculates the coordinates of the resection point but uses on-board software for this.

6.8   Control networks

After studying this section you should be aware that traversing is only one of several methods that can be used for providing survey control on site. You should also be familiar with the concept of working from the whole to the part and have an understanding of the terms primary and secondary control.

Extending control into networks

As an alternative to traversing, the positions of control points for a construction project can be fixed using a network.

A *triangulation network* consists of a series of single or overlapping triangles as shown in Figure 6.39, where the vertices of each triangle are the positions of control points. As shown in earlier sections of this chapter, it is necessary to know the angles formed by these triangles and the lengths of the sides of the triangles so that the coordinates of the points can be computed. In this case, position is determined by measuring all the angles in the network and by measuring the length of a baseline such as AB in Figure 6.39. Starting at this baseline, application of the sine rule in each triangle throughout the network enables the lengths of all the triangle sides to

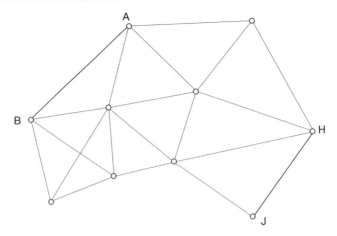

Figure 6.39 ● Triangulation network.

be calculated. To check the calculation of distance, another baseline is measured (say HJ in Figure 6.39) and this is compared with its calculated distance.

A *trilateration network* also takes the form of a series of single or overlapping triangles, but in this case position is determined by measuring all the distances in the network instead of all the angles. To enable the control point coordinates to be calculated, the measured distances are combined with angle values derived from the side lengths of each triangle.

A *combined network* (or simply a *network*) is a control scheme in which a mixture of both angles and distances are measured, usually with a total station. Figure 6.40 shows a small scheme consisting of six control points and the differences between a traverse and a network.

Although control networks could be made up entirely from a chain of single triangles, it is often better to use more complicated figures such as braced quadrilaterals and centre-point polygons, as shown in Figure 6.41. Clearly, compared with traversing or a network consisting of simple triangles, these figures involve taking far more measurements and the subsequent computation of position with them is much more complicated. However, the advantage of incorporating figures more elaborate than simple triangles in a control scheme *strengthens the network* by increasing the

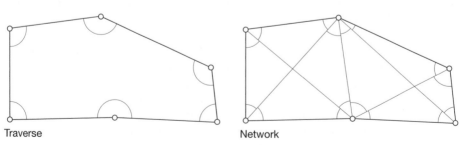

Traverse   Network

Figure 6.40 ● Comparison of traverse and network observations for the same set of control points.

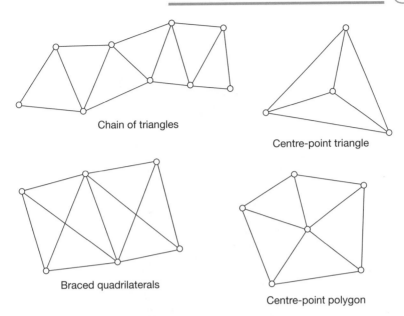

Chain of triangles

Centre-point triangle

Braced quadrilaterals

Centre-point polygon

Figure 6.41 ● Network figures.

number of redundant measurements taken, which enables more checks to be applied to the measurements and a better estimate of the accuracy of the results to be obtained.

When calculating the coordinates of points in a network, a special procedure known as *least squares* is used. If set up correctly, this produces a unique set of coordinates for the control points no matter how complicated the network. In addition to computing the coordinates for the network, it is also capable of providing a complete analysis of the positional accuracy of each coordinated point. This information can be used to check that the survey has met its specification and to detect errors in the observations. To compute coordinates and make full use of the advantages of least squares, a computer and software are required because the complicated nature of the calculations makes a manual calculation very difficult and time-consuming. Although it is possible to develop in-house software for network analysis and computation of coordinates by least squares it is seldom worth doing so, as there are many reasonably priced commercial packages available – see Further reading and sources of information at the end of this chapter. Least squares computations are also discussed in more detail in Section 9.5.

On small construction sites, traversing is a very popular method for locating the positions of control points. However, as the size of the site increases, a network of some sort is usually required to locate the positions of control points because these are better at maintaining accuracy over larger areas. Some control schemes are networks combined with traversing – an example of this is the control that was established for the construction of Munich Airport. Shown in Figure 6.42, this consists of a network of points with an average distance of about 1 km between

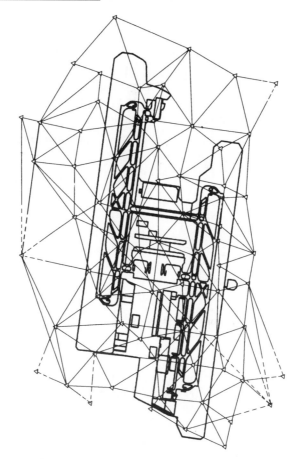

Figure 6.42 ● Control network for Munich Airport (courtesy Leica Geosystems).

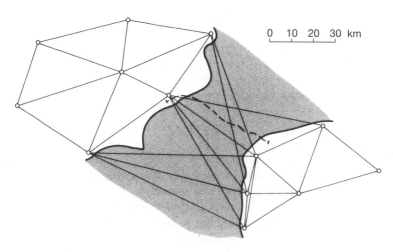

0 10 20 30 km

Figure 6.43 ● Control network for Channel Tunnel.

stations in which all the angles and distances were observed to obtain as many redundant measurements as possible. As can be seen, this network incorporates many braced quadrilaterals, centre-point polygons and overlapping figures to strengthen it. For setting out purposes, a much higher density of control points was required and the positions of these were obtained by link traversing between control points on the network to give a point-to-point separation for setting out of about 60 m. The advantage of this system was that the control points required for setting out were only surveyed at the time when they were needed, depending on the work in progress.

This approach to control surveys follows the well-established method in surveying of *working from the whole to the part*. In this, primary control is established first from which secondary control is fixed. *Primary control* consists of a framework of points which are surveyed as accurately as possible – the number of these is kept to a minimum to reduce the chance of unacceptable errors occurring and then propagating through a network. *Secondary control* is not as accurate and uses the primary control as reference points – this is either a higher density network or a series of traverses measured between the primary control points. The secondary control can then be used for mapping, detail surveying, setting out and so on. This process ensures that errors throughout a survey are kept to a minimum, especially when it covers a large area. Primary and secondary control and the specifications for these are discussed further in Sections 9.4 and 11.3.

The primary network surveyed for the Channel Tunnel project during the 1980s is shown in Figure 6.43. This network is clearly much larger than the one used at Munich Airport and many special techniques were used to measure all the angles and distances involved. From this, secondary and site control was established for construction work in England and France but using a common coordinate grid. Even though traditional methods of measurement were employed on this project (including theodolite and electronic distance measurement), a network of this size would be observed using GPS methods today. However, the principle of working from the whole to the part still applies, and primary control is established first for a GPS control survey followed by secondary control.

Reflective summary

With reference to control networks, remember:

— As well as traversing, the positions of control points on site can also be fixed using a network.

— The advantage of a network is that it enables a better accuracy than traversing to be achieved as the size of the construction site gets larger. The disadvantages are that more fieldwork is required and a computer and special software are needed for calculating the coordinates of the control points.

— For small construction sites and projects that extend to about a kilometre, use traversing for establishing control. For medium-sized projects that extend for a few kilometres in size, a network can be used. For any site bigger than this, use GPS for establishing control. Clearly, these boundaries can overlap depending on the nature of the project.

— Whatever method is used to establish survey control on site, working from the whole to the part is essential. Always use primary and secondary control to maintain accuracy.

Exercises

6.1 Explain how rectangular coordinates are converted into their polar equivalents in a rectangular → polar conversion.

6.2 Explain how a bearing and distance can be converted into rectangular coordinate differences using a polar → rectangular conversion.

6.3 Describe how a total station can be set up to measure coordinates.

6.4 Explain what a coordinate transformation is and why these are needed in surveying.

6.5 What is the purpose of a reconnaissance in a control survey?

6.6 How is it possible to check the measured angles in a polygon and link traverse?

6.7 List the various ways in which a control point can be marked – which of these are considered to be short duration and long duration?

6.8 What is three-tripod traversing and why is it used?

6.9 Discuss the various options available for defining a north direction on a coordinate grid.

6.10 What is the fractional linear misclosure for a traverse and how is it calculated?

6.11 Describe how a total station can be used for traversing.

6.12 Discuss the various sources of error that can occur when measuring traverse angles and distances.

6.13 How does an intersection differ from a resection?

6.14 Discuss the importance of checking traverse and other coordinate measurements taken with a total station.

6.15 As they have not been used for some time, the coordinates of three control points A, S and T are to be checked by measuring the internal angles and lengths of the triangle AST. Using the coordinates given below, calculate the required angles and distances.

Point	mE	mN
A	1507.319	632.017
S	1635.904	725.769
T	1738.612	627.301

(Hint: To solve this problem, calculate rectangular → polar conversions along each line of the triangle. This gives the distances. To calculate the internal angles use the bearings obtained where angle A = bearing AT – bearing AS and so on)

6.16 For the polygon traverse shown in Figure 6.44, calculate

(i) the angular misclosure and adjust the angles

(ii) the bearing of line HA from the coordinates of H and A and then bearing HR

(iii) the forward bearings of all the other lines (choose the direction of computation to be HTPEWRH so that the internal angles shown are left-hand angles)

(iv) the coordinate differences of all the lines (adjust any misclosure by the Bowditch method)

(v) the fractional linear misclosure

(vi) the coordinates of the unknown stations

6.17 For the polygon traverses shown in Figure 6.45, calculate the coordinates of the unknown stations and the fractional linear misclosure.

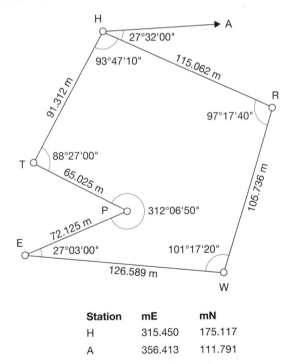

Station	mE	mN
H	315.450	175.117
A	356.413	111.791

Figure 6.44 ● Traverse diagram for Exercise 6.16.

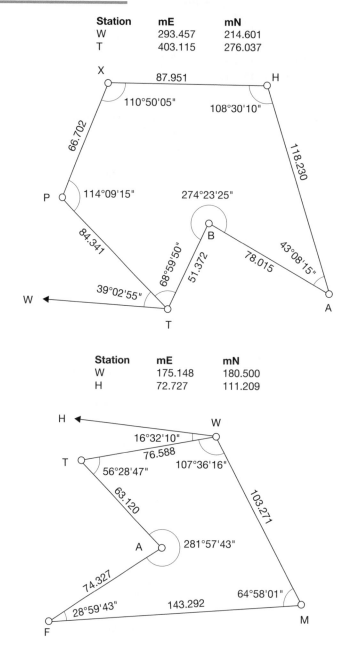

Figure 6.45 ● Traverse diagrams for Exercise 6.17.

6.18 The measured angles and distances for a link traverse MABCP are shown in Figure 6.46. The coordinates of point M are 207.120 m*E*, 526.063 m*N* and those of point P are 651.008 m*E*, 699.120 m*N*. Line MN has a whole-circle bearing of 241°17'35" and line PQ has a whole-circle bearing of 153°00'55".

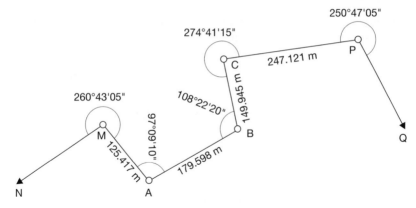

Figure 6.46 ● Traverse diagram for Exercise 6.18.

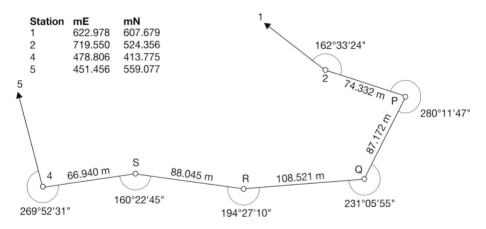

Figure 6.47 ● Traverse diagram for Exercise 6.19.

Calculate for this traverse

(i)   the angular misclosure and adjust the angles

(ii)   the forward bearings (choose the direction of computation to be MABCP so that the angles shown are left-hand angles)

(iii) the coordinate differences (adjust any misclosure equally)

(iv)   the fractional linear misclosure

(v)   the coordinates of stations A, B and C

6.19 The abstract for a link traverse 2PQRS4 is shown in Figure 6.47. Use this to calculate the coordinates of stations P, Q, R and S and the fractional linear misclosure for the traverse.

6.20 A new 100 m × 40 m warehouse is to be set out from four traverse stations A, B, C and D that were used in the site survey. However, all design coordinates are to be based on a new structural grid aligned with the warehouse and not the coordinate grid adopted for the site survey.

Using the data given below, calculate

(i)   the transformation parameters to convert coordinates from the site grid to the structural grid

(ii)   the coordinates of B and D on the structural grid

Station	Site survey		Structural grid	
	mE	mN	mE	mN
A	150.000	350.000	109.515	251.780
B	424.887	510.985		
C	467.804	288.117	329.456	14.175
D	234.100	128.848		

6.21 In order to provide extra control on a construction site, the coordinates of two targets T_1 and T_2 located at the top of nearby buildings are obtained by intersection from control points A, B and C. Using the data given below, calculate the coordinates of T_1 and T_2.

Point	mE	mN	Observed angles	
A	195.002	344.901	$T_1AB = 123°51'06"$	$T_2AB = 79°48'48"$
B	176.600	227.615	$ABT_1 = 28°01'18"$	$ABT_2 = 58°17'53"$
C	357.646	193.511	$T_1BC = 63°43'48"$	$T_2BC = 33°27'06"$
			$BCT_1 = 63°57'05"$	$BCT_2 = 68°23'45"$

6.22 A free station point P is established on site by taking observations to four control points H, L, M and N. Using the data given below, calculate the coordinates of point P.

Point	mE	mN	Angle	Observed value
H	314.202	422.788	HPL	24°28'09"
L	321.360	528.885	LPM	27°45'46"
M	414.763	566.583	MPN	49°46'28"
N	548.670	452.733		

(Hint: To solve this problem, calculate a three-point resection using control points H, M and N for the best geometry – then use point L as a check)

6.23 A distance resection is carried out at point P to two control points A and B. Calculate the coordinates of P using the data given below.

Point	mE	mN	Measured distances and angle	
A	379.588	758.723	PA = 158.635 m	PB = 208.272 m
B	411.800	572.850	APB = 60°05'00"	

(Note: The approximate position of P is 240 mE, 690 mN)

Further reading and sources of information

For specifications in control surveys and setting out, consult

BS 5964 (ISO 4463): 1996 *Building setting out and measurement* (British Standards Institution, London). BSI web site `http://www.bsi-global.com/`.

ICE Design and Practice Guide (1997) *The management of setting out in construction.* Thomas Telford, London.

The following journals include frequent articles and annual reviews of least squares software: *Civil Engineering Surveyor, Engineering Surveying Showcase* and *Geomatics World* (see Chapter 1 for further details).

Global Positioning System

 Aims

After studying this chapter you should be able to:

- Describe the development of the Global Positioning System (or GPS) and the impact this has had on site surveying

- Explain how GPS can be used to take code and phase measurements to determine position and be able to explain the difference between these

- Identify the various sources of error in GPS and explain how each of these affects the accuracy obtained

- Understand the reasons why differential and relative methods are essential for high precision surveying with GPS

- Outline the methods involved when performing static and kinematic surveys with GPS

- Distinguish between the different types of GPS receivers and systems currently available and be able to find further information to help choose one of these for engineering surveys

- Identify the main applications for GPS in civil engineering and surveying

- Keep up with developments in satellite surveying systems including GLONASS, Galileo and Network RTK

This chapter contains the following sections:

7.1   The development of GPS

7.2   Components of GPS

7.3   GPS positioning methods

7.4   Errors in GPS

7.1   The development of GPS

After studying this section you should understand the reasons why GPS was originally planned and developed. You should be able to compare and contrast GPS with conventional surveying and appreciate that to use it effectively an understanding of the operating principles of GPS is required.

NAVSTAR GPS

The Global Positioning System, or simply GPS, is something that is now familiar to most people. Today, recreational sailors and walkers can use a handheld GPS receiver, as shown in Figure 7.1, to locate their position on a chart or map, and motorists may have a dashboard mounted display for this – many different versions of these devices are now available. GPS has found widespread use in aviation, navigation, in scientific areas such as weather prediction and oceanography, and for locating features in Geographic Information Systems (GIS). It is used in site surveying for everyday tasks ranging from control surveys and setting out to machine control. For this work, GPS receivers are available that can determine coordinates at the millimetre level. This equipment is more sophisticated than a handheld receiver and is often pole-mounted, as shown in Figure 7.2.

Designed primarily for military users, GPS is managed and under the control of the United States Department of Defense (DoD). The development of GPS began in the early 1970s and evolved from the US space programme. At this time, the DoD decided that it wanted a worldwide navigation and positioning system for passive military users (passive because users should not transmit anything, as they might be detected) that could be denied to an enemy and have unlimited users. It was to

Figure 7.1 ● Garmin GPSmap 60C GPS receiver (courtesy Garmin Ltd).

Figure 7.2 ● Surveying with GPS (courtesy Topcon)

operate in real time for all military applications, varying from the infantry (a soldier with a receiver in a back pack or handheld device) through to aircraft navigation and missile guidance systems. The development of the NAVSTAR (**NAV**igation **S**atellite **T**iming **A**nd **R**anging) Global Positioning System began in 1973, with the first satellite being launched in 1978; it was fully operational by 1993. The potential for civilian use of GPS was quickly realised by the surveying and navigation professions and applications were soon recognised and hardware and software developed for these.

With GPS, it is possible to obtain a position (as three-dimensional coordinates) anywhere on the Earth at any time of the day or night in any weather conditions by receiving and processing signals from orbiting GPS satellites. To do this, the orbits of the satellites are always known and broadcast to the user. To locate position, a GPS receiver determines the distances to a number of satellites, which are the equivalent of orbiting control points; a computer then calculates the three-dimensional coordinates of the user. In some respects, GPS is like a distance resection (as described in Section 6.7) in which distances are measured at an unknown point to control points whose coordinates are known. However, GPS is very different compared to conventional (or terrestrial) surveying using levels, theodolites and total stations.

Since GPS is a passive system and users simply receive GPS signals, there is no limit to the number of GPS receivers that can use GPS signals at the same time without overloading the system – this is a similar situation to TV and radio, where millions can tune in at any time without affecting the broadcast. This is a big advantage for a large-scale position-fixing system, but because of this GPS signals are complex and receivers must be capable of collecting a large amount of information to be able to determine position.

The theoretical and scientific background to GPS and the mathematics involved to describe it are very complicated and a full knowledge of these is not needed to be able to use GPS. However, while those using it for construction work are not expected to be GPS specialists, they do need to know enough of these to assess where GPS is useful on site and what its advantages and limitations are. This chapter covers the operating principles of GPS, the equipment available for survey work and the field procedures and applications it has on site. Chapter 8 gives details of the way in which GPS coordinates are defined and how these are transformed for use on site.

Reflective summary

With reference to the development of GPS, remember:

— The operation of GPS is the responsibility of the United States Department of Defense and it is under military control at all times.

— GPS is very different from conventional (or terrestrial) surveying and new ways of surveying have to be learnt in order to use it properly.

— Although anyone who uses GPS does not have to have an expert knowledge of it, in order to understand its advantages, limitations and best applications on site some of the basics of GPS should be learnt.

7.2   Components of GPS

After studying this section you should know that the space segment of GPS is the satellite constellation and the control segment consists of a series of tracking stations and a control centre. You should be aware of the function that each of these fulfils and be aware of the information that each satellite broadcasts.

This section includes the following topics:

● GPS segments
● GPS signal structure

GPS segments

GPS consists of three segments, called the space segment (the satellites orbiting the Earth), the control segment (stations positioned around the Earth to control the satellites) and the user segment (anybody that receives and uses a GPS signal).

The *space segment* consists of a minimum of 24 satellites, all of which are in orbits of 20,200 km above the Earth with an orbital period of 12 sidereal hours. Because the sidereal day is slightly longer than the solar day, it appears to an observer on the

Earth that the satellites return to the same positions four minutes earlier each day. This has practical implications as a good satellite configuration for observations on one day may change to a less favourable one the following and subsequent days. Great care was taken when designing the GPS satellite constellation, as it is necessary to have the optimum satellite coverage on the Earth from the minimum number of satellites. To achieve the best coverage, the satellites are located in six orbital planes spaced equally around the plane of the equator but inclined at 55° to the equator.

Over the last 25 years, various generations (blocks) of GPS satellites have been launched. There are no operational *Block I satellites* today, and most currently in use are *Block II, IIA* and *IIR satellites*, where the *R* stands for *replenishment*. The fourth generation of GPS satellites are known as *Block IIF satellites*, where the *F* stands for *follow-on*.

As the accurate measurement of time is essential in GPS, each satellite is fitted with up to four atomic clocks. Block II and Block IIA satellites have caesium or rubidium clocks on board with extremely high accuracies of up to 1 in 10^{12}, whereas the later Block IIR and F satellites use highly stable hydrogen maser clocks with an even better accuracy of 1 in 10^{14}. As a comparison to everyday clocks and watches, if these were good enough to keep time to within 1 second each day, this would only be equivalent to an accuracy of 1 in 10^5.

A number of tracking stations at fixed locations around the world form the *control segment* of GPS. The Master Control Station (MCS) is located in the Consolidated Space Operations Center (CSOC) at Schriever Air Force Base near Colorado Springs in the USA.

To be able to position accurately with GPS, the exact position of each satellite has to be known at all times (remember that the satellites are in effect the GPS control points) and, despite their phenomenal accuracy, the satellite clocks do drift and they must be kept synchronised with GPS time as defined at the CSOC. As they orbit the Earth, the satellites are subjected to the varying gravitational attraction of the Earth, the attractions of the Sun and Moon, and solar radiation pressure. All of these cause the satellite orbits to change with time and these have to be measured and predicted by some means. To do this, a network of 13 tracking stations continuously monitors all of the GPS satellites in view at all times and this orbital data is relayed to the MCS, where it is used to predict future orbits for the satellites. As well as this, the clock in each satellite is also monitored and compared to GPS time to enable corrections to be computed in order to keep the satellite clocks in step with GPS time. The predicted satellite orbital positions (which are known as *ephemeris predictions*) and satellite clock corrections computed at the MCS are sent to ground antennae at the tracking stations, where this information is uploaded to the satellites every two hours and then broadcast via the GPS satellites to users of the system. This enables a GPS receiver to determine the positions of the satellites at any given time.

The *user segment* of GPS consists of anyone, civilian or military, who uses a GPS receiver to determine their position. As already noted, there is now an extensive civilian GPS community and as far as construction work is concerned, GPS is used for control surveys, mapping, setting out, machine control, monitoring and measurement of volumes.

GPS signal structure

Each satellite in the GPS constellation continuously broadcasts two electromagnetic signals that are in the L-band used for radio communication. The precise atomic clocks in the satellites have a fundamental clock rate or frequency f_0 of 10.23 MHz and the two L-band frequencies assigned to GPS are derived from the atomic clocks as L1 = 154 × f_0 = 1575.42 MHz and L2 = 120 × f_0 = 1227.60 MHz.

Both the L1 and L2 signals act as carrier waves and have information modulated onto them that can be used by a GPS receiver to determine position. Unlike total stations, which use intensity modulation for measuring distances using a laser or infrared carrier wave (described in Section 5.2), GPS uses phase modulation on the two L-band waves.

A *navigation code* or *message* is also modulated on both the L1 and L2 carriers. This has a low frequency of 50 Hz and contains various sub-frames of information, including:

- A *clock correction* – remember that the tracking stations in the ground segment measure and upload this correction to each satellite, and rather than keep altering the atomic clock to keep it in step with GPS time, a correction is applied by a GPS receiver to determine GPS time at the satellite.

- *Satellite ephemeris* or orbital information – this is obtained from the data collected by the tracking stations and a prediction computed by the MCS at Colorado Springs.

- *Atmospheric correction* data from the tracking stations.

- An *almanac* which tells a receiver how to find all the other satellites – this is not the same as the orbital information above and only gives a receiver enough information to track another satellite, not to compute its position.

- *Satellite health data* for each satellite which informs a GPS user of any malfunction on any satellite before a receiver tries to use that particular satellite.

As well as the navigation code, two binary codes are continuously modulated onto the L-band signals; these are called the *coarse/acquisition* or *C/A code* and the *precise* or *P code*. These are not the same as the data contained in the navigation code and are not uploaded to the satellites by the control segment – they are continually broadcast by each satellite and are processed by a GPS receiver to obtain the raw data for time and distance measurements. Both of these codes are known as *pseudo-random noise* or *PRN codes*, and although they appear to be random they are generated in known sequences that can be replicated. Each satellite broadcasts a unique C/A code on the L1 carrier at a rate of 1.023 million bits per second (Mbps or a frequency of f_0 ÷ 10 = 1.023 MHz), and this is repeated every millisecond. Because this gives a chip length (equivalent to 1 bit of information) of 293 m, this means that measurement of distance on the C/A code is carried out using a wavelength of 293 m. There is only one P code which has a frequency of 10.23 Mbps (frequency of f_0 = 10.23 MHz and chip length 29.3 m) which consists of 37 one-week segments. This is available on both the L1 and L2 carriers and each satellite broadcasts a different portion of the P

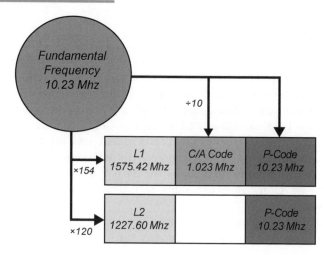

Figure 7.3 ● GPS signal structure (courtesy Leica Geosystems).

code, which it repeats every seven days starting at midnight on Saturday. GPS satellites are identified according to which portion of the P code they transmit. For example, satellite PRN 10 (or SV 10, where SV = *Space Vehicle*) has week 10 of the P code assigned to it.

Remember that GPS was primarily designed for military applications, so the P code is generally encrypted by the DoD and changed to a secret Y code that can only be read by users approved by the US military using special GPS receivers. This is known as *anti-spoofing* (*AS*), but many of the manufacturers of civilian GPS equipment have designed receivers that can access the P code and take measurements with it even when the Y code is in operation.

The relationship between the various GPS signals and the fundamental frequency is shown in Figure 7.3.

Reflective summary

With reference to GPS segments, remember:

— GPS is organised into a space segment consisting nominally of 24 orbiting satellites and a control segment consisting of an operations centre in Colorado, USA, and 13 tracking stations situated around the world.

— The GPS satellite constellation is continuously monitored, and when satellites become inoperable (usually because of old age) new ones are launched to replace these.

— When determining position with GPS, very accurate timing is required to ensure that all measurements are synchronised with GPS time. This is the reason that very accurate (and very expensive) clocks are installed in each satellite.

— The signals broadcast by each satellite contain all the information that a user needs to determine position. The signal structure is quite complicated but it is made up of two basic parts – PRN codes for determining distances and a navigation message for determining the positions of the satellites.

7.3   GPS positioning methods

After studying this section you should be aware that there are two basic methods for determining position with GPS equipment – code ranging and carrier phase measurements. You should understand how each of these is used to measure distances and you should appreciate the difference in accuracy that these are capable of producing.

This section includes the following topics:

- Code ranging
- Carrier phase measurements

Code ranging

To locate a position, a GPS receiver determines distances to a number of different GPS satellites and then uses these to compute the coordinates of the receiver antenna. In GPS, there are two *observables* that can be used to determine distance – *pseudoranges* and the *carrier phase*. Most GPS receivers are capable of determining pseudoranges by code ranging, and this provides low-accuracy, generally instantaneous, point positions with handheld and mapping-grade receivers. Carrier phase measurements are used in surveying for high-precision work using geodetic receivers.

Code ranging is the simplest form of GPS positioning and is carried out with a single receiver. To determine the distance to each satellite by code ranging, a receiver measures the time taken by the C/A code to travel from a satellite to the receiver and then calculates the distance or range R between the two as $R = c\Delta t$, where c is the signal velocity (speed of light) and Δt is the time taken for the codes to travel from satellite to receiver (called the *propagation delay*). There are some difficulties with these measurements because the process is one-way and the time at which the signal leaves the satellite must be determined as well as the time of arrival of the PRN codes at the receiver – both of these measurements are synchronised with GPS time, otherwise the transit time will be wrong and the distances will be incorrect.

To overcome these problems, the following technique has been developed for GPS code measurements. All the satellites use their on board atomic clocks to generate individual C/A codes and the relationship of these to GPS time is known. When a receiver locks onto a satellite, the incoming signal triggers the receiver to generate a C/A code identical to that transmitted by the satellite. The replica code generated by the receiver is then compared with the satellite code in a process known as *autocorrelation*, in which the receiver code is shifted in time until it is in phase, or correlated, with the satellite code. The amount by which the receiver-generated code is shifted is equal to the propagation delay between satellite and receiver. Multiplied by the speed of light, this gives the distance between satellite and receiver.

As already stated, all of the satellites are fitted with atomic clocks whose offsets from GPS time are precisely known through the navigation message. This enables the time at which the PRN codes were transmitted from a satellite to be determined almost exactly with respect to GPS time. For practical reasons, and because it would cost too much, it is not feasible to install atomic clocks in a GPS receiver. Instead, they have electronic clocks (oscillators) installed with accuracies much less than those of atomic clocks but still far better than everyday clocks. As a result, receiver clocks are not usually synchronised with GPS time and the receiver-generated replica codes will contain a clock error. Because of this, all distances (or ranges) determined by a receiver will be biased and are called *pseudoranges* – it is essential that the clock errors that cause these biases are removed from measurements.

As well as pseudorange measurements, a receiver will also process the navigation code, and from this it computes the position of each satellite it tracks at the time the measurements are taken.

To compute position, the satellites are treated as control points of known coordinates to which distances are determined using a receiver located at a point whose position is unknown. In Figure 7.4, the range from point P to any satellite S can be defined as

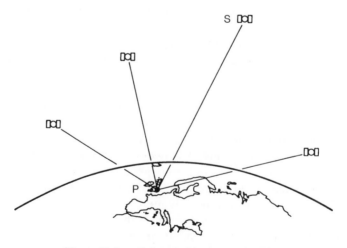

Figure 7.4 ● Point positioning with GPS.

$$R_{SP} = \rho + c\,(dt - dT)$$

where

R_{SP} = the geometric range from satellite S to P
ρ = the pseudorange measured from satellite S to P
dt = the satellite clock offset from GPS time
dT = the receiver clock offset from GPS time

Since the satellite clock offset dt can be modelled using data contained in the navigation code, this equation reduces to $R_{SP} = \rho - cdT$.

The geometric range R_{SP} is given by

$$R_{SP} = \sqrt{(X_S - X_P)^2 + (Y_S - Y_P)^2 + (Z_S - Z_P)^2}$$

in which X_S, Y_S and Z_S are the known coordinates of the satellite and X_P, Y_P and Z_P are the unknown coordinates of point P.

If observations are taken to four satellites simultaneously, four point positioning equations will be obtained as follows

For satellite 1 $\quad \sqrt{(X_{S1} - X_P)^2 + (Y_{S1} - Y_P)^2 + (Z_{S1} - Z_P)^2} = \rho_1 - cdT$

For satellite 2 $\quad \sqrt{(X_{S2} - X_P)^2 + (Y_{S2} - Y_P)^2 + (Z_{S2} - Z_P)^2} = \rho_2 - cdT$

For satellite 3 $\quad \sqrt{(X_{S3} - X_P)^2 + (Y_{S3} - Y_P)^2 + (Z_{S3} - Z_P)^2} = \rho_3 - cdT$

For satellite 4 $\quad \sqrt{(X_{S4} - X_P)^2 + (Y_{S4} - Y_P)^2 + (Z_{S4} - Z_P)^2} = \rho_4 - cdT$

These four equations can be solved to give the four unknowns: the receiver coordinates X_P, Y_P and Z_P and its clock offset dT.

As can be seen, *it is necessary to track and measure to at least four satellites in any GPS survey.* This does not, however, guarantee that it will be successful, and sometimes more are needed. The satellite constellation described earlier makes this possible by placing the satellites at high altitudes in six orbital planes so that position can be determined anywhere on the Earth's surface.

This is a simplified model of pseudoranging and it is shown in the next section that the distances determined from measurements taken to satellites are subject to several other errors as well as those due to timing.

Code ranging is also known as the *navigation solution* and this fulfils the original aims of GPS. Because it is possible to determine the receiver clock error quite easily in this process, the clocks in GPS receivers need not be as accurate as atomic clocks and this makes the receivers much less expensive and suitable for field use. In any given navigation solution, the equations are solved many times using a computer and software installed in the receiver – the operator does not have to perform any calculations and merely reads the position displayed by the receiver in real time or records this electronically.

When using GPS for point positioning, the C/A code broadcast on the L1 frequency allows access to the *Standard Positioning Service* or *SPS* which is used for most civilian applications and has a three-dimensional accuracy of about 10–20 m. To reflect the fact that GPS coordinates, like other quantities in surveying, have a measurement uncertainty associated with them, this accuracy is normally quoted at the 95% probability level. This means that 95% of the time, the results obtained from the SPS will as good as or better than 10–20 m, but there is a 5% chance they may not be. For a general discussion of errors in surveying and an explanation of the significance of probability in measurements, see Chapter 9. The P code broadcast on L1 and L2 allows access to the *Precise Positioning Service* or *PPS* which is much more accurate for point positioning. However, because of anti-spoofing only the military, equipped with special GPS receivers, can read the P code directly.

Carrier phase measurements

The accuracy of code ranging is clearly inadequate for high-precision surveying where accuracies at the centimetre level or better are required.

In Section 5.2, the method of electronic (electromagnetic) distance measurement by phase shift is described in which distance is given by $N\lambda + \Delta\lambda$ in which $N\lambda$ is the whole number of wavelengths in a distance and $\Delta\lambda$ is the fractional part of a wavelength in a distance. A total station measures the fractional component of a distance $\Delta\lambda$ by comparing the phases of signals transmitted and received by the same instrument. Although total stations are very different from GPS receivers, GPS receivers also use the phase shift method to measure distances to satellites, but this is done by comparing the phase of an incoming satellite signal with a similar signal generated by the receiver. These phase measurements are taken on the L1 and L2 carrier waves (not the modulated C/A and P codes) with a resolution of about 1%, and because the carriers have very short wavelengths of 0.19 m (L1 signal) and 0.24 m (L2 signal) it is possible to determine a range with millimetre precision.

The main problem with GPS phase measurements is determining the integer number N of carrier wavelengths between satellite and receiver. In other words, the problem is how to find N in $N\lambda + \Delta\lambda$ in order to resolve what is known as the *carrier phase ambiguity* or simply the *integer ambiguity*. This is very difficult to do because of the one-way communication between the satellites and receivers and the long distances involved, and because the satellites are always moving.

Many techniques have been developed for solving the carrier phase ambiguity both rapidly and reliably, the most successful of which use strategies based on dual-frequency (L1 and L2) phase with both C/A and P code measurements. The starting point for some ambiguity resolution techniques is for the receiver to use code ranging and a navigation solution to obtain an approximate position. Following this, the receiver removes the modulated PRN codes from the L1 and L2 signals so that it can access their carrier waves and take phase measurements on these. The integer ambiguity itself is then solved using a number of different methods. In one, the position of the receiver is estimated using carrier phase data from multiple observations – the error in the estimate is then found by checking the solution against further phase measurements taken as the satellite constellation changes over a long period of time. For a quicker solution, other strategies involve trial and error in which statistics are used to give the optimum position of the receiver from a number of different possible solutions. In all cases, the receiver software will calculate the reliability of the solution obtained but it has to be noted that the statistical tests fail sometimes so ambiguity resolution is not always possible.

Reflective summary

With reference to GPS positioning methods, remember:

— Code ranging, which is also called *point positioning*, is only capable of producing three-dimensional accuracies of about 10–20 m at present with a single receiver accessing the Standard Positioning Service. It is expected that this will improve and may eventually be in the 1–5 m range.

— Carrier phase measurements are capable of determining highly accurate distances, and therefore positions, with accuracies of between 10–50 mm. However, these are much more complicated because of integer ambiguities, coupled with the inability to access the P code directly.

— All GPS coordinates have a degree of uncertainty associated with them and all accuracies are quoted with a 95% probability.

— GPS receivers that use codes only are not very expensive but not very accurate. Receivers that use codes and carrier phase measurements are very expensive but very accurate. Both are used in surveying depending on the accuracies required.

— All measurements with GPS require that at least four satellites are tracked simultaneously in order to be able to compute a three-dimensional position.

7.4 Errors in GPS

After studying this section you should be aware of the various sources of error, and their possible magnitude, in GPS surveys. You should also know that these can either be modelled or suitable field procedures can be used to minimise or even remove them.

This section includes the following topics:

● Ionospheric and tropospheric delays

● Satellite and receiver clock errors

- Satellite orbital errors
- Selective Availability
- Dilution of position
- Multipath
- Antennae

Ionospheric and tropospheric delays

So far, it has been assumed that the measurements taken by a GPS receiver are free of error. As might be expected, this rarely happens in practice and there are several sources of error that can degrade a GPS position.

As described for total stations in Section 5.5, it is well known that the speed of electromagnetic radiation is affected by the medium through which it is travelling. GPS signals, which are in the electromagnetic part of the spectrum, must pass through the atmosphere as they approach the Earth, which causes them to be refracted. The effect of this is to slow down the L1 and L2 carriers and to introduce timing errors into pseudorange and carrier phase measurements.

The *ionosphere* is a layer in the atmosphere extending from about 50–1000 km above the Earth, and it consists of free electrons, the density of which change according to the rate at which gas molecules in the ionosphere are ionised by the Sun's ultraviolet radiation. *Ionospheric refraction* is a function of free electron density and it varies with magnetic activity, location, time of day and the direction of a GPS observation. During daylight hours, the slowing effect of the ionosphere on electromagnetic GPS signals (known as *ionospheric delay*) is much worse than at night; it is also higher from November to March than at other times of the year and can be quite severe at times of high sunspot activity, which varies on an 11 year cycle. The positional error in pseudorange measurements resulting from ionospheric delay is typically 10 m, but it can be as large as 150 m.

A property of ionospheric refraction is that it is dispersive and frequency-dependent. This means that the L1 signal is affected differently from the L2 signal, and dual-frequency receivers (those that can measure on both L1 and L2) are able to measure different arrival times for both signals. These differences can be used to calculate an estimate for the ionospheric delay, which can then be removed from observed data to give an *ionosphere-free observable*. As single-frequency receivers cannot do this, they have to model the delay using information contained in the navigation message – although this reduces the error, it does not completely remove it.

After the ionosphere, the next layers that GPS signals encounter are the stratosphere and the troposphere. The *troposphere* is the closest layer to the Earth and extends for about 10 km; above this is the *stratosphere*, which extends to 50 km, where it meets the ionosphere. As far as GPS is concerned, the effect of the troposphere and stratosphere are combined and give rise to *tropospheric refraction* and a *tropospheric delay* which gives a positional error of about 2 m for pseudoranges. Tropospheric delays are not frequency-dependent and cannot be removed using dual-frequency techniques, but even though the variability of the troposphere is greater

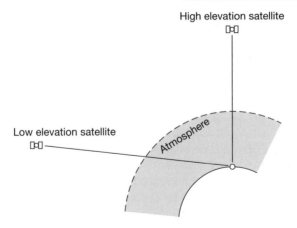

Figure 7.5 ● Effect of satellite elevation.

than the ionosphere, it is easier to model because it is closer to the Earth. To correct for tropospheric delays, two models are required because tropospheric refraction has dry and wet components – the dry component is closely related to atmospheric pressure and is easier to measure than the wet component, which depends on the amount of water vapour in the troposphere. At present, tropospheric modelling can remove almost all of the tropospheric delay for GPS measurements taken along baselines up to 20 km and where the heights at each end of the baseline are similar (for an explanation of what a baseline is, see the next section on differential and relative GPS). However, for the best possible height determination with GPS, it is very important that users are aware of the limitations of a failure to model tropospheric effects correctly. Although a dry model can remove 90% of the error, the remaining 10% wet component is very difficult to model and centimetre errors can easily propagate into computed heights. For the most accurate positioning, the wet component is parameterised and computed instead of being modelled.

A factor that is crucial to both ionospheric and tropospheric effects is the satellite elevation. Signals from low-elevation satellites will have a longer distance to travel through the atmosphere than those at high elevations, as shown in Figure 7.5. Consequently, these will experience greater ionospheric and tropospheric delays, and because it is very difficult to model the troposphere at very low elevations, pseudorange errors from this source can increase by a factor of 10. To overcome these problems, signals from low-elevation satellites are generally not used by inserting a *cut-off* or *mask angle* in the receiver software to exclude them from any processing – a typical value for this is a vertical angle of 10–15° above the observer's horizon.

Satellite and receiver clock errors

For GPS point positioning, the satellite clock error is computed using corrections broadcast in the navigation code – this reduces the error from this source to about 1 m for pseudoranges. The receiver clock error is determined by taking observations to at least four satellites and performing a navigation solution.

Both of these errors can also be eliminated by using double-differencing, a technique used when relative positioning with GPS (see the next section).

Satellite orbital errors

As already stated, the orbital data (or *ephemeris*) broadcast by each satellite is updated every two hours by the control segment of GPS. However, because the satellites are subject to variations in the parameters used to predict their orbits, the orbital data is never exact and positional errors of up to 5 m can be expected from this source for code ranging.

As a guide, inaccuracies in carrier phase measurements as a result of incorrect satellite orbits can be related to the length of a GPS baseline as follows

$$\text{baseline error} \approx \frac{d}{20,000} \times \text{orbital error}$$

where d is the baseline length in km and the baseline and orbital errors are in metres.

For example, if a baseline of 10 km is being measured and the broadcast orbital error is 2–10 m, this causes a baseline error of 1–5 mm. At 50 km, the errors are 5–25 mm.

To improve the accuracy of GPS results, as well as the real-time broadcast satellite orbital data, post-computed satellite orbital information is also available from a variety of providers. However, it would only be necessary to use this data in combination with carrier phase measurements and only when the best possible accuracy is required. For further details of these services, see Section 8.6.

Selective Availability

When designing GPS, the US DoD intended that measurements taken with the C/A code would only give an accuracy of about 150 m for point positioning and that a 15 m accuracy would only be possible by using the encrypted P code. This would stop civilian or other users from obtaining high accuracy from GPS. In practice, however, it is possible to obtain accuracies as good as 10 m even with the C/A code. This was looked on as a security risk by the DoD because GPS could be used for precise positioning by those hostile towards the USA, and led to the implementation of *Selective Availability* or *SA*. The effect of this was twofold: first, the accuracy was made worse by applying a *dither* to satellite clocks so that it was difficult to measure transit times precisely, and second, the data message was altered so that satellite orbits would be computed incorrectly (this is known as *episilon*). With SA, the accuracy of single-point positioning degrades from 10 to about 100 m.

Fortunately, SA was switched off on 2 May 2000 by presidential order. However, there is always a possibility that it could be turned on again.

Satellite geometry

In Section 6.7 it was stated that, for intersection and resection, the geometry of the control stations and unknown point should be well conditioned; otherwise the accuracy of computed coordinates could be poor. This is also true for GPS and the position of the satellites at the time of observation is a crucial factor in determining the accuracy of GPS coordinates.

As shown in Figure 7.6, if the satellites are all close to each other in the same part of the sky, this will produce a poor geometry and poor accuracy in a GPS position. On the other hand, if the satellites are spread over a large area of the sky, a good geometry is obtained and the accuracy of GPS coordinates is much better. The effect of satellite geometry is assessed in GPS using *Dilution Of Precision (DOP)* factors – a low DOP factor means that the satellite geometry is good, and a high DOP means the geometry is poor. Most receivers used for high-precision surveying are able to compute DOP values from a knowledge of the satellite almanac, the receiver position and the time of observation. GPS processing software can also be used to predict what the DOP values will be for a survey area, helping the GPS survey planning process. This is discussed further in Section 7.8.

The DOP factors used in surveying are

VDOP = Vertical Dilution Of Precision

HDOP = Horizontal Dilution Of Precision

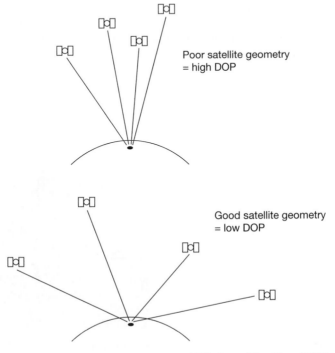

Poor satellite geometry
= high DOP

Good satellite geometry
= low DOP

Figure 7.6 ● Satellite geometry and Dilution of Position (DOP).

PDOP = Positional Dilution Of Precision

GDOP = Geometric Dilution Of Precision

All of these indicate the uncertainty in a GPS position that results from the satellite configuration and receiver position at the time of measurement. Perhaps the most useful of these is the GDOP, which indicates the degradation of the full three-dimensional position and time. Clearly, a low DOP is desirable for any survey and the best way of ensuring this is obtained is to observe as many satellites as possible, but avoiding those at low elevations.

Multipath

The antennae used with GPS receivers are designed to track satellites over wide areas and cannot be made directional. Because of this, multipath errors can occur in GPS measurements when a satellite signal does not travel directly to the receiver but is reflected off a nearby surface first (see Figure 7.7). The reflected signals can interfere with those arriving directly from other satellites and can cause errors of up to 1 m in pseudoranges. More importantly, multipath errors can also cause a receiver to lose lock on a satellite because it cannot process the reflected signal. This in turn may cause the DOP to deteriorate and may cause difficulties in resolving integer ambiguities. Because they depend on the location of a receiver in relation to its surroundings, multipath errors cannot be removed by using any differential or relative methods.

Surfaces such as tall buildings (especially those with glass sides) and chain link fences have been found to cause serious multipath errors, and one way of reducing these, especially in built-up areas, is to choose antenna locations for receivers that are as far as possible from reflective surfaces.

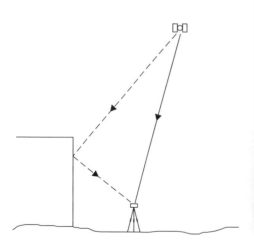

Figure 7.7 ● Multipath error.

Figure 7.8 ● Ground plane antenna (courtesy Trimble Navigation Ltd)

Figure 7.9 ● Choke ring antenna (courtesy Leica Geosystems.)

Multipath effects are worse at low elevation angles, and other ways of dealing with these are to use specially made antennae that stop reflected signals from reaching the receiver, to automatically detect them in the receiver software or to use a combination of these. An example modification to the antenna is to use a *ground plane* (Figure 7.8), which is a circular metallic disc about 500 mm in diameter that is attached underneath the antenna to prevent low-angle reflections reaching it. Another example for high-precision surveys is to use a *choke ring antenna* (Figure 7.9).

Antennae

When carrying out high-precision surveys with GPS, the antenna is usually tripod-mounted. As with total stations and theodolites, any centring and levelling of a tripod and tribrach must be done properly or errors of the same magnitude as the miscentring will be introduced into measurements.

A GPS measurement is taken to a point inside the antenna called its electrical *phase centre*. The relationship between the phase centre and the mark on the antenna which is used to measure its height above ground will be published by the manufacturer. This height difference, plus the height of the antenna above the survey point, must be known and recorded. It is vital that the make and model of the antenna being used are entered correctly into the processing software to correctly identify the position of the phase centre and that the height of the antenna above the survey point is carefully measured. Doing either of these incorrectly is one of the most common sources of error in GPS work and can cause big vertical inaccuracies.

Reflective summary

With reference to errors in GPS, remember:

— Although GPS surveys do not require any observations or readings to be taken in the same way as for other surveys, the results they produce are still subject to errors and it is important to know what these are, what their possible magnitudes are and how to control them.

— The worst sources of error in GPS are ionospheric and tropospheric delays, and these are compensated for by mathematical modelling. Although this is the subject of much research at present, by using dual-frequency measurements, ionospheric delays can be estimated quite well to give an ionosphere-free observable. Tropospheric modelling is also quite reliable for baselines up to about 20 km providing the correct model is used.

— For most GPS surveys, it is recommended that no satellites with an elevation of less than 10–15° above the horizon are used, because the signals from these experience very large ionospheric and tropospheric delays.

— Another serious source of error is a poor satellite geometry resulting in a high dilution of position. Unlike ionospheric and tropospheric delays, this is a site-dependent error and can be controlled by choosing appropriate times to take measurements.

— Multipath is another site-dependent error – this can be very difficult to detect. It is possible to perform tests to identify whether a particular site suffers from multipath, but these can be elaborate and take a long time to carry out. For these reasons, it is important to reduce multipath errors as much as possible through careful siting. If very large multipath errors are suspected, it is best to change the siting of a GPS control point or to use a total station for measurements.

— Some care is needed with the antenna in a GPS survey. It is vital to measure the height correctly and to enter the exact make and model into the receiver's processing software (sometimes called its *firmware*) so that it can determine the correct phase centre offset.

7.5   Differential and relative GPS

After studying this section you should know that the accuracy of GPS is improved considerably by using differential and relative methods. You should be aware that differential methods are used with code measurements and relative methods with phase measurements. You should have an understanding of how these methods bring about a better accuracy. Furthermore, you should appreciate that differential and relative results can be real-time or post-processed.

This section includes the following topics:

● Improving the accuracy of GPS

● Differential GPS (DGPS)

● Precise relative GPS

Improving the accuracy of GPS

In the previous section, the various sources of error that can affect GPS observations have been discussed. Combining all of these gives an accuracy for point positioning of around 10–20 m using the C/A code on the L1 frequency. For many applications requiring spatial information, the accuracy of this and carrier phase measurements can be improved by using differential and relative methods, in which two or more receivers work together. One of the receivers is located at a point whose coordinates are already known – this is called a *reference receiver* and the known point at which it is set up is called a *base station*. During a survey, the reference receiver remains at the base station and another receiver called a *rover* moves around the survey area collecting data at points whose coordinates are to be obtained.

As the reference receiver processes information from satellites, it is able to compute a position based entirely on satellite data but which includes errors due to ionospheric and tropospheric delays, clock biases, incorrect satellite orbits and other sources. These coordinates are compared with those already known and this gives a set of corrections equal to the difference between measured and known coordinates at the base station. Since the extent of most surveys is very small on a global scale and very small compared to the size of the satellite constellation, the rover will be using signals that have propagated along similar paths from the same satellites as used by the reference receiver. Consequently, it is assumed that similar errors will be obtained in the coordinates computed by the rover and reference receivers. Since the reference receiver can compute corrections for these, the positions computed by the rover can also be corrected. To do this, the corrections at the reference receiver have to be combined with the data collected by the rover and this is done either in *real time* with a radio or telecommunications link between the two or later by storing the observations and obtaining the results by *post-processing* using a computer. In summary, an improvement in the accuracy of measurements is made possible in differential methods by correlating the errors in two or more receivers assuming they are all observing to the same satellites.

In *real-time differential and relative surveying*, the data collected by the reference receiver must be continuously transmitted to the rover so that it can compute corrections. For some surveys, more than one rover may be used, and these may receive data via a radio link from a transmitter at the base station. There are many different types of radio link that can be used to broadcast corrections and the performance of these depends on a number of factors, including the frequency and power of the transmitter together with the type of antenna used and its position. Instead of a radio link, mobile phone technology can be used to transfer data from the reference receiver to each rover in a survey. When the rover has received data from the base station, it instantly computes and displays the corrected three-dimensional coordinates of the survey point.

In *post-processed differential and relative surveying*, all of the data collected by the reference receiver and rover are downloaded into a computer when the fieldwork has been completed. Using post-processing software, both sets of data are combined and the coordinates of each survey point are computed. The accuracy of post-processed results can be slightly better than that for real-time measurements, but for most applications of GPS in civil engineering and construction, the advantage of having real-time results automatically computed by a rover far outweighs any differences there are in the accuracy of the results that can be obtained.

Differential GPS (DGPS)

Differential GPS (or DGPS) is the name given to differential methods that are usually applied to code measurements only. Compared to conventional point positioning, DGPS has an accuracy of about 0.5–5 m, depending on the receiver and antenna used and how corrections are transmitted to each rover.

Although users could establish their own base stations, DGPS real-time correction services are available that make it possible for a user to carry out GPS surveys without having to do this and to use one receiver only – these are known as *Ground Based Augmentation Systems* (*GBAS*). Some GBAS use powerful radio transmitters located at permanent base stations that provide a network of beacons transmitting GPS differential correction data over a large geographical area, as shown in Figure 7.10. For example, in the UK, the *General Lighthouse Authority* (*GLA*) provides a public marine DGPS service based on 14 ground-based reference stations (for further details visit `http://www.trinityhouse.co.uk/`). These are linked to radio transmitters located at lighthouses situated at points along the coastline, plus two augmenting transmitters inland, each broadcasting over long ranges. To use this service, a rover must be fitted with a radio modem and antenna to enable it to receive the GLA correction message. The GLA services are provided free of charge, but some GBAS are operated by privately owned companies and are available by subscription only. Whatever service is used, there is a common format or standard in use for broadcast of electronically transmitted data called the *Radio Technical Commission for Maritime Services, Study Committee-104* (*RTCM SC-104*) and all receivers that want to use DGPS correction services generally receive data in this format (which is known as the *RTCM format*). However, DGPS corrections are also starting to be broadcast over the Internet.

In addition to ground-based beacon transmitters and Internet technology, there are some satellite correction services available. Known as *Satellite-Based Augmentation Systems* (*SBAS*), these use a network of ground tracking stations that collect GPS data and compute corrections – the corrections are then transmitted to SBAS satellites, which in turn

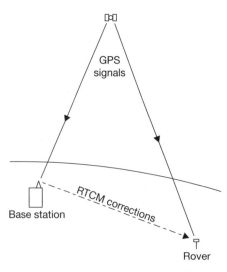

Figure 7.10 ● Differential GPS (DGPS).

broadcast corrections to any user with a receiver adapted for this. Three such systems are the European Space Agency funded *European Geostationary Navigational Overlay Service (EGNOS)* satellite system, the *Wide Area Augmentation System (WAAS)* in the USA and the *Multifunctional Transport Satellite Augmentation System (MSAS)* in Japan. These are all intended for reliable and accurate positioning of aircraft, but with the side benefit that anyone with a receiver can access the data transmitted without charge. Strictly speaking, these augmentation systems are not differential and a user computes a corrected absolute position directly from incoming GPS and SBAS signals rather than fixing position relative to a base station. There are also a number of commercially available SBAS systems.

EGNOS consists of three geostationary satellites and a network of 30 monitoring stations, four master control centres and six uplink stations. The monitoring stations collect data from GPS satellites, which is passed to the master control centres where real-time correction data for satellite orbital and clock errors and for ionospheric effects are computed. Using a telecommunications link, these are sent to the EGNOS satellites by the uplink stations. EGNOS aims to provide an accuracy of 5 m for satellite position fixing across Europe for users with an EGNOS-compatible receiver.

The WAAS is based on a network of 25 ground reference stations in the USA. Each of these receives signals from GPS satellites which are relayed to a wide area master station where a correction message is computed. This is then uplinked to two geostationary satellites, one of which is above the Pacific coast of the USA and the other above the Atlantic coast. The accuracy quoted for a WAAS receiver is again of the order of 5 m for point positioning.

MSAS also uses two geostationary satellites to provide a similar accuracy over Japan.

All SBAS can be combined with *Local Area Augmentation Systems (LAAS)*. These are similar to GBAS but are smaller and provide high-quality positional information over small areas such as airports, seaports or at other locations.

One of the major applications of real-time GBAS and SBAS has been for marine navigation both for commercial and recreational purposes. They are also used for aircraft navigation and land-based vehicle tracking, but one of the growth areas for these in surveying is in the collection of data for geographic information systems. GPS receivers made for GIS data collection are sometimes called *mapping grade receivers*. An advantage when working with real-time GIS data collection systems is that the results obtained can be verified on site, and it is possible to instantly check and plot the features and attributes surveyed on a pre-loaded map or GIS layer. This equipment is described further in Section 7.7.

Precise relative GPS

With an accuracy at the metre level, Differential GPS is still not good enough for work on site. In order to improve this accuracy, another type of differential surveying, known as *precise relative GPS,* is applied to carrier phase measurements. As with DGPS, two receivers are required for carrying out precise relative GPS with the reference receiver being located at a base station that has known coordinates. By combining the data collected at the base station with that from a roving receiver, it is possible to obtain the coordinates of the unknown points much more accurately. The methods used for obtaining the coordinates of the rovers are outlined in the following paragraphs.

The processes used in precise relative GPS enable the vector between the reference and rover receivers to be determined. This is called a *baseline* and the *baseline vector components* ΔX, ΔY and ΔZ obtained by GPS along a baseline are used to compute the unknown coordinates of the rover X_P, Y_P and Z_P using

$$X_P = X_{REF} + \Delta X \qquad Y_P = Y_{REF} + \Delta Y \qquad Z_P = Z_{REF} + \Delta Z$$

where X_{REF}, Y_{REF} and Z_{REF} are the known coordinates of the reference receiver.

For precise relative GPS, *differencing techniques* are used to remove most errors in the baseline vector components. These rely on the reference receiver and rovers collecting data at the same time and the same rate (or *epoch*) to be able to apply these.

Differencing involves forming several different types of linear combinations of satellite measurements.

A *single difference* (Figure 7.11(a)) is formed when the measurements between one satellite and two receivers are subtracted. *Since both receivers are observing the same satellite at the same time*, single differencing completely removes the satellite clock error. Also, if both receivers are within 20–30 km of each other, they will record nearly identical atmospheric and orbital errors so that these also reduced. However, single differencing increases the noise factor and measurement errors from this source by a factor of $\sqrt{2}$.

A *double difference* (Figure 7.11(b)) is formed when the measurements between two satellites and two receivers (all taken at the same time) are subtracted. This removes all the errors that single differencing does and has the added advantage that the receiver clock errors are also completely removed. This is the process used by most GPS software, as it provides a baseline solution that is free of timing errors and one in which the effects of ionospheric and tropospheric delays are reduced. Because it is two single differences, double differencing increases the noise factor and measurement errors from this by a factor of 2.

A *triple difference* (Figure 7.11(c)) is formed by differencing two double differences taken at two different times t_1 and t_2 – note that the same satellites and receivers must form these differences.

To process a baseline, a GPS receiver with the appropriate software on board or post-processing software is required. This can begin with a point position solution by measuring pseudoranges which define a search area at the receiver, which can be thought of as a three-dimensional volume of uncertainty in which the receiver position lies. Following this, carrier phase measurements take over and the receiver starts forming difference equations. A double difference solution gives the best possible accuracy, but requires that the integer ambiguity is resolved. Recall that this is N in the carrier phase measurement $N\lambda + \Delta\lambda$ between a satellite and receiver and is the integer number of complete carrier wavelengths between the receiver and

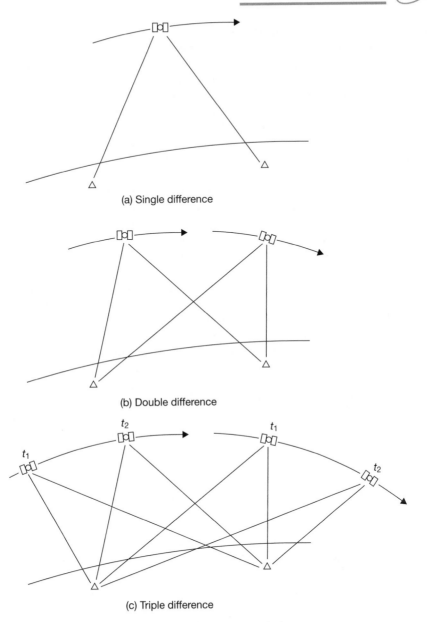

(a) Single difference

(b) Double difference

(c) Triple difference

Figure 7.11 ⬤ Differencing techniques.

satellite at the instant the receiver locks on to the satellite. This is a process of estimation and the receiver starts a search routine by computing the most likely ambiguity resolutions to all the satellites it is locked on to. This is called a *float solution* and gives an accuracy for a baseline that does not normally meet specifications. Using a second set of observations, the process is repeated and a second set of ambiguities are obtained. The two sets of ambiguities are compared and if they are not the same, the receiver keeps computing further sets until both are accepted, at which point the ambiguities are considered to be resolved. At this point, the baseline components are recomputed holding these ambiguities fixed to give what is known as the *fixed solution*, which produces baseline components of much better precision. The procedure is continually repeated and the ambiguities are constantly checked to see if they are acceptable. If, for some reason, they are not, the receiver changes to a code solution and the whole ambiguity search routine starts all over again.

In some receivers, triple differences are used to help resolve ambiguities. When these differences are formed, the integer ambiguity N is temporarily eliminated. However, the triple difference solution is not as accurate as double differencing and they are only used to obtain a first estimate of the position of a receiver and as a starting point for ambiguity resolution by the more accurate double difference solution. This method can produce a better estimate of position than a pseudorange measurement.

For any given baseline solution, the initial value of N resolved to a particular satellite does not change from the time the receiver has locked onto it, and as the satellite moves during a measurement, the receiver counts the number of whole wavelengths, or cycles, that change. Consequently, carrier phase measurements rely on the ambiguities being resolved to determine the cycle count N at lock-on together with an observed cycle count. A *cycle slip* occurs when the receiver loses lock on a satellite during a measurement and it counts the wrong number of cycles that have occurred since N was determined. These can occur when the line of sight to a satellite is obstructed or poor quality signals are received for some reason (these are signals with a low signal to noise ratio). Poor reception is commonly caused by adverse atmospheric conditions, by multipath errors and by using low elevation satellites. Because they can remove the integer ambiguities, triple differences are also used to detect cycle slips.

With a resolved ambiguity solution double differences are used to compute least squares values for the baseline vector components. A least squares solution (see Section 9.5) is necessary because even in a relatively short observation period, a considerable amount of positional information is generated. Consequently, all GPS surveying relies entirely on sophisticated software to deal with all the data collected and its processing.

Although the processes described above apply to how a receiver operates in real time, precise relative GPS can also be carried out by post-processing the results obtained. In this case, the post-processing software will follow similar procedures to resolve ambiguities and compute coordinates. By using precise relative GPS, the positional accuracy of carrier phase measurements is improved to 10–50 mm.

Reflective summary

With reference to differential and relative GPS, remember:

— The accuracy of GPS measurements is improved a significant amount by using differential and relative methods. These require a reference receiver to be located at a known point (called a base station) and a rover to be located at the point whose coordinates are to be determined. By combining the measurements from both receivers, it is possible to reduce and even eliminate some of the GPS errors.

— The results from differential and relative GPS can be real-time or post-processed. Real time requires that a radio or some other communication link is established between the receiver at the base station and each rover. Post-processing requires all the measurements to be stored on site and the position of unknown points to be computed at a later stage.

— Differential GPS (or DGPS) is a term applied to differential code measurements that can be taken with a single receiver only. This is made possible by using a real-time correction service known as an augmentation system.

— Augmentation systems can be ground-based, such as the one provided by the GLA in the UK, or satellite based, such as the EGNOS in Europe. With the proper receiver and associated software, accuracies at the metre level are possible with these.

— For precise positioning at the centimetre level with GPS, precise relative methods have to be applied to carrier phase measurements.

— Differencing techniques are used to remove errors from carrier phase measurements – triple differences are formed to help resolve integer ambiguities and double differences are formed to compute position.

7.6   Surveying with GPS

After studying this section you should be familiar with the methods that can be used with GPS to take survey measurements.

This section includes the following topics:

● Static
● Rapid static
● Kinematic surveys

Static

Although there are several methods that can be used for surveying on site with GPS, the DGPS and augmentation systems referred to in previous sections are only used in engineering surveys for small-scale mapping, in data collection for GIS and for navigation, all of which generally require a low accuracy at around the metre level. For high-precision surveying at the centimetre level, GPS methods all rely on two or more carrier-phase receivers collecting satellite data at different points.

Static surveying was the first high-precision method developed for GPS and is the standard GPS method for determining the length of baselines that are longer than 20 km.

In this, the reference receiver is located at a known control point and a rover is set up at a point whose coordinates are to be determined. Both receivers are generally tripod-mounted (see Figure 7.12), but could be on a pillar; they collect data simultaneously during a survey which must be to at least four common satellites and which can be up to several hours duration. The *observation* (or *occupation*) *time* depends on the accuracy required and this varies according to the type of receiver used and the number of satellites observed, together with their geometry and length of the baseline. Generally, as the distance between the reference and rover increases, the observation times must also increase to maintain precision. As a guide, with five satellites and a good DOP, an observation time of 30 minutes for a 50 km baseline will give a coordinate precision of around 100 mm horizontally and around twice this for heights. After 1 hour, the precision will be 50 mm and four hours may be required for a precision of 20 mm. The reason why a relatively long observation period is needed is to allow the satellite geometry to change sufficiently so that enough data is available to resolve integer ambiguities and to allow systematic errors to be removed. Single- and dual-frequency receivers can be used for static surveys, but the occupation times are generally much longer for single-frequency receivers. Although it is possible for the receivers to collect data continuously, this would cause problems with the huge amount of data that would be collected over the long observation times required. Consequently, all GPS receivers are set to record data at intervals known as *epoch rates*. For static surveys, data is usually collected with an epoch rate of 10 or 15 seconds.

All the data collected is post-processed to give the baseline vector components between the reference receiver and rover. Baselines are processed by using triple differences first to remove any cycle slips followed by double differences on the carrier phase measurements. Statistical methods are used to solve the integer ambiguities. For best results, as many errors as possible should be removed from the

Figure 7.12 ● Tripod mounted static GPS (courtesy Sokkia Ltd).

observations. To achieve this, a dual-frequency receiver should be used so that the effects of ionospheric refraction can be minimised. In addition, precise orbital satellite data should be used – this is more accurate than the broadcast data and is available through the Internet from a variety of providers. It is obtained from the tracking stations which monitor the GPS satellites, but the satellite orbits are computed after they have passed by. Because of this, precise orbital data is only available for post-processing after the fieldwork for a survey has been completed. Table 8.1 gives details of the different GPS satellite orbital products that are currently available. For further details of the methods used to process static observations, see Chapter 8.

Under ideal conditions, static GPS has an accuracy of about 10 mm + 1 ppm of the baseline length and 20 mm + 1 ppm vertically.

Rapid static

GPS is now the preferred method for control surveys on large construction sites. For these surveys, the static method described above can be used, but the rapid static method is preferred provided the baselines are not as long. This is a technique similar to static positioning but has shorter occupation times of 10–30 minutes.

As before, a reference receiver is located at a known point and the rover (or rovers) occupy the unknown points. Because observation sessions are short, good communication between the operators at the reference and rovers is required. The method relies on a faster ambiguity resolution approach, and to achieve this dual-frequency receivers must be used together with special post-processing software. In addition, a good satellite geometry (low DOP) is essential as well as observed data that is free of cycle slips, multipath and interference. Another restriction is that *baselines should not be longer than 10–20 km*, and the shorter these are, the better. All of these conditions are needed because the software has to compute each baseline observed with less information than for the static method. The epoch rate is typically 5–10 seconds.

At the longer observation times, the accuracy of rapid static surveying is of the order of 10–20 mm + 1 ppm horizontally and 20–30 mm + 1 ppm vertically. This is slightly less accurate than the static method because it is more sensitive to environmental conditions and there are limitations in the processing software which cannot model atmospheric effects as well as for static observations.

Kinematic surveys

For site work, *kinematic GPS* is used when a lot of points are to be surveyed in a relatively small area and where the accuracy requirement is not as high as for static surveys – this includes detail surveying (mapping) and construction measurements. As for all high-precision surveying with GPS, kinematic methods require a reference receiver to be located at a known position, but in this case, instead of remaining stationary, the rover is moved around the site recording position at discrete points or whilst the rover is continuously moving. This is why this method is called kinematic (or on the move) GPS surveying.

At the start of a kinematic survey, the rover has to perform an *initialisation* in order to resolve ambiguities. To do this, a method known as *kinematic on-the-fly* has been developed in which the initialisation is performed while the rover is moving. So, at the start of a survey, the operator can take measurements as soon as the equipment is turned on. The integer ambiguity is then resolved using a special processing technique, after which a back solution is carried out to compute the position of the points already surveyed up to the time the initialisation was successful. When initialising an on-the-fly survey, a minimum of five satellites must be tracked, and it is especially important that the signals obtained are free of cycle slips and multipath. When performing an initialisation, it is best to move to an area where there is a clear view of the sky. If, at any time, loss of lock occurs, the system will re-initialise automatically when the satellite coverage resumes. When initialising in this way, it must be realised that the longer the baseline, the longer it takes to fix the initial integer ambiguities.

During a *post-processed kinematic survey*, all the data collected by the reference and roving receivers is stored in a handheld computer, a controller or on board the receiver and then transferred to a host computer after fieldwork has been completed. Consequently, the results for the coordinates of all the surveyed points are not immediately available. A major disadvantage with post-processed surveys is that setting out cannot be carried out because positional data for this is required in real time.

The problems with post-processed results not being available on site are overcome by using a *real-time kinematic* (or simply *RTK*) GPS surveying system. As with all GPS surveying methods, RTK surveying requires two receivers to operate at the same time and the reference receiver is again located at a known point. However, the difference between post-processed kinematic and RTK systems is that an RTK reference receiver has a communications device attached to it (this is typically a radio or mobile phone) and it transmits all the satellite data it collects to the rovers working on site, as shown schematically in Figure 7.13. As with DGPS, the reference receiver

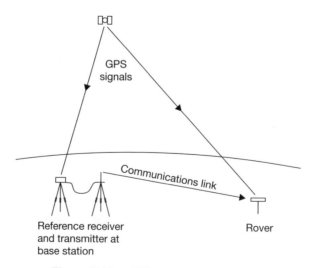

Figure 7.13 ● RTK surveying with GPS.

Figure 7.14 ● RTK base station showing reference receiver (left) with UHF transmitter (courtesy Thales).

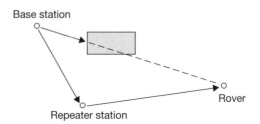

Base station
Repeater station
Rover

Figure 7.15 ● Repeater station.

and its transmitter is called a base station, a typical configuration of which is shown in Figure 7.14. Each RTK rover also has a communication link and receives all the data from the base station as well as collecting positional information from satellites. To compute position, both of these sets of data are processed instantaneously by a rover and as soon as initialisation is complete and ambiguities have been resolved, the rover can display and record coordinates and heights in real time. To check that the system has initialised correctly, it is recommended that a point with known coordinates is visited as soon as this is complete so that a comparison can be made.

To be able to work in real time, the software that would be used for post-processing has to be installed in the rover, and because it is also fitted with a communications link, RTK equipment is more expensive than kinematic post-processed systems. However, with RTK systems, it is immediately apparent if data is acceptable when setting out and for other surveys because it is possible to check work before leaving the site. Because of all the advantages of working in real time, RTK is now used for most survey work on site.

The key component of RTK surveying is the communications link between the base station and rover – many systems use UHF or VHF radio modems which are supplied as part of the equipment. The range over which RTK can be carried out using a radio link depends on the power of the transmitter at the base station. In the UK, the output power of private radio transmissions is limited to 0.5 W and this restricts the range of RTK surveys to within about a 5 km radius of the base station. A radio link should generally be line-of-sight between the base station and each rover, but on many sites this is not possible because of natural obstructions such as hills, trees and vegetation or obstructions caused by buildings and other structures. To overcome this problem, base stations should be located on high ground or the transmitter can be mounted on a pole, mast or tall building – a higher base station radio antenna (not the GPS antenna) will reduce line-of-sight problems and may increase the range of radio communication. Another means of improving the range is to use *repeater stations* – these receive a signal from the base station and then re-transmit it as shown in Figure 7.15. A further problem that can be encountered with radio communication is interference, and care must be taken to ensure that no other radio conflicts are caused by other users of the same frequencies in the survey area, especially from other RTK users. The latest technology for linking a base station to a rover is to use mobile telephones. These do not suffer the disadvantages of a radio link and can be used at ranges of up to 25 km from the base station, but are more expensive. The most common types of mobile phone technology used are GSM (*Group Spéciale Mobile*), in which the user pays for the time that the phone is connected, and GPRS (*General Packet Radio Services*), in which the user pays for the amount of data transmitted during a survey.

Kinematic GPS can be carried out by walking with the antenna mounted on a back-pack (see Figure 7.16) or by mounting it on a vehicle and with data collected at

Back-packed antenna

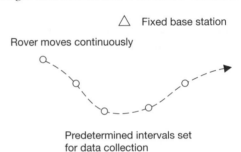

Figure 7.16 ● Back-packed GPS for kinematic surveying (courtesy Trimble Navigation Ltd).

Pole-mounted GPS

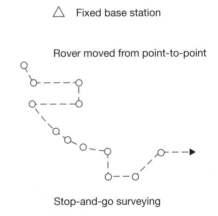

Figure 7.17 ● Pole-mounted GPS for stop-and-go surveying (courtesy Trimble Navigation Ltd).

predetermined intervals – these can as short as 0.05 s. This is a very useful technique in construction surveys for measuring longitudinal profiles and cross-sections and for obtaining data for forming digital terrain models (DTMs). The rover can also stop at distinct points and collect data for a few epochs after which a feature code can be entered into the receiver or its data collector to identify the point. To do this, the equipment is usually pole-mounted (see Figure 7.17), and this is the way in which detail surveys (mapping) are carried out with GPS. A similar method is also used for setting out and the method is sometimes called *semi-kinematic* or *stop-and-go* surveying. A typical epoch rate for detail surveying is 1 or 2 s. The accuracy of semi-kinematic surveying is typically 10–20 mm + 1 ppm horizontally and 20–30 mm + 1 ppm vertically, but this can degrade in unfavourable observing conditions.

In a kinematic survey, the reference and rover must maintain lock on the same four satellites during the observation session. If lock is maintained, the rover need only collect data over a very short time at each surveyed point, because the integer ambiguity is known from the initialisation and the rover keeps a cycle count between it and each satellite as they both move. Each new calculation of position is then simply due to changes in satellite and rover positions which only require a relatively small amount of data to compute. If lock is lost at any time and fewer than four satellites are tracked, re-initialisation of each rover is required before the survey can continue. To avoid this occurring, care must be taken in a kinematic survey not to move too close to any object that might prevent signals reaching the antenna.

In a similar manner to augmentation systems used with code-based GPS receivers, it is possible to carry out RTK surveys with a single receiver and maintain accuracy at the centimetre level. *Network RTK* (or a *Virtual Reference System*) operates using a set of reference stations that have been established at about 60–70 km intervals across a region. These continuously transmit GPS observations to a central server where a processor models and computes ionospheric, tropospheric and satellite orbital errors for the network. When a rover arrives on site at any point within the network, it sends its approximate position to the central server via a mobile phone. The server then establishes a virtual reference station in the vicinity of the rover and interpolates a set of GPS corrections for that location – these are then sent to the rover again using a cellular modem. As a result, the corrections computed by the server are transmitted to the rover as if they were coming from a reference station on site. At present, the Ordnance Survey is establishing a GPS infrastructure to provide a network RTK system in Great Britain. The network is being developed for internal usage only, but a public service is being investigated. It is also expected that an RTCM standard for communications will be released for networked RTK services.

Reflective summary

With reference to surveying with GPS, remember:

— There are three techniques available for taking measurements with GPS for site work – static, rapid static and real-time kinematic (RTK).

— *Static surveying* is used to fix the positions of control points in primary networks where the points are a long way apart and where a high accuracy is required.

— *Rapid static surveys* are similar to static surveys and are used to fix the positions of control points in secondary and other networks. The accuracy is not quite as good as that for static surveys but the method is quicker and more suited to shorter baselines. It is becoming the standard method for establishing survey control on medium to large sized construction sites.

— All the results obtained from static and rapid static surveys have to be post-processed. This involves storing all the observations taken while on site and then transferring these to a computer which is used to calculate the coordinates of all the points surveyed.

— *RTK surveying* is the preferred method for all survey work with GPS on site, mainly because all the results obtained with it are real-time – no post-processing is required. This means that measurements taken for a detail or other dimensional survey can be checked, verified and edited on site and, of course, means that it can be used for setting out.

7.7 GPS instrumentation

After studying this section you should be aware of the different types of GPS receiver that are available and their accuracies. You should also realise that for nearly all work on site, high-quality geodetic dual-frequency receivers are required.

This section includes the following topics:

● Code-based receivers
● Mapping grade receivers
● High-precision geodetic dual-frequency receivers

Code-based receivers

The purpose of this section is to offer some general information on what types of GPS receiver and system are available for engineering surveying. There are three types of receiver that can be used – code-based receivers, mapping grade receivers and geodetic receivers.

A *code-based* (or *navigation grade receiver*) is usually a handheld device familiar to walkers and other recreational users of GPS (Figure 7.1). These use code ranging techniques to determine position, and most of these will have the capability of producing real-time coordinates on a national mapping system with an accuracy of about 10 m. However, when they are used with an augmentation system, the accuracy of navigation grade receivers could be 2–3m. This type of receiver is seldom used in engineering surveying.

Mapping grade receivers

Mapping grade receivers use a combination of code ranging and DGPS to determine position, normally by using kinematic methods. The Geo Explorer from Trimble and the Leica GS20 PDM (Professional Data Mapper), shown in Figure 7.18, are specifically aimed at GIS data capture and asset location. As already stated, these are mostly code-based receivers with modifications and have accuracies ranging from sub-metre to about 5 m. At the metre level, the receiver usually operates in conjunction with an augmentation system such as EGNOS or by using RTCM broadcast differential corrections. These will enable real-time data collection to be carried out but when using RTCM corrections, an RTCM receiver, power supply and communication hub are needed to receive the correction. For the Leica GS20, all of these are fixed to a special belt which is worn by the operator (see Figure 7.19). To improve reception of the GPS signal, a separate antenna can be used which is mounted on a backpack or even on a moving vehicle. To achieve sub-metre accuracies, the results have to be post-processed or data collected using code measurements together with L1 phase measurements. Apart from GIS applications, these receivers can be used for mapping at medium and small scales and for taking preliminary measurements for determining quantities on site.

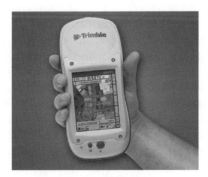 

Trimble GeoExplorer　　　　　　　　　Leica GS20 PDM

Figure 7.18 ● Mapping grade GPS receivers (courtesy Trimble Navigation Ltd and Leica Geosystems).

Figure 7.19 ● Belt-mounted RTCM correction system (courtesy Leica Geosystems).

High-precision geodetic dual-frequency receivers

The category of GPS receiver that is required for nearly all work in surveying on construction sites is a *high-precision geodetic dual-frequency receiver* (sometimes called a *survey grade receiver*). These use precise relative methods for determining position and are capable of performing static and RTK surveys with accuracies at the centimetre level.

Although the technical specifications for GPS surveying systems can appear to be very complicated, the key parameters to note for any geodetic receiver are the number of channels available, which observables are measured and what survey methods it can perform. Normally, geodetic receivers will have at least 12 channels, which means that they can track 12 satellites at the same time – some receivers enable the user to receive signals from the Russian GLONASS system (see Section 7.9). To achieve the accuracies needed for surveying, it is necessary for geodetic receivers to take a combination of C/A and P code plus carrier phase measurements – these vary from receiver to receiver, where each combination has its advantages and disadvantages (these have been discussed in Section 7.3). Dual-frequency receivers measuring on both L1 and L2 are always preferred to single frequency as this generally improves accuracy and enables a quicker and more reliable ambiguity resolution to be computed. All will be capable of performing static and fast static surveys for control work where results are post-processed using software provided by each manufacturer for use with their receivers. For on-site applications, a receiver and system with full RTK capability is required so that setting out and machine control can be carried out – this may also be needed for measuring areas, sections, volumes and for acquiring three-dimensional data for digital terrain models. Most geodetic receivers can also be used in DGPS mode and are compatible with an augmentation system – the accuracy in this mode will be suitable for mapping purposes and earth-works only and may require post-processing, but the advantage of operating in this mode is that only one receiver is required.

The main components of a geodetic receiver are the antenna which is connected to the receiver, together with a radio modem or mobile phone if in RTK mode, power supply and a controller. The antenna can be tripod-mounted for static surveys, but for RTK surveys the complete system is usually mounted on a detail pole. In this configuration, the antenna, receiver, modem and batteries can all be placed in a single housing at the top of the pole with the controller mounted at waist height on the pole – this is a type of handheld computer or data collector that is used to operate the system and to store and process observations (for a full description of these refer to Section 5.4). Alternatively, the receiver, modem and batteries are placed in a backpack and only the antenna is on a detail pole. Data can also be stored on PCMCIA or flash cards and some equipment uses camcorder-type lithium-ion batteries because these are very light. This is especially important because all the equipment is carried on the detail pole or in a backpack – these batteries have operating times of up to 15 hours for RTK work.

Figure 7.20 shows a selection of geodetic receivers for use on site. As well as the equipment shown here, a base station is also required (see Figure 7.14) consisting of a full tripod-mounted set up with radio transmitter or phone. A computer is also needed if results are to be post-processed.

A feature that is increasingly being incorporated into geodetic receivers is *Bluetooth wireless technology*. This enables portable equipment such as computers, mobile phones and other electronic devices to communicate with each other without using cables. When they are within range of each other (about 10 m), Bluetooth devices will detect each other and establish contact – if they are programmed to communicate a link is then established between them for data transfer. In order for Bluetooth devices to work alongside each other they must be capable of identifying and connecting only with other intended users, and must ignore others. To do this, a user can define a selection of other equipment as trusted devices – when any equipment fitted with Bluetooth is switched on it scans for other trusted devices within range and lists any of these it finds. If found, the user can then select one from the list and connect to it. A form of radio communication known as *spread spectrum* is used in Bluetooth technology. This enables any two devices, when they are connected, to change their frequency of communication across a wide part of the radio spectrum and at a fast rate. Because the sequence of frequencies used and the rate at which they change is known only to the two connected devices, other Bluetooth devices can even be within range of these and not interfere. Spread spectrum technology also makes Bluetooth communication very resistant to interference from other nearby electronic equipment using the same frequencies. In the UK, it is illegal for private users to use spread spectrum technology for broadcasting over wide areas, so this technology can only be used by low-power devices that are close to each other and *not* for an RTK link between a reference receiver and rover. With Bluetooth technology installed in GPS equipment, there is no need to have a cable connecting the receiver to the controller and when the controller is taken to a total station or computer, data can also be transferred to these without cables. For further details on Bluetooth visit http://www.bluetooth.com/ and for general information on specifications for infrared wireless communication, refer to the Infrared Data Association (IrDA) at http://www.irda.org/.

Bluetooth is a good example of one of the many developments taking place with GPS, and new hardware and software is being introduced all the time, for all types of

Leica GPS 1200

Sokkia Radian IS

Thales Z-MAX

Topcon HiPer

Figure 7.20 ● Geodetic receivers (courtesy Leica Geosystems, Sokkia Ltd, Thales and Topcon)

receiver. In addition, the range of equipment currently available for surveying with GPS is extensive. For information, all the manufacturers and suppliers advertise regularly and produce many brochures which are updated frequently. As well as this, because GPS is the subject of much research and development, many of the journals associated with surveying, engineering and construction publish a substantial number of scientific and other papers concerned with GPS on a regular basis. To keep up to date with this rapidly changing technology, refer to the information given at the end of the chapter.

Reflective summary

With reference to GPS instrumentation, remember:

— There are various categories of GPS receiver available. For site surveying, *mapping grade receivers* with accuracies at the sub-metre level can be used for detail surveying and data collection for earthworks, and with accuracies at about the 1–5 m level for data collection in GIS. For all other work on site, RTK *geodetic receivers* are required, which are capable of accuracies at the centimetre level.

— Geodetic equipment consists of an antenna, receiver, communication device for real-time work, controller and power supply (batteries). All of this can be tripod-mounted for static surveys, but it is mounted on a detail pole for kinematic surveys – this makes the equipment very portable and suitable for activities such as setting out.

— The specification for a geodetic receiver required for site use will vary according to the work it has to be used for, but the better (and more expensive) equipment has at least 12 channels, is dual-frequency, may have advanced data communications (such as Bluetooth) and may have a controller with a large graphic screen for editing and displaying data (this might be Windows CE-based).

— Each manufacturer of GPS equipment publishes a series of brochures with technical specifications to describe their products and these are updated frequently. For the latest information, it is best to access the companies' web sites. These are given in Further reading at the end of this chapter.

7.8 GPS in engineering surveying

After studying this section you should have some understanding of the planning and fieldwork required when using GPS for site work.

This section includes the following topics:

● First considerations
● Planning and observing control surveys
● RTK surveying with GPS

First considerations

Throughout the early years of GPS, it was only used in surveying for establishing the positions of points in control networks covering large areas and even continents, and often only for research purposes. Today, the situation has changed completely and satellite positioning systems are now used extensively in surveying and on site for an increasing number of applications. Some of these are listed in Table 7.1, where the different methods of surveying with GPS in a construction and engineering environment are identified. For all of these, the basic principles of surveying remain the same as for those used with total stations and other equipment – a control network is put in place first from which other measurements are taken.

When it is used on site, GPS equipment is often thought of as some sort of 'black box' that just needs to be switched on and it will immediately start producing coordinates, heights and other useful information at the press of a button or two.

Table 7.1 ● Applications of GPS in engineering surveying.

GPS technique	Applications
Static	Locating reference points for primary control networks, for deformation surveys and for research and scientific studies (e.g. sea level and tidal monitoring).
Rapid static	Used extensively for densifying control networks (eg adding secondary control to primary control to define a site grid). Rapid static also provides the coordinates of reference points and base stations in kinematic methods.
Real-time kinematic (RTK)	Used for data collection in topographical (detail) surveys (stop-and-go method) and for measurement surveys (continuous method where data is collected at pre-determined rates).
	Required GPS method for setting out and machine control on construction sites.

Unfortunately, this is not the case and careful thought has to be given to a number of questions when first using GPS on site. These will include whether GPS should be used at all, what equipment is suitable, what coordinate system is to be adopted and the transformations that may be needed to realise this, who will operate the equipment, whether there will be enough satellites available and so on. These and other issues are discussed in this section.

The most important question to be asked by anyone proposing to use GPS is whether it is the most appropriate method for their site. It may be useful to contact some sites that are currently using GPS to ask how it compares to total stations and other well-established methods of surveying. It is worth noting that GPS is not as accurate as total stations, theodolites and levels for measurements that are taken over relatively short distances and when features such as structural grids and lines are set out. This, plus the fact that GPS equipment is more expensive than conventional surveying equipment, may put GPS at a disadvantage for work requiring a high accuracy of less then 10 mm on construction sites. However, compared to total stations, GPS has the advantages that it only requires a single operator to use it (although some total stations are single-operator – see Section 5.3 on motorised total stations), it is much quicker if a high density of points is to be set out and it is much more cost-effective when collecting large amounts of data for detail surveys and measurement purposes, especially in sectioning and three-dimensional modelling.

Having decided to use GPS, it is recommended that each of the following are considered when choosing the equipment that will be used.

- For all engineering and construction work, carrier phase *geodetic receivers* are essential, as these will provide the best possible accuracy for measuring and setting out. These are the most expensive type of GPS receiver and some of their specifications have already been discussed in the previous section.

- *Precise relative positioning* must be used, again to provide the best possible accuracy. This requires base stations to be established, possibly with repeaters on those sites where radio reception is difficult. Access to a mobile phone network may also be required.

- When results are post-processed, a suitable *computer* must be available that can run suitable software and that can also store the large amounts of data generated by GPS surveys. Regarding data management, it is advantageous if the *data collectors/controllers* supplied with the system are compatible with other survey equipment so that the data can be integrated with that from total stations and other instruments (examples of these are given in Section 5.4 that deals with electronic data collection and storage).

- For some sites where the use of GPS is only occasional, it may be more cost-effective to *hire the equipment* required rather than purchase it or even *contract out* the GPS surveying altogether to a specialist company. This arrangement may suit those sites where total stations are already available, as a specialist company can then be employed to use GPS methods to establish a primary or secondary control network for the site, from which total stations can be used for the rest of the work.

- The issue of *training* with GPS is very important. If staff who are expected to use the equipment have little or no experience of GPS, training will have to be provided. If any results are to be post-processed, tuition in the use of the software for this will also have to be provided. Usually, training is offered by the manufacturers and suppliers of GPS equipment, and since learning how to use GPS hardware and software can be quite daunting at first, it is sensible to ask them how many training sessions they provide as part of the equipment purchase.

- As well as training, it is advisable to find out what *technical support* is offered by a manufacturer to assist with any problems that may be encountered after making a purchase. One way of obtaining an independent view of this is to ask, as suggested earlier, other users of GPS at other sites how their equipment has performed, what problems they may have encountered with it and what level of technical support and backup they have received.

- Finally in this section, it is obvious that GPS technology is developing at a fast pace – will the equipment being purchased become *out of date* quickly or will it be capable of using the extra signals proposed for GPS and the satellite coverage planned for GLONASS and Galileo? These are satellite systems that are currently being developed – see Section 7.9 for more details. Can the software installed in the controllers be modified as better ways of resolving ambiguities and modelling errors become available and can the post-processing software be updated as well?

Planning and observing control surveys

Having made the choice of equipment and supplier, GPS can be implemented on site. In common with all the other methods used in surveying, good planning is essential for GPS fieldwork. The key issues to be resolved when implementing GPS on site for control surveys are discussed in the following sections.

Control network and coordinate system

When a GPS receiver and controller records data and computes position from orbiting satellites, this can be done using a coordinate system known as the *World Geodetic System 1984* (or *WGS84*). For survey work in the UK, another coordinate system known as the *European Terrestrial Reference System 1989* (or *ETRS89*), which is a fixed definition of WGS84, should be used. Both of these are global three-dimensional Earth-centred coordinate systems (even though ETRS89 is only valid across Europe) and are very different compared to the two-dimensional plane grids used for traversing, resection and intersection described in Chapter 6. The reason it is necessary to use such coordinate systems in GPS is discussed in Chapter 8 – this also includes definitions of WGS84 and ETRS89. Although the coordinates obtained on these global and regional coordinate systems could be used for survey work, they are so different from the eastings, northings and heights that are normally encountered on site that they have to be transformed into a readily understood version for everyday work. Consequently, a decision has to be made as to what coordinate system is to be used for GPS-based surveys and how this is realised for all site work.

Sometimes the control network on site for a GPS survey consists of one reference station. It is necessary to know the WGS84 or ETRS89 coordinates of this point, and these could be determined by single-point positioning (a navigation solution, but observed over a long period of an hour or more). If the local coordinates of this point are also known or assumed, it is then possible to derive a set of transformation parameters to convert the GPS coordinates to those to be used on site. Using this reference point, further control could be added using static or rapid static methods or RTK surveys could be carried out. Either the post-processing software (for control surveys) or the receiver/controller software (for RTK surveys) will then be capable of producing or displaying site-based coordinates.

Although it is possible to carry out GPS surveys using the method described above, *it is not recommended* because:

- A control network consisting of a single point is not a strong network.

- It only provides a single point of known position for checking purposes – a site needs several of these.

- Unless special procedures are followed, the GPS coordinates of the reference point obtaining using single-point positioning will not be accurate enough.

- Most importantly, it does not produce a reliable and accurate set of parameters for transforming GPS coordinates to a site grid.

To overcome the problem of obtaining accurate GPS coordinates at one or more reference points, the Ordnance Survey has set up a system for providing these at any point in Great Britain. To do this, users define their position using what is known as the National GPS Network. This consists of a series of permanent GPS stations called *Continuously Operating Reference Stations (CORS)*, which are points located throughout Great Britain whose positions are precisely known in ETRS89. At each CORS site, the Ordnance Survey have installed a geodetic GPS receiver that continuously collects satellite data – this data is available, free of charge, from the Ordnance Survey National GPS Network web site and acts as a reference station in a static survey. A full description of the National GPS Network is given in Section 8.6, but see the Further reading section at the end of this chapter for details of how to access this web site. To obtain ETRS89 coordinates, a user places a geodetic receiver at an unknown point and takes static observations over a length of time depending on the type of receiver used and the distance from the CORS. The observed data is then combined with the reference data from the nearest CORS sites in a post-processing procedure to compute the precise coordinates of the unknown point on the ETRS89 coordinate system. This is capable of giving a position with an accuracy at the centimetre level on ETRS89 and has the advantage that only one receiver is required to do this. Several other countries have a similar setup provided by their national mapping agency for determining accurate coordinates either on WGS84 or its local equivalent (see some of the web sites listed at the end of the chapter). *Using a point or points connected to the Ordnance Survey National GPS Network is the recommended procedure for obtaining the start coordinates for any GPS control survey in Great Britain.*

The problems of using a single point and of obtaining a reliable set of transformation parameters are resolved by establishing a control scheme with more than one control point and using these to derive a more rigorous set of parameters. This requires that a set of control points is already available with positions known in the site-based coordinate system.

If the construction project is to be based on a national coordinate system, there may be control points available in the area with coordinates already known and published on this that could be used as reference stations and for deriving the transformation parameters. This has the advantage that a control network is already available and need not be surveyed. However, these points may not be sufficiently accurate, as many national surveys were observed and computed some time ago using equipment and methods that do not have the accuracies that are possible today, especially when using GPS over long baselines. Consequently, if national control points are to be used as a datum for a GPS control survey, it must be realised that the accuracy obtained will be limited to that of the original survey, however good the subsequent GPS results are. It is very probable that the accuracy of many national surveys will not meet the specifications required at the present time for setting out and other work on site. To overcome this problem in Great Britain, the Ordnance Survey has derived a set of national transformation parameters for mathematically converting observed ETRS89 coordinates directly into National Grid coordinates as an easting and northing and a sea level height based on the Ordnance Datum Newlyn and other offshore datums (further details of these transformation parameters are given in Sections 8.7 and 8.8). Large construction sites or those along linear projects such as major roads and railways may well be based on these coordinate systems. If National Grid or Ordnance Survey datum heights are to be used on a construction site or project, the transformation may be carried out at a few control points only. The national control can then be carried through the survey by using traditional methods based on, for example, total stations and levelling.

Another method of using the ETRS89 coordinates is to transform them directly into coordinates based on a local site system. To be able to do this, a set of local instead of national transformation parameters must already be available, or they have to be derived. Sometimes called *calibration*, deriving a set of parameters involves control points already defined in the site grid coordinate system. Each of these is visited and their ETRS89 coordinates determined either from the National GPS Network or by using static/rapid static methods from a reference station. By comparing the known site-based coordinates with the observed ETRS89 coordinates, it is possible to determine local transformation parameters. For this to produce a reliable set of transformation parameters, at least three, but preferably four or five, control points with known coordinates should be used. For small sites, the coordinates of control points on the site grid may have to be obtained by traversing or (for larger sites) by observing a network. If a local control network is already in place, this can be used for reference stations and for deriving transformations *provided it meets accuracy requirements*. The different procedures used for transforming GPS coordinates are discussed further in Section 8.8.

As an alternative to national or local control, the ETRS89 coordinates obtained at each control point can be used directly to define a project or site datum if GPS is to be used extensively on site. This is possible because most setting out and machine control systems today make use of graphical displays to provide positional information rather than asking the operator to process or read coordinates and it does not matter which coordinate system is used to provide this. However, this does have implications elsewhere for data management, as a project would have to be designed using ETRS89 coordinates and these would have to be known by some means at the design stage for downloading into a GPS controller ready for use on site.

Whatever reference points are used for any project, they should cover the entire area. Control may be split into primary control, which consists of a few highly accurate control points, and secondary control. A definition of these is given in Section 6.8 and the way in which they are implemented on site is also discussed in Chapter 11. Primary and secondary control should have accuracies conforming to those given in publications such as BS 5964 *Building setting out and measurement* or the ICE Design and Practice Guide *The management of setting out in construction* (see Further reading at the end of this chapter). The primary part of a network should be observed using the National GPS Network and processed separately, especially if it is a long way from the CORS. From this, a secondary network of control points can be established using rapid static observations. For those sites where GPS will be the predominant survey method, primary and secondary control locates the positions of base stations and points for checking initialisations, and this usually only requires a low density of control points. For those sites where total stations and other traditional equipment are to be used a high density of points may be required, and the secondary network must be located to ensure that all setting out and measurements have enough control available.

As can be seen, the coordinate system to be adopted and how this is to be realised is a major factor in planning a GPS survey.

Satellite coverage

When performing any GPS survey it is essential to determine which satellites will be available at a proposed control point or over a site during an observation period. In order to predict this, each manufacturer provides *mission planning software* that computes and then displays the bearing and elevation of each satellite for the times when future observations are planned. When a GPS receiver has locked onto any satellite in the GPS constellation, it decodes the current satellite almanac from the navigation message – this is a record of the approximate positions of all the satellites and it is transferred automatically to the system computer when the controller transfers observational data to it. To access the almanac, the approximate latitude, longitude and height of the proposed control point or site together with the date and time are entered into the system computer and the mission planning software will display a *satellite availability plot* as shown in Figure 7.21. On this example, the number of satellites and their GDOP/PDOP are shown for a selected location and time interval and for a mask angle of 15°. At a time of 12:30 it can be seen on the plot that a *GDOP spike* has occurred, probably because there were only four satellites available just prior to this, followed by five satellites possibly forming a poor three-dimensional geometry. Observations should not be carried out at this time, but the remainder of the planned observation period is satisfactory as the GDOP is about 3.

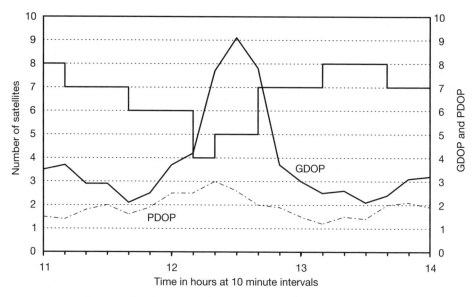

Figure 7.21 ● Satellite availability plot.

The latest generation of GPS controllers have the mission planning software installed in them. These are capable of computing and displaying satellite availability and DOP values directly without having to process these on a computer.

Site reconnaissance

At the start of the planning stage, the positions for control points are chosen with the aid of any existing maps and aerial photographs that may be available plus any information from previous surveys that may have been carried out in the area. Following this, a site reconnaissance is required to check each proposed control point for any problems.

At each control point, there should be no significant obstructions to visibility and there should be a clear view of the sky above 10–15° all round the horizon at tripod or antenna height. If there are any trees, bushes or buildings nearby, the positions of these should be surveyed (a magnetic bearing and approximate vertical angle are all that is required) and this data should be entered as an additional mask to the 10–15° already entered into the mission planning software. This will then give a more realistic satellite availability plot and DOP, as it accounts for any satellites that may be blocked by the obstructions. In the northern hemisphere, satellites cross the sky from the south so obstacles to the north can be less important, depending on how large they are.

As well as this, there should be no overhead power transmission lines or mobile phone/radio masts nearby, as these might interfere with GPS signals. Another situation to be aware of is the possibility of multipath errors, and control points should not be placed at locations where GPS signals can be reflected and received from surfaces such as buildings, parked vehicles or even water.

Following the site visits, it may be necessary to review the positions chosen for some points.

Station marking

When the positions of all the control points in a network have been finalised, some sort of station mark is put in place at each one, the design of which depends on the ground conditions and duration of the project for which they are required. Examples of these have already been given in Section 6.4 for traversing, but further suggestions can be found in BS 5964 *Building setting out and measurement*. At the time it is put in place, each point can have its approximate coordinates recorded using a handheld code-based receiver.

Taking the observations

With a control network in place fieldwork can commence, and preparation for this focuses on a number of topics. Some of these are arranging for staff to carry out the survey and for equipment to be in place and transported. Careful planning of the observation scheme is needed to ensure that enough data will be recorded to determine accurate GPS coordinates, to determine transformation parameters (when these are needed) and to ensure that there is a redundancy in the observations. When fieldwork has started, good communication between survey parties is essential and it is good practice to have a detailed observation timetable in place. GPS generates large amounts of data and data management is an issue – does each instrument have enough memory cards for this and how is data to be downloaded from the controller, stored and backed up? In addition, how and where is the data to be post-processed?

When observing a static survey, it is best to plan at least two independent sessions in which all the control points are visited twice at different times with different satellite constellations. As already noted, observation sessions must be planned well in advance to ensure that at least four or five satellites are always available and that DOP spikes are avoided. It is also worth noting that the phase centre of a GPS antenna is not a fixed point and it varies mainly according to satellite elevation. To help compensate for this effect, it is recommended that all the antennae used in a static survey are orientated in the same direction to cancel any errors. Care is also needed when measuring the height of the antenna and if different makes and models of antenna are used, it is also important to note these so that they can be accounted for at the post-processing stage.

Post-processing

All data for static and rapid static surveys are post-processed to obtain the coordinates of the surveyed points. When this is carried out, it is recommended that the software provided by the manufacturers of the equipment used during the survey is used for this. Most of the processes used in the software are fully automated and the user will merely follow on-screen instructions issued by the software to carry out data processing.

Of particular importance in post-processing is the quality of the results obtained. For most surveys, this is given as standard deviations (see Section 9.2) for each set of coordinates obtained. It is also important to choose the most appropriate method for transforming coordinates to those on a site grid and to know the accuracy of the transformation parameters derived for this.

Post-processing involves computing a set of baselines between observed points. When processing, care must be taken not to compute any *trivial baselines*. For example, control points A, B and C in a survey could be processed as baselines A–B, A–C and B–C. However, if A–B and A–C are computed first, B–C becomes a trivial baseline because it is not independent of the data already used in computing A–B and A–C. If computed, baseline B–C may introduce significant biases into the results obtained for a network. In order to close the triangle and check the results obtained, it is best to observe B–C in a separate observation session.

Further details of post-processing, especially with regard to CORS sites and coordinate transformations, is given in Sections 8.6 and 8.8.

This section is only an overview of the procedures involved in planning, observing and processing a GPS control survey – for more information on these, the references given at the end of the chapter should be consulted.

RTK surveying with GPS

In addition to control surveys, GPS is used on site for detail surveying, for providing data when areas and volumes have to be measured, in sectioning, when forming digital terrain models, and for setting out and machine control. All of these are done with an RTK system. As already noted, these operate in real time and are able to tell the operator that the system is working correctly when on site rather than at a post-processed stage. Some general information is given on RTK surveying in this section – further details of how GPS is used in each of the applications given above can be found in the subsequent chapters that deal with these topics.

To use an RTK system, it is necessary to establish a single or series of base stations on site. These must be set up at secure locations, as the reference receiver and communication link will be left unattended for long periods of time. There should be no obstructions to radio and phone communications from these to each rover. The problems this causes and how they might be overcome have already been discussed in Section 7.6. The coordinates of base stations will usually be determined by a static or rapid static survey.

Each RTK survey starts with an initialisation to resolve ambiguities. This is normally done very reliably by a receiver and its software (or firmware), but becomes increasingly difficult to do with baselines exceeding 10 km. The range over which an RTK system can operate also depends on the communications link between the receiver and the nearest base station. Compared to a radio link, the range of RTK can be extended up to 25–30 km by using a mobile phone, although the ambiguity resolution for these is a lot slower. Remember that during initialisation, five satellites must be observed at all times with a good PDOP of less than five. A receiver will normally compute a coarse solution first at the metre level, and when initialised the precision of the solution improves to the centimetre level. An audible/visual signal is issued at this point to tell the operator that the initialisation is finished and ambiguities have been resolved. Some form of display of the precision of the horizontal and vertical coordinates being obtained is also given. Following this, it is advisable to occupy an independent point whose coordinates are known to see whether these agree with those displayed by the controller – the occupation time at such a check point should be longer than for surveyed points and two to three minutes is suggested for these. This confirms that the ambiguities have been resolved correctly by the rover and that the base station has had the correct coordinates assigned to it. As can be seen, the density of control points required for reliable and efficient RTK surveying can be more than that required merely for the setting up of base stations.

It must be remembered that following initialisation, at least four common base/rover satellites must be tracked at all times and it is best to keep the rover antenna clear of any possible obstructions caused by vegetation and buildings. If loss of lock occurs, re-initialisation will automatically be carried out by the receiver, but any points surveyed in between loss of lock and re-initialisation will have a poor precision and should be re-observed. Problems with initialisations can also occur when the communications link between the base station and rover is lost for some reason.

Reflective summary

With reference to GPS in engineering surveying, remember:

— GPS is not the only surveying method used on site – total stations and other optical equipment are used just as much.

— Always ask the question 'Is GPS the right equipment for this job or will something else do it better (and maybe cheaper)?'.

— Choose your equipment carefully, taking into account the level of training and technical support on offer. Do you need to buy a GPS system, or is hiring a better option?

— Some thought must be given to the coordinate system that will be used with any GPS equipment and how this is going to be realised. This will almost certainly involve extra fieldwork and costs – if this is not done properly GPS will not give the correct results.

— A GPS control network consisting of one reference point only is not recommended. These should consist of at least three but preferably five or more points to enable reliable transformation parameters to be determined and to provide points that can be used for checking purposes.

- Like control networks measured with total stations, all observed GPS control networks must be checked by some means and must meet specifications.

- Before taking any observations with GPS it is a good idea to check, in advance, that the satellite coverage is going to be adequate, especially for programmed construction work. If the coverage is poor for some reason, an alternative survey method must be used.

- Observing a control scheme with GPS requires careful planning and organisation, especially if the work covers a large area. It is not a good idea simply to arrive on site and then start thinking about how the observations will be taken. Similarly, RTK surveys must be planned in advance.

- When GPS results are to be post-processed, it is necessary to know how to use the software provided for this – because these programs are complicated, make sure the manufacturer has provided adequate training in their use.

- The majority of GPS survey work on site is done with RTK equipment. For this to work correctly, an adequate number of base stations must be established and radio or phone communications to each rover must be maintained at all times. This will also add to the cost of GPS.

- After initialising an RTK receiver, it is best to check this has been done correctly by placing the detail pole on a known point. This should also be done at frequent intervals when setting out or taking other real-time measurements.

7.9 Developments and the future for satellite surveying

After studying this section you should be aware that GPS improvements are planned and that new satellite systems are emerging, all of which will greatly enhance satellite surveying in the future. You should also be aware that a new generation of survey instruments that combine GPS with a total station are now available. In addition, you should appreciate that augmentation systems and Network RTK will continue to develop, as will the quality and user-friendliness of GPS equipment. Finally, you should realise that much more attention has to be paid to information management on site, as GPS relies on a seamless transfer of data from the design to each RTK system in use.

This section includes the following topics:

- Improved GPS access
- GLONASS and Galileo
- GPS combined with a total station
- Other improvements
- Summary

Improved GPS access

Due to the success of GPS for civilian as well as military use, the Department of Defense is adding a second civilian frequency to the L2 signal, which will be called the L2C signal. As new satellites are launched, the C/A code will gradually become available across the GPS constellation on the L2 signal as well as the L1 signal. It is anticipated that this will be completed by 2013 and will improve the accuracy of point positioning to within 5–10 m, and for kinematic surveys will enable a better lock to be maintained on satellites once they are acquired. Another improvement planned from 2006 to 2014 is to add a third civilian frequency known as the L5 signal at 1176.45 MHz – this is going to be broadcast by each satellite eventually. It is expected that this will improve point positioning further to 1–5 m, and for RTK surveys this will enable easier ambiguity resolutions to be carried out and will provide faster recovery after signal loss when a rover passes under or close to an obstruction.

GLONASS and Galileo

As well as GPS, two other satellite systems, known as GLONASS and Galileo, will be fully operational in the future.

GLONASS is an abbreviation for the *GLObal NAvigation Satellite System* operated by the Russian Aerospace Agency for the Russian Federation. The first satellite was launched for GLONASS in 1982 and the system was at one time fully operational in 1995, when 24 satellites were in orbit. Following this, economic difficulties prevented new satellites being launched to replace those going out of service and the number of satellites has dwindled to 10 at present. However, there is a commitment from the Russian aerospace industry to restore GLONASS to full operating capability. In many respects, this is very similar system to GPS and the full constellation of 24 satellites will orbit at an altitude of 19,100 km and have a period of 11 hours 15 minutes. The planned GLONASS constellation has three orbital planes at 65° inclination to the equatorial plane of the Earth, such that five satellites will always be in view. Each satellite transmits two L-band signals, the frequency of which is different for each satellite, and instead of broadcasting satellite positions on the WGS84 coordinate system, GLONASS uses the PZ90 reference frame. Another difference between GLONASS and GPS is that different clocks at different locations are used to define time, and these are not synchronised.

For surveying, GLONASS is currently used in conjunction with GPS and not as an alternative, and despite the technical difficulties caused by the differences between GPS and GLONASS, receivers are available that are compatible with both systems. All commercially available survey receivers that use GLONASS are combined GPS + GLONASS receivers and they are essentially GPS receivers that use GLONASS as an enhancement. The advantage of using GPS + GLONASS is that there will always be more satellites available compared to using GPS on its own. By having more satellites available, much faster ambiguity resolution is made possible, reducing the time required for initialisations in RTK surveys. In addition, the better satellite coverage enables more observations to be taken over longer periods each day and allows users to attempt measurements in areas where GPS has difficulty operating when under trees, in deep valleys and in built-up areas surrounded by tall buildings (these are known as *urban canyons*).

Galileo is a new global navigation satellite system developed by the European Commission and the European Space Agency. In contrast to the military controlled GPS, Galileo will be operated by the European Union and will provide an accurate, guaranteed global positioning service under civilian control. The fully deployed system will consist of 30 satellites (27 operational + 3 spares) located in three medium Earth orbit planes with an inclination of 56° at an altitude of 23,616 km and with a period of 14 hours 22 minutes. The ground segment will consist of two Galileo Control Centres located in Europe that will control the satellites and produce updates to navigation messages. There will be a global network of 20 Galileo Sensor Stations continuously monitoring the satellites, whose data will be sent to the control centres through a communications network. The exchange of data between the control centres and satellites is accomplished using 15 uplink stations, again located around the globe. Galileo is planned to reach its full operational capability in 2008, although this might be delayed.

Galileo will be compatible with GPS and GLONASS and a user will eventually be able to determine position with the same receiver from any satellites in the three constellations. Without any doubt, a constellation of over 60 satellites will provide enormous benefits in global satellite surveying.

GPS combined with a total station

A logical progression in the development of GPS and total stations is to integrate these into one instrument. The Leica *SmartStation* shown in Figure 7.22 combines their GPS *SmartAntenna* (an antenna and receiver mounted in a single housing) with their *TPS 1200 total station* already shown in Figure 5.20 and elsewhere in the book. When the SmartStation is used for GPS measurements, the SmartAntenna is connected to the top of the TPS1200 together with an RTK communications device. In this configuration the TPS1200, through its keyboard and display, acts as if it were a GPS controller. At the start of a survey, the SmartStation uses RTK GPS to determine its coordinates, after which the SmartAntenna and communications device are

Figure 7.22 ● Leica SmartStation (courtesy Leica Geosystems).

removed. Knowing the position of the instrument, the TPS1200 is then used as a total station to take measurements either manually or in robotic mode.

When detail surveying with the SmartStation, it is not necessary to set it up over a control point; it can simply be set up anywhere convenient for taking measurements. The position of the point occupied is then determined to centimetre accuracy using RTK GPS with reference to a base station. Having determined the coordinates of the SmartStation, a second unknown point is sighted. The coordinates of this second point are not determined at this time but the total station is orientated to it and measurements are taken to the required points of detail. Following this, the SmartStation is set up at the second point and its position is now determined once again using the integrated GPS SmartAntenna. Since the coordinates of both points are now known, the correct bearing between them is determined by the SmartStation, which then re-computes the coordinates of all the detail surveyed from the first point using the correct bearing. To continue measurements, the SmartStation now works as a total station and is orientated by sighting the first point occupied.

When a detail survey is carried out conventionally, traversing or GPS might be used to determine the positions of a series of control points on site from which further observations would be taken to locate detail. This requires two separate surveys but by integrating GPS with a total station, a survey could be completed in one operation with one instrument resulting in a saving in time. In addition, given that GPS must be used to fix the position of at least two points, the technique allows a total station to be used where GPS signal reception can be unreliable (for example near to vegetation and in built-up areas). Ideally, a check should be carried out on any detail survey carried out with the SmartStation by coordinating a known position with the TPS1200 or by measuring the distance between the two RTK points with the TPS1200 and by comparing this with that obtained from their known coordinates.

When setting out with the SmartStation, it is again set up wherever convenient and coordinates are determined at this point using RTK GPS. The instrument is then moved to a second unknown point and the coordinates of this are also determined. By using the first point for orientation, setting out can then be performed with the TPS1200. In this case, fixing the position of the SmartStation with GPS is rather like using a free station point (see Sections 6.7 and 11.3), but without the need for any control to be sighted or occupied. This can have advantages when site control points become damaged or obstructed and cannot be used. Again, if GPS signal reception is intermittent or unreliable on site but can be used to fix the two set up points, it allows setting out to continue with a total station, without the need for control points. Of course, some sort of check must be made on the position of the SmartStation and its orientation by taking observations to a known point independent of the two RTK points.

The accuracy of the SmartStation is dictated by the accuracy of RTK surveying. Since this is at the centimetre level at present, the instrument is ideal for detail surveying at most scales, but may not be accurate enough for some setting out operations. Before setting out with the SmartStation, it is essential to check that measurements taken with it will meet specifications.

The reference or base station used by the SmartStation must be set up in the usual way as described in earlier sections of this chapter. Care must taken to ensure that radio or telephone communication to the base station is possible, and to help with this the SmartStation can use Bluetooth technology to connect to a mobile phone which then contacts a base station instead of using a dedicated communications device. If a network of base stations has already been established in the area, these can be used and this avoids having to set up, equip and maintain a base station. In future, such networks may be provided by government agencies or private companies (see Section 7.6 under *Kinematic surveys* and *High accuracy with a single receiver*, below, for further details). Using an already established base station would make combined GPS and total stations a very efficient way of surveying.

Clearly, the integration of GPS and total stations into one instrument represents a major step forward in surveying technology. It is expected that most of the manufacturers of GPS equipment will be producing systems similar to the SmartStation in the near future and that these will find many applications in surveying and on site.

Other improvements

High accuracy with a single receiver

A lot of development in satellite surveying at present is aimed at trying to obtain high accuracy with a single receiver. For this, augmentation systems such as EGNOS and WAAS are to play an increasing role. These already make accuracies at the metre level possible over large areas with a code-based receiver and many geodetic receivers are now also EGNOS/WAAS compatible and can produce accuracies at the sub-metre level when operated as a single receiver. The use of Network RTK is also expected to increase and the scenario where anyone can arrive on site and immediately start using an RTK system at the centimetre level without having to set up a base station is under development. As already noted, the Ordnance Survey may well be providing a public RTK service for obtaining National Grid coordinates and ODN heights in Great Britain.

Pseudolites

Devices known as *pseudolites* are also the subject of some research at present. These look much the same as a GPS receiver and antenna and can be tripod-mounted, but instead of passive signal reception, they transmit a signal identical to that of a satellite. With a series of these appropriately positioned, GPS surveys could be carried out in locations that are not possible at the moment (for example, indoors or in tunnels).

Ongoing improvements

Other improvements that make GPS systems a much more attractive solution for measurements on site today are their reliability and that they can work in harsh conditions. As with total stations and other survey equipment, the cost of GPS systems is also steadily decreasing, they are becoming easier to use and they are capable of accuracies that are more than adequate for most construction work.

Information management

Because GPS is changing the way in which data is collected and processed, an area to which site engineers are expected to become more adapted in the future will be the management of survey data as this becomes more integrated. As discussed in Chapter 1, a fully integrated surveying system consists of a surveying sensor (GPS receiver, total station, level and so on), data collection hardware and software, data communications and processing/design software, all working together. As a result, there needs to be a seamless transfer of survey data between site and office and the role of the engineer will be to organise and monitor this. Consequently, site engineers will spend less time setting out and more on ensuring that adequate survey control is provided on site. They will spend more time making sure that data is in the right place, at the right time and in the correct format and that quality and checking procedures are implemented. Because of the automation in survey equipment brought about by GPS, setting out may be carried out by other site operatives. Accordingly, it is expected that site engineers will devote more time to analysing and manipulating data and will be expected, in future, to know more about data management and computing systems than ever before.

Summary

The future for satellite surveying is guaranteed, especially now that GLONASS and Galileo are expected to become fully operational. With GPS, these are all known as *Global Navigation Satellite Systems* (*GNSS*). Alongside the development of these and the associated receiver technology to enable any user to access any satellite, the other areas of development expected in the near future are in better data communications (good examples here are the introduction of Bluetooth technology and improved mobile phone technology), in data integration and in single-receiver RTK systems.

Without any doubt, GPS has had a significant effect on how position can be determined for surveying (and many other activities) and its development is expected to continue with new technologies and applications emerging. However, despite its considerable prominence it is simply just another tool in the surveying toolbox.

Compared to total stations, theodolites, levels and tapes, GNSS are good at some survey work but not so good at others, and the site engineer must be capable of choosing the right survey equipment for each task.

Reflective summary

With reference to developments and the future for satellite surveying, remember:

— The United States Department of Defense is going to improve civilian access to GPS over the next ten years by adding the C/A code to the L2 signal and by adding a third frequency known as the L5 signal.

— Two other satellite positioning fixing systems are currently under development, known as GLONASS and Galileo. GLONASS is the Russian equivalent of GPS and Galileo is funded by the European Union. If all of these are fully operational, over 60 satellites will be available.

— Instruments that combine GPS with a total station in one instrument are now available. It is expected that these will find many useful applications on site.

— There is a lot of development taking place at present aimed at producing a satellite surveying system that can display real-time coordinates with accuracies at the centimetre level but using a single receiver.

— All of the surveying carried out on site with GPS relies on data being transferred directly from design to survey equipment. A procedure for doing this reliably and quickly must be put in place for each project where GPS is to be used. Because of this, it is expected that engineers will spend more time computing and processing GPS data leaving the setting out with it to others.

— GPS cannot be used for all the surveying required on site and it should complement other equipment and not replace all of it.

Exercises

7.1 What is the function of the control segment of GPS and how is it organised?

7.2 Describe the GPS satellite constellation and explain why it has been designed so that at least four satellites are always in view from any point on Earth.

7.3 Explain how the C/A and P codes in GPS signals differ.

7.4 Give details of the information contained in the GPS navigation message.

7.5 What are the differences between the L1 and L2 signals transmitted by GPS satellites?

7.6 Explain the difference between code ranging and carrier phase measurements.

7.7 In GPS positioning, what do the following terms mean?

observable
propagation delay
autocorrelation
pseudorange
navigation solution
standard and precise positioning services
integer ambiguity
point positioning

7.8 List the errors that can have an effect on GPS measurements with an estimate of their magnitude in code ranging.

7.9 What causes ionospheric and tropospheric delays and how can the effect of these be reduced?

7.10 Discuss the phenomenon known as dilution of position in GPS. What can be done to reduce this?

7.11 Explain what a multipath error is and suggest ways in which it can be avoided.

7.12 How does post-processing differ from real-time surveying with GPS?

7.13 What is EGNOS?

7.14 How is it possible to improve the accuracy of GPS measurements using differential and relative methods?

7.15 Explain how double and triple differences are used in precise relative measurements.

7.16 What is a cycle slip and how is it possible to detect these?

7.17 What are the differences between static and rapid static surveying?

7.18 What methods can be used to carry out kinematic surveys?

7.19 With reference to RTK surveying, what is meant by the following?

base station
initialisation
repeater station
radio modem
network RTK

7.20 On site, mapping grade and geodetic receivers can be used. What is the difference between these?

7.21 What are the main components of a geodetic receiver that could be used for setting out?

7.22 Explain how Bluetooth technology makes it possible for GPS equipment to operate without cables.

7.23 If GPS is to be used on site for the first time, what items should be discussed before any equipment is purchased?

7.24 Explain how it is possible to use National Grid coordinates and ODN heights for GPS surveys on site.

7.25 Explain how it is possible to use local grid coordinates and heights for GPS surveys on site.

7.26 What is mission planning software and how is it used?

7.27 Describe the improvements planned for GPS and discuss how these will affect point positioning and RTK surveys in the future.

7.28 Describe the GLONASS and Galileo satellite surveying systems.

7.29 Discuss the importance of good data management on site, why this is essential for GPS surveys and the role the engineer is expected to play in this.

7.30 Use a search on the Internet to see the enormous amount of information available on GPS from this source and the problems filtering it.

Further reading and sources of information

The following web sites are very useful for obtaining information on GPS and satellite surveying systems. The most important of these is

http://www.gps.gov.uk/

as it gives details of the National GPS Network plus other background information on GPS from the Ordnance Survey. When entering this for the first time, it is best to register and obtain a user name and password. This gives access to all parts of the site.

Other sites include:

http://www.ngs.noaa.gov/	(for GPS in US)
http://www.navcen.uscg.gov/gps/	(for status of GPS)
http://www.galileoju.com/	(for status of Galileo)
http://www.glonass-center.ru/	(for status of GLONASS)
http://www.esa.int/	(for information on EGNOS)
http://gps.faa.gov/	(for information on WAAS)
http://www.trinityhouse.co.uk/	(for information on the GLA)

The manufacturers' web sites are

http://www.leica.com/
http://www.sokkia.co.uk/
http://www.thalesnavigation.com/
http://www.topcon.co.uk/
http://www.trimble.com/

Some of the manufacturers have produced explanatory notes and some have online tutorials.

When any of the major Internet search engines are used to generate information on GPS, each one will produce a considerable amount of information. Some care (and time!) is needed to filter these for suitable material.

The following journals include annual reviews of GPS equipment and include frequent articles on GPS: *Civil Engineering Surveyor, Engineering Surveying Showcase, Geomatics World* and *GIM International* (see Chapter 1 for further details). Another useful journal is *GPS World*, which is published monthly by Advanstar Communications Inc. The web site for this journal is http://www.gpsworld.com/. The ASCE's *Journal of Surveying Engineering* (see Chapter 1) also includes many articles on GPS.

The following publications are also recommended:

RICS Guidance Note (2003) *Guidelines for the use of GPS in surveying and mapping.* RICS Business Services Ltd, Coventry.

ICE Design and Practice Guide (1997) *The management of setting out in construction.* Thomas Telford, London.

English Heritage (2003) *Where on earth are we? The Global Positioning System (GPS) in archaeological field survey.* http://www.english-heritage.org.uk/. Look in *Publications – Free publications* and search for 'Where on Earth are we?'; Direct download (June 2005): http://www.english-heritage.org.uk/upload/pdf/where_on_earth_are_we.pdf. [Although this is written with archaeologists in mind, it provides a very useful introduction to the concepts of GPS and its application in mapping]

Kelly, C. (2001) Setting out the Channel Tunnel rail link. *Proceedings of ICE, Civil Engineering,* **144**, 73–7. [This provides a useful insight into the implementation of GPS on a large civil engineering project]

Leick, A. (2003) *GPS Satellite Surveying,* 3rd edn. John Wiley & Sons, Chichester.

Van Sickle, J. (2001) *GPS for Land Surveyors,* 2nd edn. Ann Arbor Press, Chelsea, Michigan.

Hofmann-Wellenhoff, B., Lichtenegger, H. and Collins, J. (2001) *GPS Theory and Practice,* 5th edn. Springer-Verlag, London.

chapter 8

GPS coordinates and transformations

 Aims

After studying this chapter you should be able to:

- Define the Geoid and an ellipsoid and discuss how they help define the shape of the Earth
- Describe position as geodetic coordinates and geocentric cartesian coordinates
- Understand the differences between a GPS (ellipsoidal) height and a mean sea level (orthometric) height
- Define coordinates using Terrestrial Reference Systems (TRS) and Terrestrial Reference Frames (TRFs)
- Outline the differences between WGS84 and ETRS89 coordinates
- Describe the Ordnance Survey National Grid horizontal coordinate system and the Ordnance Datum Newlyn vertical coordinate system
- Access Ordnance Survey National GPS Network data and understand how precise ETRS89 coordinates are obtained using these
- Discuss the procedures used for changing ETRS89 coordinates into National Grid and Ordnance Datum Newlyn values using the OSTN02™ and OSGM02™ transformations
- Describe some of the procedures used for transforming ETRS89 and WGS84 coordinates into local site grid coordinates

This chapter contains the following sections:

8.1   Plane surveying and geodesy

8.2   Geoids and ellipsoids

8.1   Plane surveying and geodesy

After studying this section you should understand when plane surveying is no longer viable and the curvature of the Earth has to be accounted for when determining position. You should also appreciate how this affects the way GPS coordinates may be used with plane surveys.

Defining position for large projects and GPS surveys

In previous chapters where control surveys have been discussed, position has been defined as an easting and northing on a plane rectangular grid together with a height (or reduced level) based on an arbitrary datum or mean sea level. This is a very convenient way of defining position in surveying and even though we know the Earth is curved, a flat surface is assumed for most engineering surveys. The advantage of adopting this approach to control surveys and setting out is that it greatly simplifies the calculations required for both horizontal coordinates and heights.

As the size of a survey or construction project increases, it is no longer possible to assume a plane surface for defining position because the effect of the curvature of the Earth is too large to ignore. For normal work, this occurs when the survey is longer than about 10–15 km in any direction but nearly all engineering projects do not exceed this limit. However, because of the advent of GPS, much more use is now made of national control networks in engineering surveying, which clearly cover areas much greater than 15 km. As well as this, it is obvious that GPS is not a plane surveying system as it operates in three dimensions on a global basis. Some questions arise from this. If a plane surface is no longer acceptable or possible for defining coordinates, what surface is used to define positions for a large construction project and for satellite surveying systems? When GPS coordinates are obtained, which system do they belong to and how can they be used on site?

The answer to the first question comes from *geodesy*, which is defined as the study of the size and shape of the Earth. All surveyors and engineers involved with large-scale civil engineering projects and those using GPS need a basic understanding of the concepts of geodesy so that they are aware of how position is defined on curved surfaces. The second question is concerned with how GPS coordinates are defined and transformed into site coordinates. These and other related issues are discussed in this chapter.

Reflective summary

With reference to plane surveying and geodesy, remember:

— Plane surveying is used for most engineering surveys and is valid for sites that do not exceed 10–15 km in any direction.

— Position for major civil engineering projects that are bigger than 10–15 km has to be computed taking into account the curvature of the Earth. To understand how this can be done, it is necessary to have a knowledge of geodesy.

— Because of the introduction of GPS, regional and global control networks are in much greater use than before and this also makes it necessary for engineers and others on site to study geodesy.

8.2   Geoids and ellipsoids

After studying this section you should be aware that the shape of the Earth is approximated by the Geoid. Since this is irregular, you should understand that an ellipsoid is also chosen because it is the mathematical surface that best represents the shape of the Earth. You should also be able to define position as geodetic and cartesian coordinates on an ellipsoid.

This section contains the following topics:

● The Geoid

● Ellipsoids

● Geodetic and cartesian coordinates

The Geoid

If a plane surface can no longer be used for a survey, a different one must be chosen to represent the shape or figure of the Earth. For surveying purposes, it must be possible to base a coordinate system of some sort on this and to be able to compute position on it. Clearly, the topographical surface of the Earth is far too irregular and complicated for this purpose and a simpler shape or surface must be used.

To help imagine what this might be, it is convenient to think initially in terms of surfaces on which heights are based. In Section 2.1, height has been defined as a vertical distance measured above a surface of equal height – this is known as a level surface. A level surface is perpendicular at all points on it to the direction of gravity, and because of this it is known as an *equipotential surface*. So far in this book, only small, flat and unconnected level surfaces that extend over a construction site have been considered, but if a level or equipotential surface on a global scale is visualised, a closed three-dimensional reference surface for heights is obtained covering the whole of the Earth. By definition, this surface must always be at right angles to the direction of gravity and although this is generally towards the centre of the Earth, the direction of gravity varies because of irregularities in the density of the Earth. Consequently, any equipotential surface chosen for a global reference will be complex and bumpy, but it is possible to measure these variations to within 5–10 cm by taking precise gravity measurements and by using satellite altimetry.

Because there are an infinite number of closed level or equipotential surfaces that could be chosen as a global height reference, each corresponding to a surface of different but constant 'height', an arbitrary choice must be made to define 'zero height'. Quite naturally, mean sea level might be considered for this surface, as anyone used to dealing with maps and plans considers sea level as the height reference. On a global scale, such a surface could be visualised if all the oceans and seas were interconnected and flowing into each other and the effects of tides and winds were ignored. However, because of ocean currents and other effects, mean sea level is not exactly equipotential, and to define an exact reference surface for global heights the equipotential surface which is closest to mean sea level is used. This is known as the *Geoid*, and it is the surface that is used to represent the true shape of the Earth. It has the property that every point on it has exactly the same height everywhere in the world both on land and at sea and that it is always within about a metre of mean sea level.

When adopting a figure of the Earth, it is essential that a mathematically definable shape is used, otherwise it is very difficult to compute position on this. Unfortunately, the Geoid is a surface containing many irregularities, and although these can be measured and modelled to some extent, they make it impossible for it to be used for defining and calculating position exactly. However, the Geoid is an important surface in surveying because it is used all the time as a reference for measurements taken with equipment such as levels, theodolites, total stations or any other surveying instrument with footscrews. When these are levelled, the vertical axis of the instrument is made to coincide with the local vertical (or direction of gravity) which is normal to the Geoid. Since the local horizontal at the instrument is at right angles to this, surveying instruments always use an equipotential surface, offset from the Geoid, as a reference for taking horizontal and vertical measurements.

Ellipsoids

Since the Geoid is not a regular surface, a different figure must be used to obtain a mathematically definable shape of the Earth. Although the Earth is treated as being spherical for many purposes, it has a slightly bigger radius at the equator compared to the poles. The mathematical shape that matches this the closest is known as an *ellipsoid*. As shown in Figure 8.1, this surface is formed by rotating an ellipse about its minor axis and the parameters that define it are its semi-major axis a and semi-minor axis b which approximately coincides with the rotation of the Earth. From these, the flattening f and eccentricity e can be obtained.

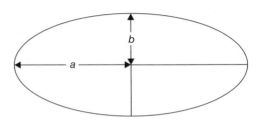

a semi-major axis
b semi-minor axis

f flattening $f = \dfrac{a-b}{a}$

e eccentricity $e^2 = \dfrac{a^2 - b^2}{a^2}$

Figure 8.1 ● Ellipsoid.

Compared to the Geoid, an ellipsoid is only a theoretical surface, but it is one on which it is possible to define and compute position exactly. When choosing an ellipsoid as a figure of the Earth, the one that is closest to the Geoid is adopted. However, because an ellipsoid cannot fit the Geoid exactly, many different ellipsoids are in use, some of which fit a region or a country well and some of which are a best fit to the whole Earth. The best-fitting global ellipsoid at present is GRS80 (Geodetic Reference System 1980) which is very similar to the WGS84 (World Geodetic System 1984) ellipsoid often used to define GPS coordinates. Due to their similar semi-major and semi-minor axes, these two ellipsoids and coordinates on them may be considered identical for most practical purposes. Historically, the ellipsoid used for mapping and coordinates in the UK is the Airy 1830 ellipsoid, which is designed to fit Great Britain only. Due to large differences between the GRS80/WGS84 and Airy 1830 definitions, the coordinates obtained for a single point on each ellipsoid will differ by many metres. Despite the existence of global ellipsoids, many others will continue to be in use for some time because they define position for other national coordinate systems.

Figure 8.2 shows a comparison of the Geoid, ellipsoid and topographical surface of the Earth.

Geodetic and cartesian coordinates

Having adopted a particular ellipsoid, it is possible to define coordinates on this. In Figure 8.3, the *geodetic latitude* ϕ_P of a point P is the angle measured between the

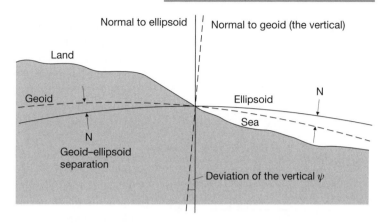

Figure 8.2 ● Comparison of the Geoid and ellipsoid.

plane of the ellipsoidal equator and the normal to the ellipsoid which extends to P above the ellipsoid. Note that this latitude is not measured from the centre of the Earth. The *geodetic longitude* λ_P is measured in the plane of the ellipsoidal equator between the meridian passing through P and the prime meridian, which is sometimes referred to as the Greenwich meridian. Both of these meridians are coincident with the minor axis of the ellipsoid, which should be aligned with the spin axis of the Earth. However, the position of the spin axis of the Earth varies as a function of time and to overcome this problem, the mean position of the poles and the spin axis of the Earth have been internationally agreed and are monitored by the *International Earth Rotation Service* (*IERS*), which is based in Paris. The IERS define the position of the *International Reference Pole* (*IRP*) and the *International Reference Meridian* (*IRM*). As a result, geodetic longitude is measured from the IRM assuming the minor axis of the ellipsoid is aligned with the IRP.

The third coordinate that defines geodetic position is known as the *ellipsoidal height* h_P. This is the distance measured along the normal above the ellipsoidal surface to point P, as shown in Figure 8.3, and is not a true height because it is not measured with reference to a level surface. It does, however, simply identify the position of a point in space above or below an ellipsoidal surface. These three coordinates will uniquely locate the position of a point, but the values obtained for ϕ_P and h_P will

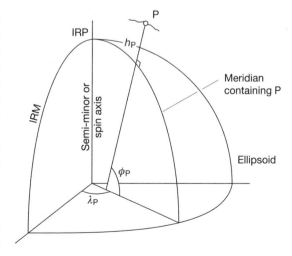

Figure 8.3 ● Geodetic coordinates.

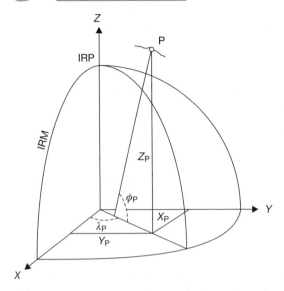

Figure 8.4 ● Geocentric Cartesian coordinates.

depend on the choice of ellipsoid. In traditional surveying over large areas, geodetic coordinates have been the classical way in which the positions of points a long way apart have been specified. To some extent they have been superseded by cartesian coordinates which are described below.

In Chapter 6, rectangular coordinates in the form of eastings and northings were introduced. If these are extended to three dimensions, a simple method for defining position is obtained that is an alternative to geodetic coordinates. In Figure 8.4, the origin of a global cartesian coordinate system lies at the centre of an ellipsoid, the IERS spin axis is the Z-axis, the X-axis is in the direction of the IRM and the Y-axis is perpendicular to both of these. Both the X-axis and Y-axis lie in the plane of the ellipsoidal equator. A point P will have coordinates X_P, Y_P and Z_P on this coordinate system, which is a *geocentric coordinate system* when its origin is located at the centre of the Earth.

Coordinates can be converted from geodetic to cartesian on the same ellipsoid using well-established formulae (see references at the end of the chapter). Note that if ellipsoids are not changed in this process, the conversions are exact. When computing these conversions the use of software is recommended, and the Ordnance Survey provides a coordinate converter, free of charge, for doing this on the National GPS Network web site (http://www.gps.gov.uk/).

The reason geocentric cartesian coordinates are now used in engineering surveying is due to GPS. Satellites in the GPS constellation have their positions determined first in terms of elliptical orbital parameters where one of the focal points of the ellipse is located at the centre of the Earth. In order to relate these to positions on the Earth, they are converted by a GPS receiver to positions on a geocentric cartesian coordinate system. A user may change these into something else, and for convenience they can be converted into geodetic coordinates or directly into other coordinate systems such as easting, northing and height based on either national or local surveys. The processes involved in converting and transforming these coordinates are much easier if they are all based on cartesian coordinate systems. When using real-time GPS on site, a user may not be aware that these conversions and transformations have been done, as the receiver firmware and controller software will have been set up to perform these automatically at each recorded or set out data point.

Reflective summary

With reference to geoids and ellipsoids, remember:

— Geodesy is the study of the shape and size of the Earth – everyone who uses GPS should have an understanding of the basics of geodesy.

— The Geoid is the surface used to approximate the true shape of the Earth. It is the equipotential surface that is closest to mean sea level around the world.

— Because the shape of the Geoid cannot be defined exactly at all points, an ellipsoid is used for defining and calculating position on the curved surface of the Earth.

— The ellipsoid that is the best fit to the Geoid is chosen for surveying and mapping. An ellipsoid could be chosen to fit a region or country, or a global ellipsoid could be used.

— Position is defined on an ellipsoid as geodetic coordinates or as cartesian coordinates.

— GPS satellites have their orbits defined on a satellite coordinate system. To relate these to positions on the Earth, they are converted into cartesian coordinates on an ellipsoid. Because both of these coordinate systems are Earth-centred and have XYZ axes, the coordinate conversions are not difficult to do.

— A GPS receiver will give position as three-dimensional global cartesian coordinates. A user must change these into something else before they can be used on site.

8.3   Heights from GPS

After studying this section you should realise that a GPS receiver gives an ellipsoidal height instead of the orthometric (or mean sea level) height used in surveying. You should also understand the difference between these and how it is possible to obtain orthometric heights from GPS measurements.

Orthometric and ellipsoidal heights

As explained in the previous section, the concept of an ellipsoidal height is misleading as it is not really a height referenced to any level surface but simply a

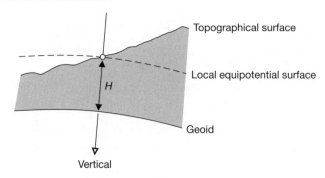

Figure 8.5 ● Orthometric height.

distance along a normal above or below the surface of an ellipsoid. In contrast, a 'mean sea level' height, which is measured when levelling or using a total station (as in trigonometrical heighting), is referenced to the Geoid and is a distance measured along the local gravity vector (the vertical) from the Geoid to the local equipotential surface passing through the point to be heighted, as shown in Figure 8.5. This measurement gives what is known as a *Geoid height* or an *orthometric height H*.

A GPS receiver first produces *XYZ* coordinates which can be converted into geodetic latitude, longitude and an ellipsoidal height. To be of use in surveying, the ellipsoidal height must be converted into an orthometric height based on the Geoid. The relationship between an ellipsoidal or GPS height *h* and an orthometric height *H* is shown in Figure 8.6, from which

$h = H \cos \psi + N$ where ψ is the deviation of the vertical

$= H + N$ if ψ is small

This gives

$H = h - N$

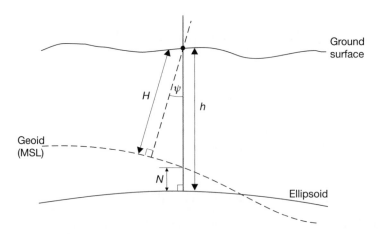

Figure 8.6 ● Relationship between ellipsoidal and orthometric heights.

where N is known as the *Geoid–ellipsoid separation*. This varies in an irregular manner depending on latitude and longitude, and to convert the GPS height to an orthometric height, a value for N is required – this is provided by a Geoid–ellipsoid separation model, more commonly termed a *Geoid model*. As can be seen, to obtain an accurate orthometric height from a GPS observation requires a good Geoid model and the development of these is of some importance in GPS coordinate systems. Because a number of different Geoid models can be used, it is necessary to state the Geoid model used when deriving orthometric heights from GPS observations. An appreciation of the uncertainty in the chosen Geoid model is also required.

Reflective summary

With reference to heights from GPS, remember:

— In surveying, all heights measured by levelling and with total stations are known as orthometric heights. These are referenced to mean sea level or a surface that is offset from this.

— GPS receivers and systems give ellipsoidal heights which are a distance above the surface of an ellipsoid – these must be converted into orthometric heights.

— The difference between an orthometric height and an ellipsoidal height is known as the Geoid–ellipsoid separation – this can be difficult to determine accurately.

— Data on Geoid–ellipsoid separations are provided by Geoid models. There are many of these in use some of which are global whilst others are only valid for a country.

— Any Geoid model has an uncertainty which needs to be taken into account.

8.4 Reference systems and reference frames

After studying this section you should realise that all surveys have a datum of some sort and that the modern name for this is a Terrestrial Reference System. You should also realise that to determine the position of any point in relation to the datum, a Terrestrial Reference Frame is needed.

Survey datums

In Chapter 6, traversing is described as a method for determining the easting and northing of a series of control points. To be able to compute these coordinates, the easting and northing of one point together with the bearing of one of the traverse lines must be known or assumed. Similarly, when levelling, the height of a TBM must be known or assumed so that the heights of other points can be determined relative to it. Both of these are examples of how origins are used to define position for a coordinate system (in the case of traversing to define two-dimensional horizontal plane coordinates and for levelling, a one-dimensional vertical height coordinate).

In order to define an origin for cartesian or geodetic coordinates, a *Terrestrial Reference System* (*TRS*) or *geodetic datum* must be established. The term 'datum' is commonly used to describe a survey origin, but Terrestrial Reference System is often used when dealing with an origin for global and satellite surveying systems.

For the coordinates used in traversing and plane surveys, a datum or reference system describes where the origin of the survey is (where easting, northing = 0) and how the *N*-axis is orientated. For cartesian coordinates, a datum defines the origin of the coordinate system (where XYZ are all zero) and in which directions the axes all point. Conventionally, a geodetic datum usually defines a set of cartesian axes together with an ellipsoid so that position can be computed as cartesian or geodetic coordinates. This will consist of eight parameters – the location of the origin in space (three parameters), the orientation of the axes (three parameters) and the size and shape of the ellipsoid (two parameters).

Since a TRS (or geodetic datum) is only a set of parameters that define a coordinate system in relation to the Earth's surface, some means of determining the position of any point in relation to a datum is needed – this is provided by a *Terrestrial Reference Frame* (*TRF*). For traversing this is the coordinates of the traverse stations, and for a levelling scheme on site it will be the heights of a series of TBMs. For GPS, the TRF is the coordinates and *velocities* (see Section 8.5) of the control points in some regional or global network. Because TRFs can include points such as traverse stations, they rely on measurements being taken to realise them (the observed traverse angles and distances in this case) and they will contain errors. Similarly, the positions and velocities of sites in a GPS network are not known exactly and the TRF defined by these also contains errors. On the other hand, a TRS is simply a set of parameters and is errorless.

Reflective summary

With reference to reference systems and reference frames, remember:

— A Terrestrial Reference System or TRS is a set of parameters defining the datum for a survey.

— For a traverse, a datum consists of an easting and northing together with a bearing, and for levelling it consists of the known or assumed height of a bench mark.

- A geodetic datum consists of eight parameters which include six parameters to define the origin and orientation of a set of three-dimensional axes plus two more to define the shape and size of an ellipsoid.

- For a traverse, the Terrestrial Reference Frame or TRF is the coordinates of the traverse stations, and for levelling it will be the heights of a series of bench marks.

- For GPS, the TRF is the coordinates and velocities of the sites in a regional or global network.

8.5   GPS coordinate systems

After studying this section you should be aware that GPS coordinates are based on a number of different datums. You should know what these datums are and how they are realised.

This section includes the following topics:

● WGS84 broadcast TRF

● The International Terrestrial Reference Frame (ITRF)

● European Terrestrial Reference Frame (ETRF89)

WGS84 broadcast TRF

WGS84 is defined by set of three-dimensional cartesian coordinates, an ellipsoid and some adopted conventions (see the National GPS Network web site for full details). As described in Chapter 7, each satellite in the GPS constellation transmits positional information enabling its coordinates on the WGS84 ellipsoid to be computed at any given time. In this case, the TRF that is used for obtaining the WGS84 coordinates of the satellites is derived from 13 military tracking stations in the ground segment of GPS whose positions are known to within about 5 cm on WGS84 – it is this network of tracking stations that define WGS84, the coordinates of which are transferred to a user through the GPS satellite constellation. Today, the broadcast TRF is mostly used for navigational positioning (single-point positioning) and different but more precise reference frames are used for all other surveying.

The International Terrestrial Reference Frame (ITRF)

The ITRF is a realisation of a datum known as the *International Terrestrial Reference System* (*ITRS*) which is provided by the IERS and not the US Department of Defense.

Compared to the broadcast WGS84 TRF, the ITRF is based on more than 500 stations at 290 sites all over the world. Because it includes many more stations and much more data, the ITRF is a more accurate network and it is produced by the civil GPS community independently of the US military. The latest version of this is called ITRF2000 and consists of a list of coordinates of each station in the network and their velocities, together with an estimation of their errors.

The *velocity* of each station is an integral part of its positional data because of tectonic plate motion, which will have an effect on global coordinates at the centimetre level. Consequently, to determine the coordinates of a station in the ITRF at a certain time (epoch), their velocities (in metres per year) have to be applied to the coordinates given by ITRF2000. If required, all of this information is available on the IERS web site (http://www.iers.org/).

European Terrestrial Reference Frame 1989 (ETRF89)

Due to the effects of tectonic plate motion mentioned above, the position of the ITRS datum is continuously changing and this makes it difficult to use in some surveying applications. To overcome this problem and still have a coordinate system that is compatible with GPS, many places in the world have defined their own GPS datum by choosing a particular epoch and fixing ITRF coordinates for points at this time. This creates a new set of cartesian axes (and datum) coincident with the ITRF at the time selected that will remain fixed for a region. However, these will gradually drift away from the ITRS axes and ellipsoid.

For Europe, a GPS datum known as the *European Terrestrial Reference System 1989 (ETRS89)* has been established that is fixed to the Eurasian tectonic plate. This coincided with the ITRF in 1989 but has been deviating from this ever since at a rate of about 25 mm per year in the UK. The TRF associated with ETRS89 is called ETRF89 and this is based on a densification, in Europe, of the ITRF network of global points. ETRF89 is a very well defined and stable reference frame that is fully compatible with WGS84. If coordinates are required on WGS84 or the ITRF at any time, the ETRS89 coordinates of any point can be converted using a transformation provided by the IERS.

In the UK, the Ordnance Survey uses ETRS89 as its datum because coordinates taken in one year will relate to coordinates taken years later. If WGS84 or ITRS coordinates were used, they would have to have a date of survey associated with them in order that they could be related. This is also the reason why ETRS89 coordinates are of importance in engineering surveying.

Reflective summary

With reference to GPS coordinate systems, remember:

— GPS coordinates are based on different datums and it is important to know which one to use.

— All satellites in the GPS constellation broadcast orbital information that enable WGS84 coordinates to be computed. This is known as the *WGS84 broadcast TRF* in which the orbital information is derived from the 13 military tracking stations in the ground segment of GPS.

— Because it is based on 500 stations worldwide, a more accurate version of the WGS84 broadcast TRF is provided by the International Earth Rotation Service and is called the *International Terrestrial Reference Frame* (ITRF). This is based on a datum known as the *International Terrestrial Reference System (ITRS)* which is produced by the civilian GPS community instead of the US military.

— Due to tectonic plate motion, the ITRS datum is moving relative to most countries. To overcome this problem in Europe, a fixed definition of the ITRF is used which is known as the *European Terrestrial Reference System* 1989 or ETRS89. This is realised through the *European Terrestrial Reference Frame 1989 (ETRF89)*.

— The Ordnance Survey now uses ETRS89 as its datum.

8.6 Ordnance Survey National GPS Network

After studying this section you should be aware that the Ordnance Survey has established an online service known as the National GPS Network to help determine precise ETRS89 coordinates in Great Britain. You should know that this consists of an Active layer of Continuously Operating Reference Stations (CORS) and a Passive layer of stations. You should also be familiar with some of the post-processing techniques used to obtain ETRS89 coordinates at the centimetre level.

This section includes the following topics:

● Ordnance Survey National GPS Network
● Processing with CORS data
● Future developments at the Ordnance Survey

Ordnance Survey National GPS Network

When surveying with GPS, a user could obtain the WGS84 coordinates of a series of points with a receiver and use these for surveying purposes. Because these would not

be very accurate, it is better to acquire ETRS89 coordinates at the centimetre level instead using the Ordnance Survey National GPS Network. This is because ETRS89 is stable in time and is more accurately defined.

The Ordnance Survey National GPS Network is a three-dimensional TRF established by the Ordnance Survey for precise surveying work with GPS and consists of two different networks or layers: the Active layer and the Passive layer.

The *Active GPS network* (or *Active layer*) currently consists of over 30 permanent primary stations located throughout Great Britain whose coordinates are known relative to the ETRS89 datum with 8 mm plan and 20 mm height accuracy. As shown in Figure 8.7, most places in the UK are within about 100 km of one of these. At each station, the Ordnance Survey has installed a geodetic dual-frequency GPS receiver that operates continuously to record positional information from orbiting satellites. Each of the stations in the Active layer is known as a *Continuously Operating Reference Station* (*CORS*). The data from these stations is collected by the Ordnance Survey and is then made available, free of charge, on the Internet

Figure 8.7 ● CORS sites and their 100 km coverage in the Ordnance Survey Active GPS Network.

through the National GPS Network web site. As shown in Figure 8.8, precise positioning can therefore be performed in precise relative mode with a single geodetic GPS receiver, instead of two, by using data downloaded from the Active stations to replace the reference or base station. With appropriate observation times, processing and GPS receiver, it is possible to obtain post-processed ETRS89 coordinates for most points in Great Britain to within 5–10 mm by using the Active layer.

The *Passive GPS network* (or *Passive layer*) is a series of about 900 secondary control stations, located throughout Great Britain at intervals of 20–35 km. These are usually ground marks positioned for easy access and each one has coordinates known in the ETRS89 reference system with a current accuracy of 55 mm in plan and 66 mm in height. They can be used as reference points for GPS positioning, but in this case the user must simultaneously occupy the Ordnance Survey Passive station and an unknown point using two receivers. Unlike the Active layer, the Passive layer is not routinely monitored and its accuracy is uncertain. When used for precise relative real-time or post-processed surveys, it is possible to obtain accuracies at about the 50–100 mm level in ETRS89 coordinates from Passive stations.

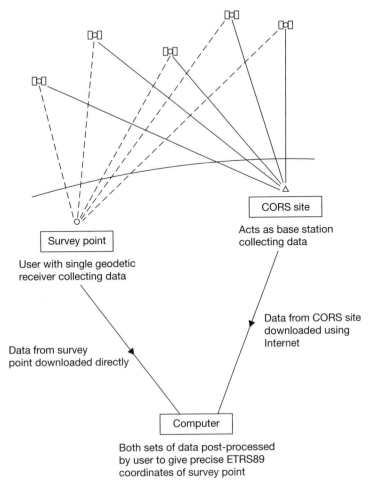

Figure 8.8 ● Surveying and processing with CORS.

Processing with CORS data

Using the Active stations of the National GPS Network, it is possible to determine the precise ETRS89 coordinates at most places in Great Britain. The data required for post-processing each CORS site is provided, for most users, by the Ordnance Survey. This is available hourly and kept for 30 days on the National GPS Network web site after which it is transferred to an archive known as the BIGF (http:// www.bigf.ac.uk/). All data of this type is available in an internationally agreed format known as the *Receiver INdependent EXchange* (*RINEX*) format. All data obtained from the CORS sites is only available at a fixed epoch rate of 15 s, but rates of 1 s can be interpolated by using software designed to do this. If a GPS surveyed point is positioned relative to either an Active or Passive station, the coordinates obtained are automatically in the ETRS89 system because both the Active and Passive stations have ETRS89 coordinates.

The advantages of using CORS are that the data from them is usually available (after a short delay) 24 hours a day every day of the year, the sites are permanent, a user only needs one GPS receiver to use them to obtain precise GPS positions and the data is available free of charge. In addition, as will be shown in the next section, the ETRS89 coordinates obtained from CORS processing can be converted directly to Ordnance Survey national control.

The disadvantages of using CORS are that Internet access is required to download the data required, difficulties may be caused by the user not having the same hardware as that at the CORS sites and the user does not have any control over the base station in a precise relative survey. Problems could arise here from different sampling rates at the CORS and those used on site plus the fact that a CORS site may not be operating at the time it is needed. However, one of the worst problems with CORS at present is that the distances to them can be excessive, giving rise to long baselines. When processing these, the height component of the ETRS89 coordinates obtained can be significantly in error because of the difficulty of removing tropospheric errors and because of the effect of *ocean tide loading*. Ocean tide loading causes variations in the position of CORS sites to occur because of the compression of the Earth's crust by tides – these effects are worse in the UK compared to other countries because it has a large tidal range and a long coastline. If not accounted for, these variations can cause errors to occur in processed coordinates of up to 10 cm vertically, but much smaller horizontally. The effect can be minimised by using GPS processing software which models the error accurately, by modifying the observation procedure to include long observation times and short baselines, or by using a good spread of CORS close to the surveyed point. Because of the difficulty of modelling some of these errors over long distances, this can limit the accuracy attainable for ETRS89 coordinates. In addition, the post-processing software may limit the length of baseline that can be processed.

In order to obtain the positions of points at the centimetre level with CORS data over baseline lengths up to 50 km (the distance to the CORS being used is up to 50 km), the following are recommended:

- Precise orbital data should be used (see Table 8.1), but remember that final orbits are not available for about 13 days.
- At least four hours' data is required at the survey point and from the CORS site.
- Great care must be taken to model tropospheric errors properly (see Section 7.4).
- The correct antenna information (type and make at the unknown point *and* the CORS) must be entered into the post-processing software, otherwise large height errors may result.
- Care must be taken to avoid high multipath environments (see Section 7.4).

To position points at the centimetre level for baseline lengths over 50 km, 24 hours' data may be required as well as specialist software.

When using CORS sites to process ETRS89 coordinates, there will be a number of different stations available for this. If these are all a considerable distance from the survey point, it is best to use only one CORS site in most cases to reduce the errors

Table 8.1 GPS satellite orbital products.

Orbit type	Availability	Updates	Precision	Baseline error at 100 km
Broadcast	Immediate		2–10 m	10–50 mm
Ultra-rapid	Immediate	Twice daily	25 cm	1 mm
Rapid	After 17 hours	Daily	5 cm	< 1 mm
Final	After 13 days	Weekly	< 5 cm	<< 1 mm

Note: Baseline error ≈ (d/20,000) × orbital error where d is the baseline length in km and the errors are in metres

caused by processing over long baselines. A second, more distant, CORS site should be used to check for any gross errors that might have occurred, but this would not be used for anything else. If there are two or three sites at about the same distance, they should all be used to process the required ETRS89 coordinates.

Future GPS developments at the Ordnance Survey

To improve the reliability of GPS positioning, the Ordnance Survey is gradually increasing the number of CORS sites throughout Great Britain. A higher density of these will help reduce the lengths of baselines that have to be processed, which in turn will help improve accuracy.

To augment the post-processing facility currently provided by the CORS sites, an RTK network (see Section 7.6) is being developed. This will include up to 110 CORS and will enable RTK positioning across most of Great Britain. At present, the network is being established for internal requirements, but the Ordnance Survey is assessing how they can make this service commercially available.

For the latest information on these developments, visit the National GPS Network web site.

Reflective summary

With reference to the Ordnance Survey National GPS Network, remember:

— The Ordnance Survey National GPS Network web site has been set up to provide users with information about GPS in Great Britain. It can be accessed by anyone and the services it provides are free of charge.

— The Active layer of the National GPS Network consists of a series of CORS situated throughout Great Britain from which it is possible to determine ETRS89 coordinates at the centimetre level – this can be done with a single geodetic receiver.

— Because the distances to the CORS can be very long, some care is needed when post-processing the baselines to these.

— The Ordnance Survey is assessing how it might introduce an RTK network that will make it possible for real-time surveying to be done across Great Britain at the 10–30 mm level with a single receiver.

8.7 Ordnance Survey National Grid and Ordnance Datum Newlyn

After studying this section you should be aware that the Ordnance Survey uses the National Grid for defining horizontal position and the Ordnance Datum Newlyn to define vertical position in Great Britain. You should be familiar with all aspects of these. You should also be aware that some problems exist when trying to use National Grid coordinates assigned to triangulation stations and Ordnance Datum Newlyn values for bench marks as some of these are no longer reliable.

This section includes the following topics:

- The National Grid
- Scale factor on the National Grid
- Grid convergence
- Ordnance Datum Newlyn (ODN)
- Using the National Grid and ODN

The National Grid

In the previous section, the method by which precise ETRS89 coordinates can be obtained has been described. These (and WGS84 coordinates) are not normally used for site surveys and setting out because they are not in a horizontal and vertical format that can be used with other equipment such as total stations and levels; they have large numerical values since they are based on global datums; and they will not be aligned with any local or site grid.

One way around this problem is to transform the GPS coordinates obtained into national coordinates. In Great Britain, this will involve changing the coordinates into values based on the National Grid and Ordnance Datum Newlyn. To understand the processes involved in doing this, the way in which position is defined on these is first described.

The National Grid is a traditional horizontal coordinate system which has been established by the Ordnance Survey for surveying and mapping in Great Britain.

The geodetic datum for the National Grid is based on 11 primary stations which define the orientation and position of an ellipsoid – the Airy 1830 ellipsoid. The TRF used to survey with (or *realise*) this datum is conventionally called OSGB36

Figure 8.9 ● Triangulation pillar.

because all the survey work that was done to determine the coordinates of points on it was carried out from 1936 to 1953. This, however, is slightly misleading because OSGB36 is both a datum and a TRF. It consists of 326 primary control points, 2695 secondary control points, 12,236 tertiary control points and 7032 fourth-order control points. These are the familiar *triangulation stations* (see Figure 8.9) that are seen on hilltops and in other prominent positions throughout the country – their name derives from the fact that most of the network was surveyed using triangulation techniques (see Section 6.8), as it was not possible to measure long distances very easily at the time.

The positions of these control points were obtained originally as latitudes and longitudes on the curved surface of the Airy ellipsoid. For convenience, these coordinates are easier to use if they are an easting and northing based on a plane rectangular grid. If this was the case, calculation of position could be carried out using plane trigonometry (as in Chapter 6) instead of much more complicated calculations on the ellipsoid. To represent ellipsoidal latitude and longitude on a curved surface as an easting and northing on a flat surface requires a *map projection*. Many different types of map projection can be used (see Further reading) and the one chosen by the Ordnance Survey for mapping and positioning in Great Britain is a *Transverse Mercator* projection. In Figure 8.10, this is shown to be a cylinder which touches the ellipsoid along a north-south line called the *central meridian*. This is then unwrapped and points on the ellipsoid are projected onto the flat surface of the unwrapped cylinder. Although Figure 8.10 shows this graphically, the geodetic coordinates of points are in fact converted to their equivalent rectangular coordinates mathematically using a set of projection formulae.

The central meridian of the Transverse Mercator projection used by the Ordnance Survey is at 2°W and this defines the direction of the *N*-axis of a plane rectangular coordinate system. The *E*-axis is at 49°N and these two axes define a rectangular grid for coordinates covering the whole of Great Britain. In order to keep all eastings positive (avoiding negative eastings west of the central meridian) and all northings (on the mainland) below 1000 kmN, this grid has a *false origin* 400 km west of the central

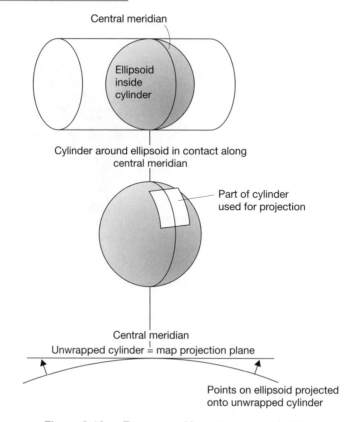

Figure 8.10 ● Transverse Mercator map projection.

meridian and 100 km north of the true origin on the central meridian at 49°N. This defines the Ordnance Survey National Grid coordinate system as shown in Figure 8.11. This is used to define horizontal position in all Ordnance Survey mapping products.

All of the constants associated with OSGB36 and the National Grid, together with the formulae used to convert coordinates from ellipsoid to projection are available on the Ordnance Survey National GPS Network web site.

Scale factor on the National Grid

When points are projected onto a plane rectangular grid from a curved ellipsoid, this will inevitably involve some distortion of the positions of points, except on the central meridian where the ellipsoid touches the cylindrical map projection. This means that the *scale distortion* on the central meridian is 1 and greater than 1 everywhere else. In this case, the *scale factor F* at any point on a Transverse Mercator projection is given by

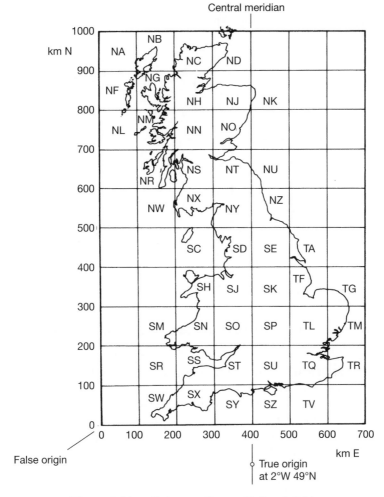

Figure 8.11 ● Ordnance Survey National Grid.

$$F = 1 + \frac{e^2}{2R^2}$$

where

e = distance of the point from the central meridian as a difference of eastings
R = the radius of the ellipsoid at the point

Since this increases as the square of the distance east and west of the central meridian, this distortion can become unacceptable at the extreme edges of the projection area. For the National Grid, the scale factor in Cornwall and the Western Isles at one side of the country and East Anglia at the other would be about 1.0008 if left unaltered. This means that a measured distance of 100.000 m would be 100.080 m on the projection, a difference of 80 mm. To reduce the scale error of

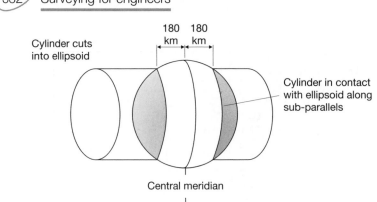

180 km   180 km

Cylinder cuts into ellipsoid

Cylinder in contact with ellipsoid along sub-parallels

Central meridian

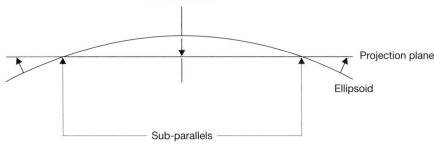

Projection plane

Ellipsoid

Sub-parallels

Figure 8.12 ● Sub-parallels on the Ordnance Survey Transverse Mercator projection.

0.0008 by half at all points on the projection, a scale factor of $1 - (0.0008)/2 = 0.9996$ is deliberately introduced to make the scale of the projection too small on the central meridian. This has the effect of halving the scale error for points at the east and west edges of the projection to 1.0004. As a result, a measured distance of 100.000 m lies between 99.960 m and 100.040 m on the projection, depending on where it is measured, with an error of 40 mm at the most. This reduction in scale error is important in mapping. By altering the scale on the projection in this way, the projection plane has been moved so that it cuts into the ellipsoid at about 180 km east and west of the central meridian, as shown in Figure 8.12. The points at which the projection is in contact with the ellipsoid are known as *sub-parallels* and the scale error along these will be zero. To calculate the scale factor at any point on the National Grid, the following formula is used

$$F = F_0 \left(1 + \frac{(E - 400{,}000)^2}{2R^2}\right) \qquad (8.1)$$

where

F_0 = the scale factor on the central meridian = 0.999 601 272
E = the National Grid easting of the point
R = the radius of the ellipsoid at the point

By using equation (8.1), the scale factor for any point can be calculated remembering that the units for E and R must be compatible. For example, given $E =$ 495 676.241 mE and taking $R = 6381$ km

$$F = 0.999\,601\,272\left(1 + \frac{(495{,}676 - 400{,}000)^2}{2(6{,}381{,}000)^2}\right) = 0.999\,7136$$

For another point with E = 182 073.450 mE

$$F = 0.999\,601\,272\left(1 + \frac{(182{,}073 - 400{,}000)^2}{2(6{,}381{,}000)^2}\right) = 1.0001842$$

When calculating position on the National Grid, the coordinate calculations given in Section 6.2 can be used but it is necessary to alter a measured distance to allow for scale factor before it is used in any calculations. This is done using

distance on the National Grid = measured distance (on the ellipsoid) × F (8.2)

For all surveys, a horizontal distance measured at ground level must be reduced to its equivalent on the ellipsoid before conversion to a length in the plane of the projection. Since the ellipsoid is close to the Geoid, which in turn is close to mean sea level, all horizontal distances measured are reduced to their equivalent values at mean sea level instead. The height or *mean sea level correction* for this is given by

$$\text{height correction} = -\frac{Dh_{\text{m}}}{R} \qquad\qquad (8.3)$$

where

D = the horizontal component of the measured distance
h_{m} = the mean height of the distance measured above mean sea level

The correction is negative unless a line below mean sea level is measured.

It is only necessary to calculate scale factors for construction projects when these are based on National Grid coordinates or another projection. By using scale factors, it has been shown that relatively small adjustments can be made to field data to enable position to be computed as plane coordinates. However, these corrections must not be overlooked, as the effect of omitting them can cause significant errors over long distances. In theory, the scale factor should be computed using the easting at each end of the line and the mean value used, but it is usually sufficiently accurate to compute it using the easting at the mid-point of the line instead. However, on long road and rail projects, a different scale factor must be computed for each 5 km section of the works.

Worked example 8.1: Use of scale factor in distance measurement

Question
Suppose a horizontal distance of 122.619 m was recorded by a total station for a line AB in a traverse that was to be computed on the National Grid. If the scale factor for AB is 0.999 631 2 and the average height of AB is 151.72 m above Ordnance Datum Newlyn (mean sea level), calculate the National Grid distance for AB.

Solution
Equation 8.3 gives (with R = 6381 km)

$$\text{height correction} = -\frac{122.619 \times 151.72}{6,381,000} = 0.0029 \text{ m}$$

and

$$\text{mean sea level distance AB} = 122.619 - 0.0029 = 122.6161 \text{ m}$$

Applying the scale factor using equation (8.2) gives

$$\text{National Grid distance} = 122.6161 \times 0.999\ 631\ 2 = \mathbf{122.571 \text{ m}}$$

Worked example 8.2: Use of scale factor in setting out

Question

In a road scheme, let the National Grid coordinates of a point on a road centre line be 612 910.741 mE, 157 062.283 mN. This is to be set out from a nearby control point with National Grid coordinates 612 963.524 mE, 157 104.290 mN. Calculate the setting out distance required for this.

Solution

The coordinates of the two points give

$$\Delta E = 612\ 963.524 - 612\ 910.741 = 52.783 \text{ m}$$

$$\Delta N = 157\ 104.290 - 157\ 062.283 = 42.007 \text{ m}$$

$$\text{National Grid distance} = \sqrt{\Delta E^2 + \Delta N^2} = 67.4584 \text{ m}$$

Substituting the mean easting into equation (8.1) gives

$$F = 1.000\ 158$$

and from equation (8.2)

$$\text{horizontal setting out distance} = \frac{\text{distance on the National Grid}}{F}$$

$$= \frac{67.4584}{1.000158}$$

$$= \mathbf{67.448 \text{ m}}$$

A height correction would have to be *added* to this if the road scheme is at an appreciable elevation.

Grid convergence

All surveys based on the National Grid produce coordinates relative to *grid north* defined by the central meridian. However, on some engineering projects, measurements are taken for a survey that give bearings relative to *true north* (or the IRP) instead and these must be converted to bearings based on grid north if projection

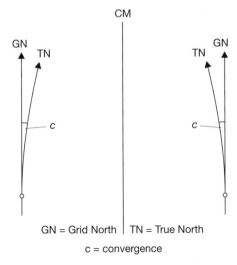

GN = Grid North | TN = True North

c = convergence

Figure 8.13 ● Grid convergence.

coordinates are to be computed. An example of this occurs in tunnelling and other underground work when a *gyro-theodolite* has been used to determine bearings for a traverse or control network. A gyro-theodolite is a north-seeking gyroscope combined with a theodolite which is capable of determining directions relative to the Earth's spin axis and true north.

Figure 8.13 shows that all lines defining grid north are parallel to the central meridian (CM), but because of the curvature of the Earth, lines that define the direction towards true north converge towards the pole and central meridian. Figure 8.14 shows how the *convergence c* is used to convert bearings from true north to grid north. Grid bearings can also be converted into their equivalent relative to true north. In this case, the grid bearing could be computed as a rectangular → polar conversion (as described in Section 6.2) using National Grid coordinates or it could be a bearing derived from measured angles that is to be checked by comparing it to a gyro-theodolite reading.

A value for the convergence can be computed using

$$c'' = \frac{e \tan \phi_G \times 206,265}{R}$$

where

e = the distance of the point from the central meridian at which the convergence is computed as a difference of eastings: this will be (400 km – E_A) or (E_X – 400 km) in Figure 8.14.

ϕ_G = the geodetic latitude of the mid-point of the line along which the bearing is computed (AB or XY in Figure 8.14)

R = the local radius of the ellipsoid at the mid-point of the line

Ordnance Datum Newlyn (ODN)

The positions of all OSGB36 control points contain only horizontal information (an easting and northing) because it was not possible to determine ellipsoidal heights easily when the observations were taken to locate these. Consequently, a different datum is used for defining heights in Great Britain.

For mainland Great Britain, this is known as the *Ordnance Datum Newlyn (ODN)* vertical coordinate system which gives heights above 'mean sea level'. This vertical

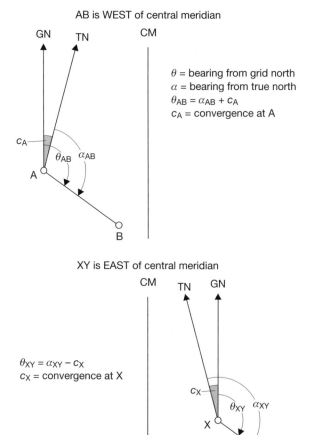

AB is WEST of central meridian

θ = bearing from grid north
α = bearing from true north
$\theta_{AB} = \alpha_{AB} + c_A$
c_A = convergence at A

XY is EAST of central meridian

$\theta_{XY} = \alpha_{XY} - c_X$
c_X = convergence at X

Figure 8.14 ● Calculation of bearings with convergence.

datum corresponds to the average sea level measured by a tide gauge at Newlyn, Cornwall between 1915 and 1921. Ideally, this should coincide with the Geoid as this is the equipotential surface that represents mean sea level. However, because of currents, tides and winds on the coast, the mean sea level obtained from the measurements at Newlyn gives a datum about 80 cm below the Geoid. Consequently, ODN is located on the equipotential surface that passes through Newlyn, which is slightly different from that of the Geoid and is offset some 80 cm below it. This difference has no effect on heights in Great Britain; it is simply the case that all ODN heights are based on a local geoid which is close enough to the Geoid and mean sea level for mapping purposes. This datum has not changed since it was adopted even though mean sea level has slowly risen since the tide gauge measurements were taken, and all heights given by the Ordnance Survey are not affected by these variations. Because of this, the correct term that should be used to describe an Ordnance Survey height on mainland Great Britain is 'orthometric height relative to ODN'. These have become a national standard in Britain and it is unlikely a new datum will be introduced. As well as ODN, there are a series of other Ordnance Survey island based datums, details of which are given on their web site.

As each country has established its vertical datum in a similar manner, national height datums are all located on slightly different equipotential surfaces, each one corresponding to mean sea level for that country. These differences are not important until the heights in one country have to be related to another. A good illustration of this is given by the Channel Tunnel project. In an ideal world, tunnelling would have proceeded from each end in England and France, and at the mid-point the two drives would have met in plan and in height. If heights based on the ODN and the French equivalent had been used for vertical control, this would not have happened because different tide gauges have been used to establish the vertical datums. These are located at Newlyn and Marseilles, on the Mediterranean coast, and the difference in the vertical datums produced by these is approximately 300 mm. This is a function of the slope of the sea surface between these two locations and not the sea slope between Dover and Calais.

The TRF that realises ODN and the other datums consists of a network of bench marks that cover the whole of Great Britain. The most accurate of these are known as fundamental bench marks and about 200 of these exist, the heights of which were obtained by a network of precise levelling lines across Britain. Using these, the TRF was densified, again by levelling, to obtain the heights of about three-quarters of a million Ordnance Survey bench marks all over the country. The majority of these lower-order bench marks can be seen cut into stone at the base of buildings and other structures (see Figure 2.3).

Using the National Grid and ODN

If a survey is to be based on the National Grid, the traditional method is to occupy a number of triangulation stations and use these for taking measurements to establish the horizontal positions of new control points or other features. To do this, the OSGB36 coordinates of the stations would be obtained from lists published by the Ordnance Survey. Fieldwork might be carried out with a total station by measuring angles and distances in a network, by using traversing, or by GPS methods. By using the National Grid in this way, it is possible to utilise a control network that covers a wide area without the need to establish a site datum.

There are a number of problems associated with this.

First, the survey that was carried out to obtain the OSGB36 coordinates of all the triangulation stations was done during the period 1936 to 1953. Since that time, survey equipment and methods have improved substantially and it has become evident from recent, more precise, measurements that there are some errors and distortions in the published coordinates of the National Grid. As well as this, many of the triangulation pillars have not been maintained for some time and their positions may have altered because of subsidence or through other disturbances.

Second, even in areas where no serious distortions have been found, the original accuracy of the network may not match that obtainable from modern equipment. Consequently, any survey adjusted to fit OSGB36 using coordinates taken from published triangulation station lists may not meet specifications even though the observations taken from them do.

Third and finally, another problem when using triangulation stations in control surveys is that they are located in remote areas, often on hilltops. This makes it difficult to include them in a network or traverse without substantially increasing the fieldwork involved.

All of these problems are overcome by using GPS methods to realise OSGB36.

On some construction projects, it is convenient to use ODN heights for vertical control, especially when the project covers a large area. Today, about half a million Ordnance Survey bench marks are still in use and it is possible to find the location and heights of these from the Ordnance Survey. As discussed in Section 2.1, most of these have not been checked by the Ordnance Survey since the 1970s and their heights can be unreliable due to possible movement of the bench marks since they were originally levelled. Because of this, the Ordnance Survey now recommends that all ODN heights are obtained by using GPS methods without reference to any of their bench marks. The advantages of this are that a reliable, uniform and accurate value for ODN heights is obtained and at any location to suit the user.

The methods by which OSGB36, ODN and the other height datums are realised with GPS are described in the next section.

Reflective summary

With reference to the Ordnance Survey National Grid and Ordnance Datum Newlyn, remember:

— The datum and TRF used to define the National Grid is called OSGB36. Traditionally, the TRF has consisted of the coordinates of a series of triangulation stations situated throughout Great Britain.

— Because position on OSGB36 is in geodetic coordinates on an ellipsoid, these are converted to an easting and a northing on a plane rectangular grid using a Transverse Mercator map projection.

— The coordinate grid defined by the Transverse Mercator projection is called the National Grid.

— Because of problems associated with errors and distortions in the coordinates of the triangulation stations used to realise OSGB36, they should *not* be used for precise engineering surveys and other site work. Instead, it is better to use GPS methods to realise OSGB36.

- When calculating position on the National Grid, it is important to take account of scale factor – if this is ignored serious errors can occur.

- ODN defines 'mean sea level' or orthometric heights in Great Britain. This corresponds to the equipotential surface passing through Newlyn, Cornwall. Other height datums exist on offshore islands.

- The TRF used to realise ODN and the other island height datums consists of all the heights of the Ordnance Survey bench marks across Great Britain.

- Because they have not been checked for a long time, some caution must be exercised when using Ordnance Survey bench marks in engineering surveying. Instead of using the published values for the heights of the bench marks, it is better to use GPS to realise ODN.

8.8   Coordinate transformations

After studying this section you should understand what a coordinate transformation is and how this differs from a coordinate conversion. You should be aware that the Ordnance Survey has produced the OSTN02 transformation for changing ETRS89 coordinates directly into a National Grid easting and northing and the OSGM02 transformation for changing ETRS89 coordinates directly into an ODN height. You should also be familiar with the methods used for transforming GPS coordinates in local surveys.

This section includes the following topics:

- Coordinate transformations
- National Grid Transformation OSTN02
- National Geoid Model OSGM02
- Transformations for local surveys

Coordinate transformations

By using the National GPS Network, it is possible to obtain the precise ETRS89 coordinates of any point in the UK by using the CORS. However, these coordinates are not convenient and have to be changed into something else – for example, a National Grid easting and northing plus ODN height or to coordinates on a local site system with arbitrary height. This is done using coordinate (or datum) transformations.

A coordinate transformation is a mathematical operation that takes the coordinates of a point in one coordinate system and changes them into the coordinates of the same point in a different coordinate system. The subject of coordinate transformations was introduced in Section 6.3, but only for simple two-dimensional problems that might involve changing coordinates between different grids on site. In contrast to these, it is possible to use transformations between different regional and global coordinate systems for GPS surveys.

There is an important difference between a coordinate transformation and a coordinate conversion. When geodetic coordinates are converted to cartesian coordinates on the same ellipsoid, the conversion is exact, as is the conversion of latitude and longitude to easting and northing via a map projection. On the other hand, coordinate transformations are not exact. The reason for this is that a coordinate transformation relies on the comparison of measured coordinates of a series of points in one coordinate system compared to the measured coordinates of the same points in the second coordinate system. Since measurements are subject to variation and inaccuracy, the transformation parameters obtained are dependent on whatever set of coordinates are compared. When a redundancy exists like this, it is impossible to define a transformation between two datums exactly and a best estimate is produced using least squares with a statistical assessment of how good the estimate is. The amount by which a transformation is in error depends on the distribution of the errors in the two coordinate systems that are being compared and how carefully the transformation has been designed to take account of these errors.

National Grid Transformation OSTN02

In Section 8.6, the methods by which precise ETRS89 coordinates can be obtained at any point in Great Britain using the National GPS Network were described. For some construction projects, it is convenient that these are transformed into OSGB36 coordinates and ODN heights.

By visiting a number of triangulation stations in the project area and taking GPS observations at these, users could obtain enough information to derive their own local transformation for the project area based on the published coordinates of the stations and the ETRS89 coordinates obtained at these. This could also be done using WGS84 coordinates as broadcast by the GPS satellites. In Section 8.7, the variable nature of the errors and distortions in the triangulation station reference frame used to define OSGB36 have been discussed. Because of these, it is very unlikely that a good transformation would be obtained and this method should not be used for transforming satellite coordinates into OSGB36 coordinates. Instead, a transformation provided by the Ordnance Survey for obtaining OSGB36 coordinates from ETRS89 coordinates must be used. This is known as the *OSTN02 transformation*, which is available, free of charge, on the National GPS Network using the Coordinate Converter facility – it can also be downloaded from this to a computer and incorporated into commercial software.

In summary, the recommended procedure for obtaining the National Grid coordinates of any point is as follows. First, GPS methods are used to obtain the three-dimensional ETRS89 coordinates of the point using the National GPS Network (for example, by taking GPS observations at the point and post-processing these with CORS data). Second, these coordinates are transformed directly into a National Grid easting and northing using OSTN02. OSTN02 is known as the *Definitive Transformation* because, along with ETRS89 coordinates, it defines OSGB36. It consists of a grid of ETRS89 to OSGB36 difference parameters – around two million of them – and this ensures that OSTN02 is accurate to about 10 cm. If the ETRS89 coordinates are replaced by WGS84 coordinates, it is only possible to achieve an accuracy at about the metre level. Some software packages use simpler transformations which cannot accurately model the eccentricities in OSGB36, and these are only accurate to between 2 and 20 m. It is therefore important to know exactly what transformation is being used by a software package when deriving National Grid coordinates from GPS observations.

By using this method, anyone with GPS equipment will have access to OSGB36 without having to visit any triangulation stations because these have been replaced by the National GPS Network and OSTN02 as the realisation of OSGB36. If, for some reason, a triangulation station is to be included in a survey, the archived coordinates of that point will no longer be true but they will agree to within 0.1 m of the existing coordinates re-fixed using the National GPS Network and OSTN02.

National Geoid Model OSGM02

A vertical transformation between ETRS89 coordinates and orthometric height is also available from the Ordnance Survey – this is known as *OSGM02*, which gives ODN and other orthometric heights directly from a GPS survey with an accuracy of about 0.02–0.15 m depending on location. For mainland Great Britain, the accuracy is at the 0.02 m level. In this way, ODN heights can be obtained without the need to visit any Ordnance Survey bench mark by using the National GPS Network in conjunction with OSGM02. Because of the problems with the ageing bench marks, this is recommended by the Ordnance Survey as the method for establishing any bench marks with an ODN height. GPS plus OSGM02 should be used to bring ODN into a local survey to establish a TBM. From this, levelling can then be used to get ODN heights throughout a survey.

As can be seen, the precise position of a point determined in ETRS89 using the National GPS Network can be converted into National Grid and ODN coordinates using transformations provided by the Ordnance Survey. A flow chart showing the transformation of GPS data to National Grid coordinates and ODN heights is shown in Figure 8.15. Compared to a horizontal triangulation network located mostly on hilltops and a separate vertical network of bench marks located mostly in lowland areas, the National GPS Network provides a three-dimensional TRF which unifies OSGB36 and ODN using transformation software.

When using commercial software to carry out these transformations, it is important to check it is using OSTN02 and OSGM02 and not different transformations provided by the manufacturer.

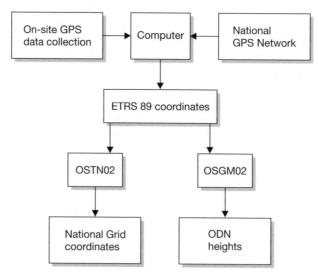

Figure 8.15 ● Transformation of raw GPS data to National Grid coordinates and ODN heights.

Transformations for local surveys

Rather than transform GPS coordinates into National Grid and ODN values, many surveys require these to be transformed directly into a local site coordinate grid. In this case, it is necessary to determine a set of local transformation parameters. There are several different methods available for doing this, most of which can be performed using software supplied by the manufacturers of GPS equipment. For plane rectangular grids, the *One Step* and *Interpolation* methods are probably the most popular for sites less than about a 10 km square because they do not require local projection and local ellipsoid parameters to be known.

The methods used to derive local transformation parameters all involve obtaining the GPS *and* local coordinates of a number of control points on site. The GPS coordinates can be obtained as ETRS89 coordinates by using the National GPS Network. This is the recommended method and has the advantages that the GPS coordinates obtained will be precisely known on the ETRS89 datum and will be consistent with other surveys carried out using this. In addition, new points can be added to a survey at a later stage even if all the original control points have been lost. As well as this, the coordinates could be transformed at any time into OSGB36 and ODN coordinates. For the best accuracy, long observation times may be necessary (over 4 hours), a geodetic GPS receiver and antenna are required and the Active stations (CORS) should be used rather than the Passive stations to ensure the best level of accuracy. The suggested procedure to follow when obtaining the ETRS89 coordinates is to observe and process each control point separately. Alternatively, these may be obtained at one point only, which could then be used as a base station for determining the coordinates of other points using static or rapid static methods. Another approach is to use raw WGS84 coordinates directly from the broadcast satellite orbital data but at a single point only. The coordinates obtained for this point could then be transferred to the other control points using this as a base station. This will, however, only produce WGS84 coordinates at about the 5 m accuracy level – it does not matter that these coordinates may be offset from WGS84 by as much as 5 m as the transformation parameters derived would take account of this. It must be noted though, that the parameters will only be valid for the WGS84 coordinates obtained at the base station (and any points these have been transferred to) and will not be compatible with any other WGS84 or ETRS89 coordinates.

The site grid coordinates required for a local transformation are obtained by traversing, observing a network and by levelling on site. Once obtained, these and the GPS coordinates are compared by the transformation software and the appropriate transformation parameters are computed. These are then stored by the system computer and used in post-processed surveys to obtain the local coordinates of any other point, or they are transferred to a GPS controller or data collector for RTK surveys.

Whatever method is used for transforming GPS coordinates into site grid coordinates, the following should be noted:

- At least four or five common control points must be used when deriving the transformation parameters. Use of a single point is not recommended.
- The resultant accuracy of the transformation should be checked to see if there are any poor control points.
- The control points used should be distributed throughout the survey area and not located in one part of the survey (see Figure 8.16).

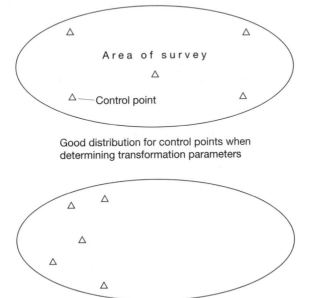

Good distribution for control points when determining transformation parameters

Poor distribution for control points when determining transformation parameters

Figure 8.16 ● Control point distribution for coordinate transformations in local surveys.

- The parameters obtained should not be used to transform any coordinates that lie outside the area enclosed by the control points.

- The Geoid–ellipsoid separation N is usually assumed to be zero at all control points in a local transformation. The parameters obtained using this assumption will only be valid if there are no gravity anomalies in the survey area.

- If WGS84 coordinates are used, the transformation parameters obtained will only be valid for the coordinates observed at the single point position (and any others derived from this). If these coordinates are changed for some reason or different control points are to be used (even in the same area), a new set of transformation parameters will have to be determined.

The procedures described above for determining transformation parameters have dealt with local sites up to about 10–15 km in extent. For much larger sites, National Grid and ODN coordinates derived from the National GPS Network are recommended.

Reflective summary

With reference to coordinate transformations, remember:

- A coordinate transformation is used to change the coordinates of points in one system to the equivalent coordinates in another system. Because of GPS, these are used routinely in surveying but without the user realising this.

- Unlike a coordinate conversion, a coordinate transformation is never exact because it relies on comparing sets of coordinates measured in two different coordinate systems.

- The OSTN02 transformation produced by the Ordnance Survey is used to transform ETRS89 coordinates directly into a National Grid easting and northing. In fact, ETRS89 coordinates + OSTN02 now *defines* the National Grid – this has replaced the triangulation stations as the realisation of OSGB36.

- An ODN and other datum heights can be obtained directly from ETRS89 coordinates using the OSGM02 transformation produced by the Ordnance Survey. This is the recommended procedure for obtaining a reliable height for a bench mark based on ODN.

- When establishing GPS control on local sites, the use of the National GPS Network and ETRS89 coordinates is advised. Try not to use WGS84 coordinates obtained from orbital data broadcast by the GPS satellites.

- Do *not* use a single-point transformation to change GPS coordinates into local site grid coordinates – it is better to derive transformation parameters by comparing coordinates at several control points.

- Be careful when computing a local transformation to ensure that the control points used are distributed across the site and not all together in one part of it.

- When computing local coordinates with a local transformation, do not do this for points well outside the area enclosed by the control points used to derive the transformation parameters.

Exercises

8.1 State the definition of geodesy and discuss the role that it plays in satellite surveying systems.

8.2 Explain how the Geoid and ellipsoid try to define the shape of the Earth.

8.3 What is an equipotential surface and why is it important in surveying?

8.4 For an ellipsoid, draw diagrams to show what the semi-major and semi-minor axes represent. State how these are related to the flattening and eccentricity.

8.5 What are the International Reference Pole and the International Reference Meridian?

8.6 Explain, with the aid of diagrams, how the position of a point can be defined as three-dimensional geodetic and cartesian coordinates.

8.7 Explain how an ellipsoidal height differs from an orthometric height.

8.8 What is the difference between a Terrestrial Reference System and a Terrestrial Reference Frame?

8.9 What is the difference between WGS84 and ETRS89 coordinates?

8.10 Describe how the Active and Passive layers of the National GPS Network differ.

8.11 When downloading data online from the National GPS Network, what are the following?

CORS
RINEX
BIGF
satellite orbital products

8.12 What special procedures should be used to maintain positional accuracy when post-processing long GPS baselines?

8.13 Describe ocean tide loading and the effect it can have on the accuracy of post-processed coordinates.

8.14 What is OSGB36 and what was the reference frame used for realising this until the arrival of GPS?

8.15 Using a map projection, coordinates on an ellipsoid can be converted into eastings and northings on a plane rectangular grid. Why is this usually done for engineering surveys?

8.16 Define the following for the Transverse Mercator map projection of Great Britain:

true origin
false origin
central meridian
convergence

8.17 What is scale factor for a map projection and how is it calculated for the Transverse Mercator map projection of Great Britain? Using Excel, plot a graph of how this varies across Great Britain.

8.18 The horizontal distance measured between two points is AB = 81.587 m. If the National Grid coordinates at point A are 395 438.009 mE, 505 854.791 mN, calculate the distance required to compute National Grid coordinates with AB when it is measured at 10 m and 1000 m above mean sea level. Assume the radius of the ellipsoid is 6380 km.

8.19 The coordinates of two control points on the National Grid are

A   640 101.358mE   306 147.630 mN
B   640 067.214 mE   306 209.052 mN

The distance between these is to be checked with a total station. What reading should be obtained for the horizontal distance AB? Assume the radius of the ellipsoid is 6380 km and that the line AB is close to mean sea level.

8.20 How does Ordnance Datum Newlyn define 'mean sea level' and how does this differ from the Geoid?

8.21 Discuss the problems that can occur when triangulation stations and Ordnance Survey bench marks are used in control surveys.

8.22 What is a coordinate transformation and how does it differ from a coordinate conversion?

8.23 What are OSTN02 and OSGM02?

8.24 When transforming GPS coordinates into site grid coordinates, why is it better to use ETRS89 coordinates instead of WGS84 coordinates?

8.25 Discuss possible sources of error and how these can be avoided when deriving local transformation parameters for GPS surveys.

8.26 Visit the Ordnance Survey National GPS Network web site and obtain a user name and password. The site contains a considerable amount of background information on GPS and its coordinate systems, all of which should be explored. It has a coordinate converter for performing a variety of coordinate conversions and transformations – these should also be looked at. It is also the primary source for downloading CORS (RINEX) data in Great Britain.

8.27 Using the National GPS Network, find and list the parameters and conventions that define GPS coordinates on WGS84.

8.28 Visit the BIGF web site for further information on this facility.

Further reading and sources of information

Most of the further reading and sources of information given at the end of Chapter 7 are also useful here.

The most important web site for this chapter is the Ordnance Survey National GPS Network which is accessed on

`http://www.gps.gov.uk/`

Other sites of interest are

`http://www.iers.org/`	for IERS information and services
`http://www.epncb.oma.be/`	for current details of ETRS89
`http://igscb.jpl.nasa.gov/`	the International GPS Service (IGS) home page
`http://www.bigf.ac.uk/`	UK site containing archived CORS data
`http://www.ngs.noaa.gov/CORS/`	for information on CORS in the USA

Some of the material presented in this chapter is based on

A guide to coordinate systems in Great Britain

and

GPS and coordinate transformations in Great Britain

Published by the Ordnance Survey on the National GPS Network.

Continuously Operating Reference Stations
CPD course notes
Stuart Edwards and Matt King, Geomatics Application Centre, School of Civil Engineering and Geosciences, University of Newcastle upon Tyne

The following books extend the coverage of geodesy, coordinate systems and map projections given here

Seeber, G. (2003) *Satellite Geodesy*, 2nd edn. Walter de Gruyter, New York.

Torge, W. (2001) *Geodesy*, 3rd edn. Walter de Gruyter, New York.

Iliffe, J. (2000) *Datums and Map Projections*. Whittles Publishing, Caithness.

Bugayevskiy, L. and Synder, J. (1995) *Map Projections: A Reference Manual*. Taylor & Francis, Basingstoke.

chapter 9

Measurements, errors and specifications

 ## Aims

After studying this chapter you should be able to:

- Discuss the reasons why the analysis of errors is important in surveying
- Describe the different types of error that can occur when surveying and give examples of how these are either eliminated or minimised
- Define what is meant by the most probable value of a quantity and give reasons why these are used in surveying
- Explain the difference between precision and accuracy in a surveying context
- Calculate most probable values and standard errors for repeated measurements in a survey
- Understand the concept of weight and how this helps to assess the precision of survey measurements
- Analyse how errors propagate through a survey
- Use various sources of information to assess specifications and tolerances for engineering surveying
- Discuss the reasons why least squares is essential for all survey analysis and adjustment

This chapter contains the following sections:

9.1   Errors and residuals

9.2   Precision and accuracy

9.3   Propagation of variances and standard errors

9.4   Survey specifications

9.1   Errors and residuals

After studying this section you should be able to describe how gross, systematic and random errors can occur in surveying and you should be familiar with some of the methods used to manage these. You should also understand why the least squares method is used to analyse survey measurements and what the significance of a most probable value is.

This section includes the following topics:

- Errors are a property of measurements
- Types of error
- Least squares estimation and most probable value

Errors are a property of measurements

In the previous chapters, the types of measurement that are fundamental to engineering surveying have been identified as horizontal distance and vertical distance (or height), together with horizontal and vertical angles. As shown throughout the book, many different techniques can be used to measure these quantities and many different instruments and methods have been developed for this purpose. Engineering and site surveying, then, is a process that involves taking observations and measurements with a wide range of electronic, optical and mechanical instruments, some of which are very sophisticated. However, even when using the best equipment and methods, it is still impossible to take observations that do not include some occasional mistakes or biases or which are completely free of small variations. These errors are very important since they are a property of all measurements and it is desirable that everyone who uses survey equipment and methods understands the errors they might be subject to and what magnitudes these might have under normal circumstances. Only by knowing the possible errors and their magnitudes and how they propagate through a measurement can the correct survey equipment and methods be chosen on site to ensure that all errors are kept within acceptable or known limits.

The use of the word *error* in surveying does not always imply that some mistake has been made, but indicates that a difference exists between the *true value T* of a quantity and any measured value of it x. Using this, an error e in an observation can be defined as

$$e = T - x$$

It is important to realise that, for surveying measurements, the true value of a quantity is usually never known, and because of this the exact error in a measurement or observation can never be known. It can also be said that every observation, no matter how carefully it is taken, is never exact and will always contain errors.

Types of error

The different types of error that can occur in surveying and other measurements are as follows.

Gross errors are often called mistakes or blunders, and they are usually much larger than the other categories of error. On construction and other sites, mistakes are often made by inexperienced engineers and surveyors who are unfamiliar with the equipment and methods they are using. As a result, gross errors are due to incompetence and also carelessness, and many examples can be given of these. Common mistakes include reading a levelling staff or tape graduation incorrectly or writing the wrong value in a field book by transposing numbers (for example 28.342 is written as 28.432). Failure to detect a gross error in a survey or in a setting out procedure can lead to serious problems on site, and for this reason it is vital that all survey work has observational and computational procedures built into it that enable mistakes to be highlighted. Examples of good practice for the elimination of gross errors are given throughout the book.

Systematic errors are those which follow some mathematical law, and they will have the same magnitude and sign in a series of measurements that are repeated under the same conditions. In other words, if the instrument, observer and surroundings do not change, any systematic errors present in a measurement will not change. If the measuring conditions change, the size of the systematic errors will change accordingly. One of the problems with systematic errors is that they are part of a measurement until something is done to remove them. In addition, they also accumulate if there is more than one present in a measurement. If an appropriate mathematical model can be derived for a systematic error, it can be eliminated from a measurement by using corrections. For example, in Section 4.3 it has been shown that the effects of a number of different factors in steel taping, such as temperature and tension, can be removed from a measurement by calculation using simple formulae (each formula is a mathematical model in this case). Another method of removing systematic errors is to calibrate the observing equipment and to quantify the error, allowing corrections to be made to further observations. In Section 5.5, it has been shown that it is often necessary to perform an electronic calibration on a total station to measure a range of instrumental systematic errors that are present in the instrument. Calibration values are then applied automatically by the total station to any subsequent measurements taken with it. A good example of a systematic error in distance measurement with a total station is the prism constant – if this is ignored or applied incorrectly, it is an

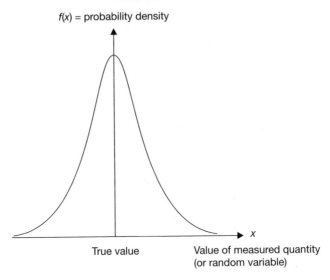

Figure 9.1 ● Normal distribution curve.

error that will be present in all readings. Similarly, failure to enter the correct phase centre offset into GPS post-processing software will result in errors in heights being obtained. Observational procedures can also be used to remove the effect of systematic errors; examples of these are to take the mean of face left and face right readings when measuring angles (to remove horizontal and vertical collimation errors from angles) and to keep the length of back sights and fore sights equal when levelling (to remove the collimation error in the level from height differences).

When all gross and systematic errors have been removed, a series of repeated measurements taken of the same quantity under the same conditions will show some variation beyond the control of the observer. These variations are inherent in all types of measurements and are called *random errors*, the magnitude and sign of which are not constant and occur by chance according to the laws of probability. Random errors cannot be removed from observations, but methods can be adopted to ensure that they are kept within acceptable limits. In order to analyse random errors or variables, statistical principles are used. In surveying it is usual to assume that random variables are *normally distributed*, as shown in Figure 9.1. This demonstrates the general laws of probability that random errors follow, which are

● Measurements that have values close to the true value occur frequently – therefore small random errors are more probable than large ones.

● Measurements that have values that are not close to the true value happen infrequently – therefore large random errors are less probable than small ones and very large errors may be mistakes and not random errors.

● Because the normal distribution curve is symmetrical, positive and negative random errors of the same size are equally probable and happen with equal frequency.

Note: *Throughout the remainder of this chapter it is assumed that all gross and systematic errors have been removed from observations and that only normally distributed random errors are being dealt with.*

Least squares estimation and most probable value

In the absence of gross and systematic errors, a random error is the difference between the true value of a quantity and an observation or measurement of that quantity. Consequently, before any random errors can be calculated for a set of observations or measurements, the true value of the observed or measured quantity must be known. Since the true value is seldom known in surveying, a quantity known as the *most probable value* (*MPV*) is calculated and used instead of the true value. If $\bar{x}$ is the most probable value of a quantity, the difference between this and a measured value x is known as *residual v*, where

$$v = \bar{x} - x$$

In surveying, a most probable value can be found using the principle of *least squares*, which states that

> The most probable value obtainable from a set of measurements of equal precision is that for which the sum of squares of the residuals is a minimum.

If a single quantity such as a distance x is measured with a tape n times, it can be shown that the least squares method gives the arithmetic mean as the MPV for the distance provided that each measurement is independent and taken under similar conditions. This could have been deduced using one of the general laws of probability: that positive and negative errors of the same magnitude occur with equal frequency.

As can be seen, the problem of not knowing true values in surveying is overcome by using MPVs and residuals to analyse the effect of random errors. However, one of the reasons that least squares has found widespread use in surveying is that it tends to produce true values from residuals even though residuals are not true observational errors. The least squares method is discussed further in Section 9.5.

Reflective summary

With reference to errors and residuals, remember:

— Three different types of error can be present in survey measurements. These are called gross, systematic and random errors. Various strategies can be put in place to deal with these.

— Since random errors cannot be determined exactly, they are replaced in survey calculations that deal with errors by residuals based on a most probable value.

— For surveying, most probable values are determined using the least
squares method.

9.2   Precision and accuracy

After studying this section you should understand the difference between precision
and accuracy. You should know that standard errors are used to measure precision in
surveying and be able to compute these. Furthermore, you should be familiar with
the concept of weight when to applied to survey measurements and how this is
linked to standard error. Finally, you should be able to compute weighted means.

This section includes the following topics:

● Definition of precision and accuracy

● Standard deviation, variance and standard error

● Weight

Definition of precision and accuracy

The terms *precision* and *accuracy* are frequently used in engineering surveying both by
manufacturers when giving specifications for their equipment and on site by
surveyors and engineers to describe how good their fieldwork is.

Precision represents the repeatability of a measurement and is concerned only with
random errors. Good precision is obtained from a set of observations that are closely
grouped together with small deviations from the sample mean $\bar{x}$ or MPV and have
the normal distribution shown in Figure 9.2(a). On the other hand, a set of observations that are widely spread and have poor precision will have the normal distribution shown in Figure 9.2(b).

In contrast, *accuracy* is considered to be an overall estimate of the errors present in
measurements, including systematic effects. For a set of measurements to be considered accurate, the MPV or sample mean must have a value close to the true value, as
shown in Figure 9.3(a). It is quite possible for a set of results to be precise but inaccurate, as in Figure 9.3(b), where the difference between the true value and the MPV is
large and probably caused by one or more systematic errors. Since accuracy and
precision are the same if all systematic errors are removed, precision is sometimes
referred to as *internal accuracy*.

For many surveying measurements, the term *relative precision* is sometimes used,
and this is the ratio of the precision of a measurement to the measurement itself. For
example, if the precision of the measurement of a distance d is s_d, the relative precision
is expressed as 1 in d/s_d (say 1 in 5000). This is the method used in Chapter 4 to define

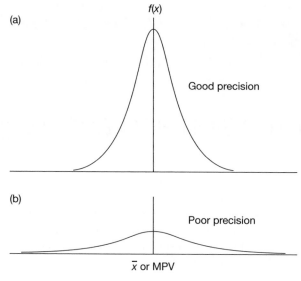

(a)

f(x)

Good precision

(b)

Poor precision

$\bar{x}$ or MPV

Figure 9.2 ● Precision.

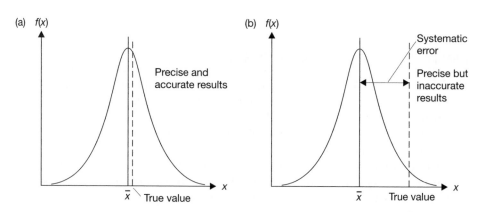

(a) f(x)

Precise and
accurate results

$\bar{x}$ True value

(b) f(x)

Systematic
error

Precise but
inaccurate
results

$\bar{x}$ True value

x

Figure 9.3 ● Accuracy and precision.

precisions in taping and in Chapter 6 to define the precision (or accuracy) of a traverse. An alternative to this is to quote relative precision in parts per million or ppm (that is 1 in 1,000,000). This is also equivalent to a precision of 1 mm per km, and both of these are used for distance measurement with total stations. The relative precision of a measurement is often calculated in surveying as soon as its precision is known, or it may be specified before starting a survey so that proper equipment and methods can be selected to achieve the stated relative precision. This is discussed in a number of chapters in the book.

In all types of surveying, attempts are always made to detect and eliminate mistakes in fieldwork, and the degree to which a survey is able to do this is a measure of its *reliability*. Unreliable observations are those which may contain gross errors without the observer knowing, whereas reliable observations are unlikely to include undetected mistakes.

Standard deviation, variance and standard error

In order to be able to compare one set of similar observations with another, the spread of a set of residuals must be assessed. To do this, they are assumed to follow a normal distribution. A normal distribution function is defined by two parameters: its *expectation* (the MPV in this case) and its *standard deviation* σ. The standard deviation is a measure of the spread or dispersion of measurements, and in Figure 9.2(a) the spread of the results is small, giving rise to a small standard deviation and good precision. Conversely, the spread of the results in Figure 9.2(b) is much greater, giving rise to a large standard deviation and poor precision. Rather than use terms like 'small' and 'large' in this context, it is obviously better to compute standard deviations, and the equation for the standard deviation of a variable x measured n times is given by

$$\sigma_x = \pm\sqrt{\frac{\sum_{i=1}^{n} v_i^2}{n}} = \pm\sqrt{\frac{\sum_{i=1}^{n}(\bar{x} - x_i)^2}{n}} \tag{9.1}$$

where

x_i is an individual measurement
$\bar{x}$ is the mean value (MPV)
v is a residual such that $v_i = (\bar{x} - x_i)$

The square of the standard deviation σ^2 is called the *variance*.

For statistical reasons, a standard deviation should be derived from a large number of observations, and since surveying measurements are usually taken in small sets, any standard deviations derived from them may be biased. For this reason, a better measure of precision is obtained for surveying observations by using an unbiased estimator for the standard deviation known as the *standard error*, s_x. This is obtained by replacing n with $(n-1)$ in equation (9.1) to give

$$\text{standard error} = s_x = \pm\sqrt{\frac{\sum_{i=1}^{n} v_i^2}{(n-1)}} = \pm\sqrt{\frac{\sum_{i=1}^{n}(\bar{x} - x_i)^2}{(n-1)}} \tag{9.2}$$

In this equation, $(n-1)$ is known as the number of *degrees of freedom* or *redundancy* and represents the number of extra measurements taken to determine a quantity. If a distance was measured 10 times, it has $(10-1) = 9$ degrees of freedom and there are nine redundant observations, since only one is required to give the distance. Without redundant observations, it is not possible to calculate standard deviations and standard errors for measured quantities in surveying. Redundant observations are also used to check fieldwork, the classic case being to measure the three angles of a triangle in a network when only two are needed to define it uniquely – the third angle is used to check that the measured angles add up to 180°.

Obviously, as the number of repetitions in a measurement increases, the difference between n and $(n-1)$ becomes smaller, as does the difference between standard deviation and standard error in equations (9.1) and (9.2). In some textbooks and software that process survey errors this difference is ignored and only standard deviations are considered.

Although the standard error s_x gives a measure of the dispersion of a set of observations, it can also be considered as the standard error of each individual measurement in the series. Also of interest is the precision of the outcome of the measurements, and the standard error of the mean (MPV) of a set of observations $s_{\bar{x}}$ is given by

$$s_{\bar{x}} = \pm \frac{s_x}{\sqrt{n}} \qquad (9.3)$$

This demonstrates a practical issue for surveying fieldwork. Suppose a quantity is to be measured using a field procedure and equipment that has a known standard error from previous work. This could be for the measurement of distance with a 30 m steel tape with a standard error of ±6 mm. If a standard error of ±3 mm is required using the same equipment and methods, equation (9.3) gives (with $s_x = \pm 6$ mm and $s_{\bar{x}} = \pm 3$ mm)

$$\pm 3 = \frac{\pm 6}{\sqrt{n}} \quad \text{or} \quad \sqrt{n} = \frac{6}{3} \quad \text{giving } n = 4$$

This shows that to double the precision, the distance would have to be measured four times and a mean computed.

The standard error for a series of measurements indicates the probability or chance that the true value for the measurements lies within a certain range of the sample mean (MPV) and it can be shown that, for the normal distribution function, there is a 68.3% chance that the true value of a measurement lies within the range $\bar{x} - s_{\bar{x}}$ to $\bar{x} + s_{\bar{x}}$. The limits or ranges within which true values are assumed to occur are called *confidence intervals*; these are shown in Table 9.1 and graphically in Figure 9.4. A reminder is given at this point that the probabilities given in Table 9.1 and Figure 9.4 assume that measurements are normally distributed and have had all gross and systematic errors removed from them.

Table 9.1 ● Probabilities associated with the normal distribution.

Probability %	68.3	90	95	95.4	99	99.7	99.9
Confidence interval	±1 $s_{\bar{x}}$	±1.65 $s_{\bar{x}}$	±1.96 $s_{\bar{x}}$	±2 $s_{\bar{x}}$	±2.58 $s_{\bar{x}}$	±3 $s_{\bar{x}}$	±3.29 $s_{\bar{x}}$

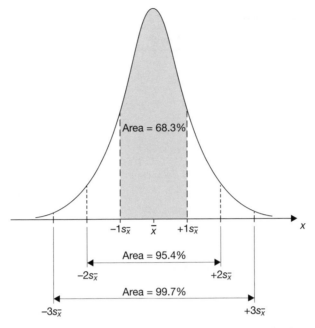

Figure 9.4 ● Confidence intervals for normal distribution.

Worked example 9.1: MPV and standard error

Question

An angle α is measured ten times with the same equipment by the same observer and the following results are obtained

| 47°56'38" | 47°56'40" | 47°56'35" | 47°56'33" | 47°56'40" |
| 47°56'34" | 47°56'42" | 47°56'39" | 47°56'32" | 47°56'37" |

Calculate the MPV of α and its standard error.

Solution

The MPV is given by

$$\bar{\alpha} = 47°56' + \frac{(38 + 40 + 35 + 33 + 40 + 34 + 42 + 39 + 32 + 37)"}{10}$$

$$= 47°56'37.0"$$

Based on the MPV, the residuals and squares of these are calculated and are listed in Table 9.2. Using these, equation (9.2) gives

$$s_\alpha = \pm\sqrt{\frac{\Sigma v_i^2}{(n-1)}} = \pm\sqrt{\frac{102}{(10-1)}} = \pm 3.4"$$

The standard error of the MPV $\bar{\alpha}$ is calculated using equation (9.3) as follows:

Table 9.2 ● Data for Worked example 9.1.

i	α_i	$v_i = (\bar{\alpha} - \alpha_i)$ sec	v_i^2 sec²
1	47°56'38"	−1	1
2	47°56'40"	−3	9
3	47°56'35"	+2	4
4	47°56'33"	+4	16
5	47°56'40"	−3	9
6	47°56'34"	+3	9
7	47°56'42"	−5	25
8	47°56'39"	−2	4
9	47°56'32"	+5	25
10	47°56'37"	0	0
		$\Sigma v_i = 0$	$\Sigma v_i^2 = 102$

$$s_{\bar{\alpha}} = \pm \frac{s_\alpha}{\sqrt{n}} = \pm \frac{3.4}{\sqrt{10}} = \pm 1.1"$$

The MPV of angle α can be quoted as

$$\bar{\alpha} = 47°56'37"\pm1.1"$$

For Example 9.1, the following should be noted:

● Since there is nothing to indicate that any individual angle is better than the others, the mean value of α is also the MPV. Should some of the observations be more reliable than the others, a weighted mean is calculated and used as the MPV (weight and weighted mean are discussed in the next section).

● When calculating residuals for angular problems, it is best to separate the seconds and calculate residuals and standard errors using these. Do *not* use whole decimal degrees, as this usually results in mistakes being made.

● The sum of the residuals of a set of observations is *always* zero – use this to check that the mean has been computed properly.

● Since only one measurement was needed to define α, there are nine redundant observations in this example.

● An alternative method for evaluating $\bar{\alpha}$ and $s_{\bar{\alpha}}$ is to use a calculator or computer. It is best to check that the standard error functions that these use are the same as those given by equations (9.2) and (9.3) before doing this.

Worked example 9.2: Rejection criteria for distance measurement

Question
Using the same tape, a setting out distance was measured 10 times by the same engineer under the same field conditions. After systematic corrections had been applied to each of the measurements, the following results were obtained (in metres):

| 23.289 | 23.290 | 23.290 | 23.289 | 23.298 |
| 23.289 | 23.288 | 23.288 | 23.291 | 23.289 |

Calculate the MPV for this distance and its standard error and confidence interval for a 95% probability.

Solution
Using similar methods to the previous example, the MPV for the distance D is given by

$$\bar{D} = 28.2 + \frac{(89 + 90 + 90 + 89 + 98 + 89 + 88 + 88 + 91 + 89)10^{-3}}{10} = 23.2901 \text{ m}$$

The residuals and their squares are given in Table 9.3, from which the standard error of a single measurement is

$$S_D = \pm \sqrt{\frac{76.9}{9}} = \pm 2.9 \text{ mm}$$

and the standard error of the MPV is

$$S_{\bar{D}} = \pm \frac{2.9}{\sqrt{10}} = \pm 0.9 \text{ mm}$$

Table 9.3 Data for Worked example 9.2.

i	D_i	$v_i = (\bar{D} - D_i)$ mm	v_i^2 mm²
1	23.289	+1.1	1.21
2	23.290	+0.1	0.01
3	23.290	+0.1	0.01
4	23.289	+1.1	1.21
5	23.298	−7.9	62.41
6	23.289	+1.1	1.21
7	23.288	+2.1	4.41
8	23.288	+2.1	4.41
9	23.291	−0.9	0.81
10	23.289	+1.1	1.21
		$\Sigma v_i = 0$	$\Sigma v_i^2 = 76.9$

This gives, for the 95% probability specified, a confidence interval for the true value of

$$\overline{D} \pm 1.96s_{\overline{D}} = 23.2901 \pm 1.96 \times 0.0009 = \mathbf{23.288 \text{ to } 23.292 \text{ m}}$$

The fifth measurement in Table 9.3 = 23.298 m. If the 95% confidence interval is to define *rejection limits* for this measurement, the 23.298 m reading would be rejected. It is very likely that this measurement is a mistake and may be a transposition error, as it could have been read as 23.289 m and recorded as 23.298 m. Although this might be the case, the reading *must not* be changed; it is either accepted unchanged or rejected. If it is rejected, the MPV and standard error are recalculated and these values are used in further computations. Without the 23.298 m observation, the MPV and standard error for D are

$$\overline{D} = \mathbf{23.289 \pm 0.0003 \text{ m}}$$

Weight

So far in this section it has been assumed that all measurements have been taken with the same precision. Although this may be the case for a lot of survey work, readings may be taken under different observing conditions with different equipment and observers. In this case, one set of results may be much more refined than another. Variations in precision are indicated by assigning different *weights* to measurements: the higher weighted the number the more precise the measurement.

The MPV of a group of measurements with varying weights is called the *weighted mean* and is given by

$$\text{weighted mean} = \frac{w_1 x_1 + w_2 x_2 + \dots + w_n x_n}{w_1 + w_2 + \dots + w_n} \tag{9.4}$$

where

measurement x_1 has weight w_1
measurement x_2 has weight w_2
. . .
measurement x_n has weight w_n

The weighted mean w will have a weight equal to the sum of all the individual weights and

$$w = w_1 + w_2 + \dots + w_n \tag{9.5}$$

Worked example 9.3: Weighted mean

Question
A distance is measured nine times with the following results (in metres):

$$L_1 = 37.379 \qquad L_2 = 37.387 \qquad L_3 = 37.392$$
$$L_4 = 37.378 \qquad L_5 = 37.370 \qquad L_6 = 37.380$$
$$L_7 = 37.369 \qquad L_8 = 37.373 \qquad L_9 = 37.365$$

Calculate the MPV for this distance when

- All the observations are equally weighted
- L_1, L_2 and L_3 have weight 1
  L_4, L_5 and L_6 have weight 2
  L_7, L_8 and L_9 have weight 3

Solution

When all the observations are equally weighted

$$\bar{L} \text{ (unweighted)} = 37.3 + \frac{(79 + 87 + 92 + 78 + 70 + 80 + 69 + 73 + 65)10^{-3}}{9}$$

$$= 22.377 \text{ m}$$

When the observations are weighted, equation (9.4) gives

$$\bar{L} \text{ (weighted)} = 22.3 + \frac{[1(79 + 87 + 92) + 2(78 + 70 + 80) + 3(69 + 73 + 65)]10^{-3}}{1(3) + 2(3) + 3(3)}$$

$$= 22.374 \text{ m}$$

Worked example 9.4: Weight and variance

Question

On a construction site, the difference in height H between two control points is measured using different equipment and observers. The results and standard errors for these measurements are $H_1 \pm s_1 = 3.539 \pm 0.005$ m and $H_2 \pm s_2 = 3.550 \pm 0.008$ m.

Calculate the most probable value of H and its standard error.

Solution

Although weight can be assigned by allocating numbers to observations, a more rigorous approach is to link weight with variance (and therefore standard error). These are related to each other as follows:

good precision = small standard error = high weight

poor precision = large standard error = low weight

This shows that weight w is inversely proportional to variance s^2 such that

$$w = \frac{1}{s^2} \qquad \text{or} \qquad w = \frac{s_0^2}{s^2} \tag{9.6}$$

where s_0^2 is called the *reference variance*, which is the variance of an observation of unit weight ($w = 1$).

To be able to solve this problem, a value for s_0^2 is required. As this has not been given, a value must be chosen, and although any value can be used, one corresponding to a standard error of an observation is often taken. In this case, $s_0^2 = 0.008^2 \ \text{m}^2$ is the value assumed for the reference variance and the weights of the two observations will be

$$w_1 = \frac{s_0^2}{s_1^2} = \frac{0.008^2}{0.005^2} = 2.56$$

$$w_2 = \frac{s_0^2}{s_2^2} = \frac{0.008^2}{0.008^2} = 1 \text{ (by definition in this example since } s_0^2 = 0.008^2 \text{ must correspond to a weight of 1)}$$

From equation (9.4) the weighted mean is

$$\overline{H} = \frac{w_1 H_1 + w_2 H_2}{w_1 + w_2} = \frac{2.56(3.539) + 1(3.550)}{2.56 + 1} = \textbf{3.542 m}$$

The weight of $\overline{H}$ is $w_{\overline{H}} = 2.56 + 1 = 3.56$ and with $s_0^2 = 0.008^2$, the variance for $\overline{H}$ is obtained from equation (9.6) as

$$s_{\overline{H}}^2 = \frac{s_0^2}{w_{\overline{H}}} = \frac{0.008^2}{3.56}$$

from which

$$s_{\overline{H}} = \pm\textbf{0.004 m}$$

When allocating weights to observations in problems such as these, it is the ratio of one weight to another that is important, not the number itself. For example, observation 1 has a precision twice as good as observation 2. These could have weights of 1 and 2 assigned to them to indicate this, but weights of 2 and 4 would produce the same outcome in any error calculations. For the problem given above, $s_0^2 = 0.005^2$ could have been chosen, or even a value of $s_0^2 = 1$. Both of these would produce exactly the same outcomes for $\overline{H}$ and $s_{\overline{H}}$. However, by choosing $s_0^2 = 0.005^2$ or 0.008^2, the arithmetic is a little easier to process.

Reflective summary

With reference to precision and accuracy, remember:

— Both precision and accuracy tend to get mixed up by everyone when surveying. Precision only refers to the behaviour of random errors, whereas accuracy defines *all* of the errors present in a survey, including any undetected gross and random errors. However, if all gross and systematic errors are removed, accuracy is the same as precision.

— The most commonly used measure of precision used in surveying is the standard error or standard deviation. Provided there is redundancy in a measurement or survey, these can be computed for any type of observation.

— Another way of indicating the precision of a measurement is to use weight. These are simply numbers that are assigned to each measurement where the ratio of one weight to another indicates the precision of one measurement compared to another.

— As with standard error and standard deviation, MPVs can be computed accounting for the different weights that might be assigned to observations – these are called weighted means.

— It is also possible to compute weighted means using standard error and standard deviation. To do this, they have to be converted into weights using a reference variance.

9.3 Propagation of variances and standard errors

After studying this section you should appreciate that errors from individual measurements will propagate through a survey according to how they are used to derive new quantities. You should also be able to calculate error propagations and understand how it is possible to use these to determine the precision of some survey methods that are in everyday use on site.

This section includes the following topics:

● Propagation of standard error
● Propagation of errors in survey methods

Propagation of standard error

Surveying measurements such as angles and distances are often used to derive other quantities using mathematical relationships: for instance, levelling heights are obtained by subtracting staff readings, horizontal distances are obtained from slope distances by calculation involving vertical angles, and coordinates are obtained from a combination of horizontal angles and distances. In each of these cases, the original measurements will be randomly distributed and will have errors, and it follows that any quantities derived from them will also have errors.

The special law of propagation of variance (standard error) for a quantity U which is a function of independent measurements $x_1, x_2, ..., x_n$, where $U = f(x_1, x_2, ..., x_n)$ is given by

$$s_U^2 = \left(\frac{\partial U}{\partial x_1}\right)^2 s_{x_1}^2 + \left(\frac{\partial U}{\partial x_2}\right)^2 s_{x_2}^2 + ... + \left(\frac{\partial U}{\partial x_n}\right)^2 s_{x_n}^2 \tag{9.7}$$

where

s_U = standard error of U

$s_{x_1}, s_{x_2}, ..., s_{x_n}$ are the standard errors of $x_1, x_2, ..., x_n$

Worked example 9.5: Standard error for the sum and the difference of two quantities

Question

Two distances a and b were measured with standard errors of $s_a = \pm 0.015$ m and $s_b = \pm 0.010$ m. Calculate the standard errors for the sum and the difference of a and b.

Solution

Using equation (9.7), the variance for $D = a + b$ is

$$s_D^2 = \left(\frac{\partial D}{\partial a}\right)^2 s_a^2 + \left(\frac{\partial D}{\partial b}\right)^2 s_b^2 = (1)^2 s_a^2 + (1)^2 s_b^2 = 0.015^2 + 0.010^2$$

and the standard error is

$s_D = \pm 0.018$ m

The variance for $D = a - b$ is

$$s_D^2 = \left(\frac{\partial D}{\partial a}\right)^2 s_a^2 + \left(\frac{\partial D}{\partial b}\right)^2 s_b^2 = (1)^2 s_a^2 + (-1)^2 s_b^2 = 0.015^2 + 0.010^2$$

and the standard error is

$s_D = \pm 0.018$ m

As can be seen, the standard error for a sum or a difference is the same.

Worked example 9.6: Standard error for the area of a building

Question

The two sides of the ground floor of a building were measured as $x = 32.00 \pm 0.01$ m and $y = 18.00 \pm 0.02$ m. Calculate the area of the building and its standard error.

Solution

The area of the building is given by

$$A = x \cdot y = 32.00 \times 18.00 = 576.0 \text{ m}^2$$

Using the law of propagation of variance

$$s_A^2 = \left(\frac{\partial A}{\partial x}\right)^2 s_x^2 + \left(\frac{\partial A}{\partial y}\right)^2 s_y^2 = y^2 s_x^2 + x^2 s_y^2$$

$$= 18.00^2 (0.01)^2 + 32.00^2 (0.02)^2$$

from which

$$s_A = \pm 0.66 \text{ m}^2$$

Worked example 9.7: Standard error for remote line measurement

Question

As it is not possible to measure the distance between two bridge piers A and B directly, this is done from observations taken with a reflectorless total station set up at point P some distance from the bridge.

Using the data given below, calculate the horizontal length of AB and its standard error.

Horizontal length PA	47.384 ± 0.005 m
Horizontal length PB	74.519 ± 0.005 m
H circle reading from P to A	$04°27'33" \pm 03"$
H circle reading from P to B	$58°44'54" \pm 03"$

Solution

Figure 9.5 shows the triangle corresponding to the data given for the question. The horizontal angle θ subtended at P by A and B is given by

$$\theta = 58°44'54" - 04°27'33" = 54°17'21"$$

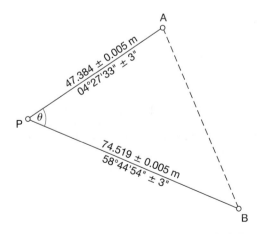

Figure 9.5 ● Data for Worked example 9.7.

The standard error of θ is given by

$$s_\theta^2 = 3^2 + 3^2 \sec^2 \qquad \text{or} \qquad s_\theta = \pm 4.24''$$

The distance AB is given by the cosine rule as

$$\begin{aligned}
AB^2 &= PA^2 + PB^2 - 2.PA.PB\cos\theta \\
&= 47.384^2 + 74.519^2 - 2(47.384)(74.519)\cos54°17'21''
\end{aligned}$$

which gives

AB = 60.632 m

The law of propagation of variance (equation 9.7) gives

$$s_{AB}^2 = \left(\frac{\partial AB}{\partial PA}\right)^2 s_{PA}^2 + \left(\frac{\partial AB}{\partial PB}\right)^2 s_{PB}^2 + \left(\frac{\partial AB}{\partial \theta}\right)^2 s_\theta^2$$

in which

$$\frac{\partial AB}{\partial PA} = \frac{PA - PB\cos\theta}{AB} = 0.064119$$

$$\frac{\partial AB}{\partial PB} = \frac{PB - PA\cos\theta}{AB} = 0.772879$$

$$\frac{\partial AB}{\partial \theta} = \frac{PA \cdot PB \sin\theta}{AB} = 47.286648$$

and

$$s_{AB}^2 = 0.064119^2(0.005)^2 + 0.772879^2(0.005)^2 + 47.286648^2\left(\frac{4.24}{206{,}265}\right)^2$$

Note that the standard error for θ **must** be converted from an angular error to a linear error by converting it into radians. This is done by multiplying it by sin 1" (= 1/206,265).

The solution of the equation for s_{AB}^2 gives

$$s_{AB} = \pm\, 0.004 \text{ m}$$

Propagation of errors in survey methods

In the previous section, a number of worked examples are given showing how individual errors propagate through a function with observed variables. By using similar methods, it is possible to study the effect of accumulating errors in survey fieldwork.

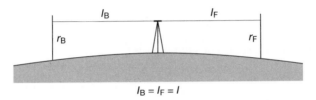

$l_B = l_F = l$

Figure 9.6 ● Propagation of errors in levelling.

Propagation of errors in levelling

Levelling is described in Chapter 2, where all of the terminology used in this section originates.

If the standard error for reading a levelling staff is s_s mm per m of the sight length and it is assumed that same sight lengths are used at every instrument setup, the standard error $s_{\Delta H}$ in the height difference ΔH obtained by levelling through a distance D can be derived.

Figure 9.6 shows a single instrument setup where a backsight staff reading r_B and a foresight r_F have been taken. With the backsight length l_B equal to the foresight length l_F, $l_B = l_F = l$ (the sight length). The height difference is given by

$$\Delta h = r_B - r_F$$

The standard error for this height difference $s_{\Delta h}$ can be obtained from

$$s_{\Delta h}^2 = s_{r_B}^2 + s_{r_F}^2$$

The standard errors in the two staff readings will be equal since the sight lengths are the same. For an individual sight length l, the standard error in a single staff reading s_r will be

$$s_r = l s_s$$

This gives

$$s_{\Delta h}^2 = 2s_r^2 = 2l^2 s_s^2$$

If n setups are required to level through the distance D, the height difference between the ends of the line is

$$\Delta H = \Delta h_1 + \Delta h_2 + \ldots + \Delta h_n$$

The variance for this sum is given by

$$s_{\Delta H}^2 = s_{\Delta h_1}^2 + s_{\Delta h_2}^2 + \ldots + s_{\Delta h_n}^2$$

Since equal sight lengths are used

$$s_{\Delta h_1} = s_{\Delta h_2} = \ldots = s_{\Delta h_n}$$

Hence

$$s^2_{\Delta H} = n s^2_{\Delta h}$$

In a distance D with n setups of sight length l

$$D = 2nl$$

and

$$s^2_{\Delta H} = \frac{D}{2l} s^2_{\Delta h} = \frac{D}{2l}(2l^2 s^2_s)$$

from which

$$s_{\Delta H} = \pm s_s \sqrt{Dl} \tag{9.8}$$

Typical values for a levelling scheme might be $s_s = \pm 0.05$ mm per m (for readings taken with an optical level), $l = 40$ m and $D = 1$ km. Substituting these into equation (9.8) gives

$$s_{\Delta H} = \pm 0.05\sqrt{1000 \times 40} = \pm 10.0 \text{ mm}$$

Although it is not realistic to expect every sight length at every setup to be equal when levelling, this example does give some idea of the accuracy that might be expected.

Compared to conventional optical levelling, a better accuracy is obtained when using a digital level and if one of these is used to take staff readings, the value of s_s changes to better than ± 0.01 mm per m. Using this value for s_s, $s_{\Delta H}$ would be ± 2 mm in the above example.

Propagation of errors in angle measurement

In Section 3.5, it is shown that angles are calculated from face left and face right readings taken with a theodolite or total station. Random errors in these are usually caused by uncertainties in the digital reading system of an electronic theodolite or are caused by differences when setting and reading a micrometer scale on an optical theodolite. As well as this, the readings will be different because of variations when the observer bisects a target.

Figure 9.7 shows an angle α as the difference between two directions d_1 and d_2 or

$$\alpha = d_2 - d_1$$

where d_1 is the mean of a face left reading d_{1L} and a face right reading d_{1R} such that

$$d_1 = \frac{d_{1L} + d_{1R}}{2}$$

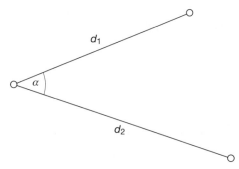

Figure 9.7 Propagation of errors in angle measurement.

If the standard error in the digital reading system or uncertainty in reading and setting a micrometer scale is s_m and that for bisecting a target is s_b the variances in d_{1L} and d_{1R} are

$$s_{d_{1L}}^2 = s_{d_{1R}}^2 = s_m^2 + s_b^2$$

This gives the variance in d_1 as

$$s_{d_1}^2 = \frac{(s_m^2 + s_b^2)}{2}$$

Similarly

$$s_{d_2}^2 = \frac{(s_m^2 + s_b^2)}{2}$$

and the variance of α is

$$s_\alpha^2 = (s_m^2 + s_b^2)$$

If α is the mean of n rounds of angles then

$$s_\alpha^2 = \frac{(s_m^2 + s_b^2)}{n} \tag{9.9}$$

When using an electronic theodolite or total station, a standard error of $s_m = \pm 5''$ is typical, and if s_b is $\pm 10''$, the propagated error for one round of angles would be

$$s_\alpha^2 = \frac{(25 + 100)}{1} = 125 \sec^2 \quad \text{or} \quad s_\alpha = \pm 11.1''$$

For two rounds $s_\alpha = \pm 7.9''$, which shows that by measuring two rounds of angles and taking the mean value an improvement in the accuracy of the measured angle is obtained as well as a check for gross errors.

In this propagation theory, no account has been taken of any systematic errors arising out of theodolite or target miscentring and from the instrument not being level. These are discussed in Section 3.6.

Propagation of errors in trigonometrical heighting

As discussed in Section 5.6, total stations are often used for trigonometrical height measurement on site. Figure 9.8 shows the situation where a total station has been centred over a control point A of height H_A and the slope distance L and vertical angle α have been measured to a detail pole held at B. The point B could be a spot height required for contouring or it could be a point to be set out. Whatever the case, the precision of the height for B is to be determined. The height of B is given by (see equation 5.2)

$$H_B = H_A + hi - hr + L \sin \alpha + (c - r)$$

where

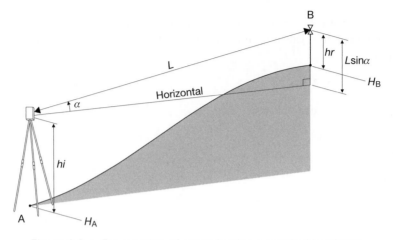

Figure 9.8 ● Propagation of errors in trigonometrical heighting.

hi = the height of the total station above A
hr = the height of the reflector above B

The term $(c - r)$ is the combined correction for curvature and refraction and for this analysis can be ignored as a systematic error. The height of B is therefore given by

$$H_B = H_A + hi - hr + L \sin \alpha$$

Applying the law of propagation of variance to this gives

$$s_{H_B}^2 = \left(\frac{\partial H_B}{\partial H_A}\right)^2 s_{H_A}^2 + \left(\frac{\partial H_B}{\partial hi}\right)^2 s_{hi}^2 + \left(\frac{\partial H_B}{\partial hr}\right)^2 s_{hr}^2 + \left(\frac{\partial H_B}{\partial L}\right)^2 s_L^2 + \left(\frac{\partial H_B}{\partial \alpha}\right)^2 s_\alpha^2$$

$$= s_{H_A}^2 + s_{hi}^2 + s_{hr}^2 + \sin^2 \alpha s_L^2 + L^2 \cos^2 \alpha s_\alpha^2$$

Let the standard errors in H_A, hi, hr and L be ±0.005 m. Under normal circumstances, α will rarely exceed ±20° and assuming the error in the measurement of this to be ±30", the error in H_B for a sighting distance of 50 m will be

$$s_{H_B}^2 = 0.005^2 + 0.005^2 + 0.005^2 + \sin^2 20° (0.005^2) + 50^2 \cos^2 20° \left(\frac{30}{206,265}\right)^2$$

giving

$$S_{H_B} = \pm 0.011 \text{ m}$$

At 100 m, $S_{H_B} = \pm 0.016$ m and at 200 m, it will be ±0.027 m.

As the sighting (slope) distance L increases, the effects of the standard errors in the height of A and in the instrument and reflector heights hi and hr, as well as the slope distance L, all decrease and an approximate value for the error in H_B is given by

$$s_{H_B} = L \cos \alpha s_\alpha = D s_\alpha$$

where D is the horizontal distance between instrument and reflector.

Propagation of angular errors in traversing

If the internal angles a_1, a_2, ..., a_n of an n-sided polygon traverse are all measured using the same equipment and methods, such that each angle has the same standard error s_a, the standard error for the sum of these angles A is determined as follows.

Applying a similar process to that given in the previous examples and with $A = a_1 + a_2 + ... + a_n$:

$$s_A^2 = \left(\frac{\partial A}{\partial a_1}\right)^2 s_{a_1}^2 + \left(\frac{\partial A}{\partial a_2}\right)^2 s_{a_2}^2 + ... + \left(\frac{\partial A}{\partial a_n}\right)^2 s_{a_n}^2$$

$$= s_{a_1}^2 + s_{a_2}^2 + ... + s_{a_n}^2$$

If

$$s_{a_1} = s_{a_2} = ... = s_{a_n} = s_a$$

then

$$s_A^2 = n s_a^2 \qquad \text{or} \qquad s_A = \pm s_a \sqrt{n}$$

This example shows that the propagated error for the sum of a set of measurements with the same standard error is proportional to the square root of the number of measurements (or observations). In elementary surveying, this is used to determine the allowable misclosures for some types of fieldwork. For example, in Section 6.5 the allowable misclosure E in the measured angles of a polygon traverse is given as $E = \pm KS\sqrt{n}$. Note the similarity in this expression to that obtained for the propagation of angles. In this case, s_a is assumed to be equal to KS and the expected propagation of this for n angles is $s_a\sqrt{n}$.

Reflective summary

With reference to propagation of variances and standard errors, remember:

— Using well-established techniques, the precision of most surveys can be determined by analysing the way in which errors propagate through them.

— For all derived quantities in a survey, it is always possible to compute a propagated variance or standard error. Whatever the type of survey or measurement that is being attempted, this should always be done to assess the precision of the outcome.

— Because the numbers that are processed in a propagation are a mix of small and large ones, some care is needed with arithmetic when processing the functions obtained in order to avoid errors occurring in the calculations.

> — If any angles are involved in a propagation, do not forget to convert their standard errors from seconds to radians – if this is not done a completely wrong answer will be obtained.

9.4   Survey specifications

After studying this section you should realise that tolerances and specifications given for surveying are important on site and that every effort must be made to meet these. You should know that advice on how to do this is available in a number of published standards and guides and that it is also possible to use error propagation theory to assess the precision of measurements required to meet specifications.

This section includes the following topics:

- Use of British and international standards in survey specifications
- ICE design and practice guide in survey specifications
- Royal Institution of Chartered Surveyors (RICS) guidelines
- Calculation of specifications

Use of British and international standards in survey specifications

In engineering surveying, it is usual to specify the precision or tolerance required for measured quantities before fieldwork commences. It is then up to those on site to decide what type of equipment and methods should be used in order to achieve these. There are a number of different ways in which this can be done, all of which are linked in some way to the theory of errors and measurements.

One method of establishing the accuracy of equipment and methods to be used on site is to consult a number of British and international standards that have been published specially for this purpose.

One of these is BS 5606 *Guide to Accuracy in Building*, which gives details of the accuracy in use of various instruments when used in engineering surveying: these are known as *A* values and are shown in Table 11.3. They can be used by anyone on site in order that equipment can be chosen which will meet specifications given for survey work. The information in BS 5606 is based on measurements carried out by the Building Research Establishment, which has performed a series of tests to determine the standard errors of various surveying activities (these include levelling as well as angle and distance measurements).

A standard of some importance for site work is BS 5964 *Building Setting Out and Measurement* (this is the same as ISO 4463). This is a wide-ranging standard that applies to all types of construction and to the control from which it is set out. It defines an accuracy measure known as the *permitted deviation* (*PD*), which is a statistical variation to take into account the normal distribution of errors. This is related to the standard deviation σ and standard error s using $PD = 2.5 \times \sigma$ or $PD = 2.5 \times s$. Since 68% is the statistical probability for the standard error of a measurement, 68% of all measurements taken for control surveys or setting out should be better than a specified *PD* divided by 2.5 to account for normally distributed errors. When working to *tolerances* on site, all measurements taken for control surveys or setting out should be three times better than the tolerance, again to account for the normal distribution of errors.

BS 5964 also includes definitions and descriptions of what primary and secondary control is for site work.

Primary control usually consists of a series of control points covering an entire site, the density of which is kept to a minimum to avoid excessive observational errors propagating through it. A primary network will be a three-dimensional closed network that is observed and adjusted independently of any other control points. A primary control network is usually required for large construction sites, long linear sites (such as rail and road works), for complicated structures and where a high accuracy is to be maintained throughout a control survey.

Secondary control is located from and adjusted to fit primary control and will usually be a higher density of points, the positions of which are located to suit site requirements. These are the control points from which most setting out is done. For most construction work, primary control may not be required and it is only necessary to establish secondary control, usually by traversing or by using GPS.

For both primary and secondary control surveys, BS 5964 gives a series of permitted deviations for the measurement of distances, angles and heights – these are summarised in Tables 9.4 and 9.5. For distances and angles, the permitted deviations for first-order primary horizontal control refer to the allowed difference between the unadjusted distances and angles measured for the network and those derived from the adjusted coordinates of control points in the network. For all other horizontal control, the permitted deviations refer to the differences allowed between a measurement taken to test the quality of a control network (called a *compliance measurement*) and that derived from the coordinates of points in the network. To interpret the values given in Table 9.4, consider the permitted deviation for distances in secondary control given as $\pm 1.5\sqrt{L}$ mm, where L is the distance in metres between the control points being checked. If $L = 100$ m, this gives a permitted deviation of ± 15 mm or a standard error of ± 6 mm for measurements. For angles at the same distance of 100 m, the permitted deviation is $\pm 0.09/\sqrt{L}$ degrees (with L again in metres), which gives a permitted deviation of $\pm 32"$ or a standard error of $\pm 13"$. For vertical control, the permitted deviations are shown in Table 9.5 – remember that these are 2.5 times the standard errors expected for levelling between these points.

Another standard that is worth considering for site work is BS 7334 (ISO 8322) *Measuring Instruments for Building Construction*. The main purpose of this standard is to enable users of surveying equipment to determine their personal accuracy for a range of surveying and setting out tasks. The results obtained from these tests would then be compared with any permitted deviations given in BS 5964 or elsewhere to

Table 9.4 ● Permitted deviations for horizontal control from BS 5964 and ICE Guide.

	BS 5964	ICE
Primary control first order		
Distances	$\pm 0.75\sqrt{L}$ mm	$\pm 0.5\sqrt{L}$ mm
Angles	$\pm \dfrac{0.045}{\sqrt{L}}$ degrees	$\pm \dfrac{0.025}{\sqrt{L}}$ degrees
Primary control second order		
Distances	$\pm 1.5\sqrt{L}$ mm	$\pm 0.75\sqrt{L}$ mm
Angles	$\pm \dfrac{0.09}{\sqrt{L}}$ degrees	$\pm \dfrac{0.032}{\sqrt{L}}$ degrees
Secondary control		
Distances	$\pm 1.5\sqrt{L}$ mm	$\pm 1.5\sqrt{L}$ mm
Angles	$\pm \dfrac{0.09}{\sqrt{L}}$ degrees	$\pm \dfrac{0.09}{\sqrt{L}}$ degrees

Note: L is the distance between the points being checked in metres.

Table 9.5 ● Permitted deviations for vertical control from BS 5964 and ICE Guide.

	BS 5964	ICE	Notes
Primary control			
Between adjacent control points (bench marks)	± 5 mm	± 5 mm	Control points less than 250 m apart
		$\pm 12\sqrt{K}$	Control points more than 250 m apart.
			K is distance between bench marks in km
Secondary control			
Between adjacent secondary control points (bench marks)	± 3 mm	± 5 mm	
Between TBMs established for structures		± 3 mm	

check that the equipment and methods chosen for a particular task are appropriate. Essentially, the test procedures in BS 7334 are based on a series of repeated measurements taken under conditions expected on site. Using the techniques given in earlier sections of this chapter, standard errors are computed using the repeated measurements from which a comparison to a permitted deviation can be made. BS 7334 is similar to ISO 17123 *Optics and Optical Instruments – Field Procedures for Testing Geodetic and Surveying Instruments.* Like BS 7334, this international standard specifies field procedures to be adopted when determining the precision of surveying instruments.

ICE design and practice guide in survey specifications

An important publication that is concerned with specifications on site is the ICE design and practice guide *The management of setting out in construction.* This gives the permitted deviations shown in Tables 9.4 for horizontal control and in Table 9.5 for vertical control. Permitted deviations are also given in the ICE guide for *tertiary control.* These are the points set out on site for construction purposes and include all the corner and centre line pegs put in place, any offset pegs, sight rails, slope rails and so on (descriptions of these are given in Chapter 11). When setting out, four categories of permitted deviation are given, and for each of these, the values given in Table 9.6 refer to the allowed difference between any calculated and check distances and angles taken from secondary control points to tertiary control points and to any measured height differences taken to check those derived from drawings.

The ICE guide also states that where tolerances are quoted, the permitted deviation for setting out (that is, for tertiary control) should be taken to be five-sixths of any tolerances specified.

Table 9.6 ● Permitted deviations for tertiary control.

Category 1 Structures

Distances	Angles	Heights
$\pm 1.5\sqrt{L}$ mm	$\pm \dfrac{0.09}{\sqrt{L}}$ degrees	± 3 mm

Category 2 Roadworks

Distances	Angles	Heights
$\pm 5.0\sqrt{L}$ mm	$\pm \dfrac{0.15}{\sqrt{L}}$ degrees	± 5 mm

Category 3 Drainage works

Distances	Angles	Heights
$\pm 7.5\sqrt{L}$ mm	$\pm \dfrac{0.20}{\sqrt{L}}$ degrees	± 20 mm

Category 4 Earthworks

Distances	Angles	Heights
$\pm 10.0\sqrt{L}$ mm	$\pm \dfrac{0.30}{\sqrt{L}}$ degrees	± 30 mm

Note: *L* is the distance between the points being checked in metres.

Royal Institution of Chartered Surveyors (RICS) guidelines

The RICS publishes some guidelines that give specifications for carrying out mapping and GPS surveys. The two guidelines published for mapping purposes are *Surveys of Land, Buildings and Utility Surveys at Scales of 1:500 and Larger, Client Specification Guidelines* and *Specification for Mapping at Scales Between 1:1000 and 1:10 000*. These provide specifications for mapping only, whereas the *Guidelines for the use of GPS in Surveying and Mapping* give specifications for all types of survey that might be carried out by GPS. As with the previous publications, these should also be considered as useful sources of information on specifications.

References for all of the British and international standards together with the ICE guide referred to in this section are given at the end of this chapter. All of these give much more information on primary, secondary and tertiary control and the specifications for these. They are also discussed further in Section 11.7 in connection with setting out. Further details of the RICS mapping and GPS guides are given in Sections 10.2 and 10.5.

Calculation of specifications

Another approach to dealing with specifications in surveying is to use error propagation theory, in which the precision required in the individual parts of a measurement can be calculated when the precision of a derived quantity is known. Some examples of these are given below.

Precision of levelling

Suppose a TBM is to be established for a construction site from an existing bench mark and that the height difference ΔH between the two bench marks is to have a specified standard error of $s_{\Delta H}$. Equation (9.8) gives

$$s_{\Delta H} = \pm s_s \sqrt{Dl}$$

and if the length between the bench marks D is known and a value for s_s is given, the maximum sighting distance l can be calculated for the levelling. Rearranging equation (9.8):

$$l = \frac{s_{\Delta H}^2}{D s_s^2}$$

If the staff readings are taken with a precision of $s_s = \pm 0.05$ mm per m, $D = 250$ m and the height difference between bench marks is to have an error of ± 0.005 m, the maximum sighting distance allowed in order to achieve this precision is

$$l = \frac{0.005^2}{250(0.05 \times 10^{-3})^2} = \textbf{40 m}$$

When developing equation (9.8) in the previous section, the variance for the height difference between two points ΔH was given as $s^2_{\Delta H} = n s^2_{\Delta h}$, which gives the standard error as $s_{\Delta H} = \pm s_{\Delta h}\sqrt{n}$, where n is the number of times the level is set up. This was used in Section 2.5 for determining the expected precision of levelling, but by using well-established values for $s_{\Delta h}$ (±5 mm was suggested as typical for this, but see Table 9.6 for different values according to the category of setting out).

Precision of angle measurement

In Section 9.3 it was shown that precision of a measured angle is given by equation (9.9) as

$$s^2_\alpha = \frac{(s^2_m + s^2_b)}{n}$$

The number of rounds n to be observed that will achieve a stated precision in the final angle α can be derived from this equation as

$$n = \frac{(s^2_m + s^2_b)}{s^2_\alpha}$$

If the theodolite or total station used for angle observations has a precision of $s_m = \pm 6"$ and targets and observers are used such that $s_b = \pm 7.5"$, the number of rounds that must be observed to achieve a precision of $\pm 5"$ in α is given by

$$n = \frac{(6^2 + 75^2)}{5^2} = 3.69$$

which indicates that **four rounds** should be taken.

Precision of trigonometrical heighting

The standard error s_{H_B} for the height of an unknown point B is derived as $s_{H_B} = L \cos \alpha s_\alpha$ in Section 9.3. If the error in the height of B is not to exceed a specified limit, it is useful to know the precision to which α must be measured so that this precision is achieved. The equation for s_{H_B} can be rearranged to give

$$s_\alpha = \frac{s_{H_B}}{L \cos \alpha}$$

With $L = 50$ m and $\alpha = 20°$, the precision to which α must be measured in order to obtain a precision in the height of B of ±0.005 m is

$$s_\alpha = \frac{0.005}{50 \cos 20°} \text{ radians} = \frac{206,265(0.005)}{50 \cos 20°} \text{ seconds} = \pm 22"$$

At 100 m for the same precision in the height of B and with $\alpha = 20°$, s_α will be ±11" and at 250 m will be ±4". Under normal circumstances, with normal equipment, it is very difficult to measure vertical angles with these precisions and this analysis shows that the precision that can be obtained for trigonometrical heighting is limited by how well the vertical angle can be measured.

Precision of slope corrections

In Chapter 4 it was shown that a slope distance L can be reduced to a horizontal distance D by measuring the slope angle θ and by applying the equation $D = L \cos \theta$. The propagated error in D for this reduction is given by

$$s_D^2 = \cos^2 \theta s_L^2 + (L \sin \theta)^2 s_\theta^2$$

Since θ is a small angle, this can be written

$$s_D^2 = s_L^2 + (L \sin \theta)^2 s_\theta^2$$

The effect of the precision of θ on the precision of D can be obtained by rearranging the equation for s_D to give s_θ as

$$s_\theta^2 = \frac{s_D^2 - s_L^2}{(L \sin \theta)^2}$$

On site, this type of slope correction is usually applied to distances measured with total stations. All of these have similar specifications and s_L has a typical value of ±3 mm for short distances. If the precision of D is specified as ±5 mm and $L = 100$ m with $\theta = 5°$, the precision required in the measurement of θ is

$$s_\theta^2 = \frac{0.005^2 - 0.003^2}{(100 \sin 5°)^2} \text{ radians}^2$$

or

$$s_\theta = \pm 1'35''$$

The angle reading system installed in most total stations is capable of measuring vertical angles with this precision.

Another method of applying slope corrections is to obtain the height difference from equation (4.2), where

$$D = L - \frac{\Delta h^2}{2L}$$

This correction is usually applied to taped distances where the height difference between the ends of a line are known. An error propagation for this equation gives

$$s_D^2 = \left(1 - \frac{\Delta h^2}{2L^2}\right)^2 s_L^2 + \left(\frac{\Delta h}{L}\right)^2 s_{\Delta h}^2$$

The term $\Delta h^2 / 2L^2$ can be ignored as it is small compared to 1 and the precision of D is given by

$$s_D^2 = s_L^2 + \left(\frac{\Delta h}{L}\right)^2 s_{\Delta h}^2$$

Assuming the precision to which the slope distance is measured is known and that the precision of D is specified, the precision required in the height difference Δh is

$$s_{\Delta h}^2 = \left(\frac{L}{\Delta h}\right)^2 (s_D^2 - s_L^2)$$

For a slope distance of 50 m, a height difference of 1 m, a taping precision of $s_L = \pm 3$ mm and a specified precision for s_D of ± 5 mm, the height difference must be measured with a precision of

$$s_{\Delta h}^2 = \left(\frac{50}{1}\right)^2 (0.005^2 - 0.003^2)$$

or

$$s_{\Delta h} = \pm \textbf{0.20 m}$$

For a height difference of 5 m, $s_{\Delta h} = \pm 0.04$ m and at a height difference of 10 m it will be ± 0.02 m. This shows that, as the height difference increases, some care is needed when measuring it (usually by levelling where the difference of two staff readings is taken to determine Δh).

Reflective summary

With reference to survey specifications, remember:

— Because of the increasing emphasis now placed on quality control, it is vital that everyone on site is aware of the need to work to specifications and tolerances when surveying and setting out.

— For construction surveys, it is strongly recommended that British and other standards are consulted for advice on how to meet and specify tolerances.

— The ICE Design and Practice Guide *The management of setting out in construction* is **essential** reading for anyone involved with control surveys and setting out on site.

— All of the various publications of the RICS that are concerned with specifications are also useful for engineering surveys – as well as published standards and the ICE guide, these are recommended reading for all those responsible for survey work in a construction environment.

— As well as consulting standards and other published material, it is also possible to assess the precision required for individual measurements in a survey. The examples given in this section show how this can be done for some of the basic techniques used in surveying but for more complicated work, it is better to use commercially produced software to predict these.

9.5   Least squares adjustment

After studying this section you should know that least squares is the most popular method used for carrying out the analysis and adjustment of survey measurements. You should also be aware of the advantages of using least squares in preference to other methods.

This section contains the following topics:

- Why least squares is used
- Quality control with least squares

Why least squares is used

At the beginning of this chapter definitions were given of the types of error that can occur in surveying measurements. The first two categories of error identified were gross errors (or mistakes), which are removed by checking fieldwork, and systematic errors which are removed by applying mathematical corrections to measurements or by adopting field procedures to cancel them. Most of this chapter has involved the analysis and discussion of random errors, which are assumed to be the only errors remaining in measurements when gross and systematic errors have been removed.

The presence of random errors in surveying measurements and observations has been shown to be due to the variation in repeated measurements of the same quantity, but their presence is also indicated by the misclosures that are usually obtained in surveying. Throughout the book, many examples are given of the way in which the random errors causing misclosures are distributed (or *adjusted*) to satisfy some mathematical or geometrical condition. In Chapter 2, levelling misclosures are distributed according to the number of times the level is set up, and in Chapter 6 the misclosure in the angles of a traverse is distributed equally to each angle and the linear misclosure is also distributed equally or by using the Bowditch rule. Although these produce acceptable solutions for the adjustment of survey measurements, they are not based on rigorous methods. Since random errors follow the laws of probability and are normally distributed, a better way of adjusting survey measurements is to use a technique that is also based on these – this is known as *least squares*.

Nowadays, all types of survey are adjusted using methods based on least squares. As discussed earlier in this chapter, adding more measurements than is needed in a survey makes the computations more complex but adds redundancy, without which an adjustment is not possible. Like all adjustment methods used in surveying, least squares also relies on redundancy in measurements, but compared with the improvised methods used for adjustment previously given, least squares will give the most probable values for each component in a survey, whether these are the heights of bench marks in a levelling network, the coordinates of traverse stations or baseline vectors in a GPS survey. For least squares to do this, however, it is emphasised once again that all gross and systematic errors must be removed from measurements before they are used in the adjustment. It is also desirable that the number of observations that are adjusted is large and that the measurements are normally distributed. Sometimes these conditions are not met, but least squares still provides the most popular and most rigorous adjustment method that is used for the majority of surveys.

The benefits of using least squares for survey analysis and adjustments are as follows.

- The theoretical basis for least squares lies in statistics and, like the random errors it deals with, it also conforms to the laws of probability. Because of this, most probable values are always obtained from a least squares adjustment.

- Standard errors or weights can be applied to all the observations taken for a survey and the effects of these can be included in the adjustment. This offers a huge advantage over other methods of adjustment.

- No matter how complicated a survey is, provided all the field data is collected and processed correctly, least squares performs a simultaneous adjustment of all the original data to produce a single solution for the survey. This will account for all of the misclosures in a survey, the precisions of the observations and any other special conditions that apply.

- Least squares performs a full error propagation on all derived quantities and provides a complete analysis of the accuracy of each point in a survey. This information can be used at the planning stage to ensure that a survey meets specifications and makes the detection of gross errors much easier.

Quality control with least squares

For all surveys, and especially on site, there is now much more emphasis on quality control. Least squares plays a significant role in this and most specifications for survey work will require a least squares adjustment of some sort to be carried out and the details included in any report that may be presented to a client for the survey.

To implement least squares at the planning stage of a survey, approximate values for all the observations in a survey are obtained (for example, by scaling angles and distances from maps and plans for a horizontal control survey) and expected precisions are assigned according to the equipment and methods that are to be used to take these. All of this data is then used in a least squares pre-analysis and the precision of the outcomes assessed and compared to the specifications for the survey. If acceptable, all the observations are taken as planned, but if not, the positions of survey points together with the precision of the equipment and methods to be used are reassessed and another least squares adjustment is performed. This process is repeated until the survey is considered to be acceptable.

Following the observation and adjustment of a survey, a further check is made to ensure that the survey is acceptable. At this stage, least squares is capable of detecting any gross, and sometimes systematic, errors that may be present in the observations. If these are suspected, it may be necessary to repeat some of the measurements and readjust.

Least squares is a technique that requires software and a computer to implement it. Although it is possible to develop 'in-house' software, many commercial packages are available and the use of these on site and elsewhere is commonplace. These are provided by all of the major manufacturers and suppliers of survey equipment, the details of which are accessed through their web sites. In addition to these, there are many third-party least squares software packages on the market, and these are periodically reviewed and regularly advertised in surveying journals. For easy reference, all of these sources of information are listed at the end of the chapter.

The application of least squares, sufficient for a thorough understanding of the subject, is beyond the scope of this book and it is not included here. However, further reading is suggested at the end of the chapter for those requiring a specialist knowledge of the subject.

Reflective summary

With reference to least squares adjustment, remember:

— The best method that can be used for survey analysis is least squares and it is recommended as an adjustment method for all survey work. The reasons for this are that it provides most probable values for all variables as well as estimates of their precisions. It is the error analysis that least squares gives that makes it very useful in surveying.

— Although least squares requires dedicated software, this is not a problem as there are many software packages to choose from, some of which can be bought with equipment from leading manufacturers and others which are produced by firms specialising in survey software.

— All GPS controllers have a least squares program already installed for real-time work, as do some data collectors, so there is often no need to purchase a computer for carrying out survey adjustments by least squares.

— That many survey procedures are quality driven, and it is often a requirement to show that a least squares adjustment has been performed on a survey to guarantee the results obtained.

Exercises

9.1 Explain the difference between gross and systematic errors, giving examples of each that occur in surveying.

9.2 What are random errors and how are they different from gross and systematic errors?

9.3 Write down the laws of probability that random errors follow.

9.4 What is a most probable value and how does it differ from a true value? How are most probable values estimated in surveying?

9.5 With reference to survey measurements and observations, what is meant by the following:

precision
accuracy
internal accuracy
relative precision
ppm
reliability

9.6 What is the difference between standard deviation and standard error? Is this important? How would you calculate the standard error for a series of repeated measurements of the same quantity?

9.7 If an error analysis is to be carried out for a survey, why is redundancy important?

9.8 Explain how a confidence interval can be used to define limits and probabilities for a measurement.

9.9 Instead of standard error, the precision of a measurement can be indicated by its weight. Describe how you would do this and explain how weight and standard error are linked.

9.10 What is the special law of the propagation of variance? Discuss the reasons why this is used extensively in the analysis of survey measurements and methods.

9.11 What British and international standards give information relating to specifications for construction surveying? If at all possible, you should try to get copies of these.

9.12 Explain what a permitted deviation is and how this is related to standard error and a tolerance.

9.13 Using BS5964 *Building setting out and measurement*, calculate permitted deviations for measured distances and angles for each of the following

– primary control (first and second order) over distances of 500 m, 1 km and 5 km

– secondary control over distances of 100–500 m at 100 m intervals

9.14 Using the ICE Design and Practice Guide *The management of setting out in construction*, calculate permitted deviations for measured distances and angles for each of the following

– primary control (first and second order) over distances of 500 m, 1 km and 5 km

– secondary control over distances of 100–500 m at 100 m intervals

– tertiary control (all categories) at distances of 25 m, 50 m and 100 m

9.15 Discuss the reasons why least squares is the most popular method in use for the analysis and adjustment of survey measurements and observations. In your answer, list the advantages that least squares has over other methods of adjustment.

9.16 Obtain copies of some survey journals and list all the least squares software packages either reviewed or advertised in these. Contact some of the sources found and obtain further information on these.

9.17 In a setting out scheme, the distance between two reference points was measured using the same steel tape by the same observers and the following ten measurements (in metres) were obtained

42.254   42.253   42.248   42.250   42.249
42.251   42.248   42.247   42.248   42.250

Calculate the most probable value for this distance and its standard error.

If a permitted deviation of 10 mm applies to this measurement, do the observers meet this specification?

9.18 An angle APB is measured by taking the following readings:

Reading along PA			Reading along PB		
°	′	″	°	′	″
00	25	36	54	17	08
00	25	22	54	16	57
00	25	25	54	17	12
00	25	46	54	17	05
00	25	18	54	16	50
00	25	29			
00	25	41			
00	25	34			
00	25	19			
00	25	09			

Calculate the most probable value of $A\hat{P}B$ and its standard error.

9.19 The height difference between two points on site is determined by taking several readings to a levelling staff held at each point in turn. Using the results given below, calculate the most probable value and standard error for the height difference.

Back sight (m)	Fore sight (m)
2.608	1.871
2.610	1.869
2.609	1.869
2.606	1.873
2.605	1.872
2.607	1.873
2.604	1.869
2.608	1.870
2.609	1.870
2.606	1.869

9.20 An angle θ is measured by three different observers with the same total station and the observations shown below were obtained. Assuming that the observations within each set are of equal precision, calculate

- which observer is the most reliable
- the number of times each observer would have to measure an angle to achieve a precision of ±3"
- the most probable value of θ and its standard error

Observation	Observer 1	Observer 2	Observer 3
1	38°42'01"	38°42'00"	38°42'01"
2	42'00"	41'59"	42'01"
3	41'59"	42'00"	41'50"
4	42'02"	41'40"	41'59"
5	42'06"	41'50"	41'58"
6	42'00"	42'02"	42'01"
7	42'18"	42'08"	42'00"
8	41'56"	41'52"	41'50"
9	41'59"	42'02"	42'01"
10	41'59"	41'59"	

9.21 Six repetitive measurements of an angle gave the following values:

12°34'17"   12°34'11"   12°34'15"
12°34'18"   12°34'18"   12°34'16"

Analyse these results using a rejection criterion of 95% probability applied to the standard error of the mean value and calculate a most probable value for the angles with an appropriate standard error.

9.22 Two angles of a triangle were measured each with a standard error of ±10". Determine the standard error in the computed third angle of this triangle.

9.23 The height of a bench mark was determined by a number of different routes, the results for these being 58.719 m ± 0.017 m, 58.713 m ± 0.032 m, 58.717 m ± 0.021 m and 58.705 m ± 0.018 m. Calculate the weighted mean for the height of the bench mark and its standard error.

9.24 The standard error for each angle observed for a six-sided polygon traverse is estimated as ±15". Determine the range of values permissible for the sum of the measured angles in this traverse at the 95% probability level. If the misclosure in the angles is limited to ±30", what is the precision required in the measurement of any individual angle, again at the 95% probability level?

9.25 Trigonometrical heighting is carried out on a coastal site to transfer heights from cliff top to beach. The following measurements are taken with a total station set at control point A located at the top of the cliff to a reflector set up at control point B on the beach.

Slope distance AB   97.328 ± 0.005 m
Vertical angle from A to B   − 15°57'05" ± 30"

The heights of the total station and reflector above points A and B are:

Height of total station above A   1.564 ± 0.005 m
Height of reflector above B   1.420 ± 0.005 m

The height of point A is 29.639 ± 0.005 m

Calculate the height of B and its standard error.

If a better accuracy was required for the height of point B, which measurement would you improve and why?

9.26 In a triangle ABC, the following measurements are taken

AB = 68.214 ± 0.008 m   BC = 52.765 ± 0.004 m $\hat{ABC}$ = 48°29'20" ± 15"

Calculate the area of the triangle and its standard error (*the area of the triangle is given by* $\frac{1}{2}$·AB·BC·sin $\hat{ABC}$).

9.27 If the distance between two primary bench marks is 200 m, what is the maximum sighting distance that should be used to level between these to ensure that the height difference between them conforms to BS 5964/ICE standards? A digital level is to be used where the standard error in reading a levelling staff is ± 0.01 mm per m of the sighting distance.

9.28 During a secondary control survey, angles are measured using a total station with a stated accuracy of ±10" and targets that are bisected also with an accuracy of ±10". If the accuracy of each angle measured is to meet BS 5964/ICE specifications, how many rounds of angles should be observed if the distances between control points are 50 m, 100 m and 200 m?

Further reading and sources of information

For information on specifications, consult:

BS 5606: 1990 *Guide to accuracy in building* (British Standards Institution [BSI], London). BSI web site http://www.bsi-global.com/.
BS 5964: 1996 (ISO 4463: 1995) *Building setting out and measurement* (British Standards Institution, London). BSI web site http://www.bsi-global.com/.
BS 7334: 1992 (ISO 8322: 1991) *Measuring instruments for building construction* (British Standards Institution, London). BSI web site http://www.bsi-global.com/.
ICE Design and Practice Guide (1997)*The management of setting out in construction*. Thomas Telford, London.

For further information on survey adjustments and least squares, consult:

Wolf, P. R. and Ghilani, C. D. (1996) *Adjustment Computations*. John Wiley & Sons, Chichester.
RICS (2005) *Reassuringly accurate*. Available as a PDF download on the Mapping and Positioning Practice Panel web page at http://www.rics.org/geo/.

For the latest information on least squares software available from survey equipment manufacturers, visit the following web sites:

http://www.leica.com/
http://www.sokkia.com/
http://www.topcon.co.uk/
http://www.trimble.com/

The following journals include annual reviews of least squares software and include frequent articles on them: *Civil Engineering Surveyor, Engineering Surveying Showcase, GIM* and *Geomatics World* (see Chapter 1 for further details).

chapter 10

Detail surveying and mapping

Aims

After studying this chapter you should be able to:

- Understand the importance of planning a detail survey and preparing its specifications well before any fieldwork begins

- Appreciate the different types of detail that exist and how symbols and abbreviations can be used in their representation

- Describe the methods used to survey detail on site using *total stations* and *GPS receivers*

- Plot a control network and add detail and contours to this to produce a survey plan

- Appreciate that although hand drawings are still undertaken it is much more common for plans to be produced using computer software packages

- Understand the processes involved in computer-aided surveying

- Appreciate the emergence of non-selective methods of detail surveying, involving *terrestrial laser scanners*, which enable data to be collected very rapidly

- Appreciate the role of surveying in the production of *Digital Terrain Models* (*DTMs*) and *Geographic Information Systems* (*GIS*)

- Describe the mapping products produced by the Ordnance Survey

- Understand the applications of *LiDAR* and *IFSAR* in mapping and terrain modelling

This chapter contains the following sections:

10.1   An introduction to plan production

10.1 An introduction to plan production

After studying this section you should understand that one of the purposes of a control network is to provide a base on which to build a survey plan. You should know what is meant by the term *scale* and you should understand that plans can be produced at a number of different scales. You should appreciate that although hand drawings are still undertaken it is much more common for plans to be produced using computer hardware and software. You should also be aware of the step-by-step procedures involved in the production of a contoured survey plan.

This section includes the following topics:

● The role of control networks in plan production

● Scale

● The predominance of computer-aided mapping methods

● The procedures involved in the production of contoured plans

The role of control networks in plan production

Previous chapters have dealt with the various methods by which networks of control points can be established. In engineering surveying, such a network is required for one of two purposes: either to use as a base on which to form a *plan of the area* in question, or to use as a series of points of known coordinates from which to *set out* a particular engineering construction. Often, these two purposes are linked in a three-stage process. First, using the control points, a contoured plan of the area is produced showing all the existing features. Second, the project is designed and superimposed on the original plan. Third, the designed points are set out on site often with the aid of the control points used to prepare the original plan. While the design stage may be undertaken by any of the engineering team, the production of the original plan and the setting out of the project are the responsibility of the engineering surveyor. Consequently, it is essential that correct surveying procedures are adopted to ensure that these two activities are carried out successfully.

The purpose of this chapter is to describe the first of these activities – the production of the original plan of an area – with particular emphasis being given to the various methods by which accurate contoured plans at the common engineering scales of between 1:50 and 1:1000 can be produced from control networks.

The second of these activities, the setting out of the project, is covered in depth in Chapter 11, where particular emphasis is given to the techniques by which control networks can be used to set out various construction projects.

Scale

All engineering plans and drawings are produced at particular scales, for example 1:500, 1:200, 1:100 and so on, in ratios of 5, 2 and 1. The scale's value indicates the ratio of horizontal or vertical plan distances to horizontal or vertical ground distances that was used when the drawing was produced.

A horizontal plan having a scale of 1:50 indicates that for a line AB

$$\frac{\text{Horizontal plan length AB}}{\text{Horizontal ground length AB}} = \frac{1}{50}$$

and, if line AB is measured as 145.6 mm on the plan then

horizontal ground length of AB = 145.16 mm × 50 = 7.280 m

The term *large-scale* indicates a small ratio such as 1:20, 1:100, whereas the term *small-scale* indicates a large ratio such as 1:25,000, 1:50,000.

On engineering drawings, scales are usually chosen to be as large as possible to enable features to be drawn as they actually appear on the ground. If too small a scale is chosen then it may not be possible to draw true representations of features and, in such cases, conventional symbols are used. This is the technique commonly adopted by the Ordnance Survey for all its mapping products.

It must be stressed that the scale value of any engineering drawing or plan must always be indicated on the drawing itself. Without this, it is incomplete and it is impossible to scale dimensions from the plan with complete confidence.

The predominance of computer-aided mapping methods

In the past twenty years there has been a fundamental change in the methods by which contoured plans and other survey drawings are produced. Traditionally, they were drawn by hand, with elegant tracings of original pencil plots being prepared by skilled draughtsmen and draughtswomen working at drawing boards. However, this is no longer the case. Hand drawings are now rarely seen in the commercial surveying world. As with many other aspects of surveying, computer-aided systems have been widely adopted and contoured plans and other survey drawings are nowadays invariably prepared on automatic multi-pen plotters using one of the many commercial surveying and mapping software packages that are currently available.

These packages and plotters enable maps and plans to be produced very quickly to a high degree of accuracy. Instead of using field sheets and books, the field survey observations are normally stored in a *database*, which can be used by the software to prepare a three-dimensional representation of the ground surface, known as a *digital terrain model* (DTM). Using the database and the DTM, plans, contour overlays, sections and perspective views can be obtained at virtually any scale in a variety of colours on the multi-pen plotter. Additional information can be added to the database in the form of layers to produce a *Geographic Information System* (GIS) containing data on underground services, geology, property boundaries, postcodes, land owners, land use, residential properties, business properties and so on. This integrated approach is now widely used in surveying and the principles on which it is based are discussed in Section 10.9.

Nevertheless, although computer-based surveying and mapping are now the norm, from an educational viewpoint, knowledge of the methods by which hand drawings are produced can greatly help the understanding of the principles on which the commercial software packages are based, since there are many similarities between the two approaches. Hence, throughout this chapter, reference is made to hand drawing methods wherever appropriate in order to enable the reader not only to gain an insight into the procedures on which computer-aided mapping methods are based, but also to appreciate the considerable advantages that such computer-based methods have over the traditional hand methods.

The procedures involved in the production of contoured plans

The procedures involved in the production of a contoured plan follow the step-by-step process outlined below. Further information about each of the steps involved can be found in the sections referenced.

● The specifications for the survey are prepared well in advance. This involves considering the purpose of the survey, its accuracy requirements and its end product and choosing suitable equipment, techniques and drawing media that will ensure that the specifications are achieved. This is discussed in Section 10.2.

● An accurate coordinate grid is established on the drawing paper or film at the required scale and the control network is plotted. How this is done on hand drawings is described in Section 10.3.

● The positions of the features in the area are located on site from the control network. The field data collected during this process is then brought back into the office, processed and plotted on the drawing. The term *detail* is used to describe the various features that are in the area. This is discussed in Section 10.4. The expression *detail surveying* or *mapping* is used to describe the process of locating or *picking up* detail on site from the control network. The most common methods for doing this nowadays are either by using GPS receivers or by using a radiation method involving total stations, both of which are discussed in Section 10.5. In addition, the recent development of laser scanners has added another powerful tool to the surveyor's detail surveying kit and these instruments are discussed in Section 10.8.

● Once all the detail has been plotted, contours can be added using the interpolation method discussed in Section 10.6.

● The plan is completed by adding a title block containing the location of the survey, a north sign, the scale, the date, the key and other relevant information. This is discussed in Section 10.7.

Reflective summary

With reference to plan production remember:

— One of the purposes of a control network is to provide a base on which the plan of an area can be formed.

— Contoured plans and other survey drawings are nowadays invariably prepared on automatic multi-pen plotters using one of the many commercial surveying and mapping software packages that are currently available.

— Although hand drawings are now rarely seen in the commercial world, knowledge of the methods by which they are produced can greatly help the understanding of the principles on which the commercial software packages are based.

— The procedures involved in the production of a contoured plan follow a step-by-step process.

10.2 Planning the survey

After studying this section, you should understand the need to plan a detail survey and to prepare its specifications well in advance of starting the actual fieldwork. You should be aware of two documents dealing with specifications for mapping and surveys that have been produced for use in the United Kingdom by the Royal Institution of Chartered Surveyors (RICS) and you should understand their purpose. You should appreciate the different types of drawing media that are available and be able to choose the most appropriate one for your requirements.

This section includes the following topics:

- Specifications for detail surveys
- Drawing paper and film

Specifications for detail surveys

Before any detail surveying field work is undertaken the first task is to prepare a *set of specifications* which set out the aims of the survey, the accuracy required, the amount and type of detail that is to be located and the intensity of any spot levels that are to be taken. This information is necessary in order that suitable equipment and techniques can be chosen to ensure that the aims of the survey can be achieved. To help with this, the Royal Institution of Chartered Surveyors (RICS) has produced two publications on the preparation of specifications for surveying and mapping and these are discussed later in this section.

Accuracy requirements

The accuracy requirements of a proposed detail survey are governed by two factors: the scale of the finished plan and the accuracy with which it can be plotted.

For *plan positions*, it is usual to assume a plotting accuracy for detail of 0.5 mm and for various scales this will correspond to certain distances on the ground. These lengths are an indication of the accuracy required at the scales in question. However, even if a plan is to be plotted at 1:500, part of it may at a later date be enlarged to 1:50, so it is always better to take measurements in the field to a greater accuracy than that required for the initial plan. A good compromise is obtained by recording all distances, where possible, to the nearest 0.01 m.

For *contours*, the vertical interval depends on the scale of the plan and suitable vertical intervals are listed in Table 10.1 for general purpose engineering surveys. Usually, the accuracy of a contour is guaranteed to one-half of the vertical interval.

All *spot levels* taken on soft surfaces (for example, grass) should be recorded to the nearest 0.05 m and those taken on hard surfaces to 0.01 m. In areas where there is insufficient detail at which spot levels can be taken, a grid of spot levels should be

Table 10.1 ● Contour vertical intervals and spot height grid sizes.

	Scale				
	1:50	*1:100*	*1:200*	*1:500*	*1:1000*
Contour vertical interval (m)	0.05	0.10	0.25	0.50	1.00
Spot height grid size (m)	2	5	10	20	40

surveyed in the field and plotted on the plan, the size of the grid depending on the scale, as shown in Table 10.1.

The amount and type of detail to be located

In addition to accuracy considerations, it is necessary to decide on the amount and type of detail that is to be located and the intensity of the spot levels that are to be taken. These decisions depend on a number of factors, for example, the *topography of the site* will influence the number of spot levels, the *purpose of the survey* will dictate the type of detail to be located and the *time available* for the work may restrict the amount that can be picked up. In addition, one very important factor that can have a considerable bearing on what is shown on the finished drawing is the budget available for its production. Since greater detail requires a greater amount of work and higher accuracy requires more expensive equipment, a survey which requires high accuracy and intensive detail will be more expensive to achieve.

Preparing the specifications

Careful planning is required in order that the accuracy and aims of the survey can be established before any fieldwork begins. This calls for the preparation of a survey specification in which the various requirements are listed. To help with this, the RICS has produced two publications on the preparation of specifications for surveying and mapping. Their titles are *Specification for Mapping at Scales between 1:1000 and 1:10000* and *Surveys of Land, Buildings and Utility Surveys at Scales of 1:500 and Larger, Client Specification Guidelines*. The purpose of the former is to provide a general technical specification for contract mapping worldwide which can be modified as required by commissioning agencies and surveyors to meet the particular needs of individual mapping projects. The purpose of the latter is to provide a standard specification for large-scale surveys. Both are relevant to engineering surveys, although the latter is more applicable since the majority of engineering plans are produced at scales of 1:500 or larger. Each takes the form of a series of sections under a range of surveying related headings. Within each section, sub-sections are provided itemising the various parameters which may be included if required. They have been set out in such a way that they can be used directly as contract documents by entering appropriate details in the spaces provided, deleting sections and numbered paragraphs which are not required, and adding appendices or additional sections and numbered paragraphs to include special requirements not covered in the documents themselves. As an example of this, Tables 10.2 and 10.3 shows sections 2.3 and 2.4, respectively, of the RICS publication *Surveys of Land, Buildings and Utility Surveys at Scales of 1:500 and Larger, Client Specification Guidelines*, which have been reproduced by kind permission of the RICS.

Although a fuller description of these specifications is beyond the scope of this book, their use is strongly recommended when planning engineering surveys. They are referenced at the end of this chapter.

Drawing paper and film

Nowadays, there is a variety of drawing media available on which to produce engineering survey plans either by hand or on computer-aided plotters. All, however, fall into one of two categories: either *paper* or *film*. Both are produced in a number of different grades, each being specified by its degree of translucency and either its weight or its thickness. Different surface finishes are also available for example, matt, semi-matt and gloss. During the course of a survey, several different types of paper or film may be used at various stages of the work. Some of the most popular types are summarised below.

Light-weight paper, 60 grammes per square metre (gsm), is an inexpensive opaque medium. However, it is easily torn, has poor dimensional stability and should be handled with care. It is ideal for pencil work and may be used for preparing an initial check plot in order to verify that the survey data is correct. Ink is possible, but is not recommended since absorption can occur, resulting in unsightly smudges.

Opaque bond paper, 90 gsm, is ideal for all pencil drawings and will also take ink. It can be used for hand drawings and computer plots. Sometimes referred to as *cartridge paper*, it usually has good dimensional stability and may be used for the production of a master pencil drawing which is later to be traced. For direct ink work, a quality must be chosen which is sufficiently high to prevent the ink from being absorbed. For both ink and pencil, its quality must be such that repeated erasures do not make the surface fibrous or cause tearing.

Tracing paper, 90 gsm, has an excellent translucency and is ideal for an ink tracing of the original pencil drawing. A high quality paper should be chosen to give good dimensional stability and to reduce the chances of tearing and creasing. This is particularly important for computer plots. An extra smooth surface will give a high standard of linework with no absorption. Tracing paper provides a good original for dyeline printing. However, with age it can become brittle and discolour.

Vellum, 80 gsm, has higher quality than tracing paper and good translucency. It is ideal for ink and does not become brittle or discolour with age. Vellum is an excellent medium for computer plots from which dyeline copies can be produced and where drawings are to be stored for a long period of time.

Polyester-based films are the top quality media available. They are translucent and are referenced by their thickness rather than their weight, for example, 75 microns. With first-class surfaces and excellent dimensional stability they are ideal for all survey drawings. They are particularly good for computer-aided plotters as they are capable of giving sharper plotted ink lines at faster speeds. Pencil and ink can be used

Table 10.2 ● Specifications for detail surveys – Planimetric Information (reproduced by permission of the RICS, which holds the copyright).

Detail to be Surveyed

2.3   Planimetric Information

The following general categories of detail shall be surveyed: (delete from the list groups of features not required)

	Permanent buildings/structures
	Temporary/mobile buildings
	Visible boundary features: walls, fences, hedges
	Road, tracks, footways, paths
	Street furniture
	Statutory Authorities' plant and service covers where visible
	Changes of surface
	Isolated trees/wooded areas/limits of vegetation
	Pitches/recreation
	Private gardens or grounds (off-site areas)
	Water features
	Earth works
	Industrial features
	Railway features with arranged access
	Other (specify)

(further information on items of detail within each of the categories listed above can be found in Annexe B of the Guidelines)

2.3.1   Accuracy

The accuracy of planimetric detail shall be such that the plan position of any well defined point of detail shall be correct to 0.3mm r.m.s.e.* at the plan scale when checked from the nearest permanent control station.

2.3.2   Obscured Ground (select the option required)

Detail which cannot be surveyed to the specified accuracy without extensive clearing shall be

either   (a)   surveyed approximately and annotated accordingly.
or   (b)   surveyed, following clearing by the Client.
or   (c)   surveyed, following clearing by the Surveyor.

---

*The r.m.s.e. (root mean square error) is sometimes used in surveying to indicate precision instead of a standard deviation.

Table 10.3 ● Specifications for detail surveys – Height Information (reproduced by permission of the RICS, which holds the copyright).

Detail to be Surveyed

2.4 Height Information

Height information shall be provided

either (a) as spot heights throughout the survey area.

or (b) by spot heights to detail specified in Annexe C, together with contours.

(Annexe C is provided in the Guidelines and lists points of detail where spot heights should be taken, for example, tops and bottoms of banks, building corners, inspection covers, gullies, ducts, conduits and so on.)

or (c) by contours only.

Sufficient levels shall be surveyed such that the ground configuration, including all discontinuities, is represented on the survey plan.

2.4.1 Spot Heights

The maximum distance between adjacent spot levels shall be .... metres. Ground survey spot levels on hard surfaces shall be correct to ±10 mm r.m.s.e*. and elsewhere to ±50 mm r.m.s.e.*, except on ploughed or otherwise broken surfaces.

The recommended maximum spacings for spot heights are:

1:500	20 metre ground spacing	(4 cm on plan)
1:200	20 metre ground spacing	(8 cm on plan)
1:100	10 metre ground spacing	(10 cm on plan)

2.4.2 Contours

Contours shall be shown at vertical intervals of .... metres.

At least 90% of all contours shall be correct to within one half of the specified contour interval. Any contour which can be brought within this vertical tolerance by moving its plotted position in any direction by an amount equal to 1/10th of the horizontal distance between the contours, or 0.5 mm at plan scale, whichever is the greater, shall be considered as correct.

2.4.3 Obstructed Ground (select the option required)

Contours which cannot be represented to the specified accuracy without extensive clearing shall be

either (a) surveyed approximately and annotated accordingly.

or (b) surveyed, following clearing by the Client.

or (c) surveyed, following clearing by the Surveyor.

*The r.m.s.e. (root mean square error) is sometimes used in surveying to indicate precision instead of a standard deviation.

on them very easily and both can be erased without leaving unsightly marks. Such films are waterproof and difficult to tear, and can be used in any reprographic process.

All paper and film can be purchased either in roll form or as packs of cut sheets. Rolls come in a variety of lengths, typically 50 m and 100 m, and a wide range of roll widths are available. Similarly, cut sheets come in packs of, typically, 100 to 250 and in a wider choice of sheet sizes. Increasingly, however, for both rolls and sheets, there is a trend towards the use of standard widths and sizes based on the dimensions recommended by the International Organization for Standardization (ISO).

International Organization for Standardization paper sizes

ISO sizes for cut sheets are based on rectangular formats which have a constant ratio between their sides of $1:\sqrt{2}$. Succeeding sizes in the range are created by either halving the longer side or doubling the shorter side. This is best illustrated by considering the most important ISO series of paper sizes, the *A-Series*.

The basic unit of the A-series is designated A0 and has an area of 1 m^2 with sides in the ration $1:\sqrt{2}$. This gives its dimensions as 841 mm × 1189 mm.

To create the next larger size, the smaller dimension is doubled and an even whole number inserted before the A0 symbol, for example.

$$2A0 = 1189 \text{ mm} \times 1682 \text{ mm} \qquad 4AO = 1682 \text{ mm} \times 2378 \text{ mm}$$

In practice, for survey drawings sizes larger than A0 are rarely used.

To create smaller sizes, the larger dimension is halved and a different number is inserted after the A, that is, A1, A2, A3 and so on. Hence A1 is half the area of A0, A2 is half the area of A1 (and a quarter of the area of A0) and so on. This enables two of each subsequent sheet size to be obtained by folding the previous size in half; for example, an A0 sheet folded in half provides two A1 size sheets. Table 10.4 lists the sizes of the A-series cut sheets commonly used for survey drawings.

Although the A-series strictly applies to cut sheets, rolls are available having widths based on the dimensions it uses. Those suitable for survey drawings are listed in Table 10.4. Each subsequent roll width is dimensioned such that two of the A-series sheet sizes can be cut from it, for example, from a roll width of 841 mm, not only A0 sized sheets (841 mm × 1189 mm) but also A1 sized sheets (594 mm × 841 mm) can be cut.

Table 10.4 ● A-series paper and film sizes.

Classification	Sheet size	Equivalent roll width
A0	841 mm × 1189 mm	841 mm
A1	594 mm × 841 mm	594 mm
A2	420 mm × 594 mm	420 mm
A3	297 mm × 420 mm	297 mm
A4	210 mm × 297 mm	210 mm

Reflective summary

With reference to planning the survey remember:

— Before any field work is undertaken for a detail survey, the first task is to prepare a *set of specifications* which set out the aims of the survey, the accuracy required, the amount and type of detail that is to be located and the intensity of any spot levels that are to be taken.

— The accuracy requirements of a proposed detail survey are governed by two factors: the scale of the finished plan and the accuracy with which it can be plotted.

— All distances, where possible, should be measured to the nearest 0.01 m, and all spot levels to the nearest 0.05 m on soft surfaces and 0.01 m on hard surfaces.

— A survey which requires high accuracy and intensive detail will be more expensive to achieve than one where a lower accuracy is specified and less detail is required.

— It is recommended that the RICS publication *Surveys of Land, Buildings and Utility Surveys at Scales of 1:500 and Larger, Client Specification Guidelines* be consulted before planning any large scale detail survey.

— There is a variety of drawing paper and film available on which to produce engineering survey plans either by hand or on computer-aided plotters.

— The most commonly used paper and film sizes are those defined by the ISO A-series.

10.3 Plotting the control network

After studying this section you should appreciate how the control network on which a detail survey is based is used as a framework for the production of the plan of the area in question. You should know how to orientate a survey so that it will fit centrally on to the drawing medium. In addition, you should be able to draw a coordinate grid on a survey drawing by hand and plot the control stations on the grid you have drawn.

This section includes the following topics:

● Computer-aided plotting compared to plotting by hand
● Orientating the survey and plot

- Plotting the coordinate grid
- Plotting the stations

Computer-aided plotting compared to plotting by hand

When a control network has been computed, as described in Chapter 6, the end-product is a set of coordinates. This is normally in the form of a list of eastings (E) and northings (N) for each of the points in the network; for example, a control station R may have coordinates (4571.56 m E, 7765.86 m N). Such coordinates have many uses in engineering surveying operations, one being to provide a framework for the production of survey drawings, either by computer-aided plotting or by hand.

For *computer-aided plotting*, the grid on which the control network is based is used by the computer to locate all the control stations and detail points in their correct E, N positions on the drawing. Different line types and symbols are then added to the points (usually in a range of colours using multiple pens) to form the finished plan and the grid itself can be added if required, although this is not necessary. Such computer-aided plotting techniques are discussed further in Section 10.9. The remainder of this section deals with the procedures involved in plotting the control network by hand since knowledge of the methods by which hand drawings are produced can greatly help the understanding of the principles on which computer-aided plotting methods are based.

When *plotting by hand*, a common mistake is to plot the stations using the angles (or the bearings) and the lengths between the stations. This is known as plotting by *angle and distance* or *bearing and distance* and involves the use of a 360° protractor and a graduated straight edge. However, the accuracy of such a method is limited, and although it can be used to plot detail (see Section 10.5) it should **never** be used to plot the control stations. Instead, the preferred and most accurate method of plotting these is to use the coordinate grid on which the computation of their coordinates was based. The procedure for this involves three steps. First, the grid on which the control network is based is carefully orientated such that the survey area falls centrally on the plotting sheet. Second, using a beam compass and a graduated straight edge, the grid lines are accurately plotted. Third, the control stations are plotted with the help of the grid and their positions carefully checked. The control stations are then used, together with the grid, to help with the plotting of the detail. The three stages involved are discussed in this section.

Orientating the survey and plot

Before commencing any plot (that is, before constructing the grid), the extent of the survey should be taken into account such that the plotted survey will fall centrally on to the drawing sheet.

In the case where the north direction is stipulated, the north–south and east–west extents of the area should be determined and the stations plotted for the best fit on the sheet.

If an arbitrary north is to be used (see Section 6.2), the best method of ensuring a good fit is to assign a bearing of 90° or 270° to the longest side. This line is then positioned parallel to the longest side of the sheet so that the survey will fit the paper properly.

Sometimes it may be necessary to set the arbitrary north to a particular direction in order to ensure that the survey will fit a particular sheet size. This will often be the case with long, narrow site surveys, where, to save paper and for convenience, the plot of the survey is to go on to a single or a minimum number of sheets or the minimum length of a roll. The boundary of the survey should be roughly sketched and positioned until a suitable fit is obtained. Again, the longest or most convenient line should be assigned a suitable arbitrary bearing.

Where the north point is arbitrary, it has to be established before the coordinate calculation takes place. In order to estimate the extent of the survey it should be sketched, roughly to scale, using the left-hand angles and the lengths between stations.

Plotting the coordinate grid

The first stage in plotting the control network is to establish a coordinate grid on the drawing medium.

Coordinated lines are drawn at specific intervals, for example, 10 m, 20 m, 50 m, 100 m in both the east and north directions to form a pattern of squares. The stations are then plotted in relation to the grid.

When drawing the grid, T-squares and set squares should *not* be used since they are not accurate enough. Instead, the grid is constructed in the following manner.

- From each corner of the plotting sheet two diagonals are drawn, as shown in Figure 10.1(a).
- From the intersection of these diagonals an equal distance is scaled off along each diagonal using a *beam compass*. This scaled distance must be large; see Figure 10.1(b).
- The four marked points on the diagonals are joined using a *steel straight edge* to form a rectangle. This rectangle will be perfectly true and is used as the basis for the coordinate grid (see Figure 10.1(c)). On all site plans and maps it is conventional to have the north point (true, magnetic or arbitrary) on the drawing such that the north direction is from the bottom to the top of the sheet and roughly parallel to the sides of the sheet. This will be achieved if the grid framework is constructed as described.
- By scaling equal distances along the top and bottom lines of the rectangle and joining the points, the vertical (E) grid lines will be formed. The horizontal (N) grid lines are formed in a similar manner using the other sides of the rectangle (see Figure 10.1(d) and (e)). All lines must be drawn with the aid of a steel straight edge and all measurements must be taken from lines AB and BC and not from one grid line to the next. This avoids accumulating errors.

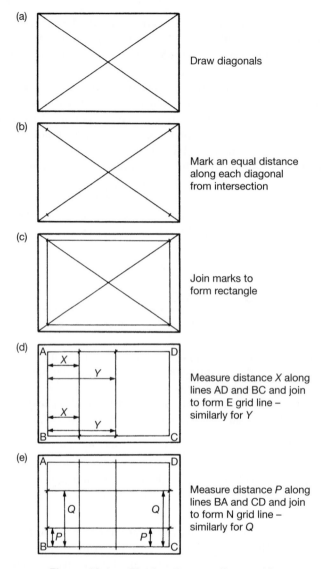

Figure 10.1 ● Plotting the coordinate grid.

● The grid lines should now be numbered accordingly. The size of a grid square on the drawing should not be greater than 100 mm × 100 mm. It is not necessary to plot the origin of the survey if it lies outside the area concerned.

Plotting the stations

Let one of the stations to be plotted have coordinates 283.62 m E, 427.45 m N and let it be plotted on a 100 m grid prepared as described above.

- The grid intersection 200 m E, 400 m N is located on the prepared grid.

- Along the 400 m N line, 83.62 m is scaled off from the 200 m E intersection towards the 300 m E intersection and point *a* is located (see Figure 10.2). Similarly, point *b* is located along the 500 m N line. Points *a* and *b* are joined with a pencil line.

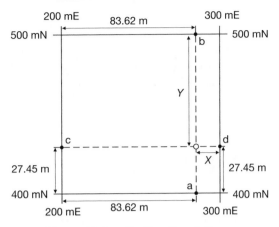

Figure 10.2 ● Plotting the stations.

- Along the 200 m E line, 27.45 m is scaled from the 400 m N intersection towards the 500 m N intersection to locate point *c*. Point *d* is found by scaling 27.45 m along the 300 m E line. Points *c* and *d* are joined.

- The intersection of lines *ab* and *cd* gives the position of the station.

- To check the plotted position, dimensions *X* and *Y* are measured from the plot and compared with their expected values. In this case *X should* equal 100.00 – 83.62 = 16.38 m and *Y should* equal 100.00 – 27.45 = 72.55 m.

- The procedure is now repeated for each of the other stations.

- When all the stations have been plotted, the lengths between the plotted stations are measured and compared with their known values.

- The control lines are added by carefully joining the plotted stations. This is to aid in the location of detail (see Sections 10.4 and 10.5).

Reflective summary

With reference to plotting the control network remember:

– The coordinates of the stations obtained during the control survey provide a framework for the production of survey drawings.

– When plotting the control stations, their positions are fixed using the coordinate grid on which the computation of their coordinates was based.

– Before beginning any plot, the extent of the survey should be taken into account such that the plotted survey will fall centrally on to the drawing sheet. It may be necessary to assign an arbitrary north direction to the grid to enable this to be achieved.

— When drawing by hand, the grid must first be plotted very accurately on the drawing medium. The positions of the control stations should then be plotted on the grid using their coordinates, *not* using the angles (or the bearings) and the lengths between the stations.

10.4 Detail

After studying this section you should understand that there are four main categories of detail. You should be aware of some of the symbols that can be used to represent detail on drawings but you should appreciate that on the large scales that tend to be used for engineering surveying drawings, symbols are not greatly used since detail can often be drawn at its correct size and shape. You should also be aware of some of the different ways by which detail can be located from control networks.

This section includes the following topics:

- Categories of detail
- Symbols
- Locating detail

Categories of detail

The term *detail* is used to describe the various features that are located in the survey area. It is a generic one that implies features not only at ground level but also those above and below ground level. Detail can be subdivided into the following four main categories:

- *Hard detail* describes well-defined features. These tend to be features which have been constructed, such as buildings, roads and walls, although some natural features such as clearly defined rocks could be given this classification. Anything with a definite edge or having a precise position that can be easily located would fall into this category.
- *Soft detail* describes features that are not well defined. These tend to be natural features such as river banks, bushes, trees and other vegetation. Anything where its edge or its precise position is in doubt would fall into this category.
- *Overhead detail* describes features above the ground, such as power lines and telephone lines. The initials O/H are sometimes added to overhead detail.
- *Underground detail* describes features below the ground, such as water pipes and sewer runs. The initials U/G are sometimes added to underground detail.

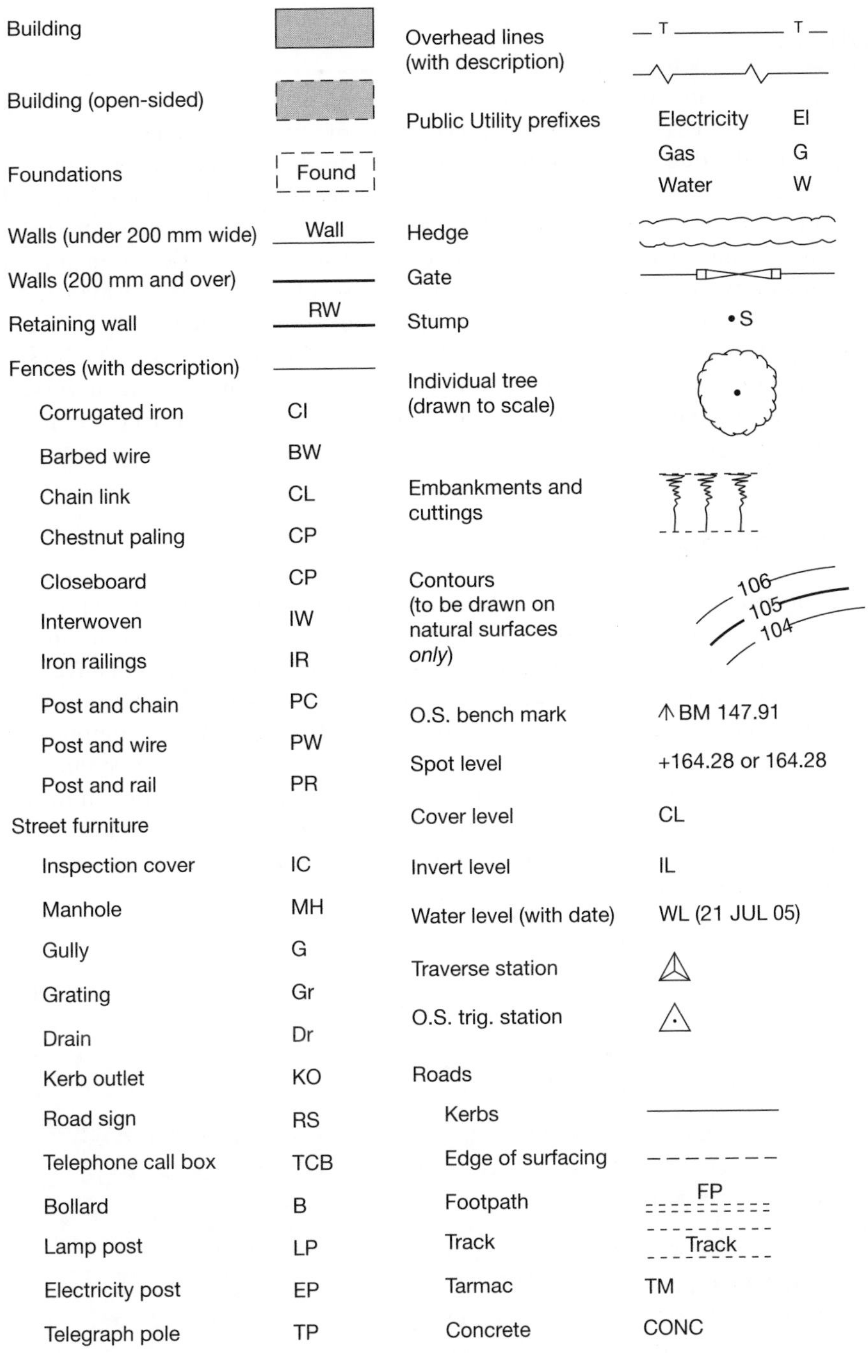

Building		Overhead lines (with description)	_ T _____ T _
Building (open-sided)		Public Utility prefixes	Electricity El
			Gas G
Foundations	Found		Water W
Walls (under 200 mm wide)	Wall	Hedge	
Walls (200 mm and over)		Gate	
Retaining wall	RW	Stump	• S
Fences (with description)		Individual tree (drawn to scale)	
Corrugated iron	CI		
Barbed wire	BW		
Chain link	CL	Embankments and cuttings	
Chestnut paling	CP		
Closeboard	CP	Contours (to be drawn on natural surfaces *only*)	106 105 104
Interwoven	IW		
Iron railings	IR		
Post and chain	PC	O.S. bench mark	⋏ BM 147.91
Post and wire	PW	Spot level	+164.28 or 164.28
Post and rail	PR	Cover level	CL
Street furniture		Invert level	IL
Inspection cover	IC	Water level (with date)	WL (21 JUL 05)
Manhole	MH	Traverse station	
Gully	G		
Grating	Gr	O.S. trig. station	
Drain	Dr		
Kerb outlet	KO	Roads	
Road sign	RS	Kerbs	
Telephone call box	TCB	Edge of surfacing	- - - - - -
Bollard	B	Footpath	FP
Lamp post	LP	Track	Track
Electricity post	EP	Tarmac	TM
Telegraph pole	TP	Concrete	CONC

Figure 10.3 ● Conventional sign list.

Symbols

On maps and plans, *symbols* can be used to represent detail. However, somewhat surprisingly, there is no universally agreed standard range of symbols for use in map and plan production and different mapping organizations tend to adopt their own conventional sign list. An example of this is provided in the United Kingdom by the Ordnance Survey, which has adopted its own extensive range of symbols for use on some of its maps. An example of a conventional sign list for use on engineering surveying drawings is shown in Figure 10.3, where a fairly comprehensive range is provided. The use of the symbols and abbreviations shown in Figure 10.3 is recommended, although it should be noted that more abbreviations than symbols are given. This is due to the fact that, at the large scales used for engineering surveys, the actual shapes of many features can be plotted to scale, and therefore do not need to be represented by symbols.

Locating detail

When undertaking a detail survey, the amount and type of detail that is located (or *picked up*) for any particular survey varies enormously with the scale and the intended use of the plan. These factors must be carefully considered in advance of any fieldwork when preparing the survey specifications discussed in Section 10.2.

Detail can be picked up from the control network by one of several methods. The most common methods employed nowadays involve the use of total stations and GPS equipment, both of which are discussed in Section 10.5. In addition, new techniques are evolving that enable enormous amounts of detail to be collected very rapidly. Foremost among these are techniques involving laser scanners, which are described in Section 10.8.

Reflective summary

With reference to detail remember:

— The term *detail* is used to describe the various features that are located in the area; there are four categories of detail: hard, soft, overhead and underground.

— On maps and plans, symbols can be used to represent detail, although at the large scales used for engineering surveys, the actual shapes of many features can be plotted to scale, and therefore do not need to be represented by symbols.

— Detail can be surveyed or picked up using total stations, GPS receivers and laser scanners.

10.5 Surveying detail using total stations and GPS equipment

After studying this section you should understand how total stations and GPS equipment can be used to locate detail from a control network and how this can be plotted. You should appreciate that both methods can produce data in a form that is ideal for use in computer-aided plotting. You should be aware that some total stations and GPS systems enable detail to be surveyed and displayed immediately, either on the instrument's own screen or on the screen of an attached control unit.

This section includes the following topics:

- Detail surveying using total stations
- Detail surveying using GPS equipment

Detail surveying using total stations

A full description of total stations is given in Chapter 5, and their application to detail surveying is given here. A total station can be used in conjunction with a prism mounted on a detail pole to locate points of detail by the *radiation* method.

The principle of radiation is shown in Figure 10.4, where a tree is to be located from a survey line AB. The plan position of the tree can be fixed by measuring the horizontal distance, r, to it from A and also the horizontal angle, θ, subtended at A by the survey line AB and the line from point A to the tree. Hence the radiation method consists of measuring horizontal angles and horizontal distances.

With reference to Figure 10.4, the total station is set up at A, the detail pole is held at the tree, and r and θ are measured. In addition, the reduced level of the point at which the detail pole is held can be obtained by measuring the vertical angle to the prism and using this in conjunction with the height of the prism and the height of the total station. Hence three-dimensional data is collected about the point being surveyed; that is, its plan position and its reduced level.

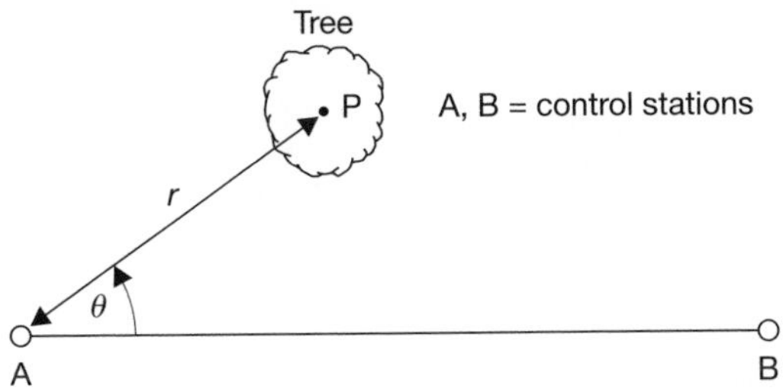

Radiation using total stations can be used very effectively for contouring, particularly in open areas where there are no points of clearly defined detail. Also, because of its high accuracy capabilities, it can be used to pick up any type of detail, either hard or soft.

Figure 10.4 ● Principle of radiation.

Traditionally, two people have been required to locate detail using total stations: one operating the total station and the other holding the detail pole. However, in the past few years, total stations have been developed which are motorized and capable of locking onto and tracking prisms on detail poles using automatic target recognition. These so-called *robotic* total stations are discussed in Chapter 5 and enable one-person operation. In this system, the instrument is set up as normal over a control point and left unattended with the operator working at the detail pole actually choosing and fixing the points of detail. The operator is armed with a detail pole fitted with a remote control unit that communicates with the instrument and allows observations to be initiated and codes and other inputs to be made via its integral keyboard.

Although the use of such robotic instruments is on the increase, two-person total station detailing is still the norm. Hence the remainder of this section assumes that two people are involved: one using the instrument and the other holding the detail pole.

Fieldwork for surveying detail using total stations

During fieldwork, the total station is set up at each point in the control network in turn and used to locate detail from them as required. Because of its inherently high accuracy, it is possible to locate detail which lies a considerable distance from the control points using this method. However, if the distance becomes too great it can be very difficult for the two people carrying out the work to communicate effectively with each other. Hence if voice communication is being used, it is recommended that sighting distances should not exceed 50 m or so. For distances greater than this the use of two-way radio communication is recommended. This, of course, is not a problem if a robotic system is being used.

Several methods of observation are possible and the following description is given as a general purpose approach only. It is assumed that the reduced levels of the control stations are known. Consider Figure 10.5, which shows a detail surveying observation being taken from a control point I to a detail point P.

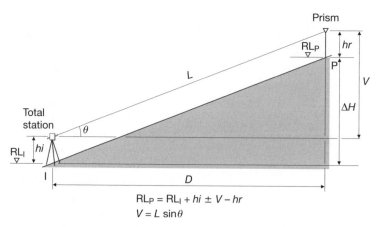

$$RL_P = RL_I + hi \pm V - hr$$
$$V = L \sin\theta$$

Figure 10.5 ● Detail surveying using a total station.

- Set the instrument up over the station mark (for example, a nail in the top of a peg) at control point I and centre and level it in the usual way. For a detail survey it is standard practice to measure horizontal and vertical angles on one face only, and hence the total station should be in good adjustment or calibration (the procedures for doing this are described in Section 5.5).

- Measure and record the height of the tilting axis (*hi*) above the station mark.

- Select a nearby control station as a reference object (RO), sight this point and record the horizontal circle reading. It may be necessary to erect a target at the RO if it is not well defined. All the detail in the radiation pattern will now be fixed in relation to this chosen direction. Some engineers prefer to set the horizontal circle to zero along the direction to the RO, although this is not essential.

- The prism mounted on the detail pole is held at the detail point being fixed (P) and observed by the total station.

- The actual observations taken at each pointing will depend on the type of instrument being used and whether they are being booked by hand or recorded automatically. In the simplest case, readings of the horizontal circle, the vertical circle and the slope distance *L* are booked or recorded. On instruments which give the horizontal distance *D* and the vertical component *V* of the slope distance directly, these two values can be booked or recorded together with the horizontal circle reading. If using an instrument which calculates and displays the coordinates (*E*, *N*) and reduced level (RL) of the point on its screen, then these can be booked or recorded directly.

- The type of detail being observed, for example, the edge of a kerb, must be booked or recorded. If hand booking and drawing is being done, the type of detail should be noted on the booking form (see later). If automatic booking and plotting is being done, a unique point number must be allocated to the detail point and a feature code must be assigned to it to so that it will be recognised by the software being used to process the data and plot the map. Feature codes can take several forms; for example, they can define the type of detail (stump, road sign, drainage gully and so on) or they can initiate an action (run a curve through the point, join to the next point, and so on). Feature codes are vital to the success of computer-aided plotting and their use is discussed in Section 10.9.

- The height of the centre of the prism (reflector) above the bottom of the detail pole (*hr*) must be booked or recorded. Since detail poles are telescopic, this height can be set as required. If hand booking and calculations are being done, these are simplified if *hr* is set to the same value as the height of the instrument above the control station *hi*. In such a case, *hi* is cancelled out by *hr* and the vertical component *V* of the slope distance *L* is equal to the difference in height between ground points I and P (see Section 5.6 for the calculations referred to here). If the observations are being recorded in a data collector or on a memory card then *hr* can be set to any convenient value. However, once it has been set, it is recommended that the height of the prism is not altered unless absolutely necessary, since every time *hr* is changed its new value must be keyed into the data collector or instrument keyboard by hand. As well as being time-consuming, this can easily be forgotten, causing errors.

● Once point P has been located the detail pole is moved to the next point and the procedure is repeated for each detail point in turn until all the observations have been completed. As far as is practicable, each of the detail points should be selected in a clockwise order to keep the amount of walking done by the person holding the detail pole to a minimum. However, if the observations are being recorded for use with a computer software mapping package it is advisable to pick up all the points needed to define a particular feature before moving on to another feature; for example, the corners of the same building, all the points along the same kerb line, all the spot heights in a field, and so on. This is discussed further in Section 10.9.

● Before packing up, the final sighting should be back to the RO to check that the setting of the horizontal circle has not been altered during observations. If it has, all the readings are unreliable and should be remeasured. Hence it is advisable, during a long series of total station readings, to take a sighting back to the RO after, say, every 10 points of detail.

Booking and calculating total station detail observations

If electronic data recorders or total stations with memory cards are being used, no booking sheets are required since the software contained in these devices will prompt the observer for all the necessary information including point numbers and feature codes. This is input directly from the built-in keyboards on either the data collector and/or the instrument.

For hand booking, however, some type of standard booking form is necessary in order that the observations are recorded accurately and neatly and can be used by others not necessarily involved with the fieldwork. Ideally, space should be provided on such a form to record the following for each point: horizontal circle, vertical circle and slope distance L, which together are known as the *raw data*. In addition, the horizontal distance D, vertical component V, coordinates E, N and reduced level RL may also be recorded. As discussed in the previous section, the type of information actually booked on the form will depend on the instrument being used and the method of plotting can also influence the type of information recorded. Great care must always be taken when booking if the raw data is not recorded, since any error in booking other values cannot later be corrected by calculation. In all cases when booking, a REMARKS column must be included in which to record details of the points being observed.

Although individuals tend to develop their own methods of booking as they gain experience, an example of a suitable form for booking and calculating radiation using total stations is shown in Table 10.5. With reference to this, the survey has been carried out using a total station from which the raw data values have been booked and used to calculate the remaining values. During the fieldwork, the height of the prism was set equal to the height of the instrument.

Table 10.5 Example booking form for radiation using a total station (all dimensions in metres unless stated otherwise).

SURVEY Flag Housing Development																		
INSTRUMENT STATION F							RL OF STATION F					47.15		OBSERVER JU				
							HEIGHT OF INSTRUMENT					1.55		BOOKER WFP				
NOTES							RL OF INSTRUMENT AXIS					48.70		DATE 28 APR 2005				
														COORDINATES (E, N)		F 719.36	911.72	
																E 724.75	1023.97	
Prism Point	Vertical Circle			Horizontal Circle			L	hr	θ			D	V	±V – hr	RL	E	N	REMARKS
	°	'	"	°	'	"			°	'	"							
E (RO)				00	00	00												Station E
1	92	35	40	08	04	27	26.248	1.55	−02	59	59	26.221	−1.188	−2.738	45.96	724.28	937.47	Edge of gravel track
2	91	09	18	51	19	48	51.753	1.55	−01	19	18	51.742	−1.043	−2.593	46.11	761.26	942.08	Post and barbed wire fence
3	89	12	55	124	44	09	21.789	1.55	+00	47	05	21.787	+0.298	−1.252	47.45	736.65	898.46	Fence meets track
4	87	35	06	143	15	17	15.044	1.55	+02	24	54	15.031	+0.634	−0.916	47.78	727.76	899.26	Hedge next to track
5	90	24	42	297	51	52	14.758	1.55	−00	24	42	14.758	−0.106	−1.656	47.04	706.66	919.24	Grass meets track
6	89	05	44	286	09	51	40.356	1.55	+00	54	16	40.351	+0.637	−0.913	47.79	681.19	924.80	Hedge next to track
7	90	05	03	293	34	11	39.471	1.55	−00	05	03	39.471	−0.058	−1.608	47.09	683.98	929.22	Opposite 6, edge of grass
8	91	21	15	305	37	53	43.886	1.55	−01	21	15	43.874	−1.037	−2.587	46.11	684.97	938.96	S Corner of store
9	91	53	08	314	25	51	37.167	1.55	−01	53	08	37.147	−1.223	−2.773	45.93	694.11	938.97	SE Corner of store
10	91	47	09	330	53	17	55.581	1.55	−01	47	09	55.554	−1.732	−3.282	45.42	694.69	961.50	NE Corner of store
E (RO)				00	00	00												
(1)	(2)			(3)			(4)	(5)	(6)									(7)

In Table 10.5, all information in columns (1) to (5) and the REMARKS column (7) is recorded in the field, the remainder being computed in the office at a later stage. The vertical circle readings entered in column (2) must be those as read directly on the instrument, reduction being carried out in column (6) where necessary (see Section 3.5).

The data shown in Table 10.5 applies to the detail survey shown in Figure 10.6. In practice, a sketch such as that shown in Figure 10.6 should be prepared to identify all the points that have been located. In addition, this sketch should indicate miscellaneous information such as types of vegetation, widths of tracks and roads, heights and types of fences and so on.

Plotting total station detail observations

If data collectors or memory cards have been used, the plotting is carried out using computer-aided methods as discussed in Section 10.9.

If the booking and calculating have been done by hand, the detail can be plotted by one of the following two methods.

Either by using the control grid to plot the rectangular coordinates of each point of detail. The RL of the point is then written alongside its plan position. In such a case, E, N and RL values are required from the booking forms.

Or by using a 360° protractor and a scale rule to plot detail points from the control stations. With reference to Figure 10.6 and Table 10.5, the procedure is as follows to plot the detail located from station F.

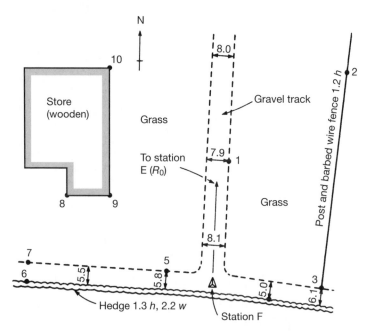

Figure 10.6 ● Example sketch for a detail survey.

- Attach the protractor to the survey plan using masking tape such that its centre is at station F and it is orientated to give the same reading to the RO, station E, as was obtained in the field on the horizontal circle of the theodolite – in this case, 00°00'00".

- Plot the positions of the horizontal circle readings taken to the detail points around the edge of the protractor. Identify each by its staff position, that is, F1, F2, F3 and so on. Since it is impossible to plot the horizontal circle readings to the same accuracy as that to which they were measured, errors can occur at this stage. These should not have a noticeable effect on soft detail. However, if any hard detail has been fixed, some adjustments to the initial plotted positions may be necessary before the plan is finalised.

- Remove the protractor and very faintly join point F to the plotted horizontal circle positions. Extend these lines.

- Using the horizontal distances (D values) from Table 10.5, measure from point F along each direction, allowing for the scale of the plan, to fix the plan positions of the points of detail.

- Write the appropriate reduced level (RL) value taken from Table 10.5 next to each point of detail.

In both methods, once the points have been plotted the detail is filled in between them with reference to the field sketches. Symbols are added as required and contours are drawn by interpolation from the plotted spot heights and levels (see Section 10.6).

Of the two methods, plotting by angle and distance is quicker but less accurate than plotting coordinates, which requires additional computations in the form of polar–rectangular conversions (see Section 6.2) if the instrument cannot provide the E, N values directly.

Real-time detail surveying using total stations

The total station radiation method described above requires the detail surveying data to be brought back into the office for subsequent analysis. This is even the case where a data collector or a memory card has been used with the total station since the data it contains must be downloaded into the appropriate software before it can be processed. However, there are some total station systems that enable the detail to be plotted instantaneously on site as it is observed. Such real-time methods have the significant advantages of not only enabling a plan of the detail survey to be produced on site but also allowing any suspect observations to be rejected and the points re-observed as required.

Real-time detail surveying using total stations has been made possible by manufacturers in one of two ways: either the total station is provided with its own large built-in screen, as in the case of the Leica TPS 1200 and the Topcon GTS-720 series of instruments (see Figure 5.36), or a control unit having a large screen can be interfaced with the total station, as in the case of the Trimble Attachable Control Unit (ACU) shown in Figure 5.29.

In both cases, integrated software enables existing map data to be uploaded and displayed on the screen. The location and orientation of the total station is obtained on site by observing to known points in the area, thereby enabling it to be correctly aligned to the existing map data on the screen. Subsequent radiation observations taken as described earlier can then be immediately processed and located on the screen by the integrated software. At the end of the session the updated map data can be transferred to office software to provide a permanent record.

Further information on total stations and their real-time surveying capabilities is given in Chapter 5.

Detail surveying using GPS equipment

A full description of the theory and equipment involved in GPS is given in Chapter 7. It can be used for several different surveying techniques, one of which is detail surveying and information on its use for such surveys can be found in a guidance note *Guidelines for the Use of GPS in Surveying and Mapping*, published by the Royal Institution of Chartered Surveyors (RICS).

For detail surveys at scales of 1:500 and larger (as specified by the RICS in their publication *Surveys of Land, Buildings and Utility Services at Scales of 1:500 and Larger, Client Specification Guidelines*, 2nd edn), the accuracy required usually means that geodetic equipment must be used, as described in Section 7.7. A kinematic survey can be carried out with one of these and data could be collected on site and then post-processed, but it is normal to use an RTK system where the results are obtained in real time. This has the advantage that a survey can be checked and edited as measurements are taken and there should be no need to revisit a site after data collection. RTK systems for detail survey are pole-mounted, lightweight and portable as the antenna, receiver, radio modem and batteries are all mounted in a single housing at the top of the pole or in a back pack. These are connected to a controller or data collector which is also fixed to the detail pole, but at a convenient height for the operator to use and view, as shown in Figure 10.7. Instructions for carrying out a survey are entered into the system via a keyboard on the controller which is also used to enter feature codes to identify each point surveyed. To assist with this, the controller may have a colour graphic screen which can also be used to edit the results as well as enter feature codes. All of the data captured and edited for the survey is also stored in the controller and this can be transferred to a computer for automatic plotting by detaching the controller from the detail pole and connecting it to a computer, by using data cards

Figure 10.7 Detail surveying using a GPS receiver (courtesy Leica Geo-systems).

or by transmitting the data by mobile phone connection directly from the site. An advantage of a detachable controller is that it can be connected to a total station as well as the GPS equipment. This can be useful in difficult areas where satellite coverage is poor, as the survey could continue with a total station taking measurements instead of GPS. Figure 10.7 shows that detail surveys by GPS are carried out as a one-person operation. Usually, a stop-and-go method of surveying is used in which the operator stops at each point to be surveyed and collects data over a number of epochs (say a few seconds), enters the feature code and then moves on to the next point.

All the receivers being used must maintain a lock on at least four satellites while the survey observations are being taken. Hence detail surveying by GPS should only be used to survey in areas where there is a good view of the sky, for example, in open areas when collecting spot heights for contouring or for calculating earthwork quantities. In urban areas, where the possibility of loss of lock is high due to tall buildings obscuring the signals, GPS may not be appropriate and another detailing method such as radiation using a total station may often be preferable.

As noted above, the GPS observations are recorded in the controller and can be processed immediately to produce a map in real time if required, depending on the type of controller being used. For example, the GPS controllers shown in Figure 10.8

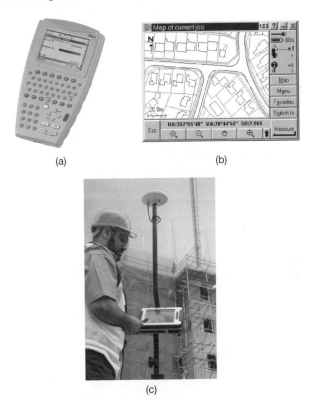

(a)   (b)

(c)

Figure 10.8 ● Graphical data collection and editing for detail surveying by GPS. (a) Leica GPS 1200 controller (courtesy Leica Geosystems); (b) Trimble ACU screen display (courtesy Trimble Navigation Ltd); (c) Sokkia Penmap Radian IS-Pod (courtesy Sokkia Ltd).

are part of an integrated graphical GPS system, which enables the GPS observations to be processed immediately to produce a map on a screen. On these, as more observations are taken they can be added to the map and any discrepancies can be detected immediately with suspect points being re-observed as required.

If real-time processing is not required, the GPS observations can be stored in raw form for subsequent downloading and post-processing.

Whichever approach is adopted, the use of booking sheets is recommended to record any relevant field notes.

As with conventional methods for detail surveys, control points must be established on site when detail surveying by GPS. These are needed as the RTK system will require not only a base station but also points with known coordinates for checking purposes and for determining transformation parameters so that site-based coordinates are obtained from GPS coordinates. The procedures required for setting up control and transforming coordinates for GPS surveys are given in Chapter 8.

The best applications for detail surveys by RTK GPS are where a lot of data is to be collected, especially on open sites where there are no overhead obstructions to limit satellite coverage or interfere with radio communications between the rover and base station. The method becomes very fast and efficient if more than one rover is used.

For detail surveys and mapping at scales of 1:1000 and smaller (as specified by the RICS in their publication *Specifications for Mapping at Scales between 1:1000 and 1:10000*, 2nd edn), data can be collected in real time by GPS using handheld mapping grade receivers in conjunction with an augmentation system such as EGNOS or WAAS. These have the advantage that they cost considerably less than geodetic receivers, they have accuracies at the sub-metre level suitable for small-scale mapping and they are less complicated than geodetic receivers. Because they are handheld or can be back-packed, they are highly mobile and are used extensively for GIS data collection. The procedure involved with these for mapping involves taking the receiver to each point of detail and recording single point positions instead of using RTK methods. A feature code would be added to the point position and all the data can be edited and stored on site as before. Augmentation systems are described in Section 7.5 and mapping grade receivers in Section 7.7. In addition to using a mapping grade receiver at small scales, a single geodetic receiver could also be used if it is EGNOS or WAAS enabled.

Reflective summary

With reference to surveying detail using total stations and GPS equipment remember:

— The total station technique is a radiation method, which involves measuring horizontal angles and horizontal distances.

— Normally, two people are required for total station detail surveying, although the development of robotic total stations has led to the evolution of one-person detail surveying.

- Total station field observations can either be booked on specially prepared field sheets or stored on a data collector or a memory card for subsequent computer-aided analysis and plotting.

- If data collectors or memory cards have been used to record the total station observations, the plotting is carried out using computer-aided methods.

- When plotting total station data by hand, field observations can either be converted into coordinates and plotted using a coordinate grid, or plotted using a protractor and a scale rule.

- Geodetic GPS receivers are required for detail surveys at scales of 1:500 or larger, whereas handheld mapping grade receivers can be used for scales of 1:1000 and smaller.

- Real Time Kinematic (RTK) GPS methods are ideal for detail surveying on open sites where there is good satellite coverage and no radio communication interference between the rover and the base station.

- RTK GPS techniques are capable of accuracies in the region of 10 mm + 1 ppm (horizontal) and 20 mm + 1 ppm (vertical).

- Detail surveying using RTK GPS can be very fast and efficient if more that one rover is used.

- Both total station and GPS detail surveying methods require control points to be established on site.

- Some total station and GPS systems enable detail to be plotted instantaneously on their screens as it is observed. These real-time methods not only enable plans to be produced immediately on site but also allow any suspect observations to be rejected and the points re-observed as required.

- The accuracy of both techniques is high and they can be used to pick up any type of detail, either hard or soft.

10.6 Contours

After studying this section you should know what contours are, why they are shown on maps and plans and what factors are considered when choosing a suitable contour interval. You should know what spot heights are, how they can be located and how contours can be interpolated from them.

This section includes the following topics:

- The role of contours
- Locating spot heights
- Interpolating contours

The role of contours

A *contour* is a line on a map or plan joining points of the same height above or below a specified datum. Contours are shown so that the *relief* (topography) of an area can be interpreted, a factor greatly used in civil engineering and construction. They can also be used to provide approximate longitudinal and cross-sectional information, as described in Section 15.3.

The difference in height between successive contours in known as the *contour interval* or *vertical interval* and this dictates the accuracy to which the ground is represented. The value chosen for any application depends on:

- The intended use of the plan.
- The scale of the plan.
- The costs involved.
- The nature of the terrain to be surveyed, that is, is it hilly or flat?

Generally a small contour interval of up to 1 m is required for construction projects, for large-scale plans and for surveys on fairly even sites. Typical contour intervals for general purpose engineering surveys are given in Table 10.1. In hilly or irregular terrain and at small scales, a larger contour interval is used. Very often, a compromise has to be reached on the value chosen, since a smaller interval requires more fieldwork, which in turn increases the cost of the survey.

Contours are obtained from *spot heights* (also sometimes referred to as *spot levels*) measured at existing detail or at points on open ground such as obvious changes of ground slope. A spot height is simply a point on the ground surface at a known height relative to some datum.

Locating spot heights

Total stations and GPS receivers are generally used to locate spot heights since they not only measure their heights but also enable their horizontal positions to be determined, which is essential if the spot heights are to be plotted on a drawing. The horizontal positions of spot heights are surveyed using total stations and GPS receivers in the same way as for other detail. Traditionally, levelling was used to measure spot heights, but this is little used nowadays since it is restricted to points whose horizontal position is already known and these may not be in the ideal positions from a contouring point of view.

In open areas where there is no detail (for example, fields, large open car parks, sites cleared prior to construction) spot heights are measured in the form of a grid pattern in which the total station reflector or GPS receiver is moved a regular distance by measuring or pacing an appropriate distance between spot height observations. Various suggestions for the spot height grid size are given in Table 10.1. For any contouring method to succeed, a sufficient number of spot heights must be recorded so that the ground surface can be accurately represented on a plan.

Interpolating contours

Spot heights acquired at existing points of detail or in a grid pattern will have height values that do not coincide with exact contour values. As a result, in order that contours can be plotted at specified intervals on a survey plan, their positions on the ground surface have to be obtained from the spot heights by interpolation. When interpolating contours it is assumed that the surface of the ground slopes uniformly between spot heights – this requires careful selection of spot height locations on the site if accurate contours are to be produced.

If the contours are to be produced using one of the many commercially available software packages, they are interpolated and plotted from the total station or GPS spot height data using the software installed on appropriate computer hardware. Such a process, from field to finished plan, is fully automated if electronic data capture and transfer is used (see Section 10.9).

If the contours are to be drawn manually, they are usually obtained using *graphical interpolation* techniques. The procedure for this is as follows:

- The horizontal positions of the spot heights are plotted on the drawing and their heights are written beside them.
- A piece of drawing film or polyester is prepared with a series of equally spaced horizontal lines drawn on it as shown in Figure 10.9(a). Every 10th line is drawn heavier than the others.
- On the plan, the film is laid in turn between pairs of adjacent spot heights. Arbitrary contour values are allocated to the lines on the film to give it a scale. In Figure 10.9, the heavier lines each represent contours at 1 m intervals and the lighter lines are at 0.1 m intervals, but different values can be assigned as required, depending on the contour interval chosen.
- The film is then rotated until the horizontal lines corresponding to the two known spot height values match the scale as shown in Figure 10.9(b).
- The positions of the 1 m contours are indicated where the heavier lines pass over the line joining the two spot heights and these positions are marked on the drawing by pressing down on the film with the point of a pencil.
- The heights of the contours are written lightly next to these positions. When all the exact contour positions have been located, those having the same height are joined by smooth curves to form the contours on the drawing.

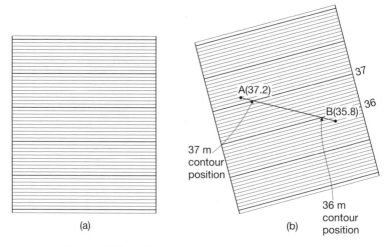

(a)   (b)

Figure 10.9 ● Graphical interpolation of contours.

Although it can be very accurate, graphical interpolation is a laborious and time-consuming method of contouring. For these reasons it is rarely used nowadays and then only for small projects.

Reflective summary

With reference to contours remember:

— A contour is a line joining points of equal height relative to some datum.

— They are shown on maps and plans to indicate the topography of the area.

— The choice of a suitable contour interval depends on the purpose and the scale of the plan, the costs involved and the nature of the terrain.

— Generally, large contour intervals are used on small-scale plans and hilly areas whereas small contour intervals are used on large-scale plans and flat areas.

— The smaller the contour interval, the greater will be the work required and the higher will be the survey cost.

— For any contouring method to succeed, a sufficient number of spot heights must be recorded so that the ground surface can be accurately represented on a plan.

- Contour lines must be interpolated from the spot height positions.

- Total stations and GPS receivers are normally used to collect spot height data for electronic transfer to commercially available mapping software packages for automatic interpolation and plotting of the contours.

- If contours are to be drawn by hand, they can be interpolated from spot heights using a graphical technique.

10.7 The completed survey plan

After studying this section you should understand the need to adopt good drawing practices during the production of the completed survey plan. You should know why it is important for the plan not only to be technically accurate but also to look professional, and you should appreciate how this can be achieved. You should be aware of the information that should be included on the finished plan.

This section includes the following topics:

- Good drawing practice
- Information to be included on the completed plan

Good drawing practice

The end-product of a detail survey is an accurate plan of the area in question at a known scale. Plans are very important documents which may be used in any contracts that are signed in connection with the construction of an engineering project in the area (see Section 11.1). Consequently, they must be accurate and must look professional. A drawing that is technically correct but which has been badly plotted, with poor line work and lettering, will probably be mistrusted and rejected. To ensure that a professional standard is achieved, great care must be taken at all stages of the work. Good quality drawing media must be used and a uniform approach should be adopted. The following are examples of this.

- All *annotation* (lettering and numbering) should be at such an orientation that it can be read without having to turn the plan upside-down. In addition, annotation of equal importance should be of equal size.

- If spot heights are to be shown, one method is to plot each as a small cross with the relevant reduced level written alongside. This will look much neater if the size and orientation of all the crosses are the same.

- Control stations are often shown in case they are needed for future use. However, the lines joining the control points are not usually shown since they do not actually exist. The only *imaginary* lines normally included on survey drawings are contours.

- If contours are included, they should normally be on *natural* surfaces only and they should not run through embankments and cuttings which have their own symbols.

- If plotting by hand, freehand must never be used on survey drawings. Straight lines should be plotted using *straight edges* such as those on steel rulers or high-quality set squares. When joining points to form curved lines, *French curves* or *flexicurves* should be used.

In practice, if a hand drawing is being produced, the original survey plan is usually prepared in pencil on good quality paper by the surveyor who undertook the fieldwork. This is known as the *master survey drawing* and usually shows all the required detail and the control information. When it has been completed, the master drawing is then traced in ink onto plastic film. Since the traced drawing will be used in reprographic processes to obtain copies for use on site, extreme care must be taken with the ink tracing, since any errors will be transferred to all the copies taken. Consequently, the job of producing the tracing is usually given to someone who has been specially trained in draughting techniques. During the tracing stage, other relevant information is added to that shown on the master drawing to produce the *completed survey plan*. The information added may have been included in the specifications as a requirement of the survey (for example, a list of the coordinates and reduced levels of the control points), or it may be necessary to help users to interpret the drawing (for example, a *key*).

If a computer-aided drawing is being produced then any additional relevant information can be programmed into the finished drawing and viewed on the computer screen before the final plan is plotted.

Information to be included on the completed plan

Whether the drawing is done by hand or by computer, the following should normally be included on the completed plan in addition to the actual surveyed area:

- A rectangular or square *border* surrounding the whole of the surveyed area. This provides a neat boundary to the drawing.

- A *title block*, within the border and running along one edge of the drawing. This is subdivided into smaller rectangles into which the additional information can be slotted. Figure 10.10 shows an arrangement of the border and title block which would be suitable for most surveys.

- The *location* of the survey. Sometimes a smaller scale *locating map* is included in the title block to show the relationship of the survey to its surrounding area.

- The *scale* of the drawing.

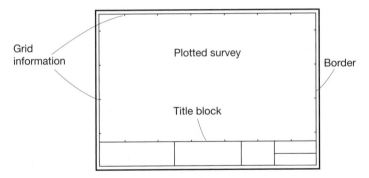

Figure 10.10 ● Border, title block and grid layout.

- The *date* of the survey.
- A *north* direction. This may be arbitrary, magnetic or true north depending on the type of survey (see Section 6.2).
- A *key* (or a *legend*) illustrating any symbols, line-types and abbreviations used.
- Details of the *control grid* used. This can either be shown in full, as a series of crosses indicating the grid intersections, or as a series of short lines along the sides of the border and title block. This last alternative is shown in Figure 10.10 and represents a good compromise in that it does not obscure the drawing but enables the grid to be reconstructed if required.
- A list of the *coordinates* and *reduced levels* of the control stations.
- The *names* of those who undertook the fieldwork and those who produced the drawing. This is useful if problems arise during its use.
- A separate box within the title block should be allocated to recording any *amendments* that have been made to the drawing. The nature of each amendment should be recorded together with its reference number or letter and the date on which it was included, for example, *Amendment A: overhead power lines added – 26 August 2005*. This is very important in engineering projects, where changes often occur as construction proceeds. Engineers and surveyors must be kept informed as drawings are amended in order that they are fully aware of any changes. Care must be taken on site to ensure that the drawings showing the latest amendments are always available.

In addition, other information may be added in the title block depending on the purpose of the survey. For example, a forestry plan could include a numbered schedule of the trees listing their types, girths and spreads, whereas a survey of underground services could include details of pipe diameters, lengths and depths.

Figure 10.11 shows a section of a survey plan and its title block containing some of the information detailed above.

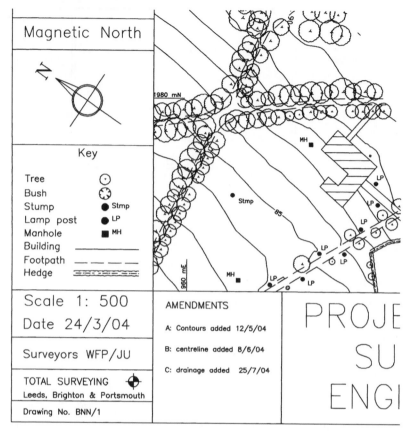

Magnetic North

Key

Tree   ⊙
Bush   ⊛
Stump   ● Stmp
Lamp post   ● LP
Manhole   ■ MH
Building   ─────
Footpath   ── ── ──
Hedge   ≈≈≈≈≈≈≈≈

Scale 1: 500

Date 24/3/04

Surveyors WFP/JU

TOTAL SURVEYING ⊕
Leeds, Brighton & Portsmouth

Drawing No. BNN/1

AMENDMENTS

A: Contours added 12/5/04

B: centreline added 8/6/04

C: drainage added 25/7/04

PROJE

SU

ENGI

Figure 10.11 ● Close-up of part of the title block from a survey drawing including some of the plan detail.

Reflective summary

With reference to the completed survey plan remember:

— The end-product of a detail survey is an accurate plan of the area in question at a known scale.

— It must be accurate and it must look professional. A drawing that is technically correct but which has been badly plotted, with poor line work and lettering, will probably be mistrusted and rejected.

— The only imaginary lines normally included on survey drawings are contours.

— In addition to the detail survey, a lot of information should be included on the plan: for example, a border, a title block, the location, the scale, the date, the north direction, a key, details of the control grid used, a list of the coordinates and RLs of the control stations used, the names of the personnel involved, details of any amendments, and so on.

10.8 Terrestrial laser scanning

After studying this section you should be aware that mapping methods involving total stations and GPS receivers are selective techniques where the person at the detail pole or receiver moves over the site being surveyed, selecting each point in turn. You should understand that new techniques have been developed in which the data required for the detail survey can be collected non-selectively in a very short period of time. You should know about one such non-selective technique, involving laser scanners, and how it can be used to produce survey maps and drawings.

This section includes the following topics:

- Selective and non-selective mapping techniques
- The principles of terrestrial laser scanning
- Data acquisition techniques
- Advantages and disadvantages of terrestrial laser scanning
- Applications of terrestrial laser scanning

Selective and non-selective mapping techniques

The total station and GPS methods of detail surveying described in Section 10.5 can be defined as *selective* techniques since they both require the operator on the detail pole or the GPS receiver to choose and visit the points to be fixed. In practice, decisions must be taken on which points to include and which points to omit and this requires the exercise of judgement by the operator. An experienced operator will be able to map a site by selecting fewer points than an inexperienced one and will choose points at more appropriate positions, particularly where contouring is to be undertaken. Hence detail surveying techniques involving total stations and GPS receivers require skills that are acquired with practice and a skilful operator will save his or her employer both time and money by collecting sufficient data in a short period of time. Nevertheless, even the most skilled operator cannot fix every point on a site, and in any selective detail surveying operation there has to be a compromise between the amount of data that can be collected and the time and money available for its collection.

Although selective methods are widely used for detail surveying, there has been considerable research in recent years into alternative techniques by which detail can be surveyed. This research has concentrated on *non-selective* methods, which locate all the visible points of detail without the need for an operator to choose or visit any of the points. One of these methods is *terrestrial laser scanning*, which is described in this section. Other methods, such as *LiDAR* and *IFSAR*, are described in Section 10.10.

The principles of terrestrial laser scanning

As its name implies, terrestrial laser scanning consists of scanning with a laser beam from the surface of the Earth. Its main areas of application are in the scanning of sites or large objects such as buildings and monuments, or small objects such as historical artefacts. An instrument known as a *laser scanner* is used to carry out the scanning. In the case of a site or a building, the scanner is taken to the feature, whereas small objects can be brought to the scanner to be measured. The latest scanners are about the size of a large total station and can be tripod mounted, as shown in Figure 10.12.

Laser scanners measure angles and distances to points on the surface of the feature and convert these into 3D information. The scanner contains a laser beam, which can be scanned at high speed and measures the distances to the points, an angle measuring system that can record the horizontal and vertical angles to the points scanned by the laser and, usually, a camera to view and record the scene being scanned. The laser beams used in current models fall into safety classes 1, 2 or 3R, depending on the manufacturer (see Section 11.5).

In operation, the scanner is set in a suitable position from which to see the feature being measured which is then scanned by the laser beam in a systematic manner. This involves the scanner taking laser distance observations to the feature at minute incremental changes of horizontal and vertical angles to build up a complete regular scan of its surface. Control is provided by including targets of known position relative

Figure 10.12 ● A laser scanner mounted on a tripod.

to a local coordinate system in the scanned area. To allow accurate identification of their centres, the targets can be scanned with very high-density point spacings.

In practice, the operator uses the camera to capture an image of the feature and then selects the area of the feature to be scanned from the image. The density of the scan to be taken is then specified either by setting the total number of points to be fixed in the horizontal and vertical directions or by allocating the point spacing. An enormous number of points on the surface of the feature can be measured. At the time of writing, data capture rates ranging from 100 points per second to 625,000 points per second are possible, with most current models operating in the range 1000–5000 points per second.

Data are captured on an attached PC and processed by commercially available software. From the resulting angle and distance data, a dense, accurate 3D (x,y,z) *point cloud* of the surface of the feature related to the local coordinate system is obtained (see Figure 10.13). As the feature is scanned, the generated 3D points are displayed instantaneously. This allows data to be easily spot-checked on site to eliminate any errors and to identify areas where more detail is required.

If necessary, the scanning process can be carried out at a number of different positions on site or around an object and the resulting individual scans can be combined together to produce one 3D digital point cloud. The digital data can then be used in a variety of ways, for example to perform field dimension checks, to create models and to generate 2D/3D computer-aided design (CAD) drawings.

Terrestrial laser scanning is a non-selective technique in that the point cloud is formed from points located in a systematic scanning process in the form of an intense angular grid. As a result of the intensive laser point bombardment that occurs during the scanning process, virtually the whole visible surface of the feature can be captured in the point cloud. Hence, when compared to total station and GPS detailing methods, the information provided by laser scanners can potentially provide a much higher level of true geometric completeness and detail for the site or feature being measured.

When planning a laser scanning survey there are a number of important factors to be taken into consideration. Information on these is given in a useful guide entitled *An Introduction to Terrestrial Laser Scanning* produced by the Royal Institution of Chartered Surveyors. Included in these factors are the need for sufficient control points to provide quality assurance checks on

Figure 10.13 ● Laser scan point cloud of the Bridge of Sighs, Cambridge (courtesy APR Services Ltd).

the point cloud, the need to locate the scanning positions such that there are no gaps in the point cloud that would lead to gaps in the survey, the need to consider the stability of the scanning locations, since scanners do not capture all the data instantaneously, and the need to be able to access the subject from all aspects in order to scan it fully.

Data acquisition techniques

In principle, laser scanners measure the distance to a target point and also the respective vertical and horizontal angles. There are three main data capture methods – phase modulation, time of flight and triangulation – all of which are outlined below.

Phase modulation

This method is similar to the phase shift method described in Section 5.2 for total stations and the transmitted laser light is intensity modulated with a sinusoidal signal. This laser light is reflected back from the object and received by a photodiode. The time of flight of the laser light from the sensor to the object and back is directly proportional to the phase difference between the transmitted and received laser light, which is proportional to the range and laser modulation frequency. The amplitude of the received diffusively reflected laser light is proportional to surface reflectivity and distance to the object.

The intensity of the reflected laser pulse is also often recorded. This provides an indication of the reflection characteristics of the surface. In some systems, the true colour of the target point is supplied via an additional passive channel. This colour information allows an instantaneous automatic texturing of the three-dimensional model.

Pulsed time of flight

In this method, the laser is transmitted in short pulses and reflected back from the object to the sensor. The time taken for the laser light to return is directly proportional to the distance travelled. This is the same process as pulsed laser distance measurement (also described in Section 5.2), but in this case the laser pulse is scanned horizontally and vertically by rotating mirrors over the scene. Again the intensity and sometimes true colour information can also be captured.

Laser triangulation

Laser triangulation systems generally consist of a laser source and either a digital camera or video recorder positioned a known distance apart. The laser beam is split into a plane of laser light, projected onto the object and swept across the field of view by an internal mirror. This laser light is reflected from the scanned surface and each scan line is captured by the digital camera. The contour of the surface is derived from the shape of the image of each reflected scan line. The colour information captured by the camera can be used for texture mapping.

Comparison of the three data acquisition methods

Each method of data acquisition has its own advantages and disadvantages. Phase modulation tends to be faster and can operate over longer ranges. For example, Riegl's LPM-2k laser scanner has a range of 10–2500 m and a quoted accuracy of 50 mm. In contrast, time of flight systems tend to operate over shorter ranges but have a higher quoted accuracy. For example, Leica's HDS2500 scanner has a range of 1.5–100 m and a quoted accuracy of ±6 mm. Triangulation systems tend to be the most accurate but can only be used at close range. For example, the Minolta VI910 has a range of 0.6–2.5 m and sub-millimetre accuracy.

Advantages and disadvantages of terrestrial laser scanning

Terrestrial laser scanning has numerous advantages over total station and GPS detail surveying techniques. It is a non-contact measurement method and therefore can be used to measure sites such as roads or railways with minimal disruption. In addition, the operation of laser scanners is independent of daylight, so they can measure in complete darkness. A vast number of points can be rapidly captured reducing on-site time and virtually eliminating costly site re-visits to gather more detail. However, the volume of data can also be a problem as high-specification hardware has to be used to process and model the data. This can be addressed by applying automated point thinning algorithms to reduce the size of the data.

There are a number of problems that occur with all laser scanner systems. They are line of sight dependent so there may be areas of an object that are hidden and unable to be scanned readily. Furthermore, divergence of the laser beam increases with range; that is, the laser beam spreads out over a distance. This makes the accurate capture of fine detail such as pipes and stairs impractical at longer distances. Another problem occurs with dark and transparent objects (e.g. marble and shiny dark objects) because laser light is absorbed more than reflected by such objects. For example, marble will cause subsurface scattering resulting in a degradation of the quality of the range data. However, this can be improved by altering the surface, for example by dusting or painting. A small amount of noise is also inherent internally in all laser scanners.

Laser scanners need a lot of power and, at the time of writing, lead acid batteries are required. This is a problem for all laser scanner manufacturers and is likely to remain so until new battery technology arrives.

For further information on the capabilities and costs of terrestrial laser scanners, the reader is recommended to study surveying journals such as *Civil Engineering Surveyor*, *Engineering Surveying Showcase* and *Geomatics World*, which produce annual reviews of this type of surveying equipment.

Applications of terrestrial laser scanning

Terrestrial laser scanners can be used to record virtually any site, building or object and are ideal for the creation of Digital Terrain Models. In civil engineering they can be used to provide as-built records of structures such as bridges and tunnels, they can be used to provide accurate dimensions of inaccessible features such as building façades and quarry faces and they can be used to record hazardous areas such as polluted sites or waste tips. In the wider context, they can be used in a variety of professions, for example to record architectural and historical buildings and artefacts for heritage preservation, to provide data for crime scene analysis and to model complex power and process plants (see Figure 10.14).

Figure 10.14 ● Laser scan point cloud of Plant pipework (courtesy APR Services Ltd).

Reflective summary

With reference to terrestrial laser scanning remember:

— Since it is a non-selective technique there is no need for an operator to choose or visit any of the points that are being surveyed.

— It involves taking laser distance observations to a feature at minute incremental changes of horizontal and vertical angles to produce a dense 3D point cloud of its surface.

— The data generated from a terrestrial laser scan can be used to perform field dimension checks, create models and generate 2D/3D computer-aided design (CAD) drawings.

— When compared to total station and GPS detailing methods, terrestrial laser scanning can potentially provide a much higher level of geometric completeness and detail.

— It is a non-contact measurement method that is independent of daylight but dependent on having a line of sight.

— A vast number of points can be captured very rapidly by terrestrial laser scanners but the sheer volume of data can cause processing problems.

- Virtually any site, building or object can be recorded using terrestrial laser scanning techniques, but accurate capture of fine detail at longer distances may not be possible due to the divergence of the laser beam.

- Information on the factors to be considered when planning a terrestrial laser scanning survey is given in the Royal Institution of Chartered Surveyors guide *An Introduction to Terrestrial Laser Scanning*.

10.9 Computer-aided survey mapping and its applications

After studying this section you should understand that survey drawings are almost invariably produced nowadays using commercially available software packages in conjunction with automatic multi-pen plotters. You should appreciate that computer-aided survey mapping follows a three-stage process, namely acquiring the data, processing the data to give the required information and then outputting this information in the required format. You should understand the importance of assigning codes to detail points. You should appreciate how the processed survey data is held in a database and how specific types of detail points can be stored in layers, which can be turned on and off as required. You should know that the output can be in either graphical or textual form and you should be aware of the different types of automatic plotters that are available. You should understand what a digital terrain model (DTM) is and how it can be formed from the processed field observations using either a grid-based or triangulation-based technique. You should be aware of some of the applications of DTMs in engineering surveying. You should understand what a geographic information system (GIS) is and how it is normally based on a computer-generated map.

This section includes the following topics:

- Introduction to computer-aided survey mapping methods
- Survey mapping software
- Digital Terrain Models
- Geographic Information Systems

Introduction to computer-aided survey mapping methods

There has been a fundamental change in the past twenty years or so in the methods by which contoured plans and other survey drawings are produced. The skilful but time-consuming process of hand drawing has been almost totally superseded by computer-aided plotting techniques, with drawings nowadays being produced very quickly and to a high degree of accuracy using *automatic multi-pen plotters* driven by one of the many commercially available *surveying and mapping software packages*.

Using these packages, the data collected on site is stored in the computer in the form of *databases* from which three-dimensional representations of the ground surface, known as *digital terrain models* (*DTMs*), can be generated. Together these devices can be used to generate a wide range of drawings and views for use in the design and setting-out procedures. By adding additional layers of information, such as the location of underground services, the boundaries of properties, the purposes for which land is being used, and so on, the database can be extended to form a *Geographic Information System* (*GIS*), which can be used by a wide range of disciplines and organisations.

A seamless approach is adopted in which the data collection capabilities of total stations and GPS receivers are linked directly to the software packages. The data is collected on site using these instruments, stored in their memory systems and then downloaded into desktop PCs or notebook computers loaded with the appropriate surveying software. Using the software, the data is analysed to enable the required drawings to be produced on the computer screen with hard copies being obtained by linking the computer to one of a wide range of automatic multi-colour plotters. These are capable of plotting on sheets up to A0 size in a fraction of the time required for hand methods and to a much higher degree of accuracy.

As a result of such developments, engineers and surveyors are now presented with a vast array of computer hardware and survey mapping and design software from which to choose. These have revolutionised surveying activities with computers and software being increasingly involved at all stages of survey work from the initial data collection on site, through all the analysis and design procedures to the final output of drawings, setting-out data and other numerical information. The following sections discuss the role of software in *survey mapping* and in the production of DTMs and GIS.

Survey mapping software

There are now many commercially produced computer software packages available for the production of engineering surveying drawings. Given the wide choice, it is not possible to give a detailed review of individual packages in a general textbook such as this. For further information on their capabilities and costs, however, the reader is recommended to study surveying journals such as *Civil Engineering Surveyor*, *Engineering Surveying Showcase* and *Geomatics World*, which produce annual reviews of the latest software packages (for further details see Chapter 1).

Nevertheless, given the widespread use of such packages, it is essential that the concepts on which they are based are understood together with the field methods and computing techniques they involve. Fortunately, many of the packages are based on similar principles and utilise similar techniques. Normally, a three-stage

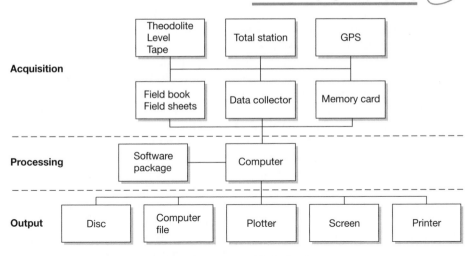

Figure 10.15 ● Computer-aided plotting process.

process is involved as illustrated in Figure 10.15. First, the raw data is *acquired* by taking field measurements. Second, the raw data is input to the computer where it is *processed* to give the required information. Third, the required information is *output* in a suitable format. These three stages are outlined below.

Acquisition of the raw data

The raw data is normally acquired using a range of surveying equipment as indicated in Figure 10.15. Traditionally, all observations were recorded by hand on field sheets or in field books. Nowadays, observations are recorded electronically either on data collectors or on memory cards plugged into total stations or using GPS receivers. The computer can accept data either from field sheets or from electronic sources although it is really designed for use with the latter, the data from which can be very quickly transferred into the computer through a standard RS232 or USB interface. If field sheets have been used, the observations must be carefully keyed in by hand and this can be very time-consuming as well as a potential source of error.

When observations are being recorded electronically, the most important aspect of this occurs during the acquisition stage. Each point surveyed in the field is given a unique number, the angle and distance observations taken to it are recorded and a *feature code* is assigned to it which indicates the type of detail being observed (tree, fence, building and so on). The feature code is vital to the success of the subsequent processing and output stages. Each feature code is recognised by the software which acts accordingly and uses it to build up the plan of the area. Feature codes can take several forms. Some simply assign a particular symbol and/or annotation to a point while others cause interactions between two or more points. An example of the former would be the code *tree* which would cause a tree symbol to be plotted at the point in question while an example of the latter would be the code *fence*, which would cause a predetermined line type to be joined between the point in question and the previous point having the code *fence*. Although the observations are recorded electronically, the use of field sketches to illustrate any unusual areas or features is strongly recommended. In practice, it is advisable to observe all the points on one particular feature before starting to pick up another feature. For example, all the points on a kerb line should be picked up using the code *kerb* before moving on to record all the spot heights in a field using the code *sh*. This avoids having to change the coding on the total station too often. Once all the observations and feature codes have been logged, they are downloaded into the computer ready for processing.

When traditional field sheets are being used, the feature codes must be recorded by hand together with the observations. Again, field sketches are strongly recommended. These can then be used to help with the interpretation of the field observations and codes when they are input by hand to the computer via its keyboard.

The acquisition stage is the most important part of the whole process. The quality of the field observations and feature coding will greatly influence the outcome of the survey. If the acquisition stage is carried out properly with all the correct codes being assigned to the points, the subsequent processing and output stages should proceed quickly without any problems. Consequently, the onus is on the surveyor and engineer to *get it right on site*, and this requires a thorough knowledge of the coding system being used. With practice, the feature codes can soon be mastered and considerable satisfaction can be gained from completing the fieldwork and watching the computer produce the finished plan with a minimum of editing and correction being required.

Processing the raw data

When they are downloaded or keyed into the computer, the software package stores the raw data (point numbers, field observations and feature codes) in a computer file. This can be viewed, printed and edited by the user as required before any processing of the data is initiated. At this stage, any changes can be made, although a file containing the original data is normally always preserved in the computer for legal purposes.

Once the data has been edited it is then automatically processed by the software. Usually this involves a number of steps.

● The three-dimensional coordinates (E, N and RL) for each point of detail surveyed in the field are computed.

● The coordinates, point number and feature codes are then stored in a *database* which can be accessed as required.

● In order to enable a plot to be created, the feature codes are checked against those contained in the software library. The software comes ready supplied with a pre-programmed library containing a range of feature codes. This is normally sufficient for the needs of most users, although users can create their own codes, if required, and store them in the library. During the checking process, if the codes are present in the library the software activates them, but if they are not present, an error message is presented on the screen.

Having processed the data in this way, it is now possible to view the detail points on the computer screen either in *textual form*, for example as a list of point numbers, *E*, *N*, RL and feature codes or in *graphical form* as a plot. Both can be displayed on the screen and edited as required with any selected section of the database being viewed. Either the keyboard or a mouse can be used to change, add or erase information. Some systems utilise *touch screens* activated by special pens. As new points are added, coordinates are generated and appropriate codes are assigned. All changes are recorded in the database. Annotation can be added at any size, at any rotation and in any position on the drawing. Plots of any part of the database can be generated at any scale and in a wide range of colours. Title blocks, borders, keys, grids, north signs and so on can all be created within the software as a series of different *planforms* and added to the plots, as required.

The software packages have a wide range of facilities. They are supplied with libraries of symbols, line types, control commands and so on. These can themselves be amended and extended to allow users to create their own libraries and feature codes. The technique of *layering* is one very important facility. In this, either at the coding stage or during subsequent editing, particular features can be placed in their own unique layer; for example all spot heights could be placed in a *spot height layer*, all trees in a *tree layer* and so on. The software allows layers to be turned on or off as required. This enables users to set up layers of detail and allows them to select and plot only those of particular interest. For example, a plot showing only the detail contained in the road network layer and the underground services layer may be required. This is easily achieved by turning off all the unwanted layers and turning on all the required layers. Figure 10.16 shows a plot in which the spot heights layer has been turned on together with several of the detail layers.

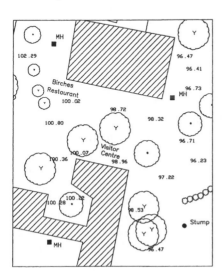

Figure 10.16 ● Section of an uncontoured plan showing spot heights and several detail layers.

Contours can also be added and placed in a *contour layer*. These are generated using the RL information stored in the database and involve the creation of a *surface* and a DTM which is discussed later in this section. Before the contours are formed, the user chooses those points from the database that are to be included in the surface and stipulates *boundaries* on the surface to define areas over which contours should not be drawn. Normally, contours are only shown on natural surfaces and they should not cross embankments or cuttings (which have their own distinct symbol). Features which may influence the shape of the contours can also be specified using *breaklines*; for example details of any ridges in the area or other lines defining changes of slope. Once the points to be included in the surface have been defined, the computer creates a DTM which approximates the shape of the actual ground surface. How this is done is described later in this section. The accuracy of the DTM depends greatly on the number, accuracy and location of the points used in its formation. When the DTM has been created, the required contour interval is defined by the user and the computer interpolates the positions of the contours and displays them on the screen. Further editing is possible to allow different colours and different contour intervals to be used. Once it is acceptable, it can be incorporated into a plot to provide a contour overlay. Figure 10.17 shows the contoured version of Figure 10.16 in which the spot height layer has been turned off.

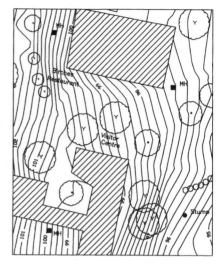

Figure 10.17 ● Section of a contoured plan with the spot height layer turned off.

Output of the required information

After all the editing has been carried out and all the contours have been generated, the information required from the survey can be obtained. Usually, this is either in graphical form as drawings produced on a plotter or in textual form as a listing of data on a printer or on a screen. However, output can also be in the form of data files stored on disc. Most of the software systems can generate files in a number of different formats. This enables data to be transferred to another computer which is either loaded with similar software or software requiring a different data format, for example, in the form of a drawing exchange file (DXF) as used by a number of CAD software packages.

The usual output, however, is a plot and, as with the software packages, there is now a wide range of plotters from which to choose. All enable multi-coloured drawings to be produced on either paper or film. The actual mode of plotting usually involves pens, although some use inkjets while others use thermal electrostatic techniques. The most accurate and the most expensive are the *flatbed* plotters, on which, as the name implies, the drawing medium is laid flat while the plot is produced. In these, the plotting mechanism moves over the paper which remains stationary at all times. The most popular for the majority of surveying applications is the *rolling drum* type, in which both the plotting mechanism and the paper move. Figure 10.18 shows such a plotter. They are manufactured in a range of sizes up to A0 (see Table 10.4) and have the advantages of being cheaper and usually not occupying as much space as an equivalent sized flatbed. Smaller *table-top* plotters are also available, usually at A2 or A3 size.

Figure 10.18 ● Rolling drum plotter.

Digital terrain models

As discussed earlier in this section, computers have been readily adopted in many surveying activities. However, as well as improving techniques which existed before, they have also caused techniques to evolve which would not have been possible but for the computers themselves. One example of this is in the subject of *terrain modelling*, which is the technique of trying to represent the natural surface of the Earth as a mathematical expression. In engineering surveying it has a large number of applications including contouring and highway design.

The name often given to a mathematical representation of part of the Earth's natural surface is a *digital terrain model* or *DTM*, since the data is stored in digital form, that is, E, N and RL. However, there are a number of other names which have been applied to such a representation, for example digital elevation model (DEM) and digital ground model (DGM). In fact, each applies to a slightly different type of surface representation. If the exact definitions are studied, the one most applicable to engineering surveying is *digital terrain model* because this is considered by most people to include both planimetric data (E, N) and relief data (RL, geographical elements and natural features such as rivers, ridge lines and so on). Since this is the exact type of information collected during a detail survey, the term DTM is used throughout this book.

Earth surface data for the formation of DTMs can be acquired in a number of ways, usually involving either ground-based methods or airborne techniques. Photogrammetry using aerial photography, LiDAR and IFSAR is particularly well suited to obtaining three-dimensional and geographical/natural information from the air over large areas where ground techniques would become laborious. However, ground methods are ideal for creating DTMs of smaller areas and the computer-aided techniques discussed earlier in this section involving total stations and GPS receivers are widely employed for this purpose and the relatively new technique of laser scanning is increasingly being used for the generation of DTMs. In addition, many of the commercially available surveying software packages have the capability of producing DTMs.

Once the field observations have been processed and stored in a database by the software, as described earlier, a DTM can normally be formed from them using one of the following two techniques.

- *Grid-based terrain modelling*

  In this technique, a regular square grid is established over the site and the RL values at each of the grid nodes are interpolated from the field data points.

  A grid size is chosen such that it is small enough to give an accurate representation of the irregular surface on which it is based. One disadvantage of this method is that the grid nodes do not coincide with the actual field data points. This tends to smooth out any surface irregularities and causes any contours generated from the grid to be less representative of the true surface. A further drawback is that any ridges and other changes of slope which have been carefully surveyed in the form of a string feature on site cannot be accurately reproduced on the grid. This will also affect the shape of any contours generated.

- *Triangulation-based terrain modelling*

  This method uses the actual field survey points as node points in the DTM. The software joins together all the data points as a series of non-overlapping contiguous triangles with a data point at each node to produce a *Triangulation Irregular Network* (TIN).

  Such a technique has none of the disadvantages of the grid-based method outlined above. A much truer representation of the surface is obtained and features which have been carefully surveyed as strings, such as the tops and bottoms of embankments, are faithfully reproduced and can be taken into consideration if contours are generated. It is also possible to set up areas of the surface from which contours can be excluded; this is not possible with the grid-based system.

In addition to the generation of contours, DTMs have numerous applications in engineering surveying. They can be viewed from different angles and presented as *wireframe views* which can highlight areas of specific interest. Examples of grid-based and TIN-based wireframe views are shown for the same area in Figures 10.19(a) and (b). Textures and shading can be applied to give them a more realistic appearance and features such as trees, hedges, walls, fences and buildings can be added. Alternatively, aerial photographic images can be draped over them and contours can be added if required, as shown in Figures 10.19(c) and (d). Once a DTM has been created, volumes of features such as lakes, spoil-heaps, stockpiles and quarries can easily be obtained and longitudinal and cross-sections quickly produced. This is discussed further in Section 15.7. DTMs are also widely used in highway design, as discussed in Section 13.5.

(a)
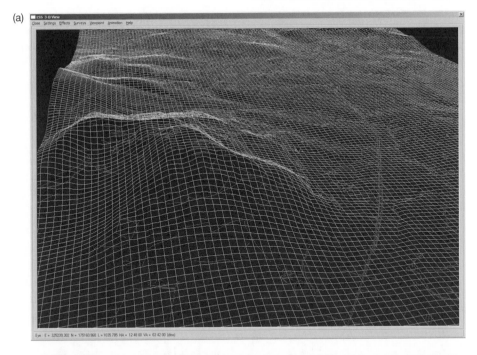

(b)

Figure 10.19a,b ● (a) Wireframe of the DTM showing a 50 m grid. Also shown are the DTM breaklines (roads, banks, walls, hedges and fences); (b) wireframe of the DTM showing the TIN.

(c)

(d)

Figure 10.19c,d ⬤ (c) Orthorectified aerial imagery draped over the DTM; (d) orthorectifed aerial imagery draped over the DTM and showing contours (all images from LSS copyright McCarthy Taylor Systems Ltd).

Geographic Information Systems

A standard definition of a *Geographic Information System* (*GIS*) is that it is a system for the collection, storage, retrieval, manipulation, analysis and presentation of geographic information. The term *geographic information* includes any data that has some form of spatial or geographical reference that enables it to be located in two- or three-dimensional space. While these definitions are accurate they can appear obscure, and a simpler but less precise definition of a GIS is that it is a computer-based system that contains a wide range of measurement-based interrelated information that can be input, linked, selected, displayed and output as required. In effect, it is a computer database that contains a wide range of measurement-related data.

Originally, before the advent of computers, geographical data was invariably presented in the form of a map. Sometimes, however, it was very difficult to show all the information on a single map, so a system of *overlays* was developed. In this, a base map was prepared showing the main features on the ground surface in an area and then a series of transparent overlays was prepared such that each overlay showed a particular item or subject of interest. For example, if the base map showed the road network in an area, an overlay could be prepared to show the drainage system that lay beneath the road network. As many overlays as required could be prepared to show different features in the area. Of course, some information could not be shown in a map-like way. For example, the condition of the road surface in a particular area is best explained in a written form rather than in a drawn form. Hence, traditionally, written reports (*descriptive* information) would accompany the overlayed maps (*spatial* information) to provide a complete picture of the problem being considered.

This principle of combining descriptive and spatial information forms the basis of geographic information systems. Using very sophisticated computer software packages, geographic information systems can be set up to contain a vast range of inter-linked spatial and descriptive information in themed layers that can be retrieved, manipulated, analysed and presented as required.

Components of a GIS

In simple terms, a GIS can be considered to contain the following components.

- A reference framework such as a map or a network of coordinated points, lines and areas stored in digital form on a computer.
- A means of inputting additional data to the computer, which is relevant to the reference framework.
- A means of linking this additional data to the reference framework.
- A means of analysing and manipulating both the additional data and the reference framework.
- A means of outputting required information in suitable formats.

A GIS can be an extremely powerful tool and it differs from other computer programs such as spreadsheets, statistical packages or drawing packages because it enables spatial operations to be performed on the data. It integrates common data-base operations such as query and statistical analysis with the unique visualisation and geographical analysis benefits offered by maps. The power of a GIS is particularly noticeable when the quantity of data is too large to be handled manually or in any other way. For example, when trying to make a realistic assessment of the effects of global warming, only a GIS has the power to store and analyse all the spatial and descriptive information necessary to achieve such an aim. It offers the power to create maps, integrate information, visualise scenarios, solve complicated problems, present powerful ideas and develop effective solutions.

In order to be able to function, a GIS requires hardware, software and data.

The *hardware* is the computer on which the GIS operates together with a digitiser/scanner and a plotter.

The *software* provides the functions needed for converting, manipulating, analysing and displaying the data. The data itself is stored in a database and the key components of the software are tools to convert the data into a form that is usable by the GIS, a system to enable the data to be stored in the database and retrieved from it as required, and tools to enable the data to be queried, analysed and output in a variety of formats, for example, as maps, graphs and numerical or textual tables.

The *data* forms the most important aspect of a GIS. Data represents information, of which there are two basic types: spatial and descriptive. Spatial information is inherently geographical in nature and can be represented in the GIS as either point, line or area features. Descriptive information can be represented in the GIS as text, colours, numbers and so on.

Data modelling

The computer software in GIS packages models the spatial data in one of two forms, either as a *vector* model or a *raster* model as shown in Figure 10.20. In addition, the data can be organised in the model either as a series of *layers* or as a series of *objects* in a single layer.

In the *vector* model, data is stored in coordinate form and each linear map feature is represented as a list of x,y coordinates. Objects or conditions in the real world are represented by the points and lines that define their boundaries, much as if they were being drawn on a map. The position of each object is defined by its placement in a map space that is organised by a coordinate reference system. In vector representation, every position in the map space has a unique coordinate value.

In the *raster* model the data is pixel based, that is, grid cells are used to define features or attributes. In this model, space is subdivided into uniformly sized cells (usually square) setting up a series of rows and columns. The row and column positions of the cells occupied by geographical objects or conditions define their location. The area that each cell represents defines the spatial resolution available. The position of each geographical feature is recorded to the nearest cell because positions are defined by row and column numbers. The cell value reports a condition and that condition pertains to the whole cell, and the units of the raster model, unlike the vector model, do not correspond directly to the spatial entities that they represent in the real world.

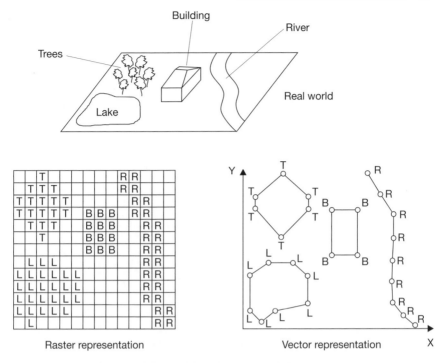

Figure 10.20 ● Vector and raster GIS modelling.

Of the two types of model, the vector model is better at handling discrete data, (for example, clearly defined features such as buildings, roads and enclosed control networks), whereas the raster model is better at handling continuous data (less clearly defined features such as forests, soil surfaces and polluted coastal regions). The vector model shows features in a map-like way, but has a more complex data structure. The raster model is better where different sets of data are being overlaid, but it can produce graphical output which is less aesthetically pleasing because boundaries tend to have a blocky appearance due to the cell structure of the model.

Data are stored by the computer in a digital database as a series of files that can contain either spatial or descriptive information about the features in the reference framework as shown in Figure 10.21. The ability of a GIS to link these types of data and maintain the spatial relationship between the features in the reference framework makes it unique from other types of database.

GIS output and applications

In engineering surveying the types of output that can be produced by a GIS include *graphical data*, such as drawings showing underground services, longitudinal sections and cross-sections, *numerical data*, such as the number of drainage gullies in a street, and *descriptive data*, such as the conditions of the road surfaces in a particular area. In addition, *three-dimensional computer modelling* can be performed, for example to give a

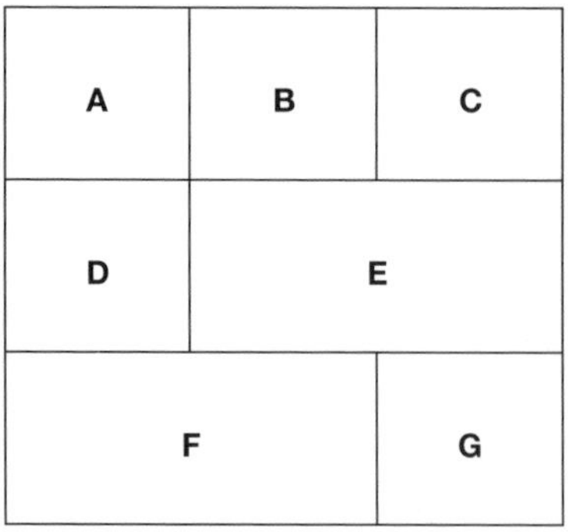

Area	Size (km²)	Trees
A	25	346
B	25	481
C	25	162
D	25	534
E	50	662
F	50	181
G	25	269

Descriptive information

Spatial information

Figure 10.21 ● Spatial and descriptive GIS data files.

drive-through visualisation of the effect a new road or a bridge will have on the environment.

In civil engineering, geographic information systems have been used to design, construct and manage complex civil engineering projects such as roads and tunnels. They help to manage assets, such as land, property and transportation infrastructure, and can be used to ensure the safety of built structures such as bridges, dams and buildings by monitoring their deformation. In a wider context, a GIS can monitor the changing environment, such as land use and climate change, as well as manage natural resources, both on land and offshore. They have also been used to plan transport and water networks and to explore for oil and gas.

GIS software packages

Every GIS is different in that each will contain its own range of data that has been input for specific purposes. However, all geographic information systems have one thing in common: they are based on some type of reference framework, usually a map, which provides a control network against which other data stored in the GIS can be referenced. Hence the starting point in the creation of any GIS is normally the input of a map, and this is usually provided using the computer-aided survey mapping methods discussed earlier in this section. As a consequence of this, several of the manufacturers of some of the currently available surveying and mapping software packages also produce GIS software packages which can be interfaced with their mapping packages to produce a seamless integration of the observed survey data with the GIS software. Information on these and other GIS packages can be found in surveying journals such as *Civil Engineering Surveyor, Engineering Surveying Showcase* and *Geomatics World*, which publish annual reviews of the latest systems.

Reflective summary

With reference to computer-aided survey mapping and its applications remember:

— Survey drawings are almost invariably produced nowadays using commercially available software packages in conjunction with automatic multi-pen plotters.

— Three stages are involved: data acquisition, data processing and data output.

— Data is normally acquired electronically either on data collectors or memory cards plugged into total stations or GPS receivers.

— Feature codes are used during the acquisition stage to define the type of detail being fixed and each detail point is given a unique number. Data is then processed using commercially available software packages that compute the three-dimensional coordinates of each point and interpret the codes. Once processed, the data is stored in a database.

— Similar types of detail, for example, trees, buildings and contours, can be stored within the database in their own specific layers that can be turned on and off as required.

— Data can be output either in *graphical form* as drawings produced on a plotter or in *textual form* as a listing of data on a printer or on a screen.

— Contours can be processed with the aid of a *digital terrain model* (DTM) formed from selected points in the database.

— A DTM includes both planimetric and relief data and can be formed from database points either using a grid-based approach or a triangulation-based approach.

— DTMs can be used to generate wireframe views, to determine volumes and to draw longitudinal and cross-sections.

— A GIS is a computer-based system that can be set up to contain a vast range of interlinked spatial and descriptive information in themed layers that can be retrieved, manipulated, analysed and presented as required.

— A GIS requires hardware, software and data. The software models the spatial data in one of two forms, either as a *vector* model or as a *raster* model. The vector model is better at handling discrete data whereas the raster model is better at handling continuous data.

— Although every GIS is different they all have one thing in common –
they are based on some type of computerised map which provides a
control framework against which all other data stored in the GIS can
be referenced.

10.10 Additional mapping systems and products

After studying this section you should be able to describe all of the Ordnance Survey
mapping products available for use in engineering surveying and construction. You
should be aware of other mapping products of use in civil engineering such as photogrammetry and orthophotography. Furthermore, you should be fully aware of
LiDAR and *IFSAR* and how these are used to produce terrain data.

This section contains the following topics:

- Alternative mapping sources
- Ordnance Survey mapping products
- Photogrammetric methods
- Orthophotography
- LiDAR
- IFSAR and NEXTMap

Alternative mapping sources

The detail surveying and plotting techniques described earlier in this chapter are
used when producing maps, plans and terrain data for most small to medium-sized
engineering projects. As described, this type of survey is carried out using total
stations and GPS, but the use of laser scanners is increasing, especially for complicated and hazardous sites. All of these methods are capable of recording three-
dimensional data at the centimetre level or even better and are usually applied to
work at engineering scales greater than 1:500. As a project gets larger or when the
accuracy requirement for mapping and terrain modelling is not so stringent,
different methods of obtaining spatial data for an engineering project or a GIS
become viable and can be more cost effective.

Ordnance Survey mapping products

One of the most popular sources of mapping data at medium to small scales in the UK
is the Ordnance Survey. To the general public, their most well-known products are

the paper maps they publish at various scales covering the whole of Great Britain. For leisure purposes, the most widely used maps are the *Explorer* and *Landranger* series at scales of 1:25,000 and 1:50,000 respectively. These are published with boundaries set by the Ordnance Survey on the National Grid; however, by specifying the coordinates or a postcode at the centre of an area it is possible to purchase a site-centred version of these using *OS Select*.

As far as engineering surveying is concerned, the Ordnance Survey maps of particular interest are those at larger scales. These include the following.

● *Superplan* plots are hard copy, graphic maps that are large-scale and either based on National Grid sheet lines or site-centred. The Ordnance Survey can deliver Superplan as customised graphic plots for anywhere in Great Britain. They are available in a wide range of paper or film sizes and can be produced at any scale from 1:100 to 1:10,000. The plot size is limited to the paper or film width (820 mm) and to a maximum length of 3 metres. As it can be site-centred, a Superplan plot can fully cover the area or site of interest, thereby avoiding the situation where several different National Grid sheets have to be purchased to obtain the coverage required. The accuracy of Superplan plots is guaranteed to that of mapping at the basic scales of 1:1250 (urban mapping: major towns and cities), 1:2500 (rural mapping: smaller towns, villages and developed rural areas) and 1:10,000 (moorland mapping: mountains, moorland and estuarine areas) and *not to the scale of the output*. The positional accuracy of all Ordnance Survey mapping, relative to the National Grid, is ±0.4 m in urban areas, ±1.1 m in rural areas and ±4.1 m in moorland areas. Part of a Superplan plot is shown in Figure 10.22.

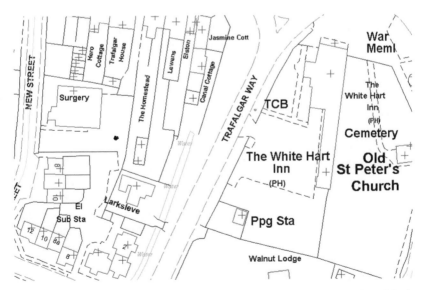

Figure 10.22 ● Ordnance Survey Superplan plot (reproduced by permission of Ordnance Survey on behalf of the Controller of Her Majesty's Stationery Office © Crown Copyright 100024463).

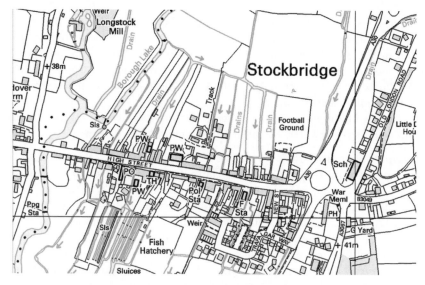

Figure 10.23 ● Ordnance Survey Landplan plot (reproduced by permission of Ordnance Survey on behalf of the Controller of Her Majesty's Stationery Office © Crown Copyright 100024463)

● *Siteplan* is a similar product to Superplan but is only available at fixed scales of 1:500 (showing 80 × 80 m of ground detail), 1:1250 (200 × 200m of ground detail) and 1:2500 (400 × 400 m of ground detail). Siteplan plots are intended for use by anyone requiring maps for planning applications.

● *Landplan* is 1:10,000 scale digitally derived colour mapping that combines both map and database information and is the largest scale mapping available from the Ordnance Survey that shows contours. It is available in National Grid format or can be site-centred by place name, postcode or National Grid coordinates and can be printed on paper or film at scales of 1:10,000 (sheet size 5 × 5 km) or 1:5000 (sheet size 3 × 3 km). An example of Landplan is shown in Figure 10.23.

All of the products described above are also available in digital format. *Superplan Data* is the equivalent of a Superplan plot but is supplied by email as a zip file which can be extracted in a DXF format that is compatible with most CAD systems. The advantages of data supplied in a digital format are that, using suitable software, the map can be combined with other topographical data, it can be edited and annotated to suit a user's requirements and can be merged into reports and printed as many times as required within the copyright license agreement purchased from the Ordnance Survey. *Siteplan Data* is the digital version of a Siteplan plot and is supplied as a raster TIFF file which can be downloaded directly to most desktop graphic software applications. Similarly, Landplan is also available in a digital format.

OS MasterMap is a digital map and database covering the whole of Great Britain. Data sets can be delivered online over the Internet or can be supplied on CD and DVD. It is based on National Grid coordinates and includes detailed topographical and geographical information in four themed layers. The *Topography Layer* is a digital

Figure 10.24 ● Topography layer in Ordnance Survey MasterMap (reproduced by permission of Ordnance Survey on behalf of the Controller of Her Majesty's Stationery Office © Crown Copyright 100024463)

large-scale database showing all features on the topographical landscape, as shown in Figure 10.24. It is based on the Ordnance Survey's 1:1250 urban mapping, the 1:2500 rural mapping and 1:10,000 moorland mapping. The themes it is broken down into include such items as buildings, roads, tracks and paths, railways, water features, structures and height. All of these can be looked at separately or in any combination and with the other MasterMap layers. The *Imagery Layer* is an aerial map of Great Britain and consists of a set of orthophotos covering the whole of the country. Orthophotography and its applications in civil engineering and construction are described in a following section. The *Integrated Transport Network (ITN) Layer* provides an overview of the transport infrastructure in the UK. At present, this consists of a Roads Network theme showing all roads in the country, but others for railways and navigable waterways are planned. The *Address Layer* provides precise coordinates of postal addresses for most residential and commercial properties in Great Britain. Clearly, this is a very comprehensive database which is capable of providing detailed information for GIS and which has applications in engineering surveying.

Other Ordnance Survey products that are useful in project planning and evaluation are the *1:10,000* and *1:25,000 Scale Raster* digital data. These are scanned images of the Landplan and Explorer mapping which provide a background map that can be overlaid with other information. They are available in black and white or colour formats.

For further details of all Ordnance Survey mapping products, refer to their web site at http://www.ordnancesurvey.co.uk/.

Photogrammetric methods

Most Ordnance Survey products are based on photogrammetric mapping using aerial photographs. Although these products are extensive and cover the whole of Great Britain at various scales, they are not always suitable for some construction projects. The main reasons for this are their positional accuracy and lack of sufficient height information at large scales. Because of this, it may be necessary to obtain topographical and terrain data for large construction projects through a company specialising in aerial photogrammetry. A survey can then be carried out to meet the exact requirements and specifications for a project. This is one of the preferred methods for obtaining height information and for terrain modelling on large sites, but photogrammetry is best suited for mapping at scales that typically vary from 1:1000 to 1:50,000. At scales larger than this, ground surveys by total station and especially GPS become more economic. Information on companies that provide photogrammetric services can be found in the various survey journals published in the UK or by using the Internet.

Orthophotography

When aerial photographs are taken, although they appear to show features on the ground in their true positions, they contain a number of distortions caused by the inevitable tilt of the aerial camera and also from relief displacement. To overcome this problem, the photographs are viewed as overlapping pairs in a photogrammetric workstation to produce a DTM or scaled map with contours. However, by using special digital processing techniques, it is also possible to remove the errors on single aerial photographs to produce what are known as *orthophotos*. These have the appearance of an aerial photograph, but any distances, areas and angles measured on the photograph will be correct. Compared to a map, an orthophoto shows every feature on the ground instead of an interpretation of this but does not provide any height information. If required, this can be determined separately and shown as contours on the photograph. Advantages of orthophotos are that they are a relatively inexpensive way of obtaining up-to-date information of large areas, they are available in digital and hard copy format, they can be combined with other data and are easy to maintain and update. In engineering surveying, their main applications are for presentational purposes and for providing base maps for GIS.

Although orthophotos can be produced to order at any location and scale by many of the companies specialising in photogrammetry, there are some data sets that are available off the shelf in the UK. *OS MasterMap* described earlier contains a national imagery layer of orthophotography covering the whole of Great Britain at 0.25 m ground resolution in full colour. This is updated every three years in urban areas and every five years in the more remote rural areas depending on flying conditions. This imagery layer can be used in conjunction with other layers in MasterMap to assess, for example, the impact of a proposed development on its surrounding environment. *MAPS (Map Accurate Photographic Survey)* is another source of orthophotography produced by an organisation known as UK Perspectives. This is a national digital aerial photographic data set that is compatible with Ordnance Survey 1:1250 scale mapping and consists of orthophotography referenced to the National Grid covering the whole of the UK. It is available as hard copy photographic prints ranging from the photographic scale of 1:10,000 to a maximum scale of 1:800. Alternatively, it can be supplied in various digital formats in tile sizes of 5 km^2 or 1 km^2 at 0.25 m ground resolution.

When producing orthophotos, it is necessary to create a DTM which is then combined with a digital image of each photograph to remove the distortions it has due to changes in ground height. When ortho-imagery is combined with a DTM, full topographical data becomes available. The DTMs created for OS MasterMap and UK Perspectives are available as separate products as follows.

- The Ordnance Survey *Land-Form PROFILE* is a 1:10,000 scale digital height dataset that is available as a DTM covering the whole of Great Britain. Heights have been derived for this on a 10 m grid with an accuracy of 2.5 m in urban and rural areas and with an accuracy of 5 m in remote moorland and mountainous areas. The DTM produced by UK Perspectives for producing their MAPS ortho-imagery is supplied as heights with an accuracy at the metre level on a 10 m grid across the whole of the UK.

- NEXTMap Britain is another DTM providing heights nationwide at the metre level in Great Britain. This uses radar instead of aerial photographs to obtain terrain data and is described in a later section.

All of these DTMs are referenced to the National Grid and, with heights at the metre level, they are particularly useful for preliminary surveys, for feasibility studies and for assessing the environmental impact of civil engineering projects. They have also found widespread use in GIS.

LiDAR

As an alternative to aerial photography, other airborne technologies are available for producing digital terrain data in civil engineering. One of these is known as *LiDAR* (**Light Detection And Ranging**).

Components of LiDAR

A LiDAR system consists of three main hardware components: a laser rangefinder, GPS and an inertial navigation system, all of which are built into a fixed wing aircraft or helicopter (see Figure 10.25). The laser is mounted on the aircraft such that it points vertically downwards during normal level flight. In a similar manner to pulsed laser distance measurement described in Section 5.2, the time it takes a laser pulse to travel from the aircraft to the ground surface and back is measured very precisely and converted into height knowing the velocity of the pulses. The data collection rate can be very fast, as the laser can be fired at rates of over 10,000 pulses per second. Because the laser is highly collimated into a small diameter beam and is

Helicopter with LiDAR head mounted underneath and video cameras on side booms

Helicopter interior showing LiDAR console

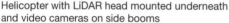

Figure 10.25 ● LiDAR system (courtesy Fugro Inpark and QinetiQ).

sufficiently powerful to penetrate vegetation, a single pulse can also be split into multiple returns, each corresponding to a different surface. In this way, LiDAR is capable of surveying a canopy and ground surface simultaneously. Rather than measure a single ground profile along the flight path of the aircraft, the laser beam is deflected through an angle of about 20° by a rotating mirror and a band or swathe is scanned along the flight path instead. As might be expected, the attitude of the aircraft is continually changing during a survey and the laser beam will not always be pointing vertically downwards. To overcome this problem, an inertial navigation system is used to measure the roll, pitch and yaw of the aircraft at all times so that the elevation data from the laser can be corrected for these. The three-dimensional position of the aircraft is determined using differential GPS where equipment installed in the aircraft works in conjunction with base stations established at known points on the ground. For the best possible accuracies, dual frequency geodetic receivers are required and the aircraft must keep within about 20 km of a base station.

Data acquisition and processing

With a fixed wing aircraft, LiDAR data is usually collected at altitudes of 300 m and upwards for wide area coverage, but with moderate sub-metre accuracy. However, for engineering surveys the best results are obtained by mounting the LiDAR system in a helicopter and flying at lower altitudes. This application of LiDAR has been developed for corridor mapping of services and utilities such as roads, railways and electricity transmission lines and for coastal surveys. It is also used by organisations such as the Environment Agency in the UK in their National Flood Mapping Program. In this configuration, and at a typical operating height of 150 m and speed of 15 m s^{-1}, LiDAR is capable of surveying in a swathe of about 50 m at a density of about 20–25 points per square metre and an accuracy of 0.05–0.15 m for plan and height.

A LiDAR system produces an enormous amount of terrain data and is the equivalent of an airborne laser scanner. Not surprisingly then, raw LiDAR data is a very dense non-selective laser point cloud where each point surveyed has three-dimensional coordinates. To obtain these coordinates, the GPS and inertial navigation system data is post-processed separately and then merged with the data from the laser. This produces LiDAR data with coordinates on the same system as the GPS base station, which will usually be provided by the Ordnance Survey National GPS Network. Terrain data that is produced in this way is known as *geo-referenced* data because it has coordinates that are tied to some geographical reference (the National Grid and Ordnance Datum Newlyn in most cases).

The post-processing of LiDAR data requires very careful analysis because the raw data will include every feature the laser encounters, whether this is the true ground surface or not. When carrying out a detail survey using a total station or GPS, a judgement is made as to what points are included in the survey and a detail pole is placed at discrete points to locate the required detail and contours. In a similar manner, a LiDAR processor must also be able to make judgements as to where the true surface in a point cloud lies and must have a good understanding of the LiDAR method to appreciate its limitations. To help with the interpretation of the point cloud, the LiDAR aircraft or helicopter is fitted with digital still or video cameras (see Figure 10.25) that record high-resolution images of the ground that are synchronised with the laser data. Because the images recorded by these cameras are geo-referenced to the GPS and inertial navigation data, they can also be used to produce orthophotos as separate products. Taking into account all of the operating and processing features of LiDAR, the accuracy of the data obtained is a function of the flying height, the diameter of the laser beam, the quality of the GPS and inertial navigation system data and the capabilities of the post-processing procedures used.

Data quality control and presentation

As with all survey methods, quality control of LiDAR data is essential and is carried out by a number of different methods. The absolute accuracy of the data produced is verified using *GPS ground truth data*, in which a series of ground-based targets are over-flown within the LiDAR scan. The positions of these are then carefully coordinated using the LiDAR processing software and the coordinates obtained are compared with those obtained by static GPS. Systematic errors can be caused by instabilities in the GPS and inertial navigation systems and temperature fluctuations in the laser rangefinder. Errors from these sources can be assessed by analysing the results obtained from the overlap between adjacent flight lines as any step in the two data sets would indicate the presence of a systematic error. Another way of checking LiDAR data is to visit the site in areas where data interpretation has been difficult, particularly below vegetation.

The way in which LiDAR data is presented depends almost entirely on the end user. In some cases, all that is required is a DTM which is exported directly to a CAD system for further processing. However, the data can be presented in many different formats to suit any application, some of which will be for engineering purposes, some for asset management and others for the development of a GIS.

Advantages and disadvantages of LiDAR

The clear advantages of LiDAR are that it is capable of carrying out remote airborne surveys covering large areas very quickly and that it is also capable of surveying complicated features such rail or power lines much more quickly than by ground-based methods. The data output is high density, is geo-referenced and has an accuracy at about the 0.1 m level for three-dimensional coordinates. One of the disadvantages at present is that raw LiDAR data is in a format that is unfamiliar to most users and needs to be presented differently to be of use elsewhere. In addition, the task of post-processing and interpreting the raw data to produce a true ground model is a compli-cated and time-consuming process. Another factor to be borne in mind is the size of the files produced by LiDAR surveys, which are typically 40 MB per km of flight for processed data – for a long linear asset survey this could exceed 10 GB, plus the storage space needed for imagery and video files. However, computers are available that can handle these storage requirements, although users may need to upgrade their system before they are able to process LiDAR files. A further drawback of LiDAR is that the laser cannot penetrate cloud and rain, which makes the method weather-dependent.

As LiDAR is still developing, it is expected that the problems of data interpretation and presentation will diminish, leading to a greater acceptance of this as a data collection technique. Since it is a remote method capable of accuracies at the centi-metre level, its use in engineering surveying is expected to increase.

IFSAR and NEXTMap

Another airborne technology that is gaining in popularity for terrain modelling is known as *IFSAR*, which is an acronym for *InterFerometric Synthetic Aperture Radar.*

In this system, two radar antennae are mounted in an aircraft a distance apart known as the interferometric baseline. During a survey, the positions of these two antennae and the baseline relative to the ground must be precisely determined and this is achieved, as with LiDAR, by post-processing GPS and inertial navigation data from equipment mounted in the aircraft. In flight, one of the antennae transmits X-band radar pulses towards the ground. These are reflected back to the aircraft and both antennae act as receivers of the return pulses. Because the radar looks to the side of the aircraft and the two antennae (receivers) are a small distance apart, there is a phase difference between the reception of the radar pulses at each antenna. This difference, together with the post-processed GPS and inertial data, can be analysed to give the required terrain data.

Compared to LiDAR, the aircraft used for IFSAR fly at high altitudes of over 5000 m at speeds in excess of 500 km per hour. This enables them to acquire data at rates of up to 100 km² per minute for wide area coverage. Unlike conventional photogram-metry and LiDAR, which use reflected light or passive sensing to provide informa-tion, IFSAR uses microwave or active sensing. The advantages of this are that data can be captured by day or night and in cloudy and rainy conditions – these are signifi-cant for the UK where daytime air traffic is heavy and cloud cover can be a problem. A disadvantage of radar is that it cannot penetrate vegetation, so any areas under trees and canopies are hidden from view.

When processing raw IFSAR data, a digital surface model (DSM) is produced first. This is the surface the radar comes into contact with and includes the tops of large trees, buildings, towers and any other features it encounters. Although DSMs are useful for some applications, most users require a DTM or 'bald earth' ground model. These are obtained from a DSM by using a computer program to digitally remove all of the surface features. When the DTM has been produced, it is also possible to produce a set of *orthorectified radar images* from the IFSAR data that are fully geo-referenced (these are the radar equivalent of an orthophoto). Taken together, all of these are capable of providing medium-scale topographical information covering wide areas.

One of the commercial companies currently engaged with IFSAR is Intermap Technologies, whose *NEXTMap Britain* project provides a national DTM covering the whole of the UK. This IFSAR product provides heights on a 5 m grid with an accuracy of 0.5 m for the southeast of England and 1 m in all other areas. The horizontal accuracy is 2.5 m. This, as well as orthorectified radar images of the whole of the country, can be bought in digital or hard copy format. The Environment Agency has also purchased the NEXTMap DTM, which will be merged with their LiDAR data in the National Flood Mapping Program.

Reflective summary

With reference to additional mapping systems and products, remember:

— Alongside detail surveying and mapping by total station, GPS and laser scanning, there are several other ways in which terrain data can be acquired for a construction project.

— The Ordnance Survey produces a range of paper and digital maps at varying scales covering the whole of Great Britain, but check the accuracy before using these to see if they meet specifications.

— Photogrammetry and orthophotography can be useful in engineering surveys. Photogrammetry is the standard method for mapping large projects, whereas orthophotographs are useful for planning and presentations.

— LiDAR is an airborne technology that is capable of mapping large areas at high speed and density with an accuracy of about 0.1 m. It is particularly useful for remote surveys of roads, railways and transmission lines and for flood plain mapping.

— IFSAR is another airborne mapping system. Although it also has the advantage of being able to survey large areas rapidly and IFSAR products are available off the shelf, the method is not as accurate as LiDAR.

Exercises

10.1 With the aid of illustrations, explain how an accurate coordinate grid should be drawn on a rectangular piece of cartridge paper.

10.2 Describe the different types of drawing paper and film that are currently available giving examples of their use in the production of surveying drawings.

10.3 Explain the principle on which the ISO *A-series* of paper is based.

10.4 Discuss the role of an *arbitrary north direction* when deciding on the best orientation for a survey drawing.

10.5 Explain what is meant by the term *detail* when carrying out the survey of an area.

10.6 With the aid of an illustration, explain what is meant by the term *radiation* when applied to detail surveying.

10.7 List the types of information that should normally be included on a completed survey drawing in addition to the actual surveyed area.

10.8 Explain the role of *codes* when picking up detail using total stations fitted with memory cards.

10.9 Explain how *layers* can be used in computer-aided survey mapping.

10.10 Discuss in detail the three-stages involved in computer-aided survey mapping; that is, data acquisition, data processing and data output.

10.11 With reference to the *Guidelines for the Use of GPS in Surveying and Mapping*, published by the Royal Institution of Chartered Surveyors (RICS), discuss the role of GPS in the acquisition of data for detail surveying purposes.

10.12 Briefly describe the principles by which data can be collected using *laser scanning* systems and *LiDAR*.

10.13 Describe the basic components and the mode of operation of a terrestrial laser scanner.

10.14 Discuss the role of surveying in the production of *Digital Terrain Models* and *Geographic Information Systems*.

10.15 Describe the difference between *grid-based terrain modelling* and *triangulation-based terrain modelling* and discuss the advantages and disadvantages of each.

10.16 Define the basic components of a *Geographic Information System*.

10.17 Describe how spatial data can be modelled in GIS packages by using either a vector model or a raster model and discuss the advantages and disadvantages of each, giving examples of the types of data for which each is best suited.

10.18 Describe the following mapping products available from the Ordnance Survey:

Superplan
Siteplan

Landplan
OS MasterMap

10.19 Explain how an orthophoto differs from a conventional aerial photograph.

10.20 Describe the way in which a LiDAR system produces terrain data giving an indication of the accuracy which can be achieved.

10.21 List the advantages and disadvantages of LiDAR surveys.

10.22 Describe IFSAR and explain how it differs from LiDAR.

Further reading and sources of information

For the latest information on preparing survey specifications, consult:

The Royal Institution of Chartered Surveyors (1988) *Specification for Mapping at Scales Between 1:1000 and 1: 10 000*, 2nd edn. RICS Books, London.
The Royal Institution of Chartered Surveyors (1996) *Surveys of Land, Buildings and Utility Services at Scales of 1:500 and Larger, Client Specification Guidelines*, 2nd edn. RICS Books, London.

For the latest information on surveying detail using GPS, consult:

ICE Design and Practice Guide (1997)*The management of setting out in construction*. Thomas Telford, London.
The Royal Institution of Chartered Surveyors (2003) *Guidelines for the Use of GPS in Surveying and Mapping*. RICS Books, London.
Rixon, S. and Sharp, M. (2004) *RICS Guidelines for the Use of GPS in Surveying and Mapping*: a working review. *Geomatics World*, **12**(3), 34–5.

For the latest information on computer-aided surveying software, consult:

Fort, M. J. (2003) Surveying for geomatics software. *Engineering Surveying Showcase 2003*, No. 2, pp. 33–49. http://www.pvpubs.com/.
Software Systems, Electronic Surveying Supplement, Civil Engineering Surveyor (2004) Institution of Civil Engineering Surveyors, pp. 58–63; http://www.ices.org.uk/.
Thinking about software? (2004) *Engineering Surveying Showcase 2004*, No. 2, pp. 38–51. http://www.pvpubs.com/.

For the latest information on terrestrial laser scanning, consult:

The Royal Institution of Chartered Surveyors (2004) *An introduction to terrestrial laser scanning*. Downloadable from the RICS website at http://www.rics.org/.
Terrestrial laser scanners for survey (2004) *Engineering Surveying Showcase 2004*, No. 1, pp. 34–5. http://www.pvpubs.com/.
Uff, J., Barber, D. and Mills, J. (2004) On the workbench: software for laser scanning. *Engineering Surveying Showcase 2004*, No. 2, pp. 25–9. http://www.pvpubs.com/.

For the latest information on LiDAR, consult:

Schnurr, D. and Partridge, D. (2004) Making sense of LiDAR. *Engineering Surveying Showcase 2004*, No. 1, 42–5. http://www.pvpubs.com/.

chapter 11

Setting out

 Aims

After studying this chapter you should be able to:

- Understand the roles of the various different types of personnel who are involved in the setting out process
- Understand the aims of setting out
- Refer to the different types of plans that may be used during the setting out process
- Appreciate the good working practices that should be undertaken in order that the aims of setting out can be achieved
- Understand the procedures required to ensure that the horizontal and vertical control requirements of the setting out operations can be met
- Set out design points on site by a number of methods including angle (bearing) and distance, intersection, measurements from a baseline and using GPS receivers
- Undertake first-stage setting out operations such as setting out a pipeline and setting out a building to ground floor level
- Apply horizontal and vertical control techniques to second-stage setting out operations such as locating formwork, establishing column positions, controlling verticality, transferring height from floor to floor, establishing pile positions and setting out bridges
- Appreciate the applications of laser instruments in surveying and setting out operations
- Understand how earthmoving machinery being used in setting out operations can be controlled using total stations, GPS and lasers
- Appreciate the need for quality assurance and accuracy in surveying and setting out and be able to refer to some of the current British standards and other publications covering these topics

This chapter contains the following sections:

11.1 An introduction to setting out

After studying this section you should understand that setting out is simply one application of surveying and involves many of the instruments and techniques used in surveying. You should be aware of the different types of personnel who are involved in the setting out process and how they interact with each other. You should understand that the responsibility for setting out on site lies with the Contractor.

This section includes the following topics:

● Definition of setting out
● The personnel involved in setting out and construction

Definition of setting out

A definition often used for setting out is that it is the reverse of surveying. What is meant by this is that whereas surveying is the process of producing a plan or map of a particular area, setting out begins with the plan and ends with the various elements of a particular engineering project correctly positioned in the area. This definition can be misleading, since it implies that setting out and surveying are opposites. This is not true. Most of the techniques and equipment used in surveying are also used in setting out and it is important to realise that setting out is simply one application of surveying.

A better definition of setting out is provided by the International Organization for Standardization (ISO) in its publication ISO 7078: 1985 *Building Construction – Procedures for Setting Out, Measurement and Surveying – Vocabulary and Guidance Notes*, which states that

> Setting Out is the establishment of the marks and lines to define the position and level of the elements for the construction work so that works may proceed with reference to them. This process may be contrasted with the purpose of Surveying which is to determine by measurement the positions of existing features.

Attitudes to setting out vary enormously from site to site and insufficient importance is attached to the process, as it tends to be rushed, often in an effort to keep ahead of the contractor's workforce. This can lead to errors, which in turn require costly corrections.

Fortunately, in recent years, greater emphasis has been placed on the need for good working practices in setting out and some of these are discussed in Section 11.2. In addition, there are now a number of national and international standards specifically dealing with the accuracy requirements of setting out and the techniques that should be employed in order to minimise errors and ensure that the construction process proceeds smoothly. Further information on these is given in the Section 11.7.

However, although progress is being made in the production of standards, the main problems of the lack of education in and the poor knowledge of suitable setting out procedures still remain. Good knowledge is vital, since despite the lack of importance often placed upon it, setting out is one of the most important stages in any civil engineering construction. Mistakes in setting out can cause abortive work and delays which leave personnel, machinery and plant idle, resulting in additional costs.

In order to fully appreciate the processes in setting out a useful starting point is to consider the personnel who are involved and these are discussed below.

Personnel involved in setting out and construction

The *Client, Employer* or *Promoter* is the person, company or government department that requires the particular scheme (the *Works*) to be undertaken and finances the project. Often, the Employer has no engineering knowledge and therefore commissions an *Engineer* (possibly a firm of Consulting Engineers or the City Engineer of a Local Authority) to provide the professional expertise. A formal contract is normally established between these two parties.

It is the responsibility of the Engineer to investigate the feasibility of the proposed project, to undertake site investigation and prepare various solutions for the Employer's consideration. Ultimately, the Engineer undertakes the necessary calculations and prepares the drawings, specifications and quantities for the chosen scheme. The Engineer also investigates the likely costs and programme for the project.

The *calculations* and *drawings* give the form and nature of construction of the Works. The *quantities* are used as a means of estimating the value of the project, for inviting competitive tenders for the project and, ultimately, as a basis for payment as the job is executed. The *specifications* describe the minimum acceptable standards of materials and workmanship included in the project. The *programme* identifies the overall time for completion of the project.

When these documents are complete, the project is put out to tender and contractors are invited to submit a price for which they will carry out the Works described. A *Contractor* is chosen from the tenders submitted and a contract is formed between the Employer and the Contractor.

Hence three parties are now involved: the Employer, the Engineer and the Contractor.

Although the Engineer is not legally a party to the contract between the Employer and the Contractor, the duties of the Engineer are described in the contract. The job of the Engineer is to act as an independent arbiter and ensure that the Works are carried out in accordance with the drawings, specifications and other conditions as laid out in the contract. The Employer is rarely, if ever, seen on site. The Engineer is represented on site by the *Resident Engineer* (RE). The Contractor is represented on site by the *Agent*.

The responsibilities of the Resident Engineer, Engineer, Agent, Contractor and Employer are described in the document known as the *Conditions of Contract*. Every scheme has such a contract and a number of different ones are available. Two of the most commonly used are the *ICE Conditions of Contract*, which is sponsored by the Institution of Civil Engineers, the Association of Consulting Engineers and The Civil Engineering Contractors Association, and the *JCT* (*Joint Contractors Tribunal*) *Standard Form of Building Contract*, which is sponsored by the Association of Consulting Engineers, the British Property Federation, the Construction Confederation, the Local Government Association, the National Specialist Contractors Council, the Royal Institute of British Architects, the Royal Institution of Chartered Surveyors and the Scottish Building Contract Committee Limited.

The RE is in the employ of the Engineer, who has overall responsibility for the contract. Many responsibilities are vested in the RE by the Engineer. The RE is helped on site by a staff which can include assistant resident engineers and clerks of works.

The Agent, being in the employ of the Contractor, is responsible for the actual construction of the Works. The Agent is a combination of engineer, manager and administrator who supervises assistant agents and site foremen who are involved in the day-to-day construction of the Works.

Many large organisations employ a Contract Manager who mainly supervises financial dealings on several contracts and is a link between head office and site.

As regards setting out, the Resident Engineer and the Agent usually work in close cooperation and they have to meet frequently to discuss the work. The Agent undertakes the setting out and it is checked by the Resident Engineer. Good communication is essential, since although the Resident Engineer checks the work, the setting out is the responsibility of the Contractor and the cost of correcting any errors in the setting out has to be paid for by the Contractor, provided the Resident Engineer has supplied reliable information in writing. If unreliable written information is given, the responsibility for correcting any errors in setting out reverts to the Employer. The whole question of responsibility for setting out will be covered by the formal contract used in the scheme. This will contain a definitive section on setting out, for example, Clause 17 of the *ICE Conditions of Contract Measurement Version*, 7th edn, takes the form of three statements. These are reproduced below by kind permission of the Institution of Civil Engineers.

(1) The Contractor shall be responsible for the true and proper setting out of the Works and for the correctness of the position levels dimensions and alignment of all parts of the Works and for the provision of all necessary instruments, appliances and labour in connection therewith.

(2) If at any time during the progress of the Works any error shall appear or arise in the position levels dimensions or alignment of any part of the Works the Contractor on being required so to do by the Engineer shall at his own cost rectify such error to the satisfaction of the Engineer unless such error is based on incorrect data supplied in writing by the Engineer or the Engineer's Representative in which case the cost of rectifying the same shall be borne by the Employer.

(3) The checking of any setting-out of any line or level by the Engineer or the Engineer's Representative shall not in any way relieve the Contractor of his responsibility for the correctness thereof and the Contractor shall carefully protect and preserve all bench-marks sight rails pegs and other things used in setting out the Works.

It is essential, therefore, that setting out records, to monitor the progress, accuracy and any changes from the original design, are kept by both the Engineer and the Contractor as the scheme proceeds. These can be used to settle claims, to provide the basis for amending the working drawings and to help in costing the various stages of the project.

Further detailed information on the topics discussed in this section can be found in some of the publications listed in the Further reading and sources of information section at the end of this chapter.

Reflective summary

With reference to an introduction to setting out remember:

— Setting out is simply one application of surveying.

— Most of the techniques and equipment used in surveying are also used in setting out.

— Mistakes in setting out can be costly.

- For setting out to be undertaken successfully good working practices should be employed.

- There are three parties involved in the construction procedures: the Employer, the Engineer and the Contractor.

- Although the Engineer checks the work, the setting out is the responsibility of the Contractor.

- The cost of correcting any errors in the setting out has to be paid for by the Contractor, provided the Engineer has supplied reliable information in writing.

11.2 The aims of setting out

After studying this section you should know that the aims of setting out are to position the Works in their correct relative and absolute positions, and to ensure that they proceed smoothly and that their costs are minimised. You should appreciate that the chances of the aims being achieved will be greatly enhanced if suitable control methods are employed, if the correct plans are available and if good working practices are adopted. You should be familiar with the different types of plans that are associated with the setting out process and you should understand some of the good working practices that should be employed.

This section includes the following topics:

- Achieving the aims
- Plans and drawings associated with setting out
- Good working practices when setting out

Achieving the aims

There are two main aims when undertaking setting out operations:

- The various elements of the scheme must be correct in all three dimensions both relatively and absolutely; that is, each must be its correct size, in its correct plan position and at its correct reduced level.

- Once setting out begins it must proceed quickly and with little or no delay in order that the Works can proceed smoothly and the costs can be minimised. It must always be remembered that the Contractor's main commercial purpose is to make a profit. Efficient setting out procedures will help this to be realised.

In practice, there are many techniques which can be used to achieve these aims. However, they are all based on the following general principles, which are discussed further in Section 11.3.

● Points of known plan position must be established within or near the site from which the design points can be set out in their correct plan positions. This involves *horizontal control* techniques.

● Points of known elevation relative to an agreed datum are required within or near the site from which the design points can be set out at their correct reduced levels. This involves *vertical control* techniques.

● Accurate methods must be adopted to establish design points from this horizontal and vertical control. This involves *positioning techniques*.

In addition, the chances of achieving the aims and minimising errors will be greatly increased if the setting out operations are planned well in advance. This requires a careful study of the drawings for the project and the formulation of a set of good working practices. These are discussed below.

Plans and drawings associated with setting out

Before any form of construction can begin, a preliminary survey is required. This may be undertaken by the Engineer or a specialist team of land surveyors and the result will be a contoured plan of the area at a suitable scale (usually 1:500 or larger) showing all the existing detail. As discussed in Chapter 10, it is usually prepared from a network of control stations established around the site. These stations are often left in position to provide a series of horizontal and vertical control points which may be used to help with any subsequent setting out. This first plan is known as the *site* or *survey* plan.

The Engineer takes this site plan and uses it for the design of the project. The proposed scheme is drawn on the site plan and this becomes the *layout* or *working* drawings. All relevant dimensions are shown on these and a set of documents giving technical details about the project is included. These form part of the scheme when it is put out to tender. The Contractor who is awarded the job will be given these drawings.

The Contractor uses these layout drawings to decide on the location of the horizontal and vertical control points in the area from which the project is to be set out and on the positions of site offices, stores, access points, spoil heaps and so on. All this information, together with the angles and lengths necessary to relocate the control points should they become disturbed is recorded on a copy of the original plan and forms what is known as the *setting out plan*.

As work proceeds, it may be necessary to make amendments to the original design to overcome unforeseen problems. These will be agreed between the Resident Engineer and the Agent. Any such alterations are recorded on a copy of the working drawings. This copy becomes the *latest amended drawing* and should be carefully filed for easy access. It is essential that the latest version of any drawing is always used, particularly if setting out operations are to be undertaken. It is also important to keep the drawings which show the earlier amendments; they may be needed to resolve a dispute or for costing purposes. When the scheme is finally completed, the drawing which shows all the alterations that have taken place during the course of the Works becomes the *as-built drawing* or *record drawing.*

Good working practices when setting out

The basic procedures involved in setting out utilise conventional surveying instruments and techniques and, given sufficient practice, an engineer can become highly proficient at undertaking setting out activities. Unfortunately, this is not sufficient if the aims stated earlier are to be achieved. There is more to setting out than simply using equipment, and many of the problems that occur are often due to lack of thought rather than lack of technical competence.

During the setting out and construction of a scheme a number of difficulties will inevitably arise. These can be concerned with such diverse matters as site personnel, equipment, ground and weather conditions, changes in materials, design amendments and financial constraints. Most will be unforeseen, but an experienced engineer will always expect some unplanned events to occur and will take steps to minimise their effects as and when they happen. Setting out is no exception to the vagaries of construction work and anyone given the task of undertaking such operations should be equally prepared for the unexpected. Of course, it is not possible to be ready for every eventuality, but by adopting a professional approach and a series of good working practices most problems can be overcome.

Emphasis has, so far, been placed on having to deal with difficulties. Although these do arise, the majority of the setting out activities on site would normally be expected to proceed without any problems. This, of course, is one of the aims stated earlier. However, such trouble-free progress does not happen by accident and the chances of mistakes and errors occurring will be greatly diminished if, as before, a series of good working practices is carefully followed.

While it is impossible to discuss all the procedures that should be adopted when setting out, the working practices given below cover most of the important considerations and it is strongly recommended that they be followed.

Keep careful records

Always record any activities in writing and date the entries made. Get into the habit of carrying a notebook and/or diary and record work in it at the end of each day (not a few days later) when it is still fresh in the mind. Anything which has an influence on the Works should be noted including the names of all the personnel involved. If requested to carry out a particular piece of work, note it in the diary and ask the person who gave the instruction to sign to confirm what has been agreed. Once it has been completed to the agreed specification, a second signature should be requested.

Try to be neat when keeping records and using field books and sheets. A dispute may not arise until months after the work was done and recorded. It is essential that the records can be fully understood in such a case, not only by the person who did the recording (who may have been transferred to another project), but also by someone who has no previous knowledge of the work. A good test is to allow a colleague to review your notebook/diary to see if it is clear.

Adopt sensible filing procedures

As work proceeds, the quantity of field books, booking forms and other setting out documents will grow quickly. These are often the only record of a particular activity, and as such could be called upon to provide evidence in the case of a dispute. They are also used to monitor the progress of the Works and to help the Quantity Surveyors with their costings. Consequently, they are extremely important and should be carefully stored in such a way that they are not only kept safe but also can be easily retrieved when requested. If the records of a number of different jobs are being stored in the same site office, great care must be taken to ensure that they are not mixed up. Different filing cabinets and plan chests clearly marked with the relevant job name should be used for each. Once a file or drawing has been consulted, it should be returned *immediately* to its correct place in order that it can be easily located the next time it is required.

Look after instruments and use them safely

Surveying instruments are the tools of the trade when setting out. Modern equipment is very well manufactured and can achieve very accurate results if used properly and regularly checked. No instrument will perform well, however, if it is neglected or treated badly. If an automatic level is allowed to roll about in the back of a Land Rover, for example, even if it is in its box, it should not come as too much of a surprise if the compensator is found to be out of adjustment. Similarly, if a theodolite is carried from one station to another while still on its tripod any jarring will be transferred to the instrument, which could affect its performance. Should the person carrying it trip and fall over the instrument could be ruined altogether.

All instruments must be treated with respect and should be inspected and checked both before work commences and at regular intervals during the work; ideally once per week when used daily and at least once every month if used occasionally. In the case of total stations, theodolites and levels, permanent tests and calibrations should be carried out and the instruments checked to ensure that all the screws, clamps and so on are functioning correctly. The purpose of testing equipment is to find out if it is in correct adjustment. Instruments which are found to be out of adjustment or beyond calibration should normally be returned to the company surveying store or to the hirer for repair. There is not usually the time or the appropriate facilities on site actually to carry out any adjustments. Other equipment, such as tapes, ranging rods and tripods, should be kept clean and oiled where necessary. If the end comes off a tape or it is badly kinked, the whole tape should be thrown away – the small cost of a new one is nothing to the costs that may arise from errors caused by incorrectly allowing for its shortened length.

If equipment gets wet it should be dried as soon as possible. Optical instruments should be left out of their boxes in a warm room to prevent condensation forming on their internal lenses. Electronic equipment, although water-resistant, is not water-proof. Should it become wet, it can start to behave unpredictably and give false readings. In such a case it is advisable to dry it off and also leave it overnight in a warm dry place out of its box. It should then be checked to ensure that it is working properly before it is returned to site.

Common sense should be shown when using equipment on site. A tripod can be left set up above a station, but its instrument should be detached and put back into its box if it is not to be used for a while. This will prevent accidental damage should one of the tripod legs be knocked by a passing vehicle. Ranging rods are not javelins and should not be thrown. Levelling staves should only be used fully extended if absolutely necessary, as their centre of gravity is much higher in such circumstances, making them difficult to handle, particularly in windy conditions. Great care should also be exercised when using levelling staves near overhead power lines.

The consequences of loss of time due to badly adjusted or damaged equipment can be extremely serious. Expensive plant and personnel will be kept idle, the programme will be delayed and material such as ready mixed concrete may be wasted. The expression *time is money* is one of the overriding considerations on site. It is essential that no time is wasted as a result of poor equipment.

Further information on the care of instruments can be found in some of the standards listed in the Further reading and sources of information section at the end of this chapter.

Check the drawings

Before beginning any setting out operations, care must be taken to ensure that the correct information is at hand. Much of this will be obtained from the drawings for the scheme and it is essential that these are checked for consistency and completeness. It is not unusual for errors to be present in the dimensions quoted or for critical dimensions to be omitted. The first step, therefore, is to study the plans very carefully, abstracting all relevant information that will be needed for the setting out operations. Should any errors or omissions be found, these must be reported immediately in writing in order that corrections can be made. It is also essential to ensure that the latest versions of the drawings are being used. A logical plan storage system must be adopted to ensure that previous versions are not used by mistake. The various different types of plans and drawings that are associated with setting out were discussed earlier in this section.

Walk the site

Even if all the drawings are correct and the relevant setting out information can be obtained, the topography and nature of the site may hamper construction. Initially, therefore, it is essential to walk over the whole of the site and carry out a reconnaissance. The setting out engineer must become very familiar with the area and this cannot be done from inside the site office. Any irregularities or faults in the ground surface which may cause problems should be noted and any discrepancies between the site and the drawings should be reported in writing.

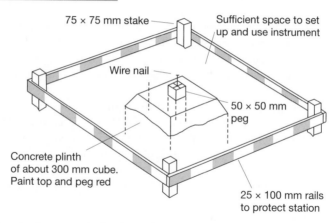

Figure 11.1 ● A primary control point and its protection (from the CIRIA book *Setting Out Procedures*, 1997).

Fix the control points

During the reconnaissance, any existing horizontal and vertical control points should be inspected and suitable positions for any new points temporarily marked with ranging rods or wooden pegs. Once they have been finalised they can be permanently marked. Many different types of marker can be used; for example, an iron bar set in concrete at ground level is ideal. In addition, there are a number of commercially available ground markers ranging from plastic discs to elaborate ground anchors. These are described in Section 6.4.

Ideally, all control points should be placed well away from any traffic routes on site and all must be carefully protected. Figure 11.1 shows a primary control point and its protection, which takes the form of a small wooden barrier completely surrounding the point and painted in very bright colours – for example, red and white stripes. Such barriers are not meant to prevent points from being deliberately disturbed, but rather to serve as a warning to let site personnel know where the points are in order that they can be avoided. There is nothing more frustrating than to spend several days establishing control only to find that half the points have been accidentally disturbed. Careful planning coupled with a thorough knowledge of the site will help to avoid such occurrences.

Inspect the site regularly

As work progresses, the RE and contractor should inspect the site daily for signs of moved or missing control points. A peg, for example, may be disturbed and replaced without the engineer being informed. Points of known reduced level should be checked at regular intervals, preferably at least once a week, and points of known plan position should be checked from similar points nearby.

Work to the programme

The detailed programme for the Works should be posted in the form of a bar chart on the wall of the site office. Using this, the Contractor should plan the various setting out operations well in advance and execute them on time to prevent delays. It is not always advisable to work too far in advance of the programme since points established at an early stage may be disturbed before they are required. Any agreed changes to the programme should be recorded immediately on the chart.

Work to the specifications

In the contract documents, details will be given of the various tolerances which apply to the different setting out operations. It is essential that everyone becomes familiar with these and works to them throughout the project. Suitable techniques and equipment must be adopted to ensure that all specified tolerances are met.

Maintain accuracy

Once the control framework of plan and level points has been established, the principle of *working from the whole to the part* should be adopted. In practice this means that all design points must be set out from the points in the control framework and not from other design points which have already been set out. This approach, which is discussed further in Section 11.3, avoids any errors in the setting out of one design point being passed to another.

Check the work

Each setting out operation should incorporate a checking procedure. A golden rule is that *work is not completed until it has been checked.* However, it is not advisable simply to carry out the same operation in exactly the same manner on two separate occasions. The same errors could be made a second time. Instead, any check should be designed to be completely *independent* from the initial method used, for example:

- Points fixed from one position should be checked from another and, if possible, from a third.
- If the four corners of a building have been established, the two diagonals should be measured and checked.
- All levelling runs should start and finish at points of known reduced level.
- Once a distance has been set out it should be measured twice as a check, once in each direction.
- Points set out by intersection should be checked by measuring the appropriate distances.

Communicate

Lack of communication is one of the main causes of errors on construction sites. All those involved with setting out must understand exactly what has to be done *before* going ahead and doing it.

In many cases, verbal communication will be perfectly acceptable. However, for matters which may be disputed, such as an agreed change in working procedures, the discovery of a discrepancy and the acceptance of a decision, it is advisable to obtain confirmation *in writing*. Signatures should also be obtained whenever possible.

Any errors in setting out should be reported as soon as they are discovered. Prompt action may save a considerable amount of money. There is nothing to be gained from trying to hide errors. This does not remove them and they will only reappear at a later stage when dealing with them will be that much more difficult and expensive.

Reflective summary

With reference to the aims of setting out remember:

— The construction must be set out correctly in all three dimensions both relatively and absolutely and the setting out process must proceed quickly with little or no delay.

— The aims can be achieved by employing the correct horizontal and vertical control techniques and appropriate positioning techniques.

— It is essential that the latest version of any drawing is used.

— Many of the problems that occur during setting out operations are due to lack of thought rather than lack of technical competence.

— Most of these problems can be overcome by adopting a professional approach and a series of good working practices.

— Records should be kept carefully and filed sensibly and the drawings should be checked for discrepancies.

— Instruments must be treated with respect and should be checked both before work commences and at regular intervals during the work.

— Control points should be positioned and protected in sensible locations and checked regularly for signs of movement or disturbance.

— Setting out operations should be planned well in advance with reference to the programme and the specifications in order that the Works progress on schedule and the accuracy requirements are maintained.

— A setting out task is not complete until is has been independently checked.

— Lack of communication is one of the main causes of setting out errors on site.

11.3 The principles of setting out

After studying this section you should appreciate that in order for the construction being set out to be located correctly on site accurate setting out procedures must be employed. You should understand that these procedures include a range of horizontal and vertical control methods and positioning techniques. You should understand how to set out design points on site by a number of coordinate methods by using tapes, theodolites, total stations and GPS receivers. You should appreciate that although they are widely used, coordinate methods are not always the most appropriate for every situation, particularly where precise alignment is critical.

This section includes the following topics:

- Setting out procedures
- Horizontal control techniques
- Vertical control techniques
- Coordinate positioning techniques

Setting out procedures

As discussed in the previous section, one of the aims of setting out is to ensure that the various elements of the scheme are positioned correctly in all three dimensions both relatively and absolutely. Hence setting out procedures must be employed that will enable the correct horizontal and vertical positions of design points to be established. If this is to be achieved then it is essential that procedures are adopted during the setting out operations that will be of sufficient accuracy to meet the specifications given in the contract.

There are a number of setting out techniques and devices that can be used to achieve this aim. In practice, they all fall into one of three main categories: *horizontal control techniques, vertical control techniques* and *coordinate positioning techniques*.

Horizontal control techniques

In order that the design points of the scheme can be correctly fixed in plan position, it is necessary to establish points on site for which the *E, N* coordinates are known. These are *horizontal control points* and, once they have been located, they can be used with a positioning technique to set out *E, N* coordinates of the design points.

When planning the establishment of horizontal control points, two factors must be borne in mind. First, the control points should be located throughout the site in order that all the design points can be fixed from at least two and preferably three of them so that the work can be independently checked. Second, the design points must all be set out to the accuracy stated in the specifications. This latter point is vital –

accuracy must be maintained throughout the control network and this can be achieved by establishing different *levels of control* based on one of the fundamental tenets of surveying, namely *working from the whole to the part*. In practice, this usually involves starting with a small number of very accurately measured control points (known as *first level* or *primary* control), which enclose the area in question and then using these to establish *second level* or *secondary* control points near the site. Under normal circumstances, this second level is used to set out the design points. When establishing a control network, great care must be taken to ensure that the tolerances stated in the contract specifications are met. Only primary and secondary control should be used to minimize the chances of errors propagating through the network, and one design point must *never* be set out from another. In addition, following the working practices discussed in Section 11.2 will help to maintain the required accuracy.

An example of working from the whole to the part using two different levels of control is given in Figure 11.2. In this, the first level of control is provided by a traverse which is run through the site in question to provide a number of well positioned primary site control points. These in turn are used to establish a second level of control, in this case secondary site points at each end of a series of baselines which define important elements of the scheme.

On some schemes the same control points that were used in the production of the site plan prior to design work are used for setting out. If this is the case, they must be re-surveyed before setting out commences. They may have altered their position owing to settlement, heave or vandalism in the time period between the original survey and the start of the setting out operations.

Horizontal control points should be located as near as possible to the site in open positions for ease of working, but well away (up to 100 m if necessary since this is easily accommodated when using total stations) from the construction areas and traffic routes on site to avoid them being disturbed. Since design points are

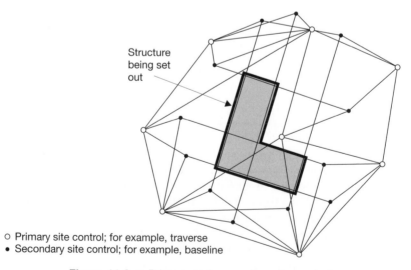

Structure being set out

○ Primary site control; for example, traverse
● Secondary site control; for example, baseline

Figure 11.2 ● Primary and secondary site control.

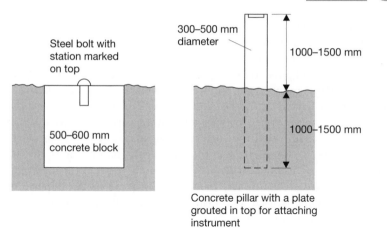

Figure 11.3 ● Permanent control points.

established from horizontal control points, design points must be clearly visible from the control points and as many should be capable of being set out from each of these.

The construction and protection of control points is very important. Wooden pegs are often used for nonpermanent stations, but they are not recommended owing to their vulnerability. Should they be the only means available, Figure 6.12 shows suitable dimensions. For longer life the wooden peg can be surrounded in concrete but, preferably, permanent stations similar to those illustrated in Figures 11.1 and 11.3 should be built. Further suggestions for the construction of control points are given in Section 6.4.

All points must be clearly marked with their reference numbers or letters and painted so that they can be easily found. They should also be surrounded by a brightly painted protective barrier to make them clearly visible to site traffic. Figure 11.1 shows a suitable arrangement.

Once established and coordinated, control points are used to set out design points of the proposed structure. They are generally used in one of the following ways.

Baselines

A *baseline* is a line running between two points of known position. Any baselines required to set out a project should be specified on the setting out plan by the designer and included in the contract between the Promoter and the Contractor. Alternatively, even if they have not been specified, they can be established by the Contractor during the setting out process as the need arises. They can take many forms; for example, they can simply join two specified points, they can run between existing buildings, they can mark the boundary of an existing development, they can be the direction of a proposed pipeline or the centre line of a new road. Given their diversity, they can also be used in a number of different ways, for example:

● Where a baseline is specified to run between two points then once the points have been established on site, the design points can be set out from the baseline by *offsetting* using tapes as shown in Figure 11.4. With reference to this, a design point D is to be set out at right angles to a baseline AB from a point C which lies at a distance y from point A. The required offset distance from C to D is x. Distances x and y will be given by the designer and will usually be *horizontal* distances. Hence, it will be necessary to apply some of the corrections discussed in Chapter 4 in order to ensure that the correct *sloping* lengths are set out on the ground surface. Worked example 11.2 at the end of this chapter shows the calculations required. The right-angle ACD (or BCD) should be set out using a theodolite as shown in Figure 11.4. Both faces should be used and the mean position taken. If only one baseline is used, extra care should be taken since there is very little check on the set out points.

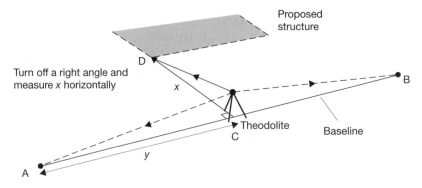

Figure 11.4 ● Setting out by offsetting from a baseline.

● Primary site control points, such as traverse stations E and F in Figure 11.5, can be used to establish a baseline AB by angle α and distance l values as shown; angle (bearing) and distance methods are discussed later in this section. Subsidiary offset lines can then be set off at right angles from each end of the baseline to fix two corners R and S of building Z as shown. Once R and S have been pegged out, the horizontal length of RS is measured and checked against its designed value. If it is within the required tolerance, points R and S can be used as a baseline to set out the corners T and U.

● Design points can be set out by taping known distances from each end of a baseline, as shown in Figure 11.6. Here, point A on building X is set out by taping dimensions 1 and 2 from the baseline and point B by taping dimensions 3 and 4. As before, the set out length of AB is then checked against its designed value and, if within tolerance, it can be used as a baseline to set out corners C and D.

● In some cases, the designer may specify a baseline that runs between points on two existing buildings. Design points are then set out from this line either by offsetting at right angles or by measuring distances from points on the line. The accuracy of this method depends greatly on how well the baseline can be estab-

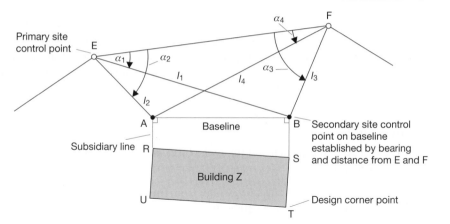

Figure 11.5 ● A baseline.

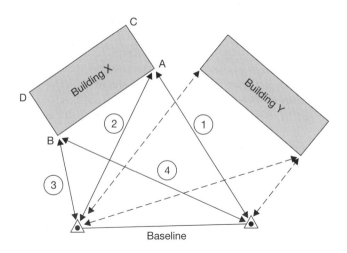

Figure 11.6 ● Setting out by taping from the ends of a baseline (based on the CIRIA book *Setting Out Procedures*, 1997).

lished and how precisely the dimensions required to set out the design points are known. If the ends of the baseline are readily accessible and the dimensions are specified by the designer then this technique can be successfully employed. However, if the ends of the baseline are inaccessible and/or the dimensions can only be obtained by scaling them from an existing plan, then this approach is unlikely to be very accurate and is not recommended unless there is absolutely no alternative. If dimensions have to be scaled from an existing plan, very great care should be taken.

The accuracy of the baselines method increases if two baselines at right angles to each other are used. Design points can be established either by measuring and offsetting from both lines, or a grid system can be set up to provide additional control points in the area enclosed by the baselines. The use of two baselines in this way leads to the use of reference grids on site.

Reference grids

A control grid enables points to be set up over a large area. Several different grids can be used in setting out.

- The *survey grid* is drawn on the survey plan from the original traverse or network. The grid points have known eastings and northings related either to some arbitrary origin or to the National Grid. Control points on this grid are represented by the original control stations.

- The *site grid* is used by the designer. It is usually related in some way to the survey grid and should, if possible, actually be the survey grid, the advantage of this being that if the original control stations have been permanently marked then the designed points will be on the same coordinate system and setting out is greatly simplified. If no original control stations remain, the designer usually specifies the positions of several points in the site grid which are then set out on site prior to any construction. These form the site grid on the ground.

  Since all positions will be in terms of the site grid coordinates, the setting out is easily achieved by measuring angles and distances (α and l values) as shown in Figure 11.7.

  The grid itself may be marked with wooden pegs set in concrete, the interval between grid points being small enough to enable every design point to be set out from at least two and preferably three grid points, but large enough to ensure that movement on site is not restricted.

- The *structural grid* is established around a particular building or structure which contains much detail, such as columns, which cannot be set out with sufficient accuracy from the site grid. An example of its use is in the location of column centres (see Section 11.4).

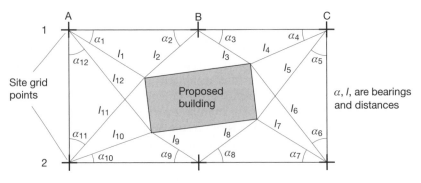

Figure 11.7 ● A site grid.

The structural grid is usually established from the site grid points and uses the same coordinate system.

● The *secondary grid* is established inside the structure from the structural grid when it is no longer possible to use the structural grid to establish internal features of the building owing to vision becoming obscured.

Note: Errors can be introduced in the setting out each time one grid system is established from another; hence, wherever possible, only one grid system should be used to set out the design points.

Offset pegs

Whether used in the form of a baseline or a grid, the horizontal control points are used to establish design points on the proposed structure. For example, in Figure 11.7 the proposed corners of a building have been established by angle and distance methods from a site grid. However, as soon as excavations for the foundations begin, the corner pegs will be lost. To avoid having to re-establish these from the horizontal control points, extra pegs known as *offset pegs* are located on the lines of the sides of the building but offset back from the true corner positions. Figure 11.8 shows these offset pegs in use, and the offset distance should be great enough to avoid them being disturbed during excavation.

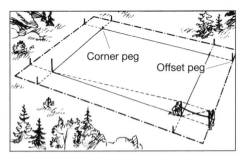

Figure 11.8 ● Use of offset pegs.

These pegs enable the corners to be re-established at a later date and are often used with *profile boards* in the construction of buildings. Offset pegs can be used in all forms of engineering construction to aid in the relocation of points after excavation.

Vertical control techniques

In order that design points on the Works can be positioned at their correct levels, *vertical control points* of known elevation relative to some specified vertical datum must be established on the site. In Great Britain, a vertical datum commonly used is *Ordnance Datum Newlyn* (ODN) and all the levels on a site that use this will be reduced to a nearby *Ordnance Survey Bench Mark* (OSBM) or GPS ODN point. The actual OSBM or point used will be agreed in writing between the Engineer and the Contractor. Alternatively, an arbitrary datum can be used. Whatever datum is adopted, a *master bench mark* (MBM) is established and this is used for two main purposes:

● First, to establish points of known reduced level near to and on the elements of the proposed scheme. These are known as *transferred* or *temporary bench marks* (TBMs). Although TBMs are often located in the new positions on the scheme, any existing horizontal control stations can be used as TBMs providing they have been permanently marked.

● Second, if there are other OSBMs nearby, their reduced levels are checked with reference to the MBM and in the case of any discrepancy, their amended values are used. This ensures that the overall vertical control remains with the MBM.

Once they have been established, the vertical control points are used to define reference planes in space, parallel to and usually offset from selected planes of the proposed construction. These planes may be horizontal (for example, a floor level inside a building) or inclined (for example, an embankment slope in earthwork construction).

As with horizontal control, it is essential that the principle of *working from the whole to the part is adopted* in order that *accuracy can be maintained*. In practice this means ensuring that all vertical design points are set out either from the MBM or from a nearby TBM, and not from another vertical design point which has been established earlier. This prevents an error in the reduced level of one design point being carried forward into that of another.

Transferred or temporary bench marks (TBMs)

The positions of TBMs should be fixed during the initial site reconnaissance so that their construction can be completed in good time and they can be allowed to settle before levelling them in. For this reason, permanent, existing features should be used wherever possible. In practice, 20 mm diameter steel bolts, 100 mm long, driven into existing door steps, ledges, footpaths, low walls and so on are ideal.

Any TBM constructed on the side of a wall should be such that the base of a levelling staff will always be at the same reduced level every time it is placed on the mark. For this reason an etched or scribed horizontal line is not recommended since it can be difficult always to return the base of the staff to exactly the same position. Instead, a bolt fitted to a piece of angle iron should be attached to the wall, as shown in Figure 11.9. This provides an excellent permanent point on which to rest the staff. Where

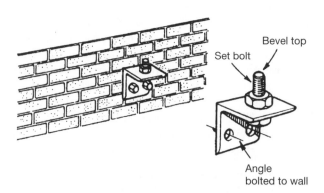

Figure 11.9 ● A TBM on the side of a wall (from the CIRIA book *Setting Out Procedures*, 1997)

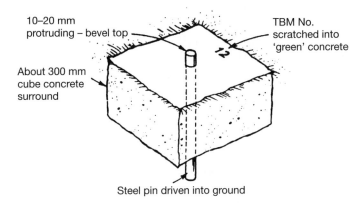

10–20 mm protruding – bevel top

TBM No. scratched into 'green' concrete

About 300 mm cube concrete surround

Steel pin driven into ground

Figure 11.10 ● A TBM at ground level (from the CIRIA book *Setting Out Procedures*, 1997).

TBMs are constructed at ground level on site, a design similar to that shown in Figure 11.10 is recommended.

Each TBM is referenced by a number or letter on the site plan and the setting out plan, and those that could be potentially damaged by site traffic should be protected since re-establishment can be time-consuming. While it is impossible to prevent them being deliberately damaged, it should be possible to protect them and highlight them in such a way that they are clearly visible and can be avoided. A suitable arrangement for doing this is shown in Figure 11.11.

Any TBMs set up on site must be levelled with reference to the agreed MBM. It is vital that the agreed datum is used since the design levels are usually based on this.

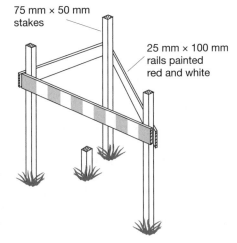

75 mm × 50 mm stakes

25 mm × 100 mm rails painted red and white

Figure 11.11 ● TBM protection.

There should never be more than 80 m between TBMs on site and the accuracy of levelling should be within the following limits:

site TBM relative to the MBM ± 0.005 m

spot levels on soft surfaces relative to a TBM ± 0.010 m

spot levels on hard surfaces relative to a TBM ± 0.005 m

Because TBMs are vulnerable, they must be checked by relevelling at regular intervals, and as soon as the project has reached a suitable stage, TBMs should be established on permanent points on the new construction. To avoid confusion, all the TBMs should be clearly marked on a copy of the site plan, together with their

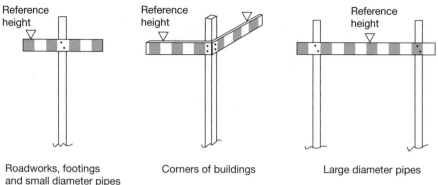

Roadworks, footings
and small diameter pipes

Corners of buildings

Large diameter pipes

Figure 11.12 ● Examples of sight rails.

reduced levels, and this should be displayed in the site office. Any redundant TBMs should be removed from the site as soon as possible

Sight rails

These consist of a horizontal timber cross piece nailed to a single upright or a pair of uprights driven into the ground. Figure 11.12 shows several different types of sight rail.

The upper edge of the cross piece is set to a convenient height above the required plane of the structure, usually to the nearest 100 mm, and should be at a height above ground to ensure convenient alignment by eye with the upper edge. The level of the top edge of the cross piece is usually written on the sight rail together with the length of traveller required. Travellers are discussed below in the following section. It is also possible to have double sight rails.

Sight rails are usually offset 2 or 3 metres at right angles to construction lines to avoid them being damaged as excavation proceeds. This is shown in Figure 11.13; Worked example 11.4 at the end of this chapter shows the calculations involved.

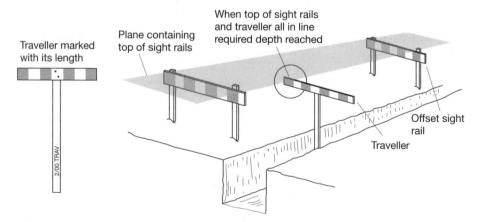

Figure 11.13 ● Sight rails and traveller used to control the excavation of a trench.

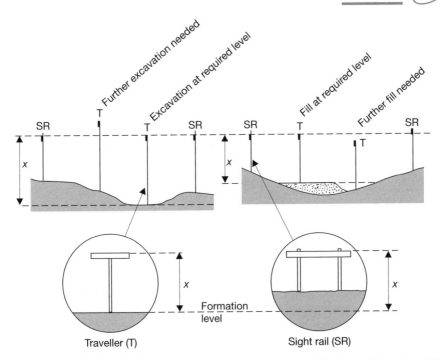

Figure 11.14 ● Sight rails being used with travellers to control the formation of a cutting and an embankment.

Travellers and boning rods

A *traveller* is similar in appearance to a sight rail on a single support and is portable. The length from its upper edge to its base should be a convenient dimension to the nearest half metre.

Travellers are used in conjunction with sight rails. The sight rails are set some convenient value above the required plane and the travellers are constructed so that their length is equal to this value. As excavation proceeds, the traveller is sighted in between the sight rails and used to monitor the cutting or filling. Excavation or compaction stops when the tops of the sight rails and the traveller are all in line.

Figure 11.13 shows a traveller and sight rails in use in the excavation of a trench and Figure 11.14 shows the ways in which travellers and sight rails can be used to monitor cutting and filling in earthwork construction.

There are several different types of traveller. Free-standing travellers are frequently used in the control of superelevation on roads, a suitable foot being added to the normal traveller, as shown in Figure 11.15. Pipelaying travellers and the use of travellers in the form of boning rods are discussed later in this section.

Slope rails or batter boards

For controlling side slopes in embankments and cuttings, sloping rails are used. These are known as *slope rails* or *batter boards*.

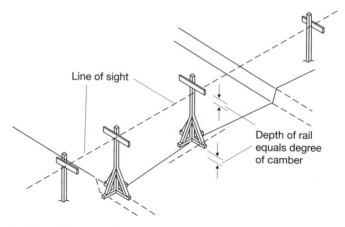

Figure 11.15 ● A free-standing traveller being use to monitor superelevation.

For an *embankment*, the slope rails usually define planes parallel to but offset some vertical distance from the proposed embankment slopes, as shown in Figure 11.16. In addition, they are usually offset a horizontal distance of at least 1 m from the toe of the embankment to prevent them from being covered during the filling operations. Travellers are always used in conjunction with the slope rails to monitor the formation of the embankments.

For a *cutting*, the slope rails can be set to define either the actual slope of the cutting, as shown in Figure 11.17, or a parallel offset slope for use in conjunction with a short traveller, as shown in Figure 11.18. Both methods are satisfactory but each has its limitations. If the exact slope is defined, it is not possible to erect an

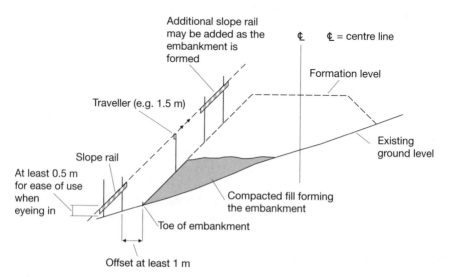

Figure 11.16 ● A slope rail being used to monitor the construction of an embankment in conjunction with a traveller.

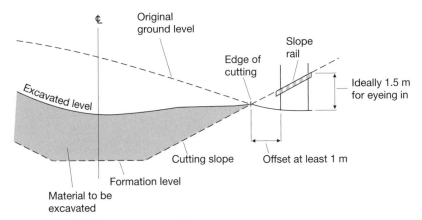

Figure 11.17 ● A slope rail being used to monitor the excavation of a cutting without a traveller.

additional slope rail as the excavation proceeds. If a parallel offset slope is defined, the height at which the slope rail must be fixed will increase by an amount equal to the length of the traveller used and this may make the operation of viewing along the slope rail very difficult to accomplish. In both methods, the wooden stakes supporting the slope rail are usually offset a horizontal distance of at least 1 m from the edge of the proposed cutting to prevent them being disturbed during excavation.

All relevant information is usually marked on the slope rails: chainage of centre line, distance from wooden stakes to centre line, length of traveller, side slopes and so on.

During the setting out, the positions of the toes of the embankments and the edges of the cuttings must be fixed in order that the wooden stakes onto which the slope

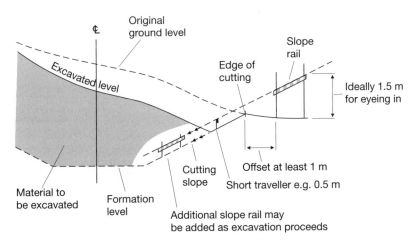

Figure 11.18 ● A slope rail being used to monitor the excavation of a cutting with a traveller.

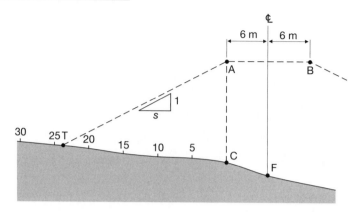

Figure 11.19 ● Locating the toe of an embankment.

rails are to be attached can be located in their correct positions. One method of doing this is as follows.

Consider Figure 11.19 in which the toe, T, of an embankment is to be fixed. The top of the embankment is to run from point A to point B and is to have a width of 12 m. Point C is on the existing ground surface directly below point A and the centre line is defined by point F, which is also on the existing ground surface. The sides of the embankment are to slope at 1 in s. The procedure is as follows.

● From the road design, obtain the reduced level of point A.

● Peg out point C by measuring a distance of 6 m horizontally from F at right angles to the centre line.

● Peg out points at 5 m horizontal distance intervals from point C along the line FC produced. Locate sufficient points to ensure that the toe T will fall between two of them.

● Measure the reduced level on the ground surface at the first 5 m peg.

● Calculate the proposed reduced level (RL) on the embankment slope directly above this point from

$$RL \text{ at } 5 \text{ m point} = RL_A - \left(\frac{5}{s}\right)$$

● Compare the values of the reduced levels measured and calculated at the 5 m peg.

● If the ground level is lower than the calculated proposed level then the toe of the embankment is further than 5 m from point C. Move to the 10 m peg and measure the RL of the ground surface. Calculate the proposed reduced level on the embankment slope directly above this point from

$$RL \text{ of } 10 \text{ m point} = RL_A - \left(\frac{10}{s}\right)$$

Compare the two RLs. If the ground level is still lower than the proposed level repeat this step for the 15 m peg. Continue moving from one peg to the next until the ground level is higher than the proposed level.

- Once the ground level at a 5 m peg is measured to be higher than the proposed level, the toe of the embankment has been passed and its position is somewhere between this 5 m peg and the previous one.
- To locate the exact position of the toe, return to the previous peg and repeat the process but advancing forward in 1 m intervals.
- Once the ground level is equal to the proposed level within 50 mm, point T has been located and a peg should be hammered into the ground at this point. As a precaution, the distance along the ground surface from point T to the centre line peg F should be measured and recorded in case the toe peg is disturbed.

On first reading, the above procedure appears to be rather slow and laborious. In practice, however, this is not the case and an experienced engineer can very quickly locate embankment toes by this method. Often it is not necessary to work from point C in 5 m intervals. If the cross-sectional drawings are available, good estimates of the positions of toes can be obtained and these will indicate the best location for the pegs set out near to the toe: for example, the first at 20 m and then every 2 metres.

Although the above procedure has concentrated on locating an embankment toe, a similar technique can be used to locate the edge of a cutting.

Once the toes and edges have been located, the wooden stakes which are to carry the slope rails can be hammered into the ground at offset horizontal distances from these as shown in Figure 11.20. The next stage is to calculate the required reduced levels at which the top edges of the slope rails must be fixed on the wooden stakes. In practice, nails are hammered into the stakes at the required levels and the rails are attached with their top edges butted up against them.

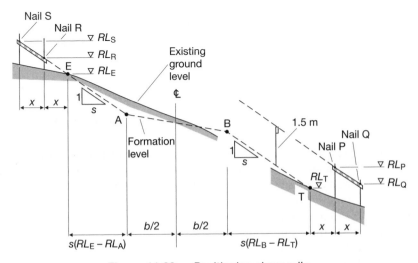

Figure 11.20 ● Positioning slope rails.

For an *embankment*, assuming that a 1.5 m traveller is to be used as shown on the right-hand side of Figure 11.20, the reduced levels at which two nails P and Q should be placed on the wooden stakes is obtained as follows. It is assumed that the RL at the toe of the embankment (RL_T) is known.

$$RL_Q = RL_T - \left(\frac{2x}{s}\right) + 1.5$$

$$RL_P = RL_T - \left(\frac{x}{s}\right) + 1.5 = RL_Q + \left(\frac{x}{s}\right)$$

Once it has been calculated, RL_Q should be compared with the reduced level of the ground directly below point Q to ensure that the difference is at least 0.5 m to enable the slope rail to be sighted along without too much difficulty. If it is less than 0.5 m, a longer traveller should be used.

Worked example 11.5 shows the calculations involved when setting out slope rails to control an embankment.

For a *cutting*, as shown on the left-hand side of Figure 11.20, the reduced levels at which two nails R and S should be placed on the wooden stakes is obtained as follows. It is assumed that the RL at the edge of the embankment (RL_E) is known and that a traveller is not being used.

$$RL_S = RL_E + \left(\frac{2x}{s}\right)$$

$$RL_R = RL_E + \left(\frac{x}{s}\right) = RL_S - \left(\frac{x}{s}\right)$$

Finally, the tops of the stakes are levelled and the values obtained are compared with the reduced levels calculated above for the nails. This gives the required distances to be measured down from the tops of the stakes and nails P, Q, R and S are hammered into them at these levels. The slope rails are then attached with their top edges butted up against these nails. If it is found that some of the wooden stakes are not long enough, it will be necessary to add extension pieces to them and then attach the slope rails to these.

Profile boards

These are very similar to sight rails but are used to define corners or sides of buildings.

It has already been shown that offset pegs are used to enable building corners to be relocated after foundation excavation. Normally a *profile board* is erected near each offset peg and used in exactly the same way as a sight rail, a traveller being used between profile boards to monitor excavation. Figure 11.21 shows profile boards and offset pegs at the four corners of a proposed building.

The arrangement shown in Figure 11.21 is quite an elaborate one and a simpler, more often used, type of corner arrangement is shown in Figure 11.22. Nails or sawcuts are placed in the tops of the profile boards to define the width of the

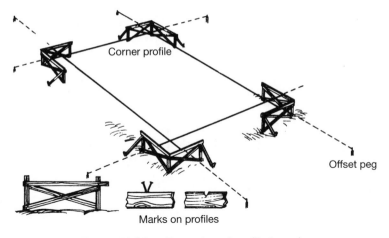

Figure 11.21 ● Examples of profile boards.

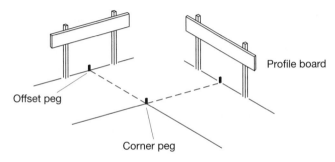

Figure 11.22 ● Profile boards at a corner of a building.

foundations and the line of the outside face of the wall. String is stretched between opposite profile boards to guide the width of cut while the traveller is used to control the depth of cut.

A variation on corner profiles is to use a *continuous profile* all round the building set to a particular level above the required structural plane. Figure 11.23 shows such a profile with a gap left for access into the building area.

The advantage of a continuous profile is that the lines of the internal walls can be marked on the profile and strung across to guide construction.

Another type of profile is a *transverse profile*, and this is used together with a traveller to monitor the excavation of deep trenches as shown in Figure 11.24.

Profile boards and their supports are normally made

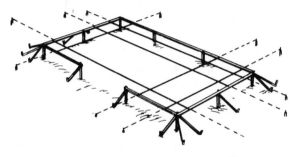

Figure 11.23 ● A continuous profile.

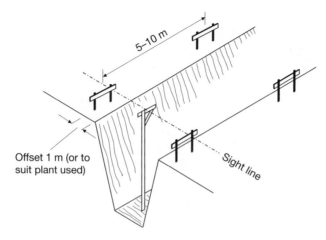

5–10 m

Offset 1 m (or to
suit plant used)

Sight line

Figure 11.24 ● Transverse profiles.

from timber. Sledge-hammering wooden stakes into the ground and nailing on cross pieces can be dangerous, especially in hard or difficult terrain. Great care must be taken to ensure that no injuries occur. It is strongly recommended that steel-toe-capped boots are worn by those involved in this type of work and that no one should be asked to hold a stake in place by hand while it is being hammered into the ground. Instead, it is not too difficult to manufacture a simple grip holder out of a piece of reinforcing rod and to use this to support the stake while it is being hit.

Worked example 11.6 shows the calculations involved when setting out profile boards to control the corner of a building.

Coordinate positioning techniques

As discussed in Section 11.2, one of the main aims of setting out is to ensure that the design points of the scheme are located in their correct plan positions. Depending on the equipment available, there are a number of different coordinate-based methods which can be adopted to ensure that this aim is achieved.

For setting out by coordinates to be possible, a control network consisting of coordinated points (with heights) must be established on site. All of the procedures for providing control using a total station are described in Chapter 6 and with GPS in Chapter 7. When choosing the locations for control points on site, some forward planning is needed to ensure that enough control is available for setting out the project and that any possible problems with control points becoming obscured as construction proceeds are avoided. It must also be possible to check work by sighting key design points from more than one control point. It is also important that any control survey carried out for site work meets the specifications for the project. Guidelines for these and how to check control points are given in the ICE Design and Practice Guide *The Management of Setting Out in Construction*.

All of the coordinate methods of setting out require the positions for all the design points to be set out to be known on the site coordinate grid – these are usually generated at the design stage of a project using office software and are stored in some digital format.

Setting out by theodolite and tape

To set out using coordinates by theodolite and tape, one of the following procedures is used.

- *Angle (bearing)* and *distance* from two control points. With reference to Figure 11.25(a), point A can be set out from a control point S by one of two methods:

  Either Using an inverse calculation, determine the horizontal length l (= SA) and the whole circle bearings (WCBs) of ST and SA from the coordinates of S, T and A as described in Section 6.2. Calculate α from α = WCB (ST) – WCB (SA). With the theodolite set up at S, sight T and set the horizontal circle to read zero along this direction. Now rotate the telescope through angle α to fix the direction to A and measure l along this direction to fix the position of A. This is known as setting out by *angle and distance*.

  Or Compute l, WCB (ST) and WCB (SA) as for the first method. Sight T from S and set the horizontal circle of the theodolite to read the WCB of ST. Rotate the telescope towards point A until the WCB of SA is read on the horizontal circle. The telescope line of sight is now defining the direction to A and the exact position of A can be fixed by measuring a horizontal distance l along this direction. This is setting out by *bearing and distance*.

- *Intersection* with two theodolites, from four control points using angles or bearings only. Intersection is shown in Figure 11.25(b) and Worked example 11.3 shows the calculations involved.

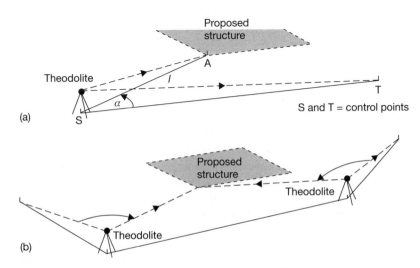

Figure 11.25 ● Positioning techniques using theodolites and tape.

- *Offsetting* from one or more baselines using a theodolite and tape, as discussed earlier in this section and shown in Figure 11.4, the offsets either being calculated from the coordinates of the ends of the baselines and the design point coordinates or being specified in the form of horizontal distances as in Worked example 11.2.

Full details on how to set up and use a theodolite are given in Chapter 3 and all of the methods used for measuring distances with tapes are described in Chapter 4. These must be read in conjunction with the three techniques given above. The following points are specific to setting out by theodolite and tape and must also be taken into consideration, where appropriate.

- All angles must be set off using a correctly calibrated theodolite; otherwise both faces should be used and the mean position taken.

- Since the design dimensions will be in the horizontal plane, any distance set out along the ground surface with a tape must allow for the slope of the ground. This requires the computation of the slope distance (see Section 4.3). If the angle of slope is required it can be measured using the theodolite, taking readings on both faces and the mean value used. If using a fibreglass tape a standardisation correction may also be necessary and if using a steel tape, standardisation, temperature, tension and sag corrections may be necessary depending on the accuracy required. All of these corrections are discussed in Section 4.3.

- The setting out must be checked. For *bearing and distance,* the setting out should be repeated from another control point, for example by placing the theodolite at T and using S as the reference point in Figure 11.25(a). For *intersection,* another control point should be used which introduces a check bearing or angle. For *offsetting,* the setting out should be repeated using a second baseline. If only one baseline is used, extra care should be taken, since there is very little check on the set-out points.

- When offsetting, right angles should be set out by theodolite and the angle turned on both faces using opposite sides of the horizontal circle to remove eccentricity and graduation errors, for example, on face right use 0° to 90° and on face left use 180° to 270°. The mean of the four pointings is the correct angle.

- To locate each design point, a 50 mm square cross-section wooden peg should be driven into the ground at the point and the exact design position marked on top of the peg with a fine tipped pen. A nail is then hammered into the peg at this point.

- In general on site, for distances up to one tape length (30 m or 50 m), tapes can be used very effectively. They are very cheap and if used properly can be very accurate. However, for distances in excess of one tape length great care must be taken to ensure that errors do not occur at the changeover point from one tape length to the next. If there are a lot of design points to be set out where the distances involved are longer than one tape length, it is better to use a total station or GPS for the setting out.

When setting out using coordinate-based methods with theodolites and tapes, the situation may arise where there are no nearby control points available for this. This

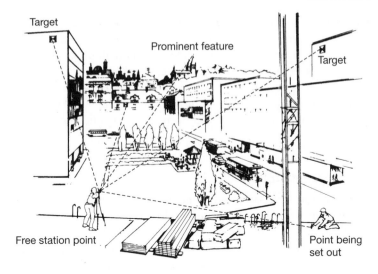

Target

Prominent feature

Target

Free station point

Point being
set out

Figure 11.26 ● Use of a free station point.

problem can be overcome by establishing a *free station point* at any convenient place
for setting out. This technique is shown in Figure 11.26 and is essentially a resection
as described in Section 6.7. Free station points are particularly applicable to large
sites where the coordinates of prominent features and targets on nearby buildings or
parts of the construction are known. The procedure for setting up a free station point
is as follows.

● The theodolite is set up at some suitable place in the vicinity of the points which
are to be set out. This gives rise to the term *free station*, since the choice of the
instrument position is arbitrary.

● An angular resection is carried out to fix the position of the free station point.
Preferably, observations should be taken to more site control points than the
minimum for checking purposes.

● The coordinates of the free station point are calculated.

Following this, setting out continues as before and the required design points are
set out using the theodolite set at the free station point.

If free station points are to be used widely on a particular site, it is essential that
there are a sufficient number of well-established control points around the site to
enable enough obstruction free sightings to be achieved while construction
proceeds. It is also possible to establish free station points using total stations. The
procedure for doing this is discussed in the next section.

Although setting out can be carried out using theodolites, tapes (and levels) in
what might be called traditional methods, a lot of work on site is done using total
stations and GPS equipment. When setting out by the so-called traditional methods,
direct measurements of angle and distance are taken to position structures and other
works from nearby control points or from baselines. Following this, offsets and
profiles are put in place to define the main lines of a building and provide vertical

control for second stage setting out. All of the procedures for doing this are described in various sections throughout this chapter. Despite their popularity on site, these well-established methods have the disadvantages that the horizontal and vertical components of setting out have to be done separately (levelling must be used for any heighting), they can be time-consuming if a lot of points have to be set out, and they require at least two people to do the setting out. Total stations and GPS can overcome these problems and offer a different approach to setting out that can be more efficient and cost effective.

Setting out by total station

To use a total station for setting out, it must be levelled and then centred over a control point in the same way as for a theodolite (the procedure for doing this is described in Section 3.5). This must be done carefully, otherwise all the subsequent readings taken with the instrument will not give correct results no matter how well they are taken. In difficult locations and environments (for example soft ground, hot sunshine) it is necessary to check the levelling and centring of the total station at frequent intervals and to re-set these, if required. These and other sources of possible errors and problems when using a total station are discussed in more detail in Section 5.5.

Having set up the total station, it has to be orientated horizontally to the site coordinate system and it may also have to be orientated vertically.

For *horizontal orientation*, the coordinates of the control point at which the instrument is set up are entered into the total station. An adjacent control point is then chosen as a reference point (usually called the reference object or RO) and the coordinates of this are also keyed in. A file containing the data for the site control points may have already been loaded into the total station or its data collector and it may only be necessary to identify control points in this process by their names or numbers to obtain their coordinates. To orientate the total station, the RO is sighted and the horizontal circle orientation programme automatically computes the bearing from the total station to the RO and sets this into the reading system and displays it on the screen (an explanation of horizontal orientation is also given in Section 6.2). Whatever direction the total station is now pointed in, it will display a bearing in the site coordinate system. It is also possible to occupy an unknown point and to obtain the coordinates of this by performing a distance resection. This allows free station points to be used on site, which can be more convenient than relying on a fixed control network. The fieldwork and calculations required for a distance resection are described in Section 6.7. Following the resection, the horizontal circle is orientated by sighting an RO.

Whatever method is used to coordinate and orientate the total station, a check must be made on these by sighting an independent control point to confirm that the bearing and distance to this agrees with the check measurements taken with the total station. It is also good practice to re-check the RO at frequent intervals, especially if the total station is set up over the same point for a long time.

For *vertical orientation*, the height of collimation of the total station has to be determined. If the height of the control point at which the total station is set is known, this is entered into the instrument or is already stored in a file containing the control point data. The height of the instrument above the ground mark is measured with a pocket tape and this is also keyed in. These two are added by the total station to give the height of collimation. If vertical orientation is required at a free station point, a control point of known height is sighted and the height of collimation determined using trigonometrical methods (trigonometrical heighting is described in Section 5.6). As with horizontal orientation a check on vertical orientation is carried out, again by sighting an independent station.

Once they have been orientated, total stations can be used for setting out horizontal positions either using the coordinates of the points to be set out directly or using bearing and distance values calculated from these coordinates. These two approaches are considered below.

● *When the coordinates of the point to be set out (the design point) are used*, these are usually contained in a file, together with the coordinates of the control points for the project, and this is downloaded to the total station or its controller/data collector before work commences. Alternatively, these can be keyed in manually if required. Using the coordinates of the point at which the total station is set up together with those of the RO, the horizontal circle is then orientated to the coordinate grid so that it displays bearings directly. The setting out mode is now selected, the design point to be set out is identified and its code or coordinates are entered into the total station. This enables the total station to calculate the bearing and distance to the design point from its coordinates. There is no need for the observer to do any inverse calculations with the coordinates as the total station does this automatically. The instrument then displays the difference between the calculated and the measured bearings. The telescope is rotated until this difference is zero and it is then pointing in the required direction for setting out the design point. Following this, a detail pole is located on the line of sight as near to the required distance to be set out as possible – once aligned, the prism is sighted and the *horizontal* distance to it measured (remember that coordinates give the horizontal distance between points and care must be taken to set the correct measurement mode on the total station to ensure the *slope* distance is *not* displayed). The difference between this and the value calculated for the horizontal distance by the total station is displayed. By moving the prism backwards or forwards along the line as required, and taking further measurements to it, this difference is reduced to zero (within an acceptable limit) to locate the design point. A pocket tape is very useful for helping to move the detail pole (prism) a particular amount to a revised position after each measurement of distance is taken. An optical guidance system (see Section 5.3) may also be useful for locating the prism but two-way radios with hand signals between instrument and reflector are often used. On some instruments, the coordinate differences between measured and calculated values are displayed instead of a bearing and distance, and the reflector is moved and measurements taken until these are zero within acceptable limits.

● *If the bearing and distance to be set out are known*, these can also be used for setting out. They are entered into the total station and, as soon as the appropriate key(s) are pressed to activate this setting out mode, the instrument once again displays the difference between the entered and measured bearing values. In order to set the required direction for setting out, the telescope is rotated until a difference of zero is displayed. With a detail pole located on the line of sight, the horizontal distance to it is measured and the difference between this and the value entered into the total station is displayed. This is reduced to zero by moving the prism backwards or forwards along the line as required. If the bearing and distance cannot be entered into the total station for some reason, setting out can still be carried out by treating the total station as if it was a theodolite. In this case, the telescope is turned until the required bearing is displayed on the horizontal circle and the distance is measured along this direction to locate the design point. To do this, the prism is moved backwards and forwards along the line as necessary until the required horizontal distance value is displayed.

As with traditional methods of horizontal setting out with theodolite and tape, it is critical that checks are applied to all points set out by a total station. This can be done by repeating the setting out from a different station and by taking distance measurements between set out design points with a tape and comparing them to their designed values.

All of the techniques described above are for two-dimensional (or horizontal) setting out. However, they can also be extended to set out vertical positions provided the height of collimation of the instrument is known and entered into the instrument together with the appropriate vertical design data.

Setting out using robotic and motorised total stations

Motorised and robotic total stations are described in Section 5.3 and many of the principles of setting out using manual total stations discussed above also apply to them, including the ability to set out vertical positions. However, such instruments offer definite advantages over the manual models – in particular, the speed at which setting out can be done and the accuracy that can be achieved are both improved.

A *motorised total station* will automatically align itself onto a design point ready for setting out when a code for the point is entered – there is no need for the operator to turn the telescope to obtain a difference of zero between displayed and computed bearings. A further improvement is offered by a *robotic total station*. This type of instrument only requires one person instead of two to perform setting out tasks. In use, the total station is set up at a control point and orientated, after which it can be left unattended, but in a secure location. All work is then carried out from the detail pole, which has a prism and control unit mounted. An example of this is shown in Figure 11.27, where a Trimble Attachable Control Unit (ACU) fitted to a detail pole is being used with a Trimble 5600 robotic total station. The code for a design point is keyed into the control unit, the detail pole is held close to the expected position for the design point and the total station is instructed, using the control unit, to search for and lock onto the prism. After lock-on, the bearing and distance are measured to the prism and the results are sent to the control unit, which guides the operator to the required position by a display on its screen – Figure 11.28 shows this for the Trimble ACU. As the detail pole is moved, the total station maintains lock and

Figure 11.27 ⬤ One-person setting out using a Trimble ACU on a pole mounted reflector in conjunction with a Trimble 5600 robotic total station (courtesy Trimble Navigation Ltd).

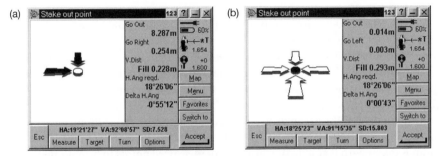

Figure 11.28 ⬤ Trimble ACU screen display fitted to a detail pole and used with a Trimble 5600 robotic total station: (a) indicating how the prism should be moved; (b) indicating that the required point has been set out (courtesy Trimble Navigation Ltd).

follows its movement. If it is set to tracking mode, the amount by which the pole has to be moved to locate the design point is continually updated on the screen, as shown in Figure 11.28(a). When the position of the design point is found, the display changes to that shown in Figure 11.28(b). Although robotic total stations are expensive, they are becoming popular with contractors on site because only one person is needed to use them and they are capable of taking rapid and accurate measurements. However, because setting out can be reduced to simply identifying point numbers and following indicators on a display, it is very important to check all setting out done with these before construction proceeds.

High precision measurement and setting out using specialised total stations

In order to meet the increasingly demanding design and construction specifications on modern buildings and structures, another application on site for total stations is in the measurement and setting out of complicated three-dimensional structures. All of the dimensions in these projects are usually specified as coordinates in a CAD model of the structure and they usually have to be set out within a few millimetres. This can be achieved using specialised total stations such as Sokkia's MONMOS system and Leica's TDM5005 and TDA5005 industrial total stations, which are described below.

The *Sokkia MONMOS system* (MONMOS = **MON**o **MO**bile 3D **S**tation) is designed for high-precision three-dimensional measurement. It consists of the Sokkia NET1200 total station and SDR4000 Control Terminal, as shown in Figure

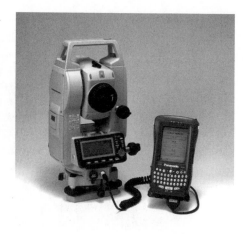

Figure 11.29 ● Sokkia MONMOS system (courtesy Sokkia Ltd).

11.29. The NET1200 is a total station capable of measuring angles with a resolution of 0.5" and precision of 1" (DIN 12857-2:1997) and distances with a resolution of 0.1 mm and precision of ±(0.6 mm + 2 ppm) to special plane reflecting targets and ±(1 mm + 2 ppm) either to a reflector or in reflectorless mode. It has an internal memory and has software installed for coordinate measurement, setting out, resection and other functions.

Although the NET1200 can be used on its own for precise coordinate measurement and setting out in the same way as most total stations, it can be set up such that any readings taken with it enable the positions of measured points to be obtained directly in some specified coordinate system. Normally, this is achieved by placing the instrument at a control point whose coordinates are known and then orientating it onto another. The advantage of the MONMOS system is that it is not necessary to do this, and the NET1200 is placed at any convenient and stable position for measurement. If existing control points are to be used, angles and distances are then measured to between two and six of these, whose coordinates are already stored in the SDR4000, which will then calculate the three-dimensional position of the NET1200 and orientate it accordingly. Alternatively, design data for a project can be stored in the SDR4000 and, instead of using control points, up to six design points on a structure can be sighted by the NET1200, which is then orientated based on their coordinates. A third alternative is to set the NET1200 to define an arbitrary coordinate system based on clearly defined points on a structure, but whose coordinates are not known – this enables MONMOS to determine relative position and dimensions.

Whatever method is used to set up and orientate MONMOS, measurement of the coordinates of any as-built point already existing on a structure or of any point to be set out are obtained by placing a reflector at the point and intersecting it using the NET1200. The angle and distance are automatically recorded and the three-dimensional coordinates of the reflector are computed and displayed. If the design data for the project is already stored in the SDR4000, an on site comparison can be made

Figure 11.30 ● Leica TDM5005 Industrial Total Station (courtesy Leica Geosystems).

between the measured and desired positions – these can be stored for later analysis or used for setting out purposes. At short distances of up to about 30–40 m it is possible to take all these measurements with MONMOS to a precision approaching ±1 mm.

The *LeicaTDM5005* and *TDA5005 Industrial Total Stations* (Figure 11.30) can also be used for three-dimensional coordinate measurement and setting out on site with a precision better than a millimetre at short range. When they are required to take real-time measurements, these instruments are used with *Leica's Axyz* software which is modular and can be used with different surveying and measuring equipment to suit different measurement requirements. Like MONMOS, these instruments can also be set up in a number of different ways to provide spatial information on site for as-built surveys or for setting out. The TDA5005 has the added advantage of being motorised and fitted with ATR (Automatic Target Recognition), which enables it to be driven automatically to a design point ready for setting out.

Both MONMOS and the Leica Industrial Total Stations are expensive and would only be used on complex construction projects where structural elements need to be positioned accurately and in three dimensions.

Setting out by GPS

For setting out by GPS, an RTK system is required consisting of two geodetic receivers working in precise relative mode. One of these will be permanently located at a base station and the other (the rover) will move around the site and take the measurements needed for positioning design points. In addition to the description of RTK methods and equipment already given in Chapter 7, this section gives details of their application in setting out.

In common with all other setting out methods, GPS is based on a control network, which must be in place before any work can start. Control points with positions defined on the site grid are needed for base stations, for determining transformation parameters when deriving site coordinates from GPS coordinates, and for checking purposes. Depending on the site, control can be local and based on an arbitrary coordinate system or it can be connected to a national system. As discussed in Sections 7.8 and 8.8, the GPS coordinates have to be transformed into site or national coordinates. For small local sites a control network consisting of at least three but preferably five points with known site coordinates and heights is required for determining transformation parameters. This can be surveyed using a total station and traverse methods, as described in Chapter 6. On large sites, whether they cover an extensive land area or are long linear sites such as those occurring on road and railway projects, site control is often based on national control. In Great Britain, the Ordnance Survey has produced two national transformation parameters for deriving National Grid eastings and northings (the OSTN02 transformation) and ODN heights (the OSGM02 transformation) from ETRS89 GPS coordinates. The way in which ETRS89 coordinates can be obtained and transformed into National Grid coordinates is discussed further in Sections 8.6 and 8.7.

In an RTK GPS system, the base station must be located at a known point. This can be an existing control point or could be a new point whose coordinates and height are obtained using either a total station or rapid static GPS methods from existing control points. The positions for base stations must be carefully chosen so that radio communications or a cell phone connection is maintained between the base station and each rover on site at all times. If the radio link or cell phone connection is lost, setting out is not possible. On difficult sites, it may be necessary to install repeater stations so that the communications link is not lost because of tall buildings, hills and other obstacles that might block it. Other important issues with regard to base stations are their security (as a GPS receiver and other hardware plus a radio transmitter will be installed and left unattended at them at all times) and the need to provide an uninterrupted power supply.

At the start of setting out, a rover must be initialised to resolve ambiguities. With an RTK system this is done whilst the rover is moving, but it requires at least five satellites to be in view for it to be performed reliably and quickly. After initialisation the rover will display real-time site grid coordinates which will change as the rover is moved – a check on these should be carried out by placing the rover on an independent point with known coordinates. If at any time a loss of lock occurs, an audible or visual warning is issued by the receiver.

For setting out to be possible with RTK GPS, a data file containing the coordinates of all the design points to be set out with the RTK system is stored in its controller. To set out a point, the setting out applications program in the controller is activated and the point identified by entering its number or code into the controller. This will then display visual symbols on its screen to help the operator locate the design point. Typical screen displays for this are shown in Figure 11.31, which shows the display for a Trimble Attachable Control Unit (ACU) fitted to a Trimble 5800 GPS rover on a

(a) 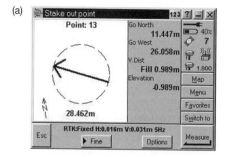 (b)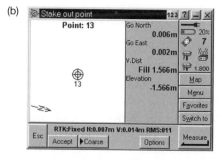

Figure 11.31 ● Trimble ACU screen display when fitted to a Trimble 5800 GPS rover: (a) indicating how the rover should be moved; (b) indicating that the required point has been set out (courtesy Trimble Navigation Ltd).

pole. This screen displays an index mark in the form of a circle containing an arrow which indicates the direction in which the pole must be moved to fix the point (Figure 11.31(a)). The operator moves the base of the pole over the surface of the ground in the direction indicated until the index mark changes to two concentric circles (Figure 11.31(b)), thereby setting out the point. Great care must be taken to keep the pole vertical and still during this process, as shown in Figure 11.32. As before with total stations, there is no need for the operator to do any inverse calculations with the coordinates.

Figure 11.32 ● Setting out using a GPS rover – it is important to keep the pole vertical and still (courtesy Leica Geosystems).

The process of setting out by real-time GPS (and total stations) relies on good data management to ensure that design data is reliably converted into a suitable digital format for transfer to survey controllers and data collectors. Because of the increasing automation in the surveying equipment used on site, this is an area in site surveying where engineers are expected to play an increasing role and where it is essential that proper training is provided by the manufacturers and suppliers of survey hardware and software. The advantages of seamless data for a construction project are enormous, and in a truly integrated survey system controllers and data collectors loaded with design data can be interchanged between total stations, GPS and other electronic surveying equipment.

The accuracy expected for setting out with RTK GPS is typically 10 mm + 1 ppm horizontally and 20 mm + 1 ppm vertically. The 1 ppm component refers to the distance to the base station and since 1 ppm is equivalent to 1 mm per km if, say, the base station is 5 km from the rover, the horizontal accuracy would be 15 mm and the vertical 25 mm. It must be realised that these figures assume good observing conditions and the accuracy can deteriorate when the satellite coverage is poor and when the satellite geometry is unfavourable. While such precisions are suitable for many setting out operations they will not be appropriate for all, and if higher precisions are specified it will be necessary to use alternative more precise techniques, such as setting out using total stations or levelling for heights. Other problems that can occur with GPS on site are multipath errors caused by plant and machinery, parked vehicles and buildings with glass or other highly reflective sides. In addition, interference with GPS signals can be caused by overhead power lines and mobile phones (not those used in the communication link to the base station). In extreme cases, where these persist, it may necessary to abandon GPS and use another method for setting out. As well as these potential difficulties, the base station and rovers must maintain a lock on at least four satellites while any setting out is being undertaken. Hence setting out by GPS should only be used to survey in areas where there is a good view of the sky. In urban areas, where the likelihood of loss of lock is high due to tall buildings obscuring the signals, it may not be possible to use GPS. This could also apply to small sites where the use of GPS may not be practical or economically viable.

The advantages that GPS has over other methods of setting out are that it is carried out by one person and it is much quicker on large-scale projects when a high density of points has to be set out and when setting out in difficult terrain. GPS is especially useful in earthworks, since the accuracy it can achieve is well matched to the accuracy normally specified for this type of work.

Further information on the use of GPS in setting out can be found in a guidance note entitled *Guidelines for the Use of GPS in Surveying and Mapping* published by the Royal Institution of Chartered Surveyors (RICS). In addition, the Institution of Civil Engineers (ICE) Design and Practice Guide entitled *The Management of Setting Out in Construction* includes a section specifically entitled 'Top tips for setting out using GPS' – an edited version of this is given in Table 11.1. It is strongly recommended that these publications by the RICS and ICE are consulted by anyone contemplating the

Table 11.1 ● Some top tips for setting out with GPS (based on information given in the Good Practice section of the ICE Design and Practice Guide *The Management of Setting Out in Construction*).

Note: It will be necessary to refer to Chapter 7 on GPS when using this table.

- A real-time RTK system is required to undertake setting out with GPS.

- GPS works on a global reference system (WGS84) and in order to operate on a local site coordinate system you must have a transformation to convert the GPS coordinates you measure to a local coordinate framework. Without this you cannot undertake setting out.

- If you require precise height measurements, particularly over longer distances, you must have determined how to model the geoid undulation over the area of your survey. Usually your GPS and controller software will provide tools to deal with this.

- At least five satellites are required for initialisation. Once this has been completed, at least four are required for setting out. Therefore you should check satellite availability in your GPS planning software or on your controller prior to fieldwork.

- Note areas of your site that are prone to multipath errors and interference errors and make others aware of these.

- For best accuracy, keep the detail pole as vertical and as still as possible.

- Make sure you use the correct height measurement with the antenna you are using. This is a common source of error.

- Treat the GPS kit with respect. Although it is more robust than optical equipment, it is considerably more expensive and must be looked after.

- Carry spare controller and receiver batteries with you at all times.

- Keep field records of all surveys and setting out. Because of the 'automatic' nature of GPS it is easy to forget about field books. These are important records and should include: details and date of the survey or setting out, theoretical and measured check point coordinates and a sketch plan of the survey.

- Don't neglect the base station – it needs to be visited regularly and its status checked.

use of GPS for setting out operations and that the reader is familiar with the contents of Chapters 7 and 8.

Suitability of total station and GPS coordinate positioning techniques

The total station and GPS coordinate positioning techniques discussed above enable points to be set out on site very accurately and quickly regardless of the terrain, and as a result they have been widely adopted on site. Their great advantage is that they can be used to set out virtually any civil engineering construction. The calculations required for this can be undertaken on a computer and the results presented either as a paper printout or uploaded into the instrument's memory for instant recall at a press of a button during the work. GPS receivers and robotic total stations have made one-person setting out a reality and even guide the operator to the exact position of the point being set out.

However, although total stations and GPS techniques are ideal for many setting out operations, they are not suitable for all. On small sites, the cost of using GPS equipment may be prohibitive and it may not be appropriate for setting out minor roads. In addition, neither total stations nor GPS receivers are best suited to situations in which precise alignment is required. For example, when setting out column centres, using total stations or GPS receivers could give unacceptable results due to their slight angular, distance or coordinate errors, causing the line joining the column centres to be outside the specified tolerance. In any alignment situation, one solution is to establish the required line using the telescope of either a theodolite or total station and then to measure along this. However, the use of two theodolites positioned at right angles, such that the intersection of their lines of sight establishes the required point, will give the best results since no distance measurement is involved. If precise reduced levels are required on site (for example, when checking the screeding on a concrete floor against a required tolerance of say ±2 mm) GPS receivers and all but the most specialized total stations would not be suitable because they could not achieve this precision. In such a case, conventional levelling would be the most appropriate technique to employ.

Reflective summary

With reference to the principles of setting out remember:

— In order to ensure that the various elements of the scheme are positioned correctly, procedures must be employed that will enable the horizontal and vertical positions of design points to be set out to the required accuracy.

— Setting out procedures fall into one of three main categories: *horizontal control methods*, *vertical control methods* and *coordinate positioning techniques*.

- Horizontal and vertical control methods are based on one of the fundamental tenets of surveying: *working from the whole to the part*. This involves the use of different levels of control to ensure that design points are always fixed from points having a higher order of precision.

- One design point should never be set out from another.

- Horizontal control points should be located as near as possible to the site in open positions for ease of working, but well away from the construction areas and traffic routes on site to avoid them being disturbed.

- Devices that can be used to establish horizontal control include *baselines*, *reference grids* and *offset pegs*.

- Devices that can be used to establish vertical control include *temporary* and *transferred bench marks*, *sight rails*, *travellers*, *boning rods*, *slope rails*, *batter rails* and *profile boards*.

- The best methods of setting out the plan positions of design points are those involving coordinates.

- Total stations and theodolites and tapes can be used to set out the plan positions of design points from a control network by *polar coordinates*, either using angles and distances or bearings and distances.

- *Intersection* methods can be undertaken using two theodolites without the need for any distance measurement equipment.

- Specialised total station systems are available where high precision is required.

- Wherever possible, each design point must be set out from at least three control points.

- The most appropriate GPS technique for setting out is a high-precision Real Time Kinematic (RTK), which is capable of achieving setting out accuracies in the region of 10 mm + 1 ppm (horizontal) and 20 mm + 1 ppm (vertical).

- Total stations and GPS techniques are suitable for many but not all setting out operations.

- The ICE and RICS have published very useful guides on the use of GPS for setting out.

11.4 Applying the principles of setting out

After studying this section you should be aware that setting out can be categorised as first stage setting out (initial site clearance and sub-surface activities) and second stage setting out (establishing the elements of the construction that are above ground level). You should understand how horizontal and vertical control together with positioning techniques can be used to set out pipelines, ground floor slabs, formwork, columns, piles and bridge abutments. You should appreciate how verticality can be controlled in multi-storey structures and you should be aware of methods by which height data can be transferred from floor to floor.

This section includes the following topics:

- Stages in setting out
- Setting out a pipeline
- Setting out a building to ground-floor level
- Setting out formwork
- Setting out column positions
- Setting out and controlling piling work
- Setting out bridge abutments
- Controlling verticality
- Transferring height from floor to floor

Stages in setting out

As the Works proceed, the setting out falls into two broad stages. Initially, techniques are required to define the site, to set out the foundations and to monitor their construction. These fall under the general heading of *first stage setting out*. Once this has been done, emphasis changes to the above-ground elements of the scheme and methods must be adopted which will ensure that they are fixed at their correct levels and positions. This falls under the general heading of *second stage setting out*. However, it must be appreciated that the division between the two stages is not always easy to define and a certain amount of overlap is inevitable.

First stage setting out

In practice, first stage setting out involves the use of many of the horizontal and vertical control methods and positioning techniques discussed in Section 11.3. The purpose of this stage is to locate the boundaries of the Works in their correct position on the ground surface and to define the major elements. In order to do this, horizontal and vertical control points must be established on or near to the site. These are then used not only to define the perimeter of the site which enables fences to be erected and site clearance to begin but also to set out critical design points on the scheme and to define slopes, directions and so on. For example, in a structural project, the main corners and sides of the buildings will be located and the required depths of dig to foundation level will be defined. In a road project, the centre line and the extent of the embankments and cuttings will be established together with their required slopes.

When the boundaries and major elements have been pegged out, the top soil is stripped and excavation work begins. During this period, it may be necessary to relocate any pegs that are accidentally disturbed by the site plant and equipment. Once the formation level is reached, the foundations are laid in accordance with the drawings and the critical design points located earlier. Setting out techniques are used to check that the foundations are in their correct three-dimensional position. The first stage ends once construction to ground floor level, sub-base level or similar has been completed. Examples of first stage setting out include setting out a pipeline and setting out a building to ground floor level, both of which are discussed later in this in section.

Second stage setting out

Second stage setting out continues on from the first stage, beginning at the ground floor slab, road sub-base level or similar. Up to this point, all the control will still be outside the main construction, for example, the pegs defining building corners, centre lines and so on will have been knocked out during the earthmoving work and only the original control will be undisturbed. Some offset pegs may remain but these too will be set back from the actual construction itself.

Second stage setting out, therefore, is concerned with the transfer of the horizontal and vertical control used in the first stage into the actual construction in order that it can be used to establish the various elements of the scheme such as internal walls and columns. It also involves transferring the control from floor to floor as the construction proceeds in order that verticality and required floor levels are maintained.

In practice, although second stage setting out does employ some of the principles of setting out discussed in Section 11.3, it sometimes requires the use of equipment that has been developed specifically for setting out purposes. Examples of second stage setting out include setting out formwork, establishing column and pile positions, setting out bridge abutments, controlling verticality in multi-storey structures and transferring height from floor-to floor, all of which are discussed later in this section.

Setting out a pipeline

The principles discussed in Section 11.3 are now considered in relation to the setting out of a gravity sewer pipeline. The whole operation falls within the category of first stage setting out.

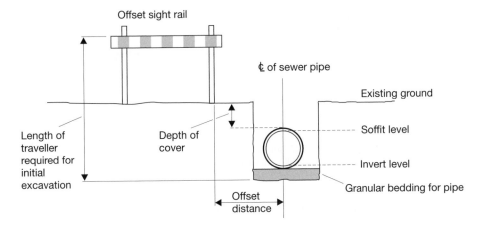

Figure 11.33 ● Sight rail for a sewer pipeline.

General considerations

Sewers normally follow the natural fall in the land and are laid at gradients which will induce a self-cleansing velocity. Such gradients vary according to the material and diameter of the pipe. Figure 11.33 shows a sight rail offset at right angles to a pipeline laid in granular bedding in a trench. Depth of cover is normally kept to a minimum, but the sewer pipe must have a concrete surround of at least 150 mm in thickness where cover is less than 1 m or greater than 7 m. This is to avoid cracking of pipes owing to surface or earth pressures.

Horizontal control

The working drawings will show the directions of the sewer pipes and the positions of manholes. The line of the sewer is normally pegged at 20 to 30 m intervals using coordinate methods of positioning from reference points or in relation to existing detail. Alternatively, the direction of the line can be set out by theodolite and pegs sighted in. *Manholes* are set out at least every 100 m and also at pipe branches and changes of gradient.

Vertical control

This involves the erection of sight rails some convenient height above the invert level of the pipe (see Figure 11.33). The method of excavation should be known in advance such that the sight rails will not be covered by the excavated material (the *spoil*). A suitable scheme for both horizontal and vertical control is shown in Figure 11.34.

Erection and use of sight rails

The sight rail upright or uprights are hammered firmly into the ground, usually offset from the line rather than straddling it. Using a nearby TBM and levelling equipment,

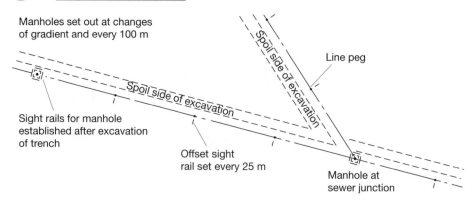

Manholes set out at changes of gradient and every 100 m

Line peg

Spoil side of excavation

Sight rails for manhole established after excavation of trench

Spoil side of excavation

Offset sight rail set every 25 m

Manhole at sewer junction

Figure 11.34 ● Layout of horizontal and vertical control for a sewer pipe system.

the reduced levels of the tops of the uprights are determined. Knowing the proposed depth of excavation, a suitable traveller is chosen and the difference between the level of the top of each upright and the level at which the top edge of the cross piece is to be set is calculated. Worked example 11.1 shows the calculations involved. Figure 11.35 shows examples of sight rails fixed in position. The excavation is monitored by lining in the traveller as shown in Figure 11.36.

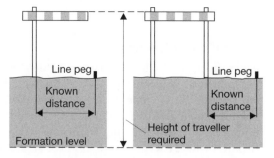

Line peg

Known distance

Formation level

Line peg

Known distance

Height of traveller required

Figure 11.35 ● Possible sight rail positions.

Where the natural slope of the ground is not approximately parallel to the proposed pipe gradient, *double sight rails* can be used as shown in Figure 11.37. Often it is required to lay storm water and foul water sewers in adjacent trenches. Since the storm water pipe is usually at a higher level than the foul water pipe (to avoid the foul water overflowing into the storm water), it is common to dig one trench to two different invert levels, as shown in Figure 11.38. Both pipe runs are then controlled using different sight rails nailed to the same uprights. To avoid confusion, the storm water sight rails are painted in different colours from the foul water ones. The same traveller is used for each pipe run. It is made with only one cross piece and is used in conjunction with the storm water or foul water sight rails as appropriate.

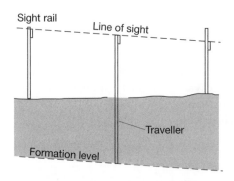

Sight rail

Line of sight

Traveller

Formation level

Figure 11.36 ● Lining in a traveller.

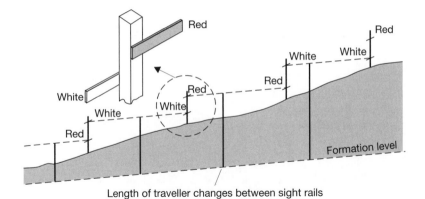

Length of traveller changes between sight rails

Figure 11.37 ● Double sight rails.

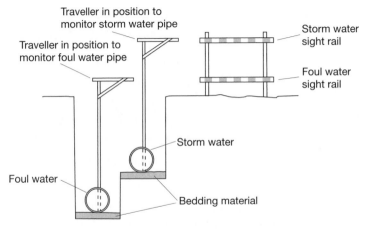

Figure 11.38 ● Setting out storm water and foul water pipes in the same trench.

Manholes

Control for manholes is usually established after the trench has been excavated, and can be done by using sight rails as shown in plan view in Figure 11.39 or by using offset pegs as shown in section in Figure 11.40.

Pipelaying

On completion of the excavation, the sight rail control is transferred to pegs in the bottom of the trench, as shown in Figure 11.41. The top of each peg is set at the invert level of the pipe and because of this such pegs are often referred to as *invert pegs*. Pipes are

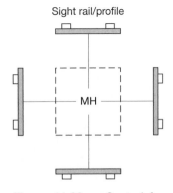

Sight rail/profile

Figure 11.39 ● Control for manholes: sight rails

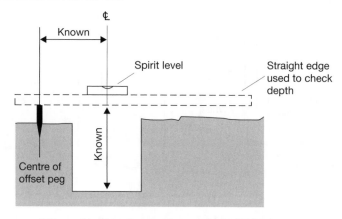

Figure 11.40 ● Control for manholes: offset pegs.

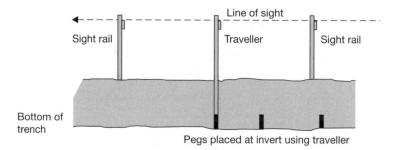

Figure 11.41 ● Setting invert pegs in a trench with a traveller.

usually laid in some form of bedding and a *pipelaying traveller* is useful for this purpose. Figure 11.42 shows such a traveller and its method of use. Pipes are laid from the lower end with sockets facing uphill. They can be bedded in using a straight edge inside each pipe until the projecting edge just touches the next forward peg or the pipelaying traveller can be used. Alternatively, three travellers can be used together, as shown in Figure 11.43. When used like this the travellers are known as *boning rods*.

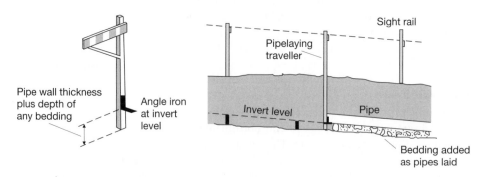

Figure 11.42 ● A pipelaying traveller and its method of use.

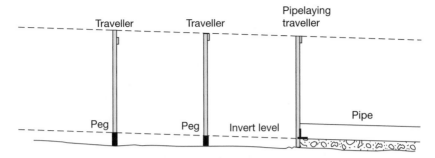

Figure 11.43 ● Boning rods.

Setting out a building to ground-floor level

This also comes into the category of first stage setting out, and the procedures required to set out the corners of a building using a theodolite and a tape are summarised below.

It is vital to remember when setting out that, since dimensions, whether scaled or designed, are almost always horizontal, slope must be allowed for in surface taping on sloping ground. The slope correction is additive when setting out, as shown in Section 4.5.

● Two corners of the building are set out from a baseline, site grid or control points using one of the methods shown in Figures 11.4, 11.5, 11.6 and 11.25.

● From these two corners, the two other corners are set out using a theodolite to turn off right angles as shown in Figure 11.44. The exact positions of each corner are then marked on the tops of wooden pegs by nails and offset pegs are established at the same time as the corner pegs (see Figure 11.8).

● The diagonals are checked as shown in Figure 11.45 and the nails repositioned on the tops of the pegs as necessary.

● Profile boards are erected at each corner or a continuous profile is used (see Figures 11.21, 11.22 and 11.23) and excavation begins. The next step is to

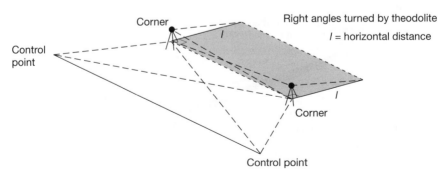

Figure 11.44 ● Setting out the corners of a building by right angles.

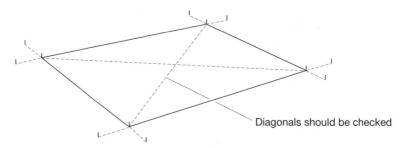

Figure 11.45 ● Checking the diagonals.

construct the foundations and these can take several forms. If concrete founda-tions have been used and a concrete ground floor slab laid, formwork is required to contain the wet concrete and this could have been set out by aligning the shut-tering with string lines strung between the profiles.

Alternatively, the four corners could be set out using total stations or GPS receivers. Using a total station, each corner would be established by bearing and distance methods from points in the control network surrounding the proposed building. If possible, each corner should be fixed from two control points and checked from a third. Using a GPS receiver, the corners would be set out directly. With both methods once the four corners have been fixed, the diagonals should be measured as a check.

Transfer of control to the ground-floor slab

Once the ground floor slab has been laid and has set, the horizontal and vertical control can be transferred to it for use in the second stage setting out operations. This is done for horizontal control by setting a theodolite and target over opposite pairs of offset pegs and using them to locate control plates on the slab, as shown in Figure 11.46. These plates are carefully marked to define the corners of the structure. For

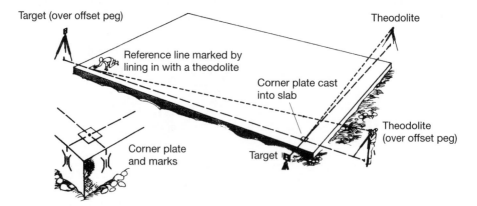

Figure 11.46 ● Transferring horizontal control to a ground floor slab.

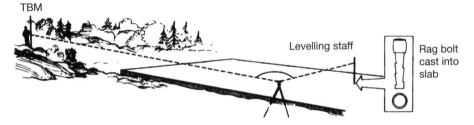

Figure 11.47 ● Transferring vertical control to a ground floor slab.

vertical control, a stable TBM is located on the slab (for example, a cast *in situ* rag bolt, as shown in Figure 11.47), and the reduced level is transferred to it by levelling from a nearby bench mark.

Setting out formwork

This is a second stage operation in which the positions where the formwork is to be located are set out with the aid of the corner plates established earlier in Figure 11.46. The formwork positions are marked on the ground floor slab with reference to the lines joining these plates, as shown in Figure 11.48. One method of marking these lines on the slab is by means of chalked string held taut and fixed at each corner position. The string is pulled vertically away from the slab and released. It hits the surface of the slab, marking it with the chalk, as shown in Figure 11.49. These slab markings are used as guidelines for positioning the formwork and should be extended to check the positioning as shown in Figure 11.50.

Setting out column positions

This is a second stage operation in which the columns should be positioned to within ±2 to 5 mm of their design position. A structural grid enables this to be achieved, and

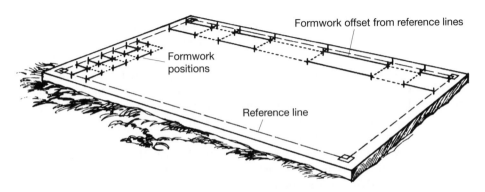

Figure 11.48 ● Setting out guidelines for formwork.

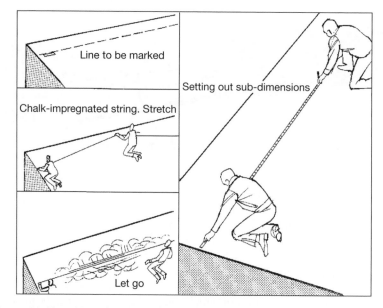

Figure 11.49 ● Use of a chalked string to mark a line on a concrete slab.

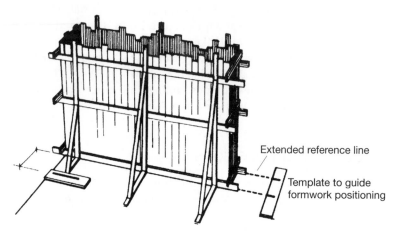

Figure 11.50 ● Positioning formwork on the guidelines.

in Figure 11.51 a grid of wooden pegs has been set out to coincide with the lines of the columns. The pegs can either be level with the ground floor slab or profile boards can be used and lines are strung across the slab between the pegs or profiles to define the column centres. If the pegs are at slab level the column positions are marked directly. If profiles are used, the lines have to be transferred to the slab surface. However they are marked, the intersections of the lines define the column centres.

Once the centres have been marked, the bolt positions for steel columns can be accurately established with a template, equal in size to the column base, placed exactly at the marked point. For reinforced concrete columns, the centres are

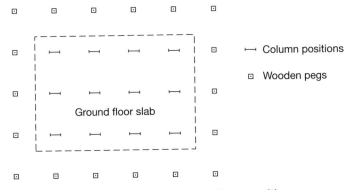

Figure 11.51 ● Setting out column positions.

established in exactly the same way, but usually prior to the slab being laid so that the reinforcing starter bars can be placed in position.

Setting out and controlling piling work

The equipment used in piling disturbs the ground, takes up a lot of space and obstructs sightings across the area. Hence it is not possible to establish all the pile positions before setting out begins, since they are very likely to be disturbed during construction. Either GPS techniques can be employed or a combination of ground-based methods (baselines plus total station or theodolite and tape positioning techniques) can be used to overcome this difficulty as follows.

● Before piling begins a baseline is decided upon and the lengths and angles necessary to set out the pile positions from each end of the baseline are calculated from the coordinates of the ends of the baseline and the design coordinates of the pile positions. The position of the baseline must be carefully chosen so that measurements taken from each end will not be hindered by the piling rig. Figure 11.52 shows a suitable scheme.

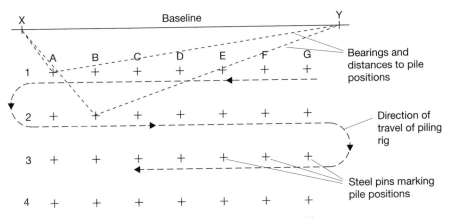

Figure 11.52 ● Setting out pile positions.

- Each pile position is set out from one end of the baseline and checked from the other. Either a total station and a detail pole or a theodolite and a steel tape are used. If the distances are longer than a tape length the total station method will be quicker and more accurate.

- The initial two or three positions are set out and the piling rig follows the path shown in Figure 11.52.

- The engineer goes on ahead and establishes the other pile positions as work proceeds.

- A variation is to use two baselines on opposite sides of the area and establish the pile positions from four positions instead of two.

Setting out bridge abutments

Structures such as bridge abutments can be set out by a combination of horizontal control methods and coordinate positioning techniques. Figure 11.53 shows the plan view of a bridge to carry one road over another. With reference to this, the procedure is as follows.

- The centre lines ℄ of the two roads are set out by one of the methods discussed in Chapters 12 and 13.

- The bridge is set out in advance of the road construction. If GPS techniques are to be used, the abutment points A, B, C and D can be set out directly. However, if total stations or theodolites and tapes are to be used then it will be necessary to establish secondary site control points around the area containing the abutments. These secondary points could either be in the form of a structural grid (see

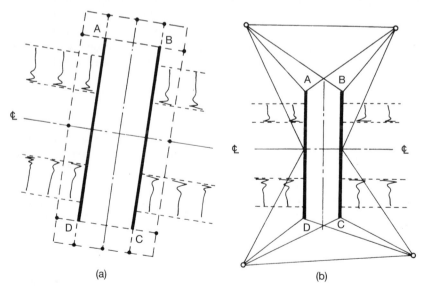

(a)   (b)

Figure 11.53 ● Setting out bridge abutments (a) using a structural grid; (b) by coordinates.

Figure 11.53(a)), which itself is set up from main site control stations by bearings and distances, or in the form of a control network from which points A, B, C and D can be set out by bearings and distances (see Figure 11.53(b)). Whichever method is used, all the points must be permanently marked and protected to avoid their disturbance during construction, and positioned well away from the traffic routes on site.

- TBMs are set up as separate levelled points or a control point can be levelled and used as a TBM.

- If a structural grid in used as shown in Figure 11.53(a), the distances from the secondary site control points to abutment design points A, B, C and D must first be calculated. They are then set out either using a theodolite to establish the directions and steel tapes to measure the distances or by using a total station. If coordinates are used as shown in Figure 11.53(b), the bearings and distances from the secondary site control points to A, B, C and D are calculated from their respective coordinates such that each design point can be established from at least two control points.

- Once points A, B, C and D have been set out, their positions should be checked by measuring between them and also measuring to them from control points not used to establish them initially.

- Offset pegs are established for each of A, B, C and D to allow excavation and foundation work to proceed and to enable the points to be relocated as and when required.

- Once the foundations are established, the formwork, steel or precast units can be positioned with reference to the offset pegs.

For multi-span bridges, the structural grid shown in Figure 11.54 can be established from the site grid or traverse stations and the centres of the abutments and piers set out from this. Since points A to P in Figure 11.54 may be used many times during the construction, they should be positioned well away from site traffic and site operations and permanently marked and protected.

Each pier can be established by setting out from its centre position (Z) using offset pegs and profiles to mark the excavation area as shown in Figure 11.55. As

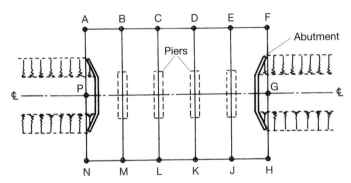

Figure 11.54 ● Setting out bridge abutments and piers from a structural grid.

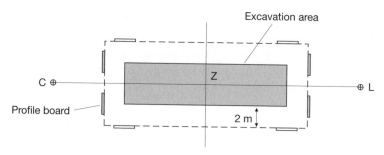

Figure 11.55 ● Setting out a bridge pier.

construction proceeds, the required levels of the tops of the piers and the subsequent deck can be established from TBMs set up nearby either by conventional levelling techniques or by using a weighted steel tape as shown in Figure 11.69.

Controlling verticality

One of the most important second stage setting out operations is to ensure that those elements of the scheme which are designed to be vertical are actually constructed to be so, and there are a number of techniques available by which this can be achieved. Several are discussed below and particular emphasis is placed on the control of verticality in multi-storey structures. In order to avoid repeating information given earlier in this chapter, the following assumptions have been made.

● Offset pegs have been established to enable the sides of the building to be re-located as necessary.

● The structure being controlled has already had its ground floor slab constructed and the horizontal control lines have already been transferred to it as shown in Figure 11.46.

The principle behind controlling verticality is very straightforward: if the horizontal control on the ground floor slab can be accurately transferred to each higher floor as construction proceeds, then verticality will be maintained. In order to ensure that tall structures are vertical and do not twist, the four points marked on the corner plates shown in Figure 11.46 should be transferred vertically from floor-to-floor as the construction proceeds.

Depending on the heights involved, there are several different ways of achieving verticality. For single-storey structures, long spirit levels can be used quite effectively, as shown in Figure 11.56. For multi-storey structures, however, one of the following techniques is preferable:

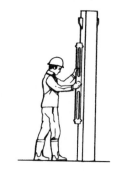

Figure 11.56 ● Plumbing a column using a long spirit level.

- Plumb-bob methods
- Theodolite methods
- Optical plumbing methods
- Laser methods

The basis behind all these methods is the same. They each provide a means of transferring points vertically. Once four suitable points have been transferred, they can be used to establish a square or rectangular grid network on the floor in question which can be used to set out formwork, column centres, internal walls and so on at that level. Plumb-bobs, theodolite methods and optical plummets are discussed below. Laser methods are discussed in Section 11.5.

Plumb-bob methods

The traditional method of controlling verticality is to use *plumb-bobs*, suspended on piano wire or nylon. A range of weights is available (from 3 kg to 20 kg) and two plumb-bobs are needed in order to provide a reference line from which the upper floors may be controlled.

In an ideal situation, the bob is suspended from an upper floor and moved until it hangs over a datum reference mark on the ground floor slab. If it is impossible or inconvenient to hang the plumb-bob down the outside of the structure, holes and openings must be provided in the floors to allow the plumb-bob to hang through, and some form of centring frame will be necessary to cover the opening to enable the exact point to be fixed. Service ducts can be used, but often these are not conveniently placed to provide a suitable baseline for control measurements. It is also not always possible to use a plumb-bob over the full height of a building owing to the need to 'finish' each floor as work progresses, for example, the laying of a concrete screed would obliterate the datum reference mark.

Unfortunately, the problem of wind currents in the structure usually causes the bob to oscillate, and the technique can be time-consuming if great accuracy is required. To overcome this, two theodolites, set up on lines at right angles to each other, could be used to check the position of the wire and to estimate the mean oscillation position. However, limited space or restricted lines of sight may not allow for the setting up of theodolites and their use tends to defeat the object of the *simple* plumb-bob.

One partial solution to dampening the oscillations is to suspend the bob in a transparent drum of oil or water. However, this tends to obscure the ground control mark being used, and if this occurs it becomes necessary either to reference the plumbline to some form of staging built around or above the drum or to measure offsets to the suspended line. This is shown in Figure 11.57, in which a freshly concreted wall is being checked for verticality. The plumb-bob is suspended from a piece of timber nailed to the top of the formwork and immersed in a tank of oil or water. Offsets from the back of the formwork are measured at top and bottom with due allowance for any steps or tapers in the wall. Any necessary adjustments are made with a push–pull prop.

Increasing the weight of the bob reduces its susceptibility to oscillations, but these are rarely eliminated completely. Plumb-bobs are very useful when constructing lift

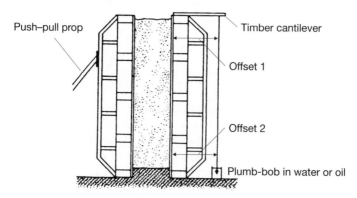

Figure 11.57 ● Use of plumb bob to control verticality of a cast-*in situ* concrete wall (reproduced from the CIRIA book *Setting Out Procedures*, 1997).

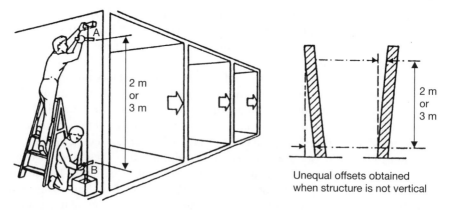

Figure 11.58 ● Use of a plumb bob to control verticality in a single-storey structure (based on BS5606:1990 with permission of the BSI).

shafts and they are ideal for heights of one or two storeys. Figure 11.58 shows a plumb-bob being used to check the verticality of a single-storey structure in which offsets A and B will be equal when the structure is in a vertical position.

The advantages of plumb-bobs are that they are relatively inexpensive and straight-forward in use. They are particularly useful for monitoring verticality over short distances, for example, when erecting triangular timber frames, a wire can be stretched across the base of the frame and a plumb-bob attached to the apex of the triangle. To ensure that it is erected in a vertical position, the frame is simply pivoted about its base until the suspended plumb-bob touches the centre of the stretched wire.

Theodolite methods

These methods assume that the theodolite is in perfect adjustment so that its line of sight will describe a vertical plane when rotated about its tilting axis. Three methods are discussed below.

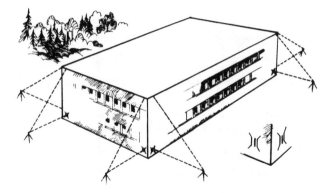

Figure 11.59 ● Transfer of control in a multi-storey structure.

Controlling a multi-storey structure using a theodolite only

The theodolite is set up on extensions of each reference line marked on the ground floor slab in turn and the telescope is sighted on to the particular line being transferred. The telescope is elevated to the required floor and the point at which the line of sight meets the floor is marked. This is repeated at all four corners, and eight points in all are transferred, as shown in Figure 11.59. Once the eight marks have been transferred, they are joined and the distances between them and their diagonal lengths are measured as checks.

If the centre lines of a building have been established, a variation of this method is to set up a theodolite on each in turn and transfer four points instead of eight, as shown in Figure 11.60. This establishes two lines at right angles on each floor from which measurements can be taken.

If the theodolite is not in perfect adjustment, the points must be transferred using both faces and the mean position used. In addition, because of the large angles of elevation involved, the theodolite must be carefully levelled and a diagonal eyepiece

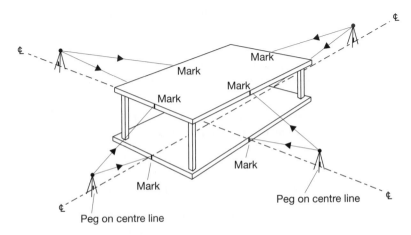

Figure 11.60 ● Transfer of centre lines.

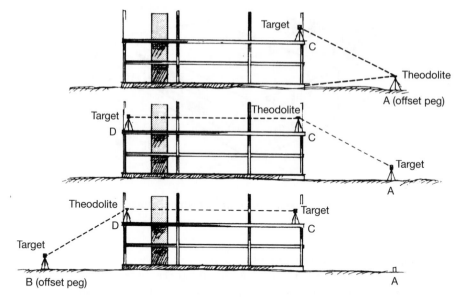

Figure 11.61 ● Transfer of control by three-tripod traversing.

attachment may be required to enable the operator to look through the telescope (see Figure 11.64).

Controlling a multi-storey structure using a theodolite and targets
In Figure 11.61, A and B are offset pegs. The procedure is as follows.

● The theodolite is set over offset peg A, carefully levelled and aligned on the reference line marked on the side of the slab (see Figure 11.46).

● The line of sight is transferred to the higher floor and a target accurately positioned at point C.

● A three-tripod traverse system is used (see Section 6.4) and the target and theodolite are interchanged. The theodolite, now at C, is sighted onto the target at A, transited and used to line in a second target at D. Both faces must be used and the mean position adopted for D.

● A three-tripod traverse system is again used between C and D and the theodolite checks the line by sighting down from D to the reference mark at B, again using both faces.

● It may be necessary to repeat the process if a slight discrepancy is found.

● The procedure is repeated along other sides of the building.

Again, the two centre lines can be transferred instead of the four reference lines if this is more convenient.

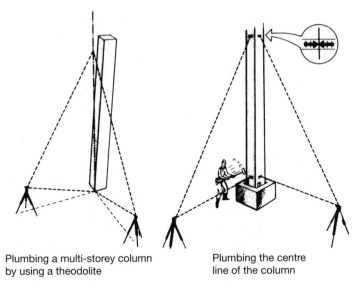

Plumbing a multi-storey column
by using a theodolite

Plumbing the centre
line of the column

Figure 11.62 ● Control of column verticality using two theodolites.

Controlling column verticality using theodolites only

Although short columns can be checked by means of a long spirit level held up against them, as shown earlier in Figure 11.56, long columns are best checked with two theodolites, as shown in Figure 11.62. Either the edges or, preferably, the centre lines of each column are plumbed with the vertical hairs of two theodolites by elevating and depressing the telescopes. The theodolites are set up directly over the necessary control lines and because of the potentially high angles of elevation, they must be very carefully levelled. The use of diagonal eyepieces is recommended. In addition, because they may not be in perfect adjustment, the verticality of the column must be checked using both faces of the theodolites.

It is not always necessary to use two theodolites, and Figure 11.63 shows the formwork for a tall column being plumbed by a single instrument set up in two positions. The theodolite is first set up on a plane parallel to but offset from one face of the formwork and sighted on suitable offset marks at the top and bottom. Ideally the theodolite should be some distance away to

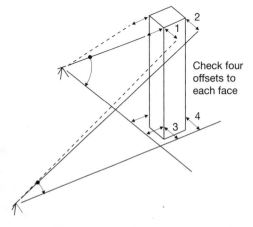

Check four
offsets to
each face

Figure 11.63 ● Control of column verticality using one theodolite (reproduced from the CIRIA book *Setting Out Procedures*, 1997).

avoid very steep angles of elevation, but this is not always possible. Observations should be taken to the top and bottom of the face of the column as a check on twisting (that is, offsets 1, 2, 3 and 4 should all be equal) and, as a further precaution, both faces of the theodolite should be used. Any discrepancy between offsets 1, 2, 3 and 4 should be adjusted and the column face rechecked. Once this face has been plumbed, the theodolite is moved and the whole procedure is repeated for the adjacent column face.

Optical plumbing methods

Optical plumbing can be undertaken in several ways. Either the optical plummet of a theodolite can be used, or the theodolite can be fitted with a diagonal eyepiece or an optical plumbing device specially manufactured for the purpose can be employed. When carrying out optical plumbing, holes and openings must be provided in the floors and a centring frame must be used to establish the exact position.

Using the optical plummet of a theodolite

The optical plummet of a theodolite provides a vertical line of sight in a downwards direction which enables the instrument to be centred over a ground mark. Optical plummets are usually incorporated into all modern theodolites, although some manufacturers now incorporate visible laser plummets instead, which serve exactly the same purpose but do not require the operator to look into a telescope. However, there are also special attachments which fit into a standard tribrach and enable high-accuracy centring to be obtained not only to reference marks below the instrument but also to control points above the instrument, for example, in the roof of a tunnel. These are *optical roof and ground point plummets* which enable centring to be achieved to ±0.3 mm over a distance of 1.5 m. On some instruments, a switch-over knob permits a selection between ground or roof point plumbings. After centring has been achieved, the plummet is replaced by the instrument or target which, by virtue of the system of forced centring, is now correctly centred. These devices are meant for short-range work only and do not provide a vertical line of sight of sufficient accuracy to control a high-rise structure.

Using a diagonal eyepiece

Diagonal eyepiece attachments are available for most theodolites. These are interchanged with the conventional eyepiece and enable the operator to look through the telescope while it is inclined at very high angles of elevation, as shown in Figure 11.64.

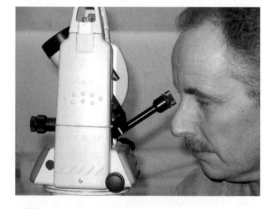

Figure 11.64 ● Using a diagonal eyepiece.

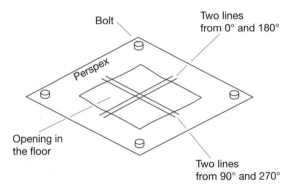

Figure 11.65 ● Perspex target.

They can be used to transfer control points upwards to special targets, either up the outside of the building or through openings left in the floors. The procedure is as follows.

● The theodolite with the diagonal eyepiece attached is centred and levelled over the point to be transferred as normal using its built-in optical plummet.

● The theodolite is rotated horizontally until the horizontal circle reads 00°.

● The telescope is tilted until it is pointing vertically upwards. If an electronic reading instrument is being used, the display will indicate when the telescope is vertical. In the case of an optical reading instrument, an additional diagonal eyepiece must be fitted to the optical reading telescope to enable it to be read.

● A perspex target, as shown in Figure 11.65, is placed over the hole on the upper floor and an assistant is directed by the theodolite operator to mark a line on the perspex which coincides with the image of the horizontal cross hair in the telescope.

● The theodolite is rotated horizontally until the horizontal circle reads 180°.

● The telescope is tilted to give a vertical line of sight once again.

● The perspex target is again viewed. If the instrument is in correct adjustment, the horizontal hair of the telescope will coincide with the line already drawn on the target. If this is not the case, the assistant is directed to mark another line on the target corresponding to the new position of the horizontal hair, as shown in Figure 11.65. A line mid-way between the two will be the correct one.

● The whole procedure is now repeated with the horizontal circle of the theodolite reading first 90° and then 270°. The mean of the two lines obtained from these values will be the correct one. The transferred point lies at the intersection of this mean line and that obtained in the 00° and 180° positions.

If care is taken with this method the accuracy can be high, with precisions of ±1 mm in 30 m being readily attainable. However, it is essential that the horizontal hair is used for alignment because its mean position in this procedure is unaffected by any non-verticality of the vertical axis, which is not the case with the vertical hair.

Using special optical plumbing devices

This is the preferred optical plumbing method and there are several variations of these available. Some can only plumb upwards, some only downwards and some both. The accuracy attainable is extremely high: for example, the Wild ZNL zenith and nadir plummet shown in Figure 11.66 is capable of achieving precisions of 1:30,000 (1 mm in 30 m) and the Wild ZL zenith plummet and the Wild NL nadir plummet are each capable of achieving precisions of 1:200,000 (0.5 mm at 100 m). Such high accuracy is due to these instruments being fitted with compensator devices similar to those used in automatic levels. They are first approximately levelled using their small

Figure 11.66 ● Wild ZNL zenith and nadir plummet (courtesy Leica Geosystems).

circular levels and their compensators then take over. This ensures a much higher degree of accuracy in the vertical line of sight than can be achieved using a theodolite fitted with a diagonal eyepiece.

In use, the plummet is first centred over the ground point to be transferred and then used with a special target in a similar way to that described above for diagonal eyepieces. Some plummets have their own centring system to enable them to be set over the point while others are designed to fit into a standard tribrach which has previously been centred over the ground mark. As with diagonal eyepieces, the control should always be transferred at the four major points of the circle, that is 00°, 90°, 180° and 270°, and the mean position used. Even if the plummet is in correct adjustment, this should still be carried out as a check. The handbook supplied with each plummet describes how it can be set to perfect adjustment.

Figures 11.67(a) and (b) show an optical plummet being used to plumb upwards to transfer control to a special centring device on the floor above. This device is fitted with an index mark which is moved as necessary by an assistant until it defines the transferred point. As it is difficult for the observer and assistant to keep in touch over a large number of storeys, the use of two-way radios is recommended. As control is transferred upwards, the perspex target is moved to the next floor, as shown in Figure 11.68.

Figure 11.67(c) shows an optical plummet being used to plumb downwards in order to transfer control up from the floor below. In this case, the optical plummet is adjusted until the reference mark on the floor below is bisected by its cross hairs. The centring device is then set in place beneath the tripod and moved until the line of sight passes through its centre.

Once at least three and preferably four points have been transferred, a grid can be established on the floor by offsetting from the transferred points. The offset points chosen for the grid should be sited away from any columns so that they can be used to fix the position of any formwork. When this has been completed, the centring devices can be removed and replaced by safety plates to avoid accidents.

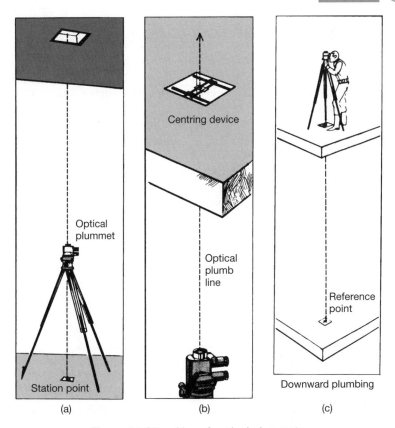

Figure 11.67 ● Use of optical plummets.

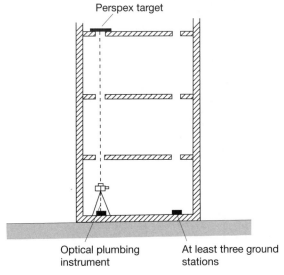

Figure 11.68 ● Plumbing a multi-storey building (reproduced from the CIRIA book *Setting Out Procedures*, 1997).

Transferring height from floor to floor

Reduced levels must be transferred several times during the second stage setting out operations as the construction proceeds from floor to floor. One method by which this can be done is to use a weighted steel tape to measure from a datum in the base of the structure as shown in Figure 11.69.

The base datum levels should be set in the bottom of lift wells, service ducts and so on, such that an unrestricted taping line to roof level is provided. The levels should be transferred to each new floor by always measuring from the datum rather than from the previous floor. This is the principle of *working from the whole to the part* as discussed in Section 11.3 and ensures that errors in transferring level to one floor are not carried over to another floor. Worked example 4.3 covers the calculations involved in this method.

Each floor is then provided with TBMs in key positions from which normal levelling methods can be used to transfer levels on each floor.

Alternatively, if there are cast-*in situ* stairs present, a level and staff can be used to level up and down the stairs, as shown in Figure 11.70. Note that both up and down levelling must be done as a check.

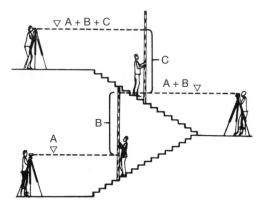

Figure 11.70 ● Transfer of height from floor to floor using levelling.

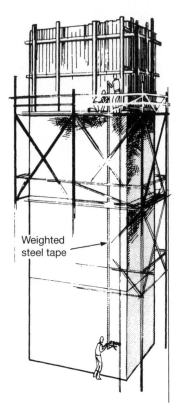

Weighted steel tape

Figure 11.69 ● Transfer of height from floor to floor using a steel tape.

Reflective summary

With reference to applying the principles of setting out remember:

— As the Works proceed, setting out falls into two broad stages: first stage and second stage.

— Both stages involve the horizontal and vertical control methods together with positioning techniques. However, second stage can also require the use of equipment that has been developed specifically for setting out purposes.

— Before the first stage can begin, horizontal and vertical control points must be established on or near the site.

— First stage setting out normally deals with defining the site, establishing critical design points on site, setting out foundations and monitoring their construction.

— Examples of first stage setting out include start-up activities such as erecting fences around the site and stripping the topsoil as well as specific projects such as setting out a pipeline and setting out a building to ground floor level.

— Second stage setting out normally deals with establishing the elements above ground level at their correct levels and positions.

— Examples of second stage setting out include transferring control into a structure, setting out formwork, establishing column and pile positions, setting out bridge abutments, controlling verticality and transferring height from floor-to-floor.

— Verticality of structures can be controlled by purpose-built equipment such as plumb-bobs and optical plumbing devices and by multi-purpose instruments such as theodolites and lasers.

11.5 Setting out using laser instruments

After studying this section you should know that the laser instruments used in surveying and setting out have beams which are either visible or invisible. You should understand the safety classification system used to categorise individual laser surveying instruments and be aware that you must not use a laser instrument if it does not have the correct labels on it. You should appreciate that there are generally two different types of laser surveying instruments, either alignment lasers or rotating lasers, each of which has been developed for different types of applications. You should understand how alignment lasers can be used to control verticality, pipelaying and tunnelling and you should appreciate the use of rotating lasers for general site levelling and setting out internal fittings.

This section includes the following topics:

- Visible and invisible beam lasers
- Safety aspects of laser beams
- Alignment lasers and their applications
- Rotating lasers and their applications

Visible and invisible beam lasers

Although a detailed description of laser techniques and equipment is beyond the scope of this book, laser instruments are now widely used in setting-out operations and a few of the more common methods are discussed in this section.

A laser generates a beam of high intensity and of low angular divergence; hence it can be projected over long distance without spreading significantly. These characteristics are utilised in specially designed laser surveying equipment and it is possible to carry out many alignment and levelling operations by laser.

There are two types of laser used in surveying equipment: those that generate a *visible beam* and those that generate an *invisible beam*.

Visible beam lasers

Originally, all visible beam lasers used in surveying equipment were produced from a mixture of helium and neon (HeNe) gas housed inside a glass tube. The beams produced had a wavelength of 632.8 nm, giving them their characteristic red colour. The need to accommodate and protect the glass tube meant that HeNe laser surveying instruments were rather cumbersome and heavy, and they also required an external power supply. In 1988, however, Toshiba produced the world's first commercially available visible laser diode using an indium gallium aluminium phosphorus (InGaAIP) diode source, which produced a red beam of wavelength of 670 nm. This effectively marked the beginning of the end of the use of HeNe lasers in surveying instruments and since then visible laser diode sources have become the norm due to their much smaller size and weight. Their power consumption is also much less than that required to generate HeNe lasers, which enables them to be operated using small built-in rechargeable or replaceable battery packs. Indeed, since their introduction in 1988, visible laser diodes have been refined to such an extent it is no longer possible to purchase a new HeNe laser surveying instrument, although many older HeNe models are still in active use.

Up to 1996, all the visible beams used in surveying instruments were red, but in that year Topcon produced the first green beam laser and nowadays both red and green beam lasers are available. The advantage of green over red is that the human eye is approximately four times more sensitive to green light than it is to red light, which means that green is about four times more visible. Hence it is much easier to see a green laser beam, particularly when using it in sunlight or in bright ambient light conditions where viewing a red beam may be difficult. Even interior work in dark rooms is improved by the brighter green beam. The main disadvantage of green beam instruments is that they are more expensive than their red counterparts.

Visible beam instruments are generally used for setting out applications in which precise alignment is required: for example, controlling verticality, pipelaying and tunnelling, and also for levelling and grading purposes. Their beams can either be detected by eye or intercepted on translucent targets. Special handheld or rod-mounted photoelectric detectors can also be used.

Invisible beam lasers

The invisible beam lasers used in surveying instruments have always been produced from semiconductor laser diodes which produce beams having wavelengths beyond those in the visible spectrum (for example 905 nm and 720 nm). Like their visible beam counterparts, invisible beam laser diodes are small, lightweight and can be operated using on-board rechargeable or replaceable batteries. However, surveying instruments incorporating them are generally restricted to levelling and grading applications, since special photoelectric detectors are always required to locate the beams. The fact that their beams are invisible makes such instruments difficult to use for alignment applications.

Safety aspects of laser beams

From a safety point of view, the lasers used in surveying are low power with outputs ranging from less than 1 to 5 mW. This range represents absolutely no hazard when the beam or its reflection strikes the skin or clothes of anyone in the vicinity. However, an output of 1 to 5 mW presents a serious hazard to the eyes and on no account should anyone look directly into the beam of a laser in this range.

Depending on their output power, some lasers are safer than others and every laser must carry a label to indicate its safety classification. In Europe and throughout much of the world (but not currently in the USA), lasers are classified according to a system specified by the International Electrotechnical Commission (IEC) in its standard IEC 60825-1: *Safety of laser products – Part 1: Equipment classification, requirements and user's guide*. The latest version of this standard, which was published in August 2001, classifies lasers according to their *accessible emission limit* (AEL). The AEL is the level of radiation to which a user is exposed when using the laser under normal operating conditions. IEC 60825-1 specifies the following seven-point classification system:

- *Class 1 lasers* are safe under reasonably foreseeable conditions of operation, including the use of optical instruments, such as a theodolite telescope, for viewing the beam directly (known as *intrabeam* viewing). *In summary, Class 1 lasers represent no risk to the eyes.*

- *Class 1M lasers* are safe under reasonably foreseeable conditions of operation, but may be hazardous if the user employs magnifying optics within the beam. Hence *intrabeam* viewing (see Class 1 above) should not be performed on a Class 1M laser. *In summary, Class 1M lasers represent no risk to naked eyes.*

- *Class 2 lasers* are safe under reasonably foreseeable conditions of operation in that the user's eyes are protected by their natural reaction to blink when presented with a bright light. This so-called *blink reflex* provides adequate protection by preventing too much laser energy from entering the eyes and damaging the retinas. This reduces any exposures to momentary ones, and as long as the 'blink reflex' is allowed to function it should provide adequate protection for intrabeam viewing, either with the naked eye or through a magnifying instrument such as a theodolite telescope. However, overriding the blink reflex must be avoided and you must not stare directly into the beam of a Class 2 laser. In addition, the beam should not be directed at other people or into areas where other people unconnected with the laser work might be present. *In summary, Class 2 lasers represent no risk to the eyes for any momentary exposures that are controlled by the blink reflex, either when viewing the beam directly or when viewing it through magnifying optics.*

- *Class 2M lasers* are safe under reasonably foreseeable conditions of operation in that like Class 2 lasers the user's eyes are protected by the blink reflex when the beam is viewed with the naked eye. However, they differ from Class 2 in that their beams may be hazardous if the user employs magnifying optics within the beam. Hence intrabeam viewing should not be performed on a Class 2M laser. As for Class 2 lasers, the beam should not be directed at other people or into areas where other people unconnected with the laser work might be present. In addition, the beam should always be terminated at a suitable non-specular (non-mirror-like) surface. *In summary, Class 2M lasers represent no risk for any momentary exposures to naked eyes that are controlled by the blink reflex.*

- *Class 3R lasers* are more powerful. The eye's blink reflex no longer affords sufficient protection, so all intrabeam viewing must be avoided either with the naked eye or through magnifying optics such as a theodolite telescope. Viewing the image of the beam on a non-reflective (diffuse) surface such as concrete, stone, wood and so on, either with the naked eye or through magnifying optics would normally not be hazardous.

  Certain precautions are necessary when using a Class 3R laser. For example, to avoid members of the public or site personnel being exposed to the beam, it should be operated well above or below eye level, the beam should not be unintentionally directed at highly reflective flat or concave surfaces such as mirrors or windows, and it should not be left unattended on site. As with Class 2 lasers, the beam should not be directed at other people or into areas where other people unconnected with the laser work might be present.

In addition, in outdoor surveying environments where people are passing by, a risk assessment should be undertaken to ascertain whether their eyes could be exposed to the beam. Such a risk is unlikely to be present when using a Class 3R pipe laser at the bottom of a manhole, but may be present when operating a total station fitted with a Class 3R laser in an open environment. If such as risk is considered to be present then it is advisable to use warning signs to alert passers-by to the fact that a Class 3R laser is being operated in the area. Any signs used should be clearly marked and easy for the general public to understand. When not in use, a Class 3R laser should be stored where unauthorised personnel cannot gain access.

Details of these and other precautions that may be necessary when using a Class 3R laser should be specified in the manufacturer's handbook accompanying the instrument. They can also be found in IEC Technical Report IEC 60825-14: *Safety of laser products – Part 14: A user's guide*, which was published in February 2004 to provide guidance on best practice in the safe use of laser products that conform to IEC 60825-1. *In summary, Class 3R lasers represent no risk to naked eyes when their beams are viewed on diffuse surfaces.*

● *Class 3B lasers* are the most powerful used in surveying equipment. Even momentary eye exposure can be hazardous, and staring at or viewing the beam optically must be avoided at all costs. The viewing of diffuse reflections represents only a low risk. There is a very small chance that skin injury may occur, but only if the beam is focused onto a minute spot. Approved safety eyewear must be worn and protective clothing may be required and the instrument must be fitted with a key control. Class 3B lasers should not be used without first carrying out a risk assessment to determine the protective control measures necessary to ensure safe operation. For surveying operations it is usual to rope off the area of operation and to erect special warning notices. Details of these and other precautions that may be necessary when using a Class 3B laser can be found in the IEC Technical Report IEC 60825-14 mentioned above for Class 3R lasers and should also be specified in the manufacturer's handbook accompanying the instrument. *In summary, Class 3B lasers represent no risk to eyes if approved safety eyewear is worn.*

● *Class 4 lasers* are the most hazardous and can cause injury not only to the eyes but also to the skin; they may also represent a fire hazard. They are not used in surveying instruments.

In the USA a comparable, but not identical, classification system is specified by the Center for Devices and Radiological Health (CDRH), which is a division of the US Food and Drug Administration (FDA). Details of this can be found in the US Federal Laser Product Performance Standard 21 of the Code of Federal Regulations (21 CFR) and is as follows: Class I, Class IIa, Class II, Class IIIa, Class IIIb and Class IV. In addition to the CDRH, the American National Standards Institute (ANSI) publishes its own slightly different laser safety classification system in its document designated ANSI Z136.1. Hence there are currently two laser classification systems in use in the USA which are slightly different from each other. Fortunately, however, the CDRH is in the process of adopting the IEC classification system and the ANSI has expressed a similar aim. Therefore within the next few years it is possible that the IEC classification system will have been adopted throughout the USA.

Dazzle

Although Class 1, 1M, 2 and 2M visible lasers all at least represent no risk for any momentary exposures to naked eyes, they can inadvertently cause alarm and possibly harm to passers by. This is due to the brightness of their beams, which may dazzle someone who is not aware that they are being used. Such distractions are potentially very dangerous if the person who is dazzled is driving a vehicle. Consequently, all visible beam lasers should be used with care and positioned well away from areas where dazzling could occur. If they have to be used near to traffic routes then it may be necessary to erect protective screens to avoid any possibility of their beams distracting drivers.

Labels required on lasers

Every laser should be clearly marked with labels indicating its safety class and any specific precautions that must be exercised by the user. When using surveying instruments incorporating lasers on site, three types of label may be required: all lasers must have an *explanatory label*, most must also have the triangular black and yellow laser sunburst *warning label* as shown in Figure 11.71, and some must have an *aperture label* placed close to the aperture through which the laser is being emitted. The aperture label shall bear the words:

Figure 11.71 ● Laser warning sign: black triangle and sunburst on a yellow background.

LASER APERTURE

or

AVOID EXPOSURE – LASER RADIATION IS EMITTED FROM THIS APERTURE

IEC 60825-1 states that the labels should be permanently fixed, legible and clearly visible during operation, maintenance or service, according to their purpose. They should also be positioned such that they can be read without the necessity for human exposure to laser radiation in excess of the maximum accessible emission level for a Class 1 laser. If these labels are missing, the laser should not be used. Instead it should be returned to the hire company or the firm from which it was purchased to have its labels replaced. The types of labels required and examples of the wording to be used on the explanatory labels for each class of laser are shown in Table 11.2.

Safety summary

Class 1 lasers are perfectly safe and, providing care is taken with their use and the instructions given on their labels are followed correctly, Class 1M, 2 and 2M lasers

Table 11.2 ● Labelling on laser instruments used in surveying and setting out.

Laser Class	Types of label required	Wording on the explanatory label
1	Explanatory	CLASS 1 LASER PRODUCT
1M	Explanatory	LASER RADIATION DO NOT VIEW DIRECTLY WITH OPTICAL INSTRUMENTS CLASS 1M LASER PRODUCT
2	Warning & explanatory	LASER RADIATION DO NOT STARE INTO BEAM CLASS 2 LASER PRODUCT
2M	Warning & explanatory	LASER RADIATION DO NOT STARE INTO THE BEAM OR VIEW DIRECTLY WITH OPTICAL INSTRUMENTS CLASS 2M LASER PRODUCT
3R	Warning, aperture & explanatory	LASER RADIATION AVOID DIRECT EYE EXPOSURE CLASS 3R LASER PRODUCT
3B	Warning, aperture & explanatory	LASER RADIATION AVOID EXPOSURE TO THE BEAM CLASS 3B LASER PRODUCT

should also be safe to use under normal operating conditions. Class 3R and Class 3B lasers require additional safety precautions to be implemented and guidance on these can be found in the IEC Technical Report IEC 60825-14: *Safety of laser products – Part 14: A user's guide.* This should also be clearly specified in the manufacturer's handbooks.

If used correctly, laser surveying instruments should not represent any significant hazard, and as is discussed in the following sections, they offer considerable savings in both the time required to complete setting out tasks and the personnel costs involved. However, should further information be required on any aspect of the safe use of lasers, full details can be found in the excellent book *Laser Safety* by Roy Henderson and Karl Schulmeister, which has been used as a source for some of the information given in this section and is referenced at the end of this chapter.

The majority of laser instruments used in surveying incorporate beams of Class 1, 2 or 3R and they fall into two main categories: either *alignment lasers* or *rotating lasers.* These are discussed in the following sections.

Alignment lasers and their applications

This category of laser instrument uses diodes to produce single red or green visible beams. For alignment purposes, these have the important advantage of producing constantly present visible reference lines that can be used without interrupting the construction works. A number of different instruments are manufactured for alignment purposes; examples include *laser theodolites, precision laser plummets, pipe lasers* and *tunnel lasers.*

Laser theodolites are a combination of a laser and a theodolite, one example being Sokkia's LDT50 laser digital theodolite shown in Figure 11.72. In this instrument a red beam of wavelength 635 nm is generated to coincide exactly with the line of collimation. On looking into the eyepiece, the observer sees a reflection of the beam which is perfectly safe. Two modes of operation are possible with the LDT50: either the beam can be focused using the telescope focusing screw as required or a parallel-sided beam can be generated. In addition, the laser power output can be set to one of two different settings, either 1 mW (Class 1) or 2.5 mW (Class 3R) to increase the working range of the instrument. For example, in cloudy daytime conditions with the laser set to 1 mW output power the working range is over 200 m, whereas when set to 2.5 mW output power and working

Figure 11.72 ● Sokkia LDT50 laser digital theodolite (courtesy Sokkia Ltd).

in dark conditions such as those found in tunnels and underground sites, the working range increases to over 400 m. In addition to the laser feature, the LDT50 has full theodolite capabilities. It can be fully transited and has an angle reading resolution of 1" with an accuracy of ±5". The LDT50 can operate continuously from an AC power supply, for more than 5 hours using its built-in rechargeable battery or for more than 30 hours using an optional external 6 V DC battery.

Laser theodolites can be used in place of conventional theodolites in almost any alignment or intersection technique and, once set up, the theodolite can be left unattended. However, since the instrument could be accidentally knocked or vibration of nearby machinery could deflect the beam, it is essential that regular checks are taken to ensure that the beam is in its intended position.

One application of laser theodolites is for *controlling verticality.* Conventional techniques for doing this were discussed in Section 11.4 and lasers present an alternative option particularly over long vertical ranges. If using a laser theodolite to control verticality, it should be set up on the ground floor slab directly over the ground point to be transferred. The beam is then projected vertically either up the outside of the building or through special openings in the floors. The essential requirement of the system is to ensure that the beam is truly vertical and, to check that this is the case, it is necessary to use four mutually perpendicular positions of the vertical telescope as described in Section 11.4.

Alternatively, verticality can be controlled using purpose built *precision laser plummets* such as Sokkia's LV1 shown in Figure 11.73. Precision laser plummets usually generate Class 2 red laser beams of wavelengths in the region of 633 nm to 635 nm from laser diodes and have measurement ranges of 100 m or more upwards and 5 m downwards. Depending on the manufacturer, they can generate a vertical beam to an accuracy of ±5" (±2.5 mm in 100 m) and a downward beam to an accuracy of ±1' (±1.5 mm in 5 m). Levelling can be provided either by a pendulum compensator system (typically over a range of ±10') or by an electronic self-levelling system

Figure 11.73 ● Sokkia LV1 precision laser plummet (courtesy Sokkia Ltd).

(typically over a range of ±3°). Precision laser plummets are compact, lightweight (2.5 kg), all-weather instruments that can operate from a range of power supplies including 110/240 V AC, replaceable alkaline batteries (for up to 80 hours) or rechargeable NiCd (nickel cadmium) batteries (for up to 30 hours).

To control verticality, the beam is intercepted as it passes the floor to be referenced by the use of plastic targets fitted in the openings or attached to the edge of the slab. The point at which the beam meets the target is marked to provide the reference. When controlling floor-by-floor construction in multi-storey buildings, intermediate targets with holes in to enable the beam to pass through can be placed on completed floors, with a solid target being used on the floor in progress. This system has the advantage that should the laser instrument be moved accidentally, the beam will be cut out by the lower targets and the operator will immediately be aware that a problem has arisen.

With conventional optical plumbing, two people are required: one to look through the instrument and one to move the target into the correct position. Such methods can cause problems of communication between the two operators. Laser plumbing has the great advantage that only the operator on the target is required, since once it has been set up, a constant visible beam will be projected for as long as is necessary. There is no danger of lack of communication and the spot can be clearly seen.

One very successful area of application of lasers is their use in *setting out pipelines*, where they eliminate the need for the sight rails and travellers discussed in Section 11.4 and enable one-person pipelaying to be undertaken. Purpose-built *pipe lasers* have been specifically manufactured for this application. Generally they fall into Class 3R, and red beam (typically 635 nm) and green beam (typically 532 nm) models are available. They are used with plastic or Perspex *targets*, as shown in Figure 11.74. In practice, the pipe laser is set up in the most appropriate location from which to set out the pipe. This can either be inside the pipe, outside the pipe, on top of the pipe or at the base of a manhole. Pipe lasers are available that will fit snugly inside a

Figure 11.74 ● A pipe laser and its target (courtesy Topcon Ltd).

150 mm diameter pipe and for use with manholes and larger diameter pipes they can be fitted to a variety of different supports. Figure 11.75 shows typical arrangements.

Pipe lasers are completely waterproof and made to be very robust. They are fully electronic self-levelling over a wide range, typically ±10% (±5.7°), and the direction

(a)

(b)

(c)

Figure 11.75 ● Various modes of operation and support for pipe lasers ((a) and (b) courtesy Trimble Navigation Ltd and (c) courtesy Leica Geosystems.

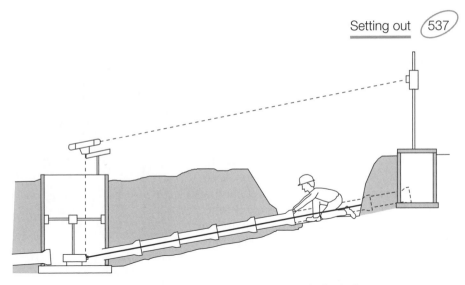

Figure 11.76 ● Principle of laser controlled pipelaying.

(line) and slope (grade) of the beam can be set using on-board controls either manually or using a wireless infra-red remote control unit. Depending on the manufacturer, line can be controlled over a range of ±15%, grade over a range of –15% to
+50% (a larger positive range is provided since pipes are usually laid uphill, as shown
in Figure 11.76) and the remote control unit has an operating range of up to 200m
from the front of the instrument and 45m from the rear. Several models are also
fitted with an automatic targeting system that finds the centre of the target up to
distances of 200 m, which is ideal for re-establishing the line quickly when returning
to a job the next day. Power can normally be supplied by one or more of the
following: from an AC source, from a car battery, from a built-in rechargeable NiMH
(nickel metal hydride) battery pack (up to 48 hours) or from replaceable alkaline
batteries (up to 70 hours).

   Before pipelaying begins, the laser is correctly set up over the required line. To
help achieve this, some pipe lasers have a visible centring point provided in the form
of a light-emitting diode (LED) to enable them to be set over the correct line plumbed
down from above. The beam is then aligned along the direction in which the pipe is
to run. This can be achieved by mounting a special sighting device directly above the
laser at the top of the manhole and by lining this in on a target positioned directly
above the required line at the next manhole, as shown in Figure 11.76. As soon as a
short length of the trench has been excavated between the two manholes, the
sighting device is tilted downwards to define the line in the trench. Once the line has
been established, the gradient of the pipe is then set on the grade indicator of the
laser.

   During pipelaying, a target is placed in the open end of the pipe length being laid,
which is then moved horizontally and vertically until the laser beam hits the centre
of the target. The pipe is then carefully bedded into position, the target is removed
and the procedure repeated. If the laser is unintentionally moved off grade, a
warning system is activated and when this happens with one of the Leica 6700 series
pipe lasers the beam blinks on and off until the unit has re-levelled itself. If this

should occur, the laser must be checked to ensure that the beam is still in the required direction and the required grade. Although pipe lasers will self-level, they will not re-level at the same height if they have been moved in a vertical direction.

Lasers have also been used very successfully for *tunnel alignment*, and several *tunnel lasers* were used to control the tunnel boring machines used on the Channel Tunnel. Tunnel lasers are made to be shockproof, waterproof and even flameproof, if required, owing to the need for them to function in the adverse conditions that can often prevail in tunnels. An example is Sokkia's SLB110 tunnel laser shown

Figure 11.77 Sokkia SLB110 tunnel laser (courtesy Sokkia Ltd).

in Figure 11.77, which generates a Class 2 red beam of wavelength 635 nm from a laser diode.

To control a tunnel drive, a laser is fixed to the wall of the tunnel on a base plate and aimed in the required direction, which is defined by a series of intermediate targets located further down the tunnel on identical base plates. The intermediate targets have holes in to allow the beam to pass, as shown in Figure 11.78. The beam is detected by special targets on the tunnel boring machine and its position on the target indicates whether or not the machine is on line. In a manual system, a bullseye target is used and the operator adjusts the controls of the machine keeping the visible laser spot on the centre of the target. In an automatic system, the beam is detected by a photoelectric sensor, which relays information about the boring machine to its control computer. The computer then automatically takes the necessary action to keep the machine on line. As the tunnel drive progresses, the laser and the targets are moved forward to the next base plates as required to maintain the line.

Figure 11.78 Controlling an automatically driven tunnel boring machine (courtesy AGL).

Rotating lasers and their applications

These instruments, which are also known as *laser levels*, generate a plane of laser light. Traditionally, this has been done by passing the laser beam through a spinning pentaprism to simulate a plane, hence the generic name *rotating lasers*. While this is still overwhelmingly the case, there are some instruments that can generate a continuous 360° laser plane without the need for any rotating parts – these are discussed later in this section.

In order to support them in use, various mountings are available for rotating lasers, ranging from tripods to special column and wall brackets.

The beams generated by rotating lasers can be either visible or invisible, depending on the manufacturer. The invisible beams are invariably Class 1, while the visible beams can be red or green, depending on the manufacturer, and are either Class 2 or Class 3R. Each invisible beam laser comes complete with its special photoelectric laser detector, which is essential to locate the beam. The rotating visible beam instruments are also normally supplied with photoelectric laser detectors, since these enable the beams to be detected to a higher accuracy than by eye. Such high accuracy is essential when levelling and grading operations are being undertaken. Figure 11.79 shows an example of a rotating beam laser being used with its sensor.

Figure 11.79 ● A rotating laser in use with its sensor (courtesy Topcon Ltd).

The majority of rotating beam instruments incorporate self-levelling devices, either in the form of a pendulum compensator system (typically over a range of ±10'), an electronic system (up to ±5°) or a liquid system (up to ±3°), although there are some manually levelled instruments available. If it is knocked accidentally out of its self-levelling range, the beam is normally cut off until it re-levels itself. Should this occur, the height of the plane must be remeasured to ensure that it is at the required value. Although the instrument will re-level, it will not necessarily re-level at the same height.

All rotating lasers can generate horizontal planes. Many can also generate vertical planes, and some, known as *grade lasers*, can generate sloping planes, either about one axis (*single grade*) or about both axes (*dual grade*). Grade lasers are used in machine control and are discussed further in Section 11.6.

On some instruments, the speed of rotation can be varied as required, from 0–600 rpm. On others, a range of pre-set speeds can be selected, for example, 0, 150, 300, 600 and 900 rpm. The fastest speed available is in the region of 1200 rpm and speeds of this magnitude are required for machine control work (see Section 11.6). The accuracy of the plane generated is normally better than ±10 seconds of arc (±5 mm in 100 m).

Depending on the instrument, power can be provided by a range of sources, ranging from 110/240 V AC to rechargeable batteries.

In use, rotating lasers are very simple to operate. To generate a horizontal plane, the instrument is attached to a tripod or other suitable support. Depending on the levelling system, it is either levelled manually or set approximately level, either using a circular level or by eye and then turned on. After a few seconds, the laser plane will start to be generated. Operating ranges up to 375 m from the laser are possible, depending on the instrument. Handheld wireless remote control units are provided with some instruments to enable the laser to be operated from a distance, in excess of 200 m in some cases.

When the instrument has been levelled and switched on, the height of any point can be determined by taking a reading on a levelling staff fitted with a photoelectric detector, as shown in Figure 11.80.

For general *levelling* work, the rotating laser is set to generate a horizontal plane at a known height and is then left unattended. Significant advantages over conventional levelling are that only one person is required, as shown in Figure 11.81, and several operators can use the same plane simultaneously, as shown in Figure 11.82. However, if more than one rotating laser is being used on a site, great care must be taken to ensure that the plane from one laser is not being sensed by the detector from

Figure 11.80 ● Rotating laser sensor fitted to a levelling staff (courtesy Leica Geosystems).

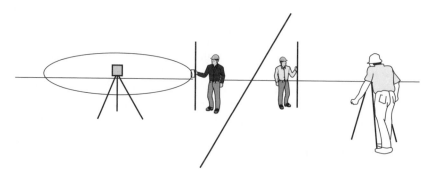

Figure 11.81 ● Single-person laser levelling compared to conventional levelling (courtesy Trimble Navigation Ltd).

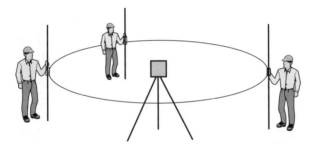

Figure 11.82 ● Multi-user laser levelling (courtesy Trimble Navigation Ltd).

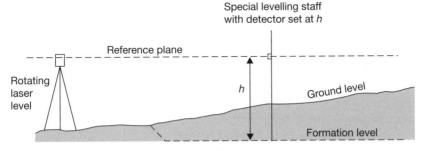

Figure 11.83 ● Setting out with a rotating laser level.

another. When used for setting out foundations, floor levels and so on, the sensor can be fixed at some desired reading and the staff used as a form of traveller, as shown in Figure 11.83, with the laser reference plane replacing sight rails. The laser sensors used for general levelling have a vertical reception length in the order or 50 mm and often have several levels of precision to which they can locate the centre of the laser plane, ranging from ±0.5 mm to ±3 mm, depending on the manufacturer.

Indicator arrows on their displays show when the beam has been located.

Lines of known level can be marked on vertical surfaces by estimating the centre of the plane as it appears on the surface. This is done by eye for visible beams and for invisible beams by holding the laser sensor against the surface on its own, without a staff, as shown in Figure 11.84.

For short-range levelling work, Pacific Laser Systems have recently introduced a laser instrument that is capable of generating a laser plane without the need for any rotating parts. This is the PLS360 model, shown in Figure 11.85, which has a range of 30 m and is offered at virtually half the cost of conventional rotating lasers. In the

Figure 11.84 ● Using a laser sensor to define a level line on a column (courtesy Leica Geosystems).

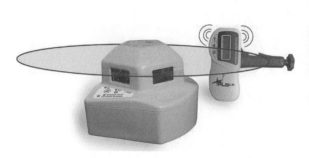

Figure 11.85 ● PLS360 non-rotating laser level (courtesy Pacific Laser Systems and Topcon Ltd).

Figure 11.86 ● An interior laser with its target being used to install an internal wall (courtesy Trimble Navigation Ltd).

PLS360, a continuous 360° beam is generated using five diodes placed at the angles of a pentagon without the need for any rotation.

Another class of instruments can generate both a single beam and a rotating beam, simultaneously, at right angles to each other. These are the *interior lasers* which generate mutually perpendicular red or green visible beams. This is done by passing a single beam through beam splitting optics that allow part of the beam to pass straight through while directing the remainder into a spinning pentaprism, which turns it through a right angle thereby simulating a plane perpendicular to the unaltered part of the beam.

Interior lasers have been specially developed for controlling the installation of mutually perpendicular internal fittings in a building, such as raised floors, partition walls and suspended ceilings. They can be mounted on a range of different supports and are designed for use without photoelectric detectors. Instead, targets such as graduated pieces of Perspex with a magnetic edge are used. These are fitted to structural members, allowing the operator to have both hands free during the installation. When installing a vertical partition wall, the laser is set to generate a vertical plane parallel to but offset a known amount from the required position of the wall. A target is then attached to the partition member being located and the operator moves it into position, finally fixing it when the beam passes through the correct graduations on the target, as shown in Figure 11.86. For a suspended ceiling, a horizontal plane is generated instead and the target is attached to the ceiling member being fitted to indicate when the correct level has been reached. For raised floors, a horizontal plane is again generated, and in this case the target is placed on top of the floor panel being fitted to indicate when it has been installed at the correct level. In all cases, the operating range to such targets is approximately 60 m and the accuracy to which fittings can be located is better than ±2.5 mm. Additional features are available on some interior lasers such as the facility to generate a laser fan to concentrate the beam over a short length, the ability to use the beam to plumb over a point and

Figure 11.87 ● Triax UL300 Universal Laser (courtesy Sokkia Ltd).

Figure 11.88 ● Topcon RT-10S Rotating Laser Theodolite (courtesy Topcon Ltd).

the option of stopping the rotating beam to provide two mutually perpendicular single beams.

Some instruments, known as *universal lasers* are multi-functional. For example, the Triax UL300, shown in Figure 11.87, generates two mutually perpendicular Class 2 visible red beam lasers. It has horizontal plane and single and dual grade capability over a radius of 150 m and can also be used for pipelaying and alignment, since the handle is detachable, enabling it to be used inside a 150 mm pipe. Topcon offers another form of universal instrument in its RT-10S model, shown in Figure 11.88, which is a rotating laser theodolite. This consists of a Class 1 rotating invisible beam laser having an operating radius of 300 m fitted on top of a 10" reading electronic theodolite. Together they can be used to generate horizontal or single grade laser planes at any slope for use in levelling and machine control operations and they can generate vertical planes in any required direction to establish ground control lines for bolts, formwork, column centres, walls and so on.

Rotating lasers are also widely used for controlling earth-moving machinery and other large construction machines for grading and screeding work and their applications in these areas are discussed in Section 11.6.

Reflective summary

With reference to setting out using laser instruments remember:

– A laser beam has high intensity and low angular divergence, which enables it to be projected over long distances without spreading significantly. These characteristics are ideal for alignment and levelling operations.

- The laser instruments used in surveying and setting out have beams which are either visible or invisible.

- All the lasers used in modern surveying instruments are generated from laser diodes.

- Lasers are classified according to a safety system specified by the International Electrotechnical Commission (IEC) in its standard IEC 60825-1.

- A user's guide to the safe use of lasers is given in IEC 60825-14.

- Every laser must be clearly marked with labels indicating its safety class and any specific precautions that must be exercised by the user. If the labels are missing the laser should not be used.

- The majority of lasers used in surveying instruments fall into safety Classes 1, 2 and 3R.

- There may still be some older instruments in use which fall into the Class 3B category.

- Under normal operating conditions, Class 1 lasers pose no hazard and Class 2 lasers present only a minimal hazard. When using lasers in either of these categories, it is normally sufficient to follow the instructions on their labels and in the manufacturer's handbook for their safe use.

- When using Class 3R and Class 3B lasers you must not only follow the instructions given on their labels and in the manufacturer's handbook but also take additional precautions as specified in IEC 60825-14.

- Surveying lasers fall into two main categories, either *alignment lasers* or *rotating lasers.*

- Alignment lasers can be used for controlling verticality, pipelaying and tunnelling, and for any application where precision alignment is critical.

- Rotating lasers can be used for general levelling work, setting out internal fittings and controlling grading and screeding operations.

- Several multi-purpose laser instruments have been developed by a number of manufacturers, for example, *digital laser theodolites, universal lasers* and *rotating laser theodolites.*

- Laser instruments enable many surveying and setting out operations to be carried out by just one person and offer not only considerable savings in the time required to complete the work but also significant improvements in the accuracy that can be achieved.

11.6 Machine control

After studying this section you should understand that earthmoving machines and other large construction plant can be controlled automatically by a number of different systems, including ones based on total stations, GPS receivers and laser instruments. You should appreciate that such systems have considerable advantages over conventional methods involving sight rails and travellers. In particular, you should be aware of the savings in time and materials and the improvements in the accuracy and uniformity of the finished surfaces that can be achieved with automatic machine control systems.

This section includes the following topics:

● Introduction to machine control systems

● Total station machine control systems

● GPS machine control systems

● Laser machine control systems

Introduction to machine control systems

Although the methods already described in previous sections of this chapter for controlling earthworks using line pegs combined with profiles, slope rails and travellers are still in use, it is possible to do this using machine control techniques. When a machine control system has been installed on site, it can be used to provide dimensional control for earthmoving and other plant, but without the need to set out anything. The advantages of this are that machine operators can work anywhere on site at any time without having to wait for profiles and rails to set out, they do not have to wait when these have to be replaced for some reason, and there is no need for anyone to check the progress of work at all times with a traveller. It is claimed that machine control improves productivity and reduces costs, especially for bulk earthmoving and grading on large projects.

The methods used in machine control are based on the use of total stations, GPS and lasers.

Total station machine control systems

A machine control system using a total station consists of a motorised total station fitted with ATR (these are described in Section 5.3) working with a dozer or grader which has a 360° prism attached to it (see Figure 11.89). The machine is also equipped with a ruggedised computer which can communicate with the total station using a radio modem or some form of telemetry link. The remaining item of equipment is a graphic display unit, mounted in the cab for the driver.

Figure 11.89 ● A 360° prism mounted on a grader for machine control with a total station (courtesy Leica Geosystems).

The software used for the design of a construction project generates a three-dimensional design surface which is usually represented as a DTM. For machine control, this data is either transferred to the computer on board the grader or dozer using data cards or it can be sent via the communications link. The total station is set up at a control point (preferably outside the working area), is orientated and has its coordinate measuring program activated (these procedures are described in Section 6.2). After this, it continuously tracks and monitors the three-dimensional position of the machine as it moves around the site. This information is also relayed to the computer on board the machine where it is combined with data measured by slope sensors.

All of this enables the three-dimensional position and orientation of the machine to be constantly updated and compared to the design surface for the project – any differences in the elevation and cross slope for the current position are calculated by the computer and these are used by the driver to steer the machine and adjust the blade. Normally, the driver will have the design profile in view at all times and can manually control the position of the blade. The display unit will show the position of the machine on a site plan together with the blade tilt and the cut or fill required (see Figure 11.90). Corrections to be applied can also be shown on light bars (see Figure 11.91). Alternatively, the differences can be used to provide automatic control of the blade at the required elevation and correct cross slope through an hydraulic system. Automatic systems are safer as all the driver has to do is concentrate on steering and not on grade adjustments at the same time. However in automatic control systems, information on grade and slope together with the blade position is displayed in the cab to assist the driver when necessary.

Because a motorised total station can track the prism mounted on the machine at all times, a machine control system based on these should be able to operate at all times. The total station can normally continue tracking through short interruptions of the ATR beam and after long interruptions, the total station will initiate a search and automatically lock onto the prism once again.

Figure 11.90 ● Machine control display units (courtesy Leica Geosystems and Trimble Navigation Ltd).

The advantages of a total station for machine control are:

● It is capable of achieving a high accuracy of ±5–10 mm and it is suitable for all types of work, particularly finished grade work.

● The total station can operate in tunnels and under cover – GPS control systems cannot be used in these conditions.

● The total station can be used for other work on site without having to be modified in any way.

Figure 11.91 ● Light bars (on left) for guiding machine (courtesy Trimble Navigation Ltd).

GPS machine control systems

GPS machine control systems are similar to those using total stations, the main difference being that RTK GPS is used to determine the position of the machine instead of a total station. As before, a computer and display unit are installed in the cab of the machine but in this case, a GPS antenna is mounted at some point on the machine, usually above the blade as shown in Figure 11.92. This is connected to a GPS receiver and the system computer. Using a computed GPS position and data collected from slope sensors, the position and height of the blade can be determined as well as its slope. The Site Vision system from Trimble uses two GPS antennae, mounted above each end of the blade, as shown in Figure 11.93. This is used with a special receiver with two inputs and because the position at each end of the blade is known, its cross slope can be computed using these, rather than from slope sensors.

Figure 11.92 ● GPS antenna mounted on a dozer blade (courtesy Topcon Ltd).

Figure 11.93 ● Dual GPS antennas mounted at each end of a grader blade (courtesy Trimble Navigation Ltd).

All of this data is again compared to the design surface loaded in the computer to give the corrections for positioning the blade and these are either viewed on the display unit for manual steering or are fed into an automatic hydraulic system.

As with any RTK GPS, a base station must be set up on the site at a known position or control point and it is essential that communication between this and the machine is maintained at all times. Although the need for a line of sight between machine and base station is not as much of a problem as with total stations, care must be taken to have a reliable communications link to the base station. Another problem to be addressed is that of transforming the GPS coordinates obtained into site coordinates used for the design. This will involve a calibration process in which the RTK system is detached from the machine and is used to take a series of measurements at control points around the site which have known coordinates and heights.

The accuracy expected for machine control with GPS is of the order of ±30 mm and it is best suited to site preparation and bulk earthworks. One of the benefits of using GPS for machine control is that one base station can control several machines simultaneously over a range of several kilometres. In addition, the same digital model used by the machine can also be loaded into a controller and attached to a pole-mounted RTK system. This can then be taken to any point on site and used not only to check the finished levels at that point but also set out grade points conventionally, as required.

Laser machine control systems

Laser machine control systems are a further application of rotating lasers and their detectors discussed in Section 11.5. They are mainly used to control *grading* operations and the principle is shown in Figure 11.94. With reference to this, a rotating laser is set to generate a plane parallel to the required formation plane and the amount of filling or cutting being undertaken by the grader is controlled by a detector mounted on the machine itself. The detectors used in machine control often have full 360° detection capability and they can be mounted on masts fitted to the blades and arms of the graders, as shown in Figure 11.95. The vertical reception length of the detectors used for machine control is much greater than that of the detectors used for general levelling and interior work. Models are available that can detect over a vertical range of 230 mm and several levels of precision can be provided, depending on the manufacturer, ranging from ±3 mm to ±25 mm.

If the required formation plane is to be horizontal this can be defined using one of the wide range of horizontal plane rotating lasers available for general site levelling work. However, if sloping planes are required, these can only be achieved using a special type of rotating laser known as a *grade laser*. Examples of this type of instrument are shown in Figure 11.96, and Figure 11.97 shows the principle of their use in controlling slopes.

All grade lasers can be set to generate a horizontal plane and a *single grade*, that is, a single slope about one axis – typically up to ±8%. Many can also generate a *dual grade*, that is, a dual slope about two mutually perpendicular axes – typically up to ±10%, although slopes of up to 50% about both axes are possible with some models. On some instruments, beam shutters are provided to ensure that the laser plane is

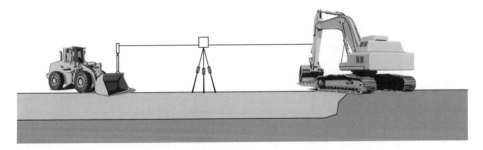

Figure 11.94 ● Principle of laser machine control.

(a)

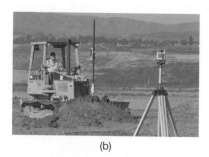
(b)

(c)

Figure 11.95 ● Controlling earthmoving and grading by laser (courtesy (a) Sokkia Ltd, (b) Topcon Ltd and (c) Trimble Navigation Ltd).

generated only in those parts of the site where it is required. This is very useful when several different lasers are being used on the same site.

Both manual and fully automatic systems are available. In a manual system, the driver is presented with some type of illuminated display, typically a series of multi-coloured vertical lights, which indicate whether the blade is too high or too low. If the centre lights are illuminated, the operation is on line, as shown in Figure 11.98. In a fully automatic system, information about the position of the beam on the detector is sent to a controlling

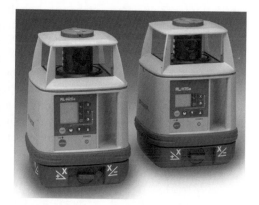

Figure 11.96 ● Grade lasers (courtesy Topcon Ltd).

computer which automatically adjusts the machine's hydraulic systems to keep the operation on line leaving the driver to concentrate on driving. Long operating ranges are possible in this type of work with some of the grade lasers having a working radius of 375 m. To help speed up operations, handheld remote control devices are

Figure 11.97 ● Principle of laser controlled grading (courtesy Topcon Ltd).

available to enable the laser to be fully controlled from some distance away – in excess of 200 m in some cases.

A range of rotation speeds is provided, and the speed used should be such that the detector does not have to hunt for the beam, thereby ensuring that the blade passes smoothly over the ground surface at all times. Generally, the faster the speed of the machine to be controlled and at longer distances, the faster should be the rotation speed of the beam.

Like total stations and GPS systems, laser grading control systems have big advantages over traditional methods. These are:

Figure 11.98 ● Machine control laser receiver with the centre lights illuminated indicating "on-grade" (Courtesy Topcon Ltd).

● They eliminate the need for sight rails and travellers.

● They prevent over-cutting or under-cutting, which saves time and materials.

● They enable the finished grade to be achieved in much fewer passes over the ground surface.

● Several machines can be controlled simultaneously.

● They are more accurate (within ±10 mm) which leads to more uniform slopes and better levels – this in turn improves drainage.

● The work can be undertaken equally well at night enabling tight schedules and weather delays to be overcome.

Concrete *floor screeding* machines can also be very accurately controlled by laser. In this application, two detectors are mounted at each end of the screed carriage,

simultaneously sensing the plane being generated by a single visible beam rotating laser. The information collected by the detectors is passed to an on-board control box which checks and rapidly adjusts the screed elevation through automatic control of the machine's hydraulics. Accuracies better than ±2 mm are possible with this system.

Reflective summary

With reference to machine control remember:

— There are several methods available, based on total stations, GPS receivers and laser instruments.

— They all eliminate the need for sight rails and travellers and offer considerable savings in costs and materials and significant increases in the accuracy and the uniformity of the finished surfaces when compared to such conventional methods.

— They can all operate at any time of day or night.

— Total station systems are capable of achieving accuracies in the region of ±5–10 mm and are suitable for all types of work, particularly finished grade work.

— GPS systems are capable of achieving accuracies in the region of ±30 mm and are suitable for site preparation and bulk earthworks.

— Laser systems are suitable for all types of work. They are capable of achieving accuracies in the region of ±10 mm on earthworks and ±2 mm on concrete screeding.

11.7 Quality assurance and accuracy in surveying and setting out

After studying this section you should appreciate that if the setting out is to proceed smoothly and meet the accuracy requirements of the project concerned then the quality of the work must be of an appropriate standard. You should be aware of the term Quality Assurance and how organisations aspire to BS 5750 accreditation to assure their customers of their quality. You will appreciate that the question of accuracy in surveying and setting out arises at all stages of the construction process. You will understand that personnel involved in engineering surveying work must know the limitations and capabilities of their surveying instruments and must adopt suitable working practices in order that the quality requirements of a project are met. You will be aware of the ICE Design and Practice Guide *The Management of Setting Out in Construction* and you will know of several British and international standards that relate to surveying and setting out operations.

This section includes the following topics:

- The need for accuracy
- Quality Assurance
- Accuracy in surveying and setting out
- British and international standards

The need for accuracy

The need for accurate and efficient setting out is paramount if civil engineering projects are to be completed on time and within the allocated budget. In order that this can be achieved, the two basic aims stated in Section 11.2 must be met, that is, the setting out must not only proceed smoothly but also meet the accuracy requirements specified in the contract. Hence the methods adopted and the equipment used must be appropriate to the needs of the project, that is, they must meet the project's quality requirements.

Due to the tighter profit margins prevailing in the construction industry being set against the need to maintain standards, there has been a marked increase in concern in recent years about the quality of work achieved on construction sites and the standards against which this quality should be assessed. This has given rise to the concept of *Quality Assurance* (QA) when undertaking surveying and setting out on construction sites, and several publications have appeared that deal specifically with this topic. In addition, a number of British Standards have been released which are directly related to quality in the form of accuracy as applied to surveying and setting out.

This section discusses the role of Quality Assurance in surveying and setting out operations and reviews those British Standards and other documents which are directly concerned with the accuracy attainable in these activities.

Quality Assurance

The term *Quality Assurance* (QA) is nowadays widely used on construction sites to indicate that certain standards of quality have been or are expected to be achieved. Unfortunately, as with many terms, it has become part of everyday jargon and those using it do not necessarily appreciate the concepts on which it is based. In general terms, QA is the creation of a fully competent management and operations structure which can consistently deliver a high-quality product to the complete satisfaction of the client. It has its origins in 1979, when, in order to provide guidelines on which a company could base a quality system, the British Standards Institute issued BS 5750 *Quality Systems*. This enabled organisations to gain BS 5750 accreditation by submitting to be assessed against the standard. Since then, thousands of companies have gained accreditation and the standard has been revised a number of times.

Identical international and European standards have been issued by the International Organization for Standardization (ISO) in its ISO 9000 series of standards and by the European Committee for Standardization (CEN) in its EN 29000 series. For example, the latest issue of the *Quality Systems* standard, BS 5750-8:1991, is identical to ISO 9004-2:1991and EN 29004-2:1993. Further information on these standards can be found on the websites of these organisations given in the Further reading and sources of information section at the end of this chapter.

BS 5750 accreditation is much sought after as a means of assuring quality to customers. In order to achieve it, companies in the UK are required to submit an application form and accompanying documentation, including a Quality Manual, for consideration by the United Kingdom Accreditation Service (UKAS), which acts on behalf of the Department of Trade and Industry (DTI). The documentation submitted must include details of the firm's policies, objectives, management structure, methods of implementing quality control, details of all quality-related management procedures and actual day-to-day working practices in order to show how quality is maintained.

From a surveying and setting out point of view, it could be argued that engineers and surveyors have always undertaken quality assurance procedures as part of their duties. Planning the work, following a programme, recording all relevant information, checking each setting out operation and working to specified tolerances are just some of the many examples of quality procedures undertaken as a matter of course during engineering surveying work. However, BS 5750 accreditation can only be obtained by submission of the correct documents for approval and it is very gratifying to note that many UK companies and contractors who specialise in surveying and setting out activities have successfully achieved this coveted status despite the considerable amount of work involved in preparing the submission.

Accuracy in surveying and setting out

The question of accuracy in surveying and setting out arises at all stages of the construction process. When the initial site survey is being carried out, the survey team must use equipment and techniques which will ensure that the plan produced shows the required detail and is within the tolerances agreed with the client. At the design stage, the designer must specify suitable tolerances which will ensure not only that the construction will function properly but also that it can be built. At the construction stage, the contractor must choose equipment and adopt working procedures which will ensure that the scheme is located correctly in all three dimensions and that the tolerances used by the designer and specified in the contract are met.

Those involved in engineering surveying work must, therefore, become familiar with the limitations and capabilities of the various surveying instruments they use in their day-to-day operations and must adopt suitable working procedures that will ensure that the tolerances specified in the contract are achieved. These requirements are directly related to the question of Quality Assurance discussed in the previous section and, in a QA-accredited surveying firm, some of their QA documents will give details of the standard working procedures used by the company in its day-to-day activities.

For those actively involved in surveying and setting out, whether preparing for QA accreditation or not, the development of suitable working procedures and the choice of appropriate equipment for a particular project are of paramount importance. This is best dealt with by the preparation of a project *Quality Plan*, which is a document setting out specific quality practices, resources and sequences of activities. Its aim is to provide a formal framework from which it will be possible to coordinate and control all elements of a project in order to meet the agreed set of standards for the project. The project Quality Plan is an organisational tool that enables the Quality Assurance criteria for the project to be met. It will give details of management responsibilities, document control, process control, inspection, testing, non-conformance and records. Setting out will be part of these. Considerable guidance on the preparation of a project Quality Plan can be found in the Institution of Civil Engineers (ICE) Design and Practice Guide *The Management of Setting Out in Construction*. Prepared by the Joint Engineering Surveying Board of the Institution of Civil Engineers and the Institution of Civil Engineering Surveyors, this excellent guide was first published in 1997 and is in the process of being updated. It discusses the overall process by which accurate and efficient setting out can be achieved and gives guidance to engineers and surveyors involved in the planning, design and execution of a project. One of its aims is to highlight the links between these processes, indicating where setting out features. The guide includes a *Quality Management* section, which first poses then answers a number of questions aimed at assisting in the compilation of a project Quality Plan. In addition, it contains sections dealing with other important aspects of setting out including the *Overall Control Framework*, *Accuracy and Precision*, *Specifications*, *Instrumentation* and *Good Practice*. It concludes with a section on the accuracy requirements of a wide range of *Specific Civil Engineering Works* and how these can be achieved. It is strongly recommended that anyone involved in the planning and execution of setting out operations consults this guide before carrying out any setting out work on site. In addition to this document, there are a number of relevant British and international standards that deal with accuracy and setting out, and these are discussed in the following section.

British and international standards

Currently, there are a number of British standards and ISO standards which relate to surveying and setting out operations. The most important ones are BS 5606, BS 5964, BS 7307, BS 7308, BS 7334 and ISO 17123. These are briefly discussed below. In those cases where the standards have been adopted by both organisations, the equivalent BS and ISO numbers are given.

BS 5606: 1990 *Guide to Accuracy in Building*

Originally published in 1978, BS 5606 was most recently revised in 1990. Its main objective is to guide the construction industry on ways to avoid problems of inaccuracy or fit arising on site. It is based on work carried out by the Building Research Establishment (BRE). Accuracies which can be achieved in practice are given and used to stress the need for realistic tolerances at the design stage. Table 11.3 has been taken directly from BS 5606 and shows the accuracy in use *A* of various pieces of equipment when used in engineering surveying by reasonably proficient operators. The values shown in the table should be used both by the designer when specifying the deviations allowed in the design in order that what is designed can actually be set out (see BS 5964 below), and by the engineer undertaking the setting out in order that equipment can be chosen which will maintain the design standards and specifications.

BS 5606 was one of the first documents produced by the British Standards Institution which dealt with the accuracy of surveying and setting out, and it remains the basic reference for this type of work. It is deliberately broad and basic in coverage, leaving guidance on more detailed and complex work to other more specific standards.

BS 5964: 1990 and 1996 *Building Setting Out and Measurement* (ISO 4463)

This standard, the first part of which was published in 1990, translates the requirements of BS 5606 to relate to the processes of setting out and measurement. It is a wide ranging standard which applies to all types of construction and to the control from which the construction is set out.

One of the most important aspects of BS 5964 is that it defines an accuracy measure known as a *permitted deviation* (*P*). This is a statistical variation taking into account the normal distribution of measurement, as discussed in Section 9.4. Since many building practices and instrument manufacturers express tolerances in terms of the statistical *standard deviation* (σ), *P* values have been related to σ values by the following equation.

$$P = 2.5 \times \sigma$$

In practice, because the standard deviation represents a probability of 68%, this equation implies that 68% of all setting out checked should be better than the permitted deviation divided by 2.5. When working to tolerances on site, the setting out should be three times better than the tolerance to take into account the normal distribution of measurements. On site, *P* is compared with the appropriate *A* values from Table 11.3 or BS 7334 (as discussed later) in order that suitable equipment is chosen to ensure that $A \leq P$.

Having defined *P* values, BS 5964 goes on to break down setting out into its component activities and assesses each in turn. Recommended practices are given for various activities and permitted deviations are defined for primary control, secondary control, setting out, levelling and plumb.

BS 7307: 1990 *Building Tolerances: Measurements of Buildings and Building Products* (ISO 7976)

This standard is published in two parts. In Part 1, alternative methods for determining shape, dimensions and deviations of building components both in the factory and on site are given. Diagrams illustrating the procedures that should be used are included, as are accuracy tables which recommend suitable equipment and its associated permitted deviations. Suitable positions for measuring points on buildings and building components are covered in Part 2 of the standard.

BS 7308: 1990 *Method for Presentation of Dimensional Accuracy Data in Building Constructions* (ISO 7737)

With the standardisation of the methods by which data is collected being covered in BS 7307: 1990, the need arises for standard formats for the presentation and processing of data. This is covered by BS 7308: 1990 which states how measured data used to check and assess accuracy should be presented. Initially, the standard defines which dimensions should be measured and then goes on to give guidance on the correct presentation of dimensional accuracy data. Blank copies of standard booking forms and tables are also included.

BS 7334: 1990/1992 *Measuring Instruments for Building Construction* (ISO 8322)

This is a very important British Standard which should be consulted by all those involved in the use of surveying instruments for surveying and setting out. Its main purpose is to enable users of equipment to determine the accuracy of the instruments they are using on site. In other words, it enables them to assess their own personal accuracy when using particular instruments for a range of surveying and setting out tasks. It is divided into eight parts; parts 1–3 were published in 1990 and parts 4–8 in 1992.

Part 1 gives the theory of how to determine the accuracy of measuring instruments. Users can employ this theory when devising test procedures for instruments not covered in other parts of the standard. Accuracy results are expressed as accuracy in use A values as discussed earlier for BS 5606. Details are also provided on how to establish that the accuracy associated with a particular surveying technique using specific equipment is appropriate to the intended measuring task. This is done by comparing the calculated A values with P values specified using BS 5964.

Part 2 gives a step-by-step guide to the observation procedures and calculation methods to be employed to determine the accuracy in use of measuring tapes.

Part 3 deals with optical levelling instruments.

Part 4 deals with theodolites.

Part 5 deals with optical plumbing instruments.

Part 6 deals with laser instruments.

Part 7 deals with instruments when used for setting out.

Part 8 deals with EDM instruments up to a range of 150 m.

Table 11.3 — Accuracy in use of measuring instruments (reproduced from BS 5606:1990 with permission of the British Standards Institution).

Measurement	Instrument	Range of deviations	Comment (see also NOTE)
T.3.1 Linear	30 m carbon steel tape for general use	±5 mm up to an including 5 m; ±10 mm for over 5 m and up to and including 25 m; ±15 mm for over 25 m	With sag eliminated and slope correction applied.
	30 m carbon steel tape for use in precise work	±3 mm up to an including 10 m; ±6 mm for over 10 m and up to and including 30 m	At correct tension and with slope, sag and temperature corrections applied.
	Electronic distance measuring (EDM) instruments (short range models) for general use	±10 mm for distances over 30 m and up to 50 m; ±(10 mm + 10 ppm)(3) for distances greater than 50 m	Accuracies of EDM instruments vary, depending on make and model of instruments. Distances measured by EDM should normally be greater than 30 m and measured from each end.
	Precise work	±(5 mm + 5 ppm)(3)	
T.3.2 Angular	Opto-mechanical (e.g. glass arc) theodolite(1) (with optical plummet or centring rod) reading directly to 20"	±20" (±5 mm in 50 m)	Scale readings estimated to the nearest 5". Mean of two sights, one on each face with readings in opposite quadrants of the horizontal circle.
	Opto-mechanical (e.g. glass arc) theodolite(1) (with optical plummet or centring rod) reading directly to 1"	±5" (±2 mm in 80 m)	Mean of two sights, one on each face with readings in opposite quadrants of the horizontal circle.
	1" opto-electronic theodolite/total station	±3" (±1 mm in 50 m)	Mean of two sights, one on each face with readings in opposite quadrants of the horizontal circle.
T.3.3 Verticality	Spirit level	±10 mm in 3 m	For an instrument not less than 750 mm long.
	Plumb-bob (3 kg) freely suspended	±5 mm in 5 m	Should only be used in still conditions.
	Plumb-bob (3 kg) immersed in oil to restrict movement	±5 mm in 10 m	Should only be used in still conditions.
	Theodolite (with optical plummet or centring rod and diagonal eyepiece)	±5 mm in 30 m(2)	Mean of at least four projected points, each one established at a 90° interval.
	Optical plumbing device	±5 mm in 100 m	Automatic plumbing device incorporating a pendulous prism instead of a levelling bubble.

Table 11.3 (continued)

	Laser upwards or downwards alignment	±7 mm in 100 m	Four readings should be taken in each quadrant of the horizontal circle and the mean value of the readings in opposite quadrants accepted. Appropriate safety precautions should be applied according to power of instrument used.
T.3.4 Levels	Spirit level	±5 mm in 5 m distance	Instrument not less than 750 mm long.
	Water level	±5 mm in 15 m distance	Sensitive to temperature variation.
	Lightweight self-levelling level	±5 mm in 25 m distance	
	Optical level		
	(a) 'builders' class	±5 mm per single sight of up to 60 m[2]	Where possible sight lengths should be equal.
	(b) 'engineers' class	±3 mm per single sight of up to 60 m[2] ±10 mm per km	
	(c) 'precise' class	±2 mm per single sight of up to 60 m ±8 mm per km	If staff readings of less than 1mm are required the use of a precise level incorporating a parallel plate micrometer is essential but the range per sight preferably should be about 15 m and should not be more than 20 m.
	Laser level (visible light source) (invisible light source)	±7 mm per single sight of up to 100 m ±5 mm per single sight of up to 100 m	Appropriate safety precautions should be applied according to power of instrument used.

[1]If a single sighting only is made when using a correctly adjusted theodolite to establish an angle the likely deviations will be increased by a factor of 3. Therefore a single sight should not be taken.

[2]Value based on measured data.

[3]Parts per million of measured distance.

NOTE: Equipment should be checked periodically according to BS 7334.

ISO 17123: 2001–2005 *Optics and Optical Instruments – Field Procedures for Testing Geodetic and Surveying Instruments*

This is the most recent ISO standard that relates to the testing of surveying instruments and at the time of writing it does not carry an equivalent British Standard number. It cancels and replaces ISO 8322, which was identical to BS 7334 discussed earlier. It consists of a series of parts, each covering a particular category of surveying equipment. It specifies field procedures to be adopted when determining and evaluating the precision (repeatability) of geodetic instruments and their ancillary equipment when used for building and surveying measurements. Primarily, these tests are intended to be field verifications of the suitability of a particular instrument for the immediate task at hand and to satisfy the requirements of other standards. They are not proposed as tests for acceptance or performance evaluations that are more comprehensive in nature. The field procedures given in ISO 17123 have been developed specifically for *in situ* applications without the need for special ancillary equipment and are purposely designed to minimize atmospheric influences. Part 1 was published in 2002, parts 2, 3 and 4 in 2001, part 6 in 2003, and parts 5 and 7 in 2005. Those already published are listed below.

Part 1: Theory
Part 2: Levels
Part 3: Theodolites
Part 4: Electro-optical distance meters (EDM instruments)
Part 5: Electronic tacheometers
Part 6: Rotating lasers
Part 7: Optical plumbing instruments

The Further reading and sources of information section at the end of the chapter lists all the British and international standards referred to above together with other publications that may be found useful by anyone concerned with the correct methods of setting out on construction sites.

Reflective summary

With reference to Quality Assurance and accuracy in surveying and setting out remember:

— The term *Quality Assurance* (QA) is used on construction sites to indicate that certain standards of quality have been or are expected to be achieved.

— In general terms, QA is the creation of a fully competent management and operations structure which can consistently deliver a high-quality product to the complete satisfaction of the client.

— Quality standards can be assessed against BS 5750 and if successful, the firm will become BS 5750 accredited.

— Many survey companies and contractors aspire to BS 5750 accreditation to prove their quality to their customers.

— The question of accuracy in surveying and setting out arises at all stages of the construction process. The original survey team, the designer and the setting out engineer must develop suitable working practices and choose suitable equipment to enable them to work within the specified tolerances.

— The development of suitable working procedures and the choice of appropriate equipment are best dealt with by the preparation of a project *Quality Plan*, which sets out specific quality practices, resources and sequences of activities to be undertaken on site, including those required for surveying and setting out operations.

— Guidance on the preparation of a project Quality Plan is given in the ICE Design and Practice Guide *The Management of Setting Out in Construction*. It is strongly recommended that anyone involved in the planning and execution of setting out operations consults this guide before carrying out any setting out work on site.

— BS 5606 gives *accuracy in use values* of various pieces of surveying equipment when used by reasonably proficient operators. These should be used both by the designer in order that what is designed can actually be set out and by the setting out engineer so that equipment can be chosen which will maintain the design standards and specifications.

— BS 5964 defines an accuracy measure known as a *permitted deviation*, which is used to check if setting out meets tolerances specified in contract documents.

— BS 7334 enables users of surveying equipment to assess their own personal accuracy when undertaking a range of surveying and setting out tasks using particular instruments.

11.8 Setting out worked examples

After studying this section you should understand how to calculate the levels at which the top edges of sight rails and profile boards must be fixed in order to control excavations in conjunction with travellers. You will be able to convert horizontal distances into ground surface distances and coordinates into angle and distance setting out data. You will understand how to calculate the positions of uprights to carry slope rails and you will be able to determine the levels at which the top edges of slope rails should be set.

This section includes the following topics:

- Setting out a pipeline using sight rails and a traveller
- Setting out by offsets from a baseline
- Setting out by intersection
- Using sight rails
- Using slope rails
- Using profile boards

Worked example 11.1: Setting out a pipeline using sight rails and a traveller

Question

An existing sewer at P is to be continued to Q and R on a falling gradient of 1 in 150 for plan distances of 27.12 m and 54.11 m consecutively, where the positions of P, Q and R are defined by wooden uprights.

Given the following level observations, calculate the difference in level between the top of each upright and the position at which the top edge of each sight rail must be set at P, Q and R if a 2.5 m traveller is to be used.

Level reading to staff on TBM on wall (RL 89.52 m)	0.39 m
Level reading to staff on top of upright at P	0.16 m
Level reading to staff on top of upright at Q	0.35 m
Level reading to staff on top of upright at R	1.17 m
Level reading to staff on invert of existing sewer at P	2.84 m

All readings were taken from the same instrument position.

Solution

Consider Figure 11.99.

Height of collimation of instrument	= 89.52 + 0.39 = 89.91 m
Invert level of existing sewer at P	= 89.91 − 2.84 = 87.07 m

This gives

Sight rail top edge level at P	= 87.07 + 2.50 = 89.57 m
Level of top of upright at P	= 89.91 − 0.16 = 89.75 m

Hence

Upright level − sight rail level	= 89.75 − 89.57 = +0.18 m

That is, the top edge of the sight rail must be fixed **0.18 m** *below* the top of the upright at P.

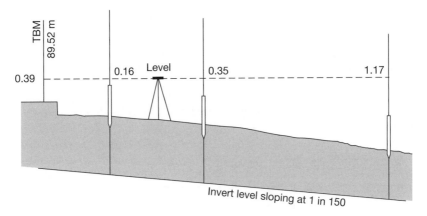

Figure 11.99

Fall of sewer from P to Q	$= -27.12 \times \left(\dfrac{1}{150}\right) = -0.18$ m

Invert level at Q	$= 87.07 - 0.18 = 86.89$ m
Sight rail top edge level at Q	$= 86.89 + 2.50 = 89.39$ m
Level of top of upright at Q	$= 89.91 - 0.35 = 89.56$ m
Upright level – sight rail level	$= 89.56 - 89.39 = 0.17$ m

That is, the top edge of the sight rail must be fixed **0.17 m** *below* the top of the upright at Q.

Fall of sewer from P to R	$= -\left(\dfrac{27.12 + 54.11}{150}\right) = -0.54$ m

Invert level at R	$= 87.07 - 0.54 = 86.53$ m
Sight rail top edge level at R	$= 86.53 + 2.50 = 89.03$ m
Level of top of upright at R	$= 89.91 - 1.17 = 88.74$ m
Upright level – sight rail level	$= 88.74 - 89.03 = -0.29$ m

That is, the top edge of the sight rail must be fixed **0.29 m** *above* the top of the upright at R.

This is achieved by nailing the sight rail to an extension piece to form a short traveller and then nailing this to the upright such that it adds 0.29 m to its height.

Worked example 11.2: Setting out by offsets from a baseline

Question

A design point Q is to be set out from a control line EF using a theodolite and a steel tape as follows:

- Measure 15.000 m horizontally from E along the line EF and establish a point R.
- From point R, turn off a right angle and measure a horizontal distance of 12.500 m to establish point Q.

The ground slopes uniformly from E to F at an angle of –02°56'40" and from R to Q at an angle of +03°41'20".

Calculate the lengths along the ground surface that must be set out from E to fix the position of point R and from R to fix the position of point Q.

Solution

A plan view of the setting out scheme is shown in Figure 11.100. From Figure 4.9(b) in Section 4.3:

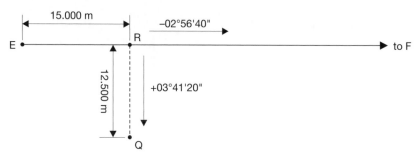

Figure 11.100

$$D = L \cos\theta$$

Hence

$$L = \frac{D}{\cos\theta}$$

Therefore

surface length ER $= \dfrac{15.000}{\cos 02°56'40"} =$ **15.020 m**

and

surface length RQ $= \dfrac{12.500}{\cos 03°41'20"} =$ **12.526 m**

Worked example 11.3: Setting out by intersection

Question

A rectangular building having plan sides of 75.36 m and 23.24 m was set out with its major axis aligned precisely east–west on a coordinate system. The design coordinates of the SE corner were (348.92, 591.76) and this corner was fixed by theodolite intersection from two stations P and Q whose respective coordinates were (296.51, 540.32) and (371.30, 522.22). All dimensions are in metres. The other corners were set out by similar methods.

When the setting out had been completed, the sides and the diagonals of the building were measured as a check. To help with this, the existing ground levels at the four corners of the proposed structure were determined by levelling as follows.

SE (156.82 m), SW (149.73 m), NE (151.45 m), NW (146.53 m)

Calculate

- The respective horizontal angles (to the nearest 20") that were set off at P relative to PQ and at Q relative to QP in order to intersect the position of the SE corner.
- The surface check measurements that should have been obtained for the four sides of the building and the two diagonals, assuming even gradients along all lines.

Solution

Consider Figure 11.101.

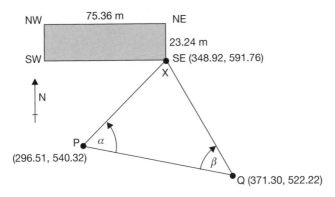

Figure 11.101

Calculation of horizontal angles α and β

Let the SE corner of the building be X.

easting of X	348.92	northing of X	591.76
easting of P	296.51	northing of P	540.32
ΔE_{PX}	+52.41	ΔN_{PX}	+51.44

Therefore from a rectangular $\rightarrow$ polar conversion:

bearing PX = 45°32'07"

easting of X	348.92	northing of X	591.76
easting of Q	371.30	northing of Q	522.22
ΔE_{QX}	-22.38	ΔN_{QX}	+69.54

From a rectangular $\rightarrow$ polar conversion:

bearing QX = 342°09'37"

easting of Q	371.30		northing of Q	522.22
easting of P	296.51		northing of P	540.32
ΔE_{QP}	+74.79		ΔN_{QP}	−18.10

Therefore from a rectangular → polar conversion:

bearing PQ = 103°36'17"

This gives

angle α = bearing PQ − bearing PX = 58°04'10"

clockwise angle to be set off at P relative to PQ = 360° − 58°04'10"

= **301°56'00"**

angle β = bearing QX − bearing QP = 58°33'20"

clockwise angle to be set off at Q relative to QP = **58°33'20"**

Both answers have been rounded to the nearest 20".

Calculation of the surface check measurements
Slope correction = $+(\Delta h^2/2L)$ (see Section 4.3) where Δh is the height difference and L the slope distance (but horizontal distance may be used without significant error).

From SE to SW corners, Δh	= 156.82 − 149.73	= 7.09	$\Delta h^2 = 50.27$
From NE to NW corners, Δh	= 151.45 − 146.53	= 4.92	$\Delta h^2 = 24.21$
From SE to NE corners, Δh	= 156.82 − 151.45	= 5.37	$\Delta h^2 = 28.84$
From SW to NW corners, Δh	= 149.73 − 146.53	= 3.20	$\Delta h^2 = 10.24$

Hence the **slope distances along the four sides** should have been as follows:

SE to SW corners $= 75.36 + \left(\dfrac{50.27}{2 \times 75.36}\right) = 75.36 + 0.33 = \mathbf{75.69\ m}$

NE to NW corners $= 75.36 + \left(\dfrac{24.21}{2 \times 75.36}\right) = 75.36 + 0.16 = \mathbf{75.52\ m}$

SE to NE corners $= 23.24 + \left(\dfrac{28.84}{2 \times 23.24}\right) = 23.24 + 0.62 = \mathbf{23.86\ m}$

SW to NW corners $= 23.24 + \left(\dfrac{10.24}{2 \times 23.24}\right) = 23.24 + 0.22 = \mathbf{23.46\ m}$

For the diagonals

Horizontal diagonals $= \sqrt{(75.36^2 + 23.24^2)} = 78.86$

From SE to NW corners, Δh = 156.82 − 146.53 = 10.29	$\Delta h^2 = 105.88$
From SW to NE corners, Δh = 151.45 − 149.73 = 1.72	$\Delta h^2 = 2.96$

Hence the **slope distances along the diagonals** should have been as follows:

$$\text{SE to NW corners} = 78.86 + \left(\frac{105.88}{2 \times 78.86}\right) = 78.86 + 0.67 = \mathbf{79.53\ m}$$

$$\text{SW to NE corners} = 78.86 + \left(\frac{2.96}{2 \times 78.86}\right) = 78.86 + 0.02 = \mathbf{78.88\ m}$$

Worked example 11.4: Using sight rails

Question

The six corners of the proposed L-shaped excavation shown in Figure 11.102 have been set out on site and offset pegs have been established to help define the sides of the excavation as shown. The horizontal dimensions of the sides and the horizontal offset distances are given in Figure 11.102.

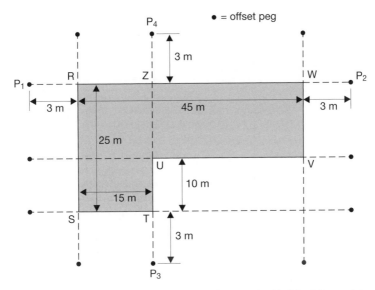

Figure 11.102 ● Plan view for Worked example 11.4 (not to scale).

The proposed formation level of the surface of the excavation at point R is 95.72 m. The surface is to fall at a slope of 1 in 150 from R to W and is to rise at a slope of 1 in 100 at right angles to the line RW. To help with the excavation, sight rails are to be erected above the offset pegs for use with a 2 m traveller. Offset pegs P_1 and P_2 define the line RW and offset pegs P_3 and P_4 define the line TU.

The reduced levels of the corner and offset pegs are shown in Table 11.4. Calculate the heights at which the top edges of the sight rails should be fixed above the tops of the offset pegs P_1, P_2, P_3 and P_4.

Solution

The best approach is to consider the line P_1RWP_2 and then to consider the line P_4UTP_3 as follows:

Table 11.4 ● Data for Worked example 11.4.

Peg R	96.86 m	Peg S	97.47 m
Peg T	96.98 m	Peg U	96.91 m
Peg V	97.61 m	Peg W	96.70 m
Peg P_1	96.95 m	Peg P_2	96.45 m
Peg P_3	97.12 m	Peg P_4	96.75 m

For the line P_1RWP_2

Figure 11.103 shows a section taken along the line P_1RWP_2. With reference to this:

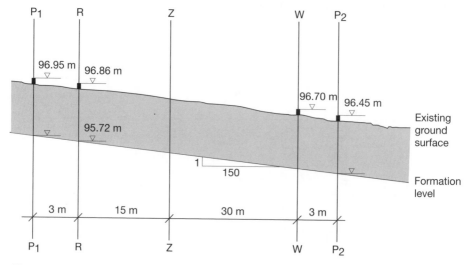

Figure 11.103 ● Section along line P_1RWP_2 for Worked example 11.4 (not to scale).

$$\text{formation level at } P_1 = 95.72 + \left(\frac{3}{150}\right) = 95.74\text{m}$$

$$\text{formation level at } P_2 = 95.72 - \left(\frac{48}{150}\right) = 95.40\text{m}$$

For offset peg P_1

required top of sight rail level = 95.74 + 2.00	=	97.74m
actual top of peg level	=	96.95m
therefore, **distance above peg P_1**	=	**0.79m**

For offset peg P_2

required top of sight rail level = 95.40 + 2.00	=	97.40m
actual top of peg level	=	96.45m
therefore, **distance above peg P_2**	=	**0.95m**

For the line P₄UTP₃

Figure 11.104 shows a section taken along the line P_4UTP_3. With reference to this, point Z is the point where the line TU produced crosses the line RW as shown in Figure 11.102 and, in order to solve this part of the question it is first necessary to calculate the formation level at point Z. With reference to Figure 11.103:

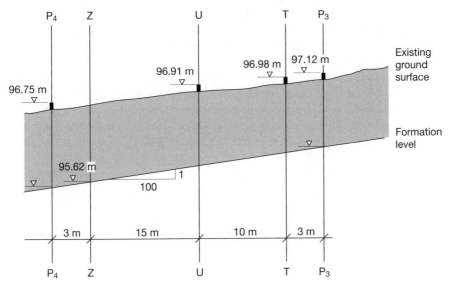

Figure 11.104 ● Section along line P_4UTP_3 for Worked example 11.4 (not to scale).

$$\text{formation level at Z} = 95.72 - \left(\frac{15}{150}\right) = 95.62 \text{ m}$$

Therefore, with reference to Figure 11.104:

$$\text{formation level at } P_3 = 95.62 + \left(\frac{28}{100}\right) = 95.90 \text{ m}$$

$$\text{formation level at } P_4 = 95.62 - \left(\frac{3}{100}\right) = 95.59 \text{ m}$$

For offset peg P₃

required top of sight rail level = 95.90 + 2.00 = 97.90 m
actual top of peg level   = 97.12 m
therefore, **distance above peg P₃**   = **0.78 m**

For offset peg P₄

required top of sight rail level = 95.59 + 2.00 = 97.59 m
actual top of peg level   = 96.75 m
therefore, **distance above peg P₄**   = **0.84 m**

Worked example 11.5: Using slope rails

Question

An embankment was constructed with a formation width of 36.00 m and a formation level of 103.59 m. The transverse slope at right angles to the centre line was 1 in 12 and the side slopes were 1 in 2. Slope rails were used with a 1.50 m traveller held vertically to monitor the formation of the embankment.

At one particular cross-section, taken at point R on the centre line, the existing ground level was 85.08 m. The slope rail on one side of the embankment was attached to two vertical wooden stakes A and B and the slope rail on the other side of the embankment was attached to two vertical wooden stakes C and D. The horizontal distances AB and CD were each 1.00 m. Stakes B and C were nearest the embankment and they were positioned such that their centres were 1.00 m away horizontally from the edges of the embankment to avoid the slope rails being covered. The tops of stakes A, B, C and D were levelled as 80.54 m, 80.81 m, 90.59 m and 89.89 m, respectively.

For this cross-section calculate:

- the slope distances that were set out along the ground surface from point P at right angles to the centre line to establish the centres of stakes A, B, C and D
- the vertical distances that were set out from the tops of the stakes A, B, C and D to fix the top edges of the sight rails in their correct positions

Solution

The embankment and slope rail positions are shown in Figure 11.105. With reference to this and also to Section 15.4 in Chapter 15, the parameters of the embankment are:

$$h = (103.59 - 85.08) = 18.51 \text{ m}; \ n = 2; \ s = 12; \ b = 18.00 \text{ m}$$

Therefore, since this is a two-level cross-section, equation (15.12) can be applied to give

$$W_G = \frac{s(b + nh)}{(s - n)} = \frac{12(18.00 + (2)18.51)}{(12 - 2)} = 66.02 \text{ m}$$

and equation (15.13) can be applied to give

$$W_L = \frac{s(b + nh)}{(s + n)} = \frac{12(18.00 + (2)18.51)}{(12 + 2)} = 47.16 \text{ m}$$

The slope distances that were set out along the ground surface from point P at right-angles to the centre line to establish the centres of stakes A, B, C and D
With reference to Figure 11.105, the horizontal distances from the centre line to the centres of the stakes are:

for stake A = W_G + 1.0 + 1.0 = 66.02 + 1.0 + 1.0 = 68.02 m

for stake B = W_G + 1.0 = 66.02 + 1.0 = 67.02 m

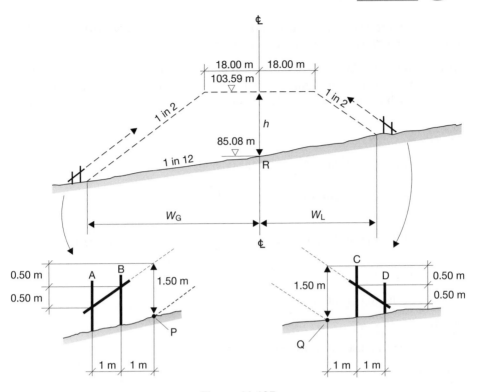

Figure 11.105

for stake C = W_L + 1.0   = 47.16 + 1.0   = 48.16 m

for stake D = W_L + 1.0 + 1.0 = 47.16 + 1.0 + 1.0 = 49.16 m

but the transverse slope = 1 in 12, hence

$$\text{transverse slope} = \tan^{-1}\left(\frac{1}{12}\right) = 04°45'49''$$

Therefore the slope distances from the centre line are:

to the **centre of stake A** = $\dfrac{68.02}{\cos 04°45'49''}$ = **68.26 m**

to the **centre of stake B** = $\dfrac{67.02}{\cos 04°45'49''}$ = **67.25 m**

to the **centre of stake C** = $\dfrac{48.16}{\cos 04°45'49''}$ = **48.33 m**

to the **centre of stake D** = $\dfrac{49.16}{\cos 04°45'49''}$ = **49.33 m**

The vertical distances that were set out from the tops of the stakes A, B, C and D to fix the top edges of the sight rails in their correct positions.

With reference to Figure 11.105, in which points P and Q represent the toes of the embankment (that is, the points where the embankment meets the existing ground surface), each of the stakes should be considered in turn starting with stake B as follows:

For stake B

RL of the top of the rail = RL_P + 1.50 –0.50

RL_P = existing RL on the centre line – $\left(\dfrac{W_G}{12}\right)$

RL of the top of the rail = $85.08 - \left(\dfrac{66.02}{12}\right) + 1.50 - 0.50 = 80.58$ m

By levelling:

RL of the top of the stake = 80.81 m

which gives

the vertical distance = (80.58 – 80.81) = –0.23 m

Therefore

the top edge of the slope rail at B must be fixed **0.23 m below** the top of the stake

For stake A

The top of the rail is 0.50 m below the top of the rail at stake B, hence:

RL of the top of the rail = 80.58 – 0.50 = 80.08 m

From the levelling:

RL of the top of the stake = 80.54 m

which gives

the vertical distance = (80.08 – 80.54) = –0.46 m

Therefore

the top edge of the slope rail at A must be fixed **0.46 m below** the top of the stake

For stake C

RL of the top of the rail = RL_Q + 1.50 – 0.50

RL_Q = existing RL on the centre line + $\left(\dfrac{W_L}{12}\right)$

$$\text{RL of the top of the rail} = 85.08 + \left(\frac{47.16}{12}\right) + 1.50 - 0.50 = 90.01 \text{ m}$$

By levelling:

RL of the top of the stake = 90.59 m

which gives

the vertical distance = (90.01 – 90.59) = –0.58 m

Therefore

the top edge of the slope rail at C must be fixed **0.58 m below** the top of the stake

For stake D
The top of the rail is 0.50 m below the top of the rail at stake C, hence:

RL of the top of the rail = 90.01 – 0.50 = 89.51 m

From the levelling:

RL of the top of the stake = 89.89 m

which gives

the vertical distance = (89.51 – 89.89) = –0.38 m

Therefore

the top edge of the slope rail at D must be fixed **0.38 m below** the top of the stake

Worked example 11.6: Using profile boards

Question
The foundations of a building are to be laid in a trench which has a uniform reduced level of 57.62 m. Corner and offset pegs have already been set out and wooden stakes are driven into the ground on either side of the offset pegs to enable profile boards to be erected. A 2.00 m traveller is to be used to monitor the excavation of the trench.

For the corner shown in Figure 11.106, the tops of stakes A, B, C and D are levelled as 59.75 m, 59.89 m, 59.78 m and 59.51 m, respectively.

Calculate the vertical distances that should be set out from the tops of the wooden stakes at A, B, C and D in order that the top edges of the profile boards are positioned correctly.

Solution

formation level of the trench = 57.62 m
traveller length = 2.00 m

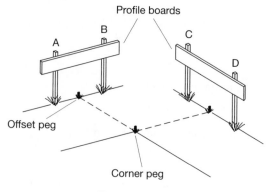

Profile boards

B C

A D

Offset peg

Corner peg

Figure 11.106

Therefore the tops of the profile boards must all be set to a reduced level of

57.62 + 2.00 = 59.62 m

For stake A

Level of top of stake   = 59.75 m
Level of top of board = 59.62 m
Difference   = +0.13 m

Therefore the top of the stake is higher than the required level of the profile board and hence the top edge of the profile board must be set **0.13 m below the top of stake A**.

Repeating the calculations for stakes B, C and D gives the following positions for the top edges of the profile boards

0.27 m below the top of stake B
0.16 m below the top of stake C
0.11 m above the top of stake D; this is done by adding an extension piece to the top of stake D to which the profile board can be attached.

Reflective summary

With reference to setting out worked examples remember:

— It is always advisable to sketch the problem and add all known data to the sketch before beginning the calculations.

— In pipeline calculations, always work from the starting point to the point in question rather than from point to point along the pipeline. This avoids carrying forward any errors.

- Ground surface dimensions will always be greater than horizontal dimensions.

- Knowledge of how to perform rectangular → polar conversions on a hand calculator is a very useful skill to learn.

- When checking square or rectangular buildings, the diagonals must always be measured. Their horizontal lengths should be equal within the specified tolerance.

- When using sight rails, slope rails and profile boards, it is their top edges that are set to known heights above the required formation levels.

- If the wooden stakes that are to carry sight rails, slope rails and profile boards are too short it will be necessary to add extensions pieces to their tops.

- A traveller 2.00 m long is much easier to use that a traveller 5.00 m long.

- The calculations should be presented in a logical sequence so that they can be easily followed for checking purposes.

- The setting out data obtained from the calculations should be presented in such a way that the person undertaking the setting out knows exactly what is required.

Exercises

11.1 State the two main aims of setting out and briefly describe the three basic principles on which setting out operations should be based in order that these aims can be achieved.

11.2 A pipeline is to be laid in a trench at a rising gradient of 1 in 150 from a point E to a point H. The horizontal distance EH is 180.00 m. Long wooden stakes E, F, G, and H have been driven into the ground at 60.00 m horizontal distance intervals on the line EH and sight rails are to be attached to them for use with a 2.00 m traveller. The tops of the stakes are levelled as follows:

E = 45.78 m   F = 45.92 m,   G = 46.15 m,   H = 46.61 m

The formation level of the bottom of the trench at E is to be 43.29 m.

Calculate the difference in level between the top of each stake and the position at which the top edge of each sight rail must be set.

11.3 A new sewer is to be laid to drain a housing development to a manhole on an existing sewer. The centres of three proposed manholes A, B and C on the new sewer have been pegged out on site and the levels of the tops of the pegs were measured as follows:

A = 98.13 m   B = 98.23 m   C = 98.68 m

The new sewer is to be constructed to fall into the manhole on the existing sewer at a gradient of 1 in 200. The horizontal distance between the manhole on the existing sewer and proposed manhole A is 80.00 m, with successive horizontal distance intervals between AB and BC being 66.00 m and 52.00 m, respectively. The cover level of the manhole on the existing sewer is 97.53 m and the formation level at the bottom of this manhole is 1.87 m below its cover level.

If a 3.00 m traveller is to be used during the construction of the trench for the new sewer, calculate the height of the sight rail required above the cover of the existing manhole and also the heights of the sight rails required above the tops of the pegs defining proposed manholes A, B and C.

11.4 Discuss the various different types of plans and drawings that may be required when undertaking setting out operations.

11.5 A rectangular building EFGH having horizontal side lengths EF = GH = 35.000 m and EH = FG = 17.500 m is to be set out on a steeply sloping site. The prevailing temperature is 10 °C.

The horizontal control is provided by a baseline XY which lies on a constant slope. The horizontal length of baseline XY is 50.000 m and its slope length is 50.543 m. Points X and Y are marked with wooden pegs each having a nail in the top.

Side EF is to be located along baseline XY such that point E is a horizontal distance of 7.500 m from X towards Y and point Q is a horizontal distance of 42.500 m from X towards Y.

The equipment available is as follows: a 5" reading electronic theodolite, a steel tape of length 50.009 m at 20 °C and 50 N tension (coefficient of thermal expansion = 0.0000112 per °C), a thermometer, tensioning gear, wooden pegs, nails and a hammer.

Describe in detail, using the required dimensions, how the exact corner positions E, F, G and H should be located.

11.6 One of the toes, T, of a proposed embankment for a new road is to be pegged out on site. The top of the embankment is to run from point A to point B and is to have a horizontal width of 12 m. The proposed reduced levels of A and B are known. Point C is on the existing ground surface directly below point A and point T is on the existing ground surface to the left of point C. The centre line is defined by point F, which is also on the existing ground surface. Only point F and other points on the proposed centre line have been pegged out. The sides of the embankment are to slope at 1 in *s*.

Describe how the position of T can be located.

11.7 A cutting having a formation width of 30.00 m and side slopes of 1 in 2 is to be formed in ground having a transverse slope at right angles to the centre line of 1 in 11.

At one particular cross-section, the cutting is to be controlled by slope rails each placed at right-angles to the centre line and carried by two wooden stakes at 1.00 m centres. The top edges of the slope rails are to be in line with the cutting slopes. Stakes A and B will carry the slope rail on the higher side of the cutting and stakes C and D will carry the slope rail on the lower side. Stakes B and C are nearest the cutting and their centres are placed 1.00 m from its edges to avoid displacement during excavation.

The existing level at the centre line is 65.32 m and it is proposed to excavate 6.98 m to formation level at the centre line. The tops of stakes A, B, C and D are levelled as 70.12 m, 69.76 m, 64.07 m and 64.21 m, respectively.

Calculate the difference in level between the top of each stake and the position at which the top edge of its slope rail must be fixed.

11.8 Discuss in detail the *personnel* who are involved in setting out. Include in your discussion details of their various roles and responsibilities.

11.9 An embankment having a formation width of 15.80 m is to be formed on ground having a transverse slope at right angles to the centre line of 1 in 8 such that the side slopes are 1 in 3. At one particular cross-section, the existing ground level and the formation level at the centre line are 211.67 m and 216.83 m, respectively.

The cross-section is to be controlled by two slope rails, one being attached to two stakes R and S at 1.00 m centres on one side of the embankment and one being attached to two stakes T and U on the other side of the embankment. Stakes R and S are on higher ground than stakes T and U. Stakes S and T are nearest the edges of the embankment and are offset a horizontal distance of 2.00 m from them. The slope rails are to be set parallel to the side slopes with their top edges a vertical distance of 1.50 m away from them. The tops of stakes R, S, T and U were levelled as 214.76 m, 214.79 m, 208.04 m and 207.22 m, respectively.

Calculate:

(i)   The slope distances at right angles to the centre line along the surface of the ground at which the centres of stakes R, S, T and U were set out

(ii)   The difference in level between the top of each stake and the position at which the top edge of its slope rail must be fixed

11.10 With the aid of illustrations, discuss the use of *sight rails, slope rails, profile boards* and *travellers* as means of providing vertical control during setting out operations.

11.11 A particular cross-section on a proposed road is part in cut and part in fill. Side slopes are to be 1 in 2 in cut and 1 in 3 in fill and the transverse slope at right-angles to the centre line is 1 in 6. The formation width of the road is to be 18.00 m; the existing ground level on the centre line is 102.72 m and the required height of fill at the centre line is 0.86 m.

The construction is to be controlled using slope rails. The slope rail on the cut side is to be in line with the required slope of the cutting and is to be attached to two stakes A and B at 1.00 m centres, the centre of B being 1.00 m away from the edge of the cutting. The slope rail on the fill side is to be parallel to but offset a vertical distance of 1.00 m from the required slope of the filling and is to be attached to two stakes C and D at 1.00 m centres, the centre of C being 1.00 m away from the edge of the filling. The tops of stakes A, B, C and D are levelled as 105.76 m, 105.81 m, 99.40 m and 99.38 m, respectively.

Calculate:

(i) The slope distances along the ground surface at right angles to the centre line at which the centres of stakes A, B, C and D must be set out

(ii) The difference in level between the top of each stake and the position at which the top edge of its slope rail must be fixed

11.12 With the aid of illustrations, discuss the various types of horizontal *control grids* that can be used during surveying and setting out operations.

11.13 Discuss the roles of *temporary* and *transferred bench marks* in surveying and setting out operations. Include illustrations showing examples of TBMs and state the accuracies to which they should be established.

11.14 A square building PQRS of horizontal side length 45.000 m is to have a ground floor that slopes uniformly at +4% from P to Q and from S to R. Sides PS and QR are to be level. Profile boards are to be erected at each corner and used with a 2.50 m traveller to monitor the excavation.

Two profile boards are to be used to monitor the excavation along side PQ. One is to be attached to two wooden stakes V and W, which straddle the line QP produced and the other to two wooden stakes X and Y, which straddle the line PQ produced. The line joining V and W and the line joining X and Y both lie at right angles to the line PQ. Line VW is offset a distance of 4.00 m from point P and line XY is offset a distance of 4.00 m from point Q.

The corner peg at point P was levelled as 45.89 m and the required depth of dig at P is 1.92 m. The RLs of the tops of the stakes at V, W, X and Y were levelled as 46.72 m, 46.58 m, 48.88 m and 48.65 m, respectively.

For each of stakes V, W, X and Y, calculate the difference in level between its top and the position at which the top edge of its profile board must be fixed.

11.15 With the aid of illustrations, discuss the use of *offset pegs* in setting out operations.

11.16 Two corners E and F of a proposed building EFGH are to be set out from two traverse stations A and B. The corners are to be set out initially by angle and distance from station A and checked by intersection from station B. The design coordinates of E and F and the traverse coordinates of A and B are based on the same coordinate system and are as follows:

Point	mE	mN
E	968.34	1217.65
F	1018.34	1217.65
A	1056.78	1269.42
B	945.31	1254.66

The ground slopes along the directions to E and F are:

A to E = $-02°17'46"$ and A to F = $-02°41'47"$

Calculate:

(i)   The clockwise horizontal angles that must be set out from A relative to the line AB and from B relative to the line BA to establish the directions to E and F

(ii)   The distances that must be measured along the ground from A to establish the positions of E and F assuming that the gradients are constant along directions AE and AF

11.17 With the aid of illustrations, discuss how the following should be undertaken:

Setting out column positions

Setting out pile positions

Transferring height from floor to floor

11.18 The corner positions of a square building ABCD were set out on a construction site from nearby horizontal control stations K and L in turn using a 1" reading total station. The designed horizontal side lengths of the building were each 30.00 m.

Initial rough setting out of one of the corners, A, enabled its approximate position to be established and the RL at this point was measured as 27.39 m. The RLs of control points K and L were known to be 25.43 m and 29.37 m, respectively.

A detail pole, set to the same height as the total station, was used to establish the precise position of point A. The other corner positions were later set out in a similar manner. The coordinates of A, K and L, which were based on the same coordinate system, were as follows:

Point	mE	mN
A	709.11	872.36
K	806.27	651.92
L	823.58	905.43

Once the setting out had been completed, reduced levels of 27.18 m, 26.27 m and 26.11 m were taken at points B, C and D, respectively. In addition, the slope lengths along diagonals AC and BD were measured with a steel tape as 42.439 m and 42.443 m, respectively. The ground surface sloped uniformly along the diagonals.

Calculate:

(i)　The horizontal angles that were set out from K relative to the line KL and from L relative to the line LK to establish the directions to point A

(ii)　The slope distances that were set out on the total station in turn along these directions to fix the positions of point A

(iii)　The errors in the diagonals AC and BD

11.19 Discuss, in detail, the *working practices* that should be adopted both before and during setting out operations in order to ensure that constructions are positioned correctly and that work proceeds without delays.

11.20 An oil drilling rig R can be seen from two horizontal control stations A and B. Using a total station, the following measurements are taken to a set of prisms located 1.563 m above the top of the rig.

At station A

Height of tilting axis above A = 1.528 m
Mean horizontal circle reading to B = 20°17'56"
Mean horizontal circle reading to R = 93°41'22"
Mean vertical angle to R = −02°09'18"
Mean slope distance to R = 1245.681 m.

At station B

Height of tilting axis above B = 1.499 m
Mean horizontal circle reading to A = 09°28'46"
Mean horizontal circle reading to R = 312°14'21"
Mean vertical angle to R = −01°57'42"
Mean slope distance to R = 1419.306 m

The coordinates and reduced levels of A and B are as follows:

A　6193.45 mE, 7091.22 mN, 117.65 m
B　7281.56 mE, 6811.93 mN, 119.42 m

Calculate:

(i)　The coordinates of the drilling rig R

(ii)　The reduced level of the top of the rig, ignoring the effects of curvature and refraction

11.21 With the aid of illustrations, describe in detail *four* methods by which *verticality* can be controlled in structures.

11.22 Two design points E and F are to be set out by intersection from the ends of a baseline XY using two 1" reading theodolites. The coordinates of the points, which are based on the same coordinate system, are as follows:

E	1192.41 mE, 1336.78 mN	F	1207.66 mE, 1108.25 mN
X	1063.17 mE, 1214.89 mN	Y	1241.11 mE, 1235.46 mN

Calculate the horizontal angles that must be set out at X relative to the line XY and at Y relative to the line YX to establish the positions of E and F.

11.23 Describe the current classification system used to categorise the safety of lasers in surveying and setting out instruments. Include details of the precautions that should be undertaken when using such instruments on site.

11.24 (a) Explain the difference between *fixed beam* lasers and *rotating beam* lasers and describe *one* instrument from each category.

(b) Discuss the advantages and disadvantages that laser instruments have compared to conventional surveying instruments when used in setting-out operations.

11.25 With the aid of illustrations, discuss how laser instruments can be used in the following operations:

Controlling pipelaying

Controlling tunneling

Fixing internal walls and suspended ceilings

Controlling earth-moving machinery

11.26 A straight tunnel is to be driven between stations P and S which are on opposite sides of a hillside. Stations Q and R were located such that P can be seen from Q and S can be seen from R with Q and S being inter-visible. The following observations were made:

Clockwise horizontal angles	Horizontal distances (m)
PQR = 78°14'26"	AQ = 126.59
QRS = 81°16'45"	QR = 915.88
	RS = 109.32

The coordinates of P are 1716.32 mE, 6897.14 mN and bearing PQ is 172°18'39".

Calculate:

(i) The clockwise horizontal angle to be set off at P from PQ to the centre line of the tunnel

(ii) The clockwise horizontal angle to be set off at S from SR to the centre line of the tunnel

(iii) The horizontal length of the tunnel if P is located a horizontal distance of 45.19 m from one end of the tunnel and S is located a horizontal distance of 87.44 m from the other end. The tunnel falls within the line PS.

11.27 Discuss the suitability of total stations and GPS receivers for setting out coordinate positions on site. Include in your discussion the advantages and disadvantages of such systems when compared to traditional setting out techniques.

11.28 A rectangular building PQRS (lettered clockwise) having plan sides of PQ = RS = 40.000 m and QR = SP = 25.000 m is to be set out using polar methods from two traverse stations H and J such that side RQ has a bearing of 00°00'00" on a local coordinate grid system. The coordinates of building corner R and those of traverse stations H and J are on the local coordinate grid system and are given below:

Point	mE	mN
Building corner R	550.00	650.00
Station H	479.32	623.77
Station J	611.89	563.21

(i)   Calculate the clockwise horizontal angles and horizontal distances to be set off at station H relative to the line HJ in order to establish corners R and S

(ii)   Calculate the clockwise horizontal angles and horizontal distances to be set off at station J relative to the line JH in order to check the positions of corners R and S as established from H

(iii)   Describe a procedure for setting out the other corners of building PQRS and their associated offset pegs starting from corners R and S

11.29 Describe, in detail, *three* different systems by which construction machinery can be automatically controlled on site.

11.30 Discuss the need for quality assurance in surveying and setting out operations with particular reference to the accuracy requirements of the various stages of the construction process. Include in your discussion a description of a *Quality Plan* and an explanation of its purpose.

Further reading and sources of information

For the latest information on the personnel involved in setting out and construction, consult

ICE Conditions of Contract Measurement Version, 7th edn (1999). Thomas Telford, London.
JCT 98 (1998) *Standard Form of Building Contract*. RIBA Publications, London. 1998 edition with amendments up to 2003.

For the latest information on setting out procedures, consult

Sadgrove, B. M. (1997) *Setting-out Procedures*, 2nd edn. Construction Industry Research and Information Association (CIRIA), London. http://www.ciria.org.uk/.

ICE Design and Practice Guide (1997) *The Management of Setting Out in Construction*. Thomas Telford, London.

RICS Guidance (2003) Note *Guidelines for the Use of GPS in Surveying and Mapping*. RICS Books, London.

For the latest information on laser safety, consult

IEC 60825-1: 1993+A1:1997+A2:2001, Ed 1.2 (2001-08), *Safety of Laser Products – Part 1: Equipment Classification, Requirements and User's Guide*, International Electrotechnical Commission, 3 rue de Varembé, Geneva, Switzerland, August 2001. http://www.iec.ch/.

IEC 60825-14: Ed 1.0 (2004-02), *Safety of Laser Products – Part 14: A User's Guide*, International Electrotechnical Commission, 3 rue de Varembé, Geneva, Switzerland, Feb 2004. http://www.iec.ch/.

R. Henderson and K. Schulmeister (2004) *Laser Safety*. Institute of Physics Publishing, Bristol.

For the latest information on quality assurance and accuracy, consult

BS 5606 (1990) *Guide to Accuracy in Building*. British Standards Institution (BSI), London. http://www.bsi-global.com/.

BS 5750-8:1991 (1991) *Quality Systems. Guide to Quality Management and Quality Systems Elements for Services*. British Standards Institution, London.

BS 5964 (1990 and 1996) *Building Setting Out and Measurement*. British Standards Institution, London.

BS 7307 (1990) *Building Tolerances: Measurement of Buildings and Building Products*. British Standards Institution, London.

BS 7308 (1990) *Methods for Presentation of Dimensional Accuracy Data in Building Construction*. British Standards Institution, London.

BS 7334 (1990, 1992) *Measuring Instruments for Building Construction* Parts 1, 2 and 3 – 1990, Parts 4, 5, 6, 7 and 8 – 1992. British Standards Institution, London.

EN 29004-2:1993 (1993) *Quality Systems. Guide to Quality Management and Quality Systems Elements for Services*. European Committee for Standardisation. http://www.cenorm.be/.

ISO 4463-1 1989: *Measurement Methods for Building – Setting Out and Measurement – Part 1: Planning and Organization, Measuring Procedures, Acceptance Criteria*. International Organization for Standardization (ISO), Geneva. http://www.iso.ch/.

ISO 7078: 1985 *Building Construction – Procedures for Setting Out, Measurement and Surveying – Vocabulary and Guidance Notes*. International Organization for Standardization, Geneva.

ISO 8322: 1991 *Building Construction – Measuring Instruments – Procedures for Determining the Accuracy in Use*. International Organization for Standardization, Geneva.

ISO 17123: 2001–2005 *Optics and Optical Instruments – Field procedures for testing geodetic and surveying instruments*. International Organization for Standardization, Geneva.

chapter 12

Circular curves

 Aims

After studying this chapter you should be able to:

- Differentiate between different types of horizontal and circular curves
- Understand the terminology and geometry of circular curves
- Calculate through chainage values along the centre lines of circular curves
- Design curves of constant radii to join straight sections of roads and railways
- Appreciate the use of computer software packages in the design of circular curves
- Set out the centre lines of circular curves on site using a variety of different methods
- Choose the most appropriate method of setting out for different situations where circular curves are used
- Plot the centre line of a circular curve on a drawing

This chapter contains the following sections:

12.1 Horizontal curves

12.2 Circular curves and their geometry

12.3 Through chainage

12.4 Designing circular curves

12.5 Introduction to setting out horizontal curves on site

12.6 Setting out circular curves on site by traditional methods

12.7 Setting out circular curves on site by coordinate methods

12.8 Plotting the centre lines of circular curves on drawings

12.1 Horizontal curves

After studying this section you should have an appreciation of the different types of horizontal curves that are used in the design of roads and railways.

Types of horizontal curves

In the design of roads and railways, straight sections of road or track are connected by curves of constant or varying radius, as shown in Figure 12.1. The purpose of the curves is to deflect a vehicle travelling along one of the straights safely and comfortably through the angle θ to enable it to continue its journey along the other straight. For this reason, θ is known as the *deflection angle*.

The curves shown in Figure 12.1 are *horizontal curves* since all measurements in their design and construction and considered in the horizontal plane. The two main types of horizontal curves are:

- *Circular curves*, which are curves of constant radius as shown in Figure 12.1(a).
- *Transition curves*, which are curves of varying radius as shown in Figure 12.1(b).

Usually, some combination of straights, circular curves and transition curves is used in the final design. Together, they form what is known as the *horizontal alignment*. This

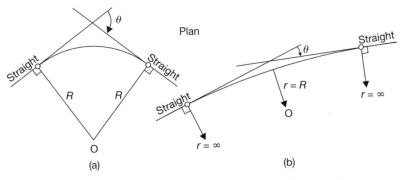

Figure 12.1 ● Horizontal curves: (a) circular curve; (b) transition curve.

chapter covers the design and setting out of circular curves, and Chapter 13 covers transition curves.

Reflective summary

With reference to horizontal curves remember:

— The purpose of horizontal curves is to ensure that vehicles travel safely from one straight section of road or track to another.

— Circular curves are just one type of horizontal curve.

— The term *horizontal alignment* is used to describe the combination of horizontal curves and straights used in the design.

12.2 Circular curves and their geometry

After studying this section, you should be aware of the different types of circular curve that can be used when designing roads and railways, and you should be familiar with their terminology and geometry. You should also understand that much of the terminology used when referring to circular curves also applies to transition curves.

This section includes the following topics:

● Types of circular curve
● Terminology used in circular curves
● Formulae used in circular curves
● Important relationships in circular curves

Types of circular curve

There are three basic types of circular curve: *simple curves, compound curves* and *reverse curves*, all of which can be referred to either as *radius curves* or *degree curves*.

Simple circular curves

A simple circular curve consists of one arc of constant radius R, as shown in Figure 12.1(a). These are the most commonly used types of circular curve.

Compound circular curves

These consist of two or more consecutive simple circular curves of different radii without any intervening straight section. A typical two-curve compound curve is shown in Figure 12.2, where a curve of radius R_1 joins a curve of radius R_2. The object of such curves is to avoid certain points, the crossing of which would involve great expense and which cannot be avoided by a simple circular curve.

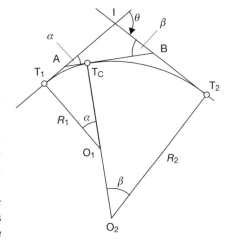

Figure 12.2 ● Compound curve.

Nowadays they are uncommon since there is a change in the radial force (as defined in Section 13.1) at the *common tangent point* T_C where one curve meets another, as shown in Figure 12.2. The effect of this, if the change is marked, can be to give a definite jerk to the passengers, particularly in trains. To overcome this problem, either *very large radii* should be used to minimise the forces involved or *transition curves* (see Chapter 13) should be used instead of the compound curve.

If a compound curve such as that shown in Figure 12.2 is being considered then the best approach is to treat it is as if it were two consecutive simple circular curves, that is, T_1T_C and T_CT_2. This is done by introducing a *common tangent line* AB, which passes through the common tangent point T_C. This creates angles α and β, which are the deflection angles for the simple circular curves T_1T_C and T_CT_2, respectively, where $(\alpha + \beta) = \theta$. Having established α and β, the radii of curvature R_1 and R_2 can be chosen and the curves designed as described in Section 12.4. With reference to Figure 12.2, $T_1A = AT_C$ and $T_CB = BT_2$ but AT_C does not equal T_CB.

Reverse circular curves

These curves consist of two consecutive curves of the same or different radii without any intervening straight section and with their centres of curvature falling on opposite sides of their common tangent point. They are much more common than compound circular curves and, like them, can be used to avoid obstacles. More usually, however, they are used to connect two straights which are very nearly parallel and that would otherwise require a very long simple circular curve.

A typical reverse circular curve is shown in Figure 12.3. In order to connect the two straights T_1I_1 and T_2I_2, it is necessary to introduce an additional straight I_1I_2. During the design process, several additional straights could be tried until a suitable solution is found and common tangent point T_C is chosen.

Once point T_C has been selected, the reverse curve can be considered as two separate simple curves with no intermediate straight section, that is, T_1T_C and T_CT_2. With reference to Figure 12.3, $T_1I_1 = I_1T_C$ and $T_CI_2 = I_2T_2$, but I_1T_C does not necessarily equal T_CI_2.

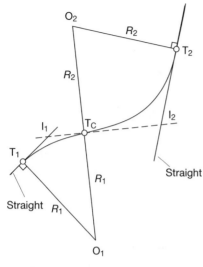

Figure 12.3 ● Reverse curve.

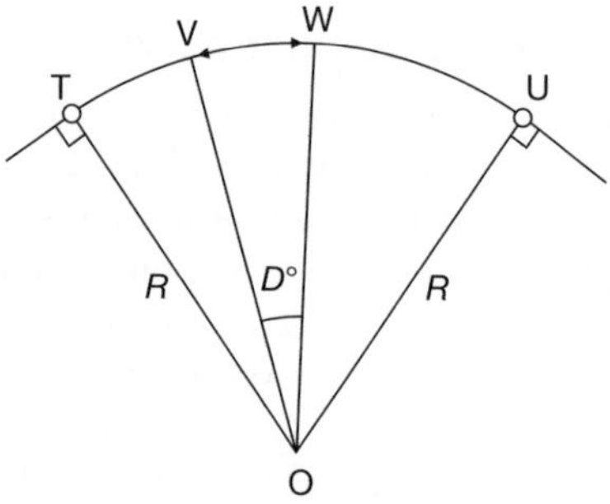

Figure 12.4 ● Degree curve.

Radius and degree curves

A circular curve can be referred to in one of two ways:

● In terms of its radius, for example, a 750 m curve. This is known as a *radius curve*.

● In terms of the angle subtended at its centre by a 100 m arc, for example a 2° curve. This is known as a *degree curve*, an example of which is shown in Figure 12.4.

In Figure 12.4, arc VW = 100 m and subtends an angle of $D°$ at the centre of curvature O. The curve TU is, therefore, a $D°$ degree curve. The relationship between radius curves and degree curves is given by the formula

$$DR = \frac{18{,}000}{\pi} \tag{12.1}$$

in which D is in degrees and R in metres.

A curve of radius 1500 m is equivalent to

$$D° = \frac{18{,}000}{1500\pi} = \frac{12}{\pi} = 3.820°$$

That is, a 1500 m radius curve = a 3.820° degree curve.

Terminology used in circular curves

Although circular curves are straightforward in nature, much of their terminology also applies to transition curves and it is vital, therefore, that a good understanding

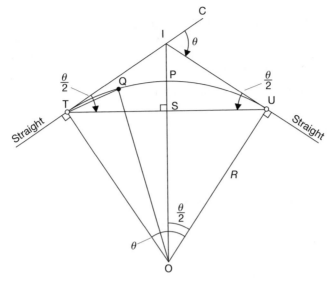

Figure 12.5 ● Circular curve geometry.

of the terminology used in circular curves is attained before proceeding to study transitions. A simple circular curve is shown in Figure 12.5 and, with reference to this, some of the terms and symbols commonly used in circular curve geometry are defined as follows:

- I is the *intersection point* of the two *straights* TI and IU
- TPU is a *circular curve* which runs around the arc from T to U
- The *length of the circular curve* around the arc TPU = L_C
- T and U are the *tangent points* to the circular curve
- TI and UI are the *tangent lengths* of the circular curve
- P is the *mid-point* of the *circular curve* TPU
- *Long chord* = TSU
- S is the *mid-point* of the long chord TSU
- *Deflection angle* = θ = *external angle* at I = angle $\hat{CIU}$
- *Intersection angle* = $(180° - \theta)$ = *internal angle* at I = $\hat{TIU}$
- *Radius of curvature* of the circular curve = R
- *Centre of curvature* of the circular curve = O
- Q is *any point* on the circular curve TPU
- *Tangential angle* = for example, angle $\hat{ITQ}$ = the angle from the tangent length at T (or U) to any point on the circular curve
- The *mid-ordinate* of the circular curve = PS
- *Radius angle* = angle $\hat{TOU}$ = *deflection angle* $\hat{CIU}$ = θ
- *External distance* = PI

Formulae used in circular curves

From the geometry of Figure 12.5, the following formulae can be derived for use in the design and setting out procedures:

- *Tangent lengths IT and IU*: in triangle IUO

$$R\tan\left(\frac{\theta}{2}\right) = \frac{IU}{IO} = \frac{IU}{R}$$

Hence $IU = IT = R\tan\left(\frac{\theta}{2}\right)$ (12.2)

- *External distance, PI*: in triangle IUO

$$\cos\left(\frac{\theta}{2}\right) = \frac{R}{IO}$$

or $IO = \dfrac{R}{\cos(\theta/2)} = R\sec\left(\dfrac{\theta}{2}\right)$

But $PI = OI - OP = OI - R$

Hence $PI = R\sec\left(\dfrac{\theta}{2}\right) - R = R\left[\sec\left(\dfrac{\theta}{2}\right) - 1\right]$ (12.3)

- *Mid-ordinate, PS*: in triangle TSO

$$\cos\left(\frac{\theta}{2}\right) = \frac{OS}{OT}$$

or $OS = OT\cos\left(\dfrac{\theta}{2}\right) = R\cos\left(\dfrac{\theta}{2}\right)$

But $PS = OP - OS = R - R\cos\left(\dfrac{\theta}{2}\right) = R\left[1 - \cos\left(\dfrac{\theta}{2}\right)\right]$ (12.4)

- *Long chord, TU*: in triangle USO

$$\sin\left(\frac{\theta}{2}\right) = \frac{US}{UO} = \frac{US}{R}$$

or $US = R\sin\left(\dfrac{\theta}{2}\right)$

But $TU = US + TS$ and $US = TS$

Hence $TU = 2R\sin\left(\dfrac{\theta}{2}\right)$ (12.5)

The *length of the circular curve, L_C,* can be obtained from one of two formulae, depending on whether the curve is a radius curve or a degree curve:

For a curve of radius R:

$$L_C = R\theta \text{ metres} \tag{12.6}$$

where R is in metres and θ is in radians.

For a $D°$ degree curve:

$$L_C = 100\left(\frac{\theta}{D}\right) \text{ metres} \tag{12.7}$$

where θ and D are in the same units, either degrees or radians.

Important relationships in circular curves

In Figure 12.5, triangle ITU is isosceles, therefore angle $\hat{ITU}$ = angle $\hat{IUT}$ = $(\theta/2)$. Using this, the following definition can be given with reference to Figure 12.6:

> *The tangential angle, α, at T to any point, X, on the curve TU is equal to half the angle subtended at the centre of curvature, O, by the chord from T to that point.*

Similarly, with reference to Figure 12.7:

> *The tangential angle, β, at any point, X, on the curve to any forward point, Y, on the curve is equal to half the angle subtended at the centre by the chord between the two points.*

Another useful relationship is illustrated in Figure 12.8, which is a combination of Figures 12.6 and 12.7.

- From Figure 12.6, angle $\hat{TOX} = 2\alpha$; hence angle $\hat{ITX} = \alpha$.
- From Figure 12.7, angle $\hat{XOY} = 2\beta$; hence angle $\hat{AXY} = \beta$.

Therefore, in Figure 12.8, angle $\hat{TOY} = 2(\alpha + \beta)$ and it follows that angle $\hat{ITY} = (\alpha + \beta)$. In words this can be stated as:

> *The tangential angle to any point on the curve is equal to the sum of the tangential angles from each chord up to that point.*

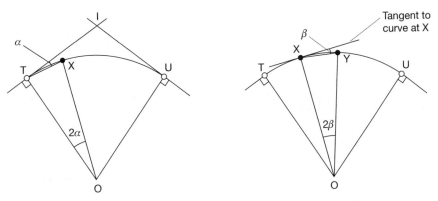

Figure 12.6   **Figure 12.7**

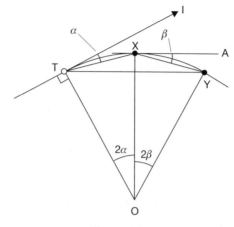

Figure 12.8

The relationships illustrated in Figures 12.6, 12.7 and 12.8 are used when setting out circular curves by the method of tangential angles (see Section 12.6).

Reflective summary

With reference to circular curves and their geometry remember:

— There are three basic types of circular curve: *simple curves*, *compound curves* and *reverse curves*, all of which can be referred to either as *radius curves* or *degree curves*.

— It is important that an understanding of the terminology used in circular curves is attained since much of it also applies to other types of horizontal curves, such as transition curves.

— Specific formulae can be derived for some of the circular curve parameters, which can be used to help both with the design and setting out of the curves.

12.3 Through chainage

After studying this section you should have an understanding of the concept of through chainage and how it can be used to reference the positions of points on the centre lines of long narrow constructions such as roads and railways. You should be able to calculate the through chainages of the tangent points of circular curves and have an appreciation of how the chainage of any point on the centre line of a circular curve can be determined.

Concept of through chainage

Through chainage, which is often just referred to as *chainage*, is simply a distance and is usually given in metres. It is a measure of the length from the starting point of the scheme to the particular point in question and is used in road, railway, pipeline, tunnel and canal construction as a means of referencing any point on the centre line.

Figure 12.9 shows a circular curve, of length L_C and radius R running between two tangent points T and U, which occurs in the centre line of a new road. As shown in Figure 12.9, chainage increases along the centre line and is measured from the point Z at which the new construction begins. Z is known as the *position of zero chainage*.

Chainage continues to increase from Z along the centre line until a curve tangent point such as T is reached. At T, the chainage can continue to increase in two directions, either along the curve (that is, from T towards U) or along the straight (that is, from T along the line TI produced). Hence intersection point I and tangent points T and U can all have chainage values.

Tangent point T is known as the *entry tangent point* because it is the point at which the curve is *entered* in the direction of increasing chainage and tangent point U is known as the *exit tangent point* because it is point at which the curve is *exited* in the direction of increasing chainage.

At the beginning of the design stage, when only the positions of the straights will be known, chainage is considered along the straights. However, once the design has been completed and the lengths of all the curves are known, the centre line becomes the important feature and chainage values must be calculated from the position of zero chainage along the centre line only. This is done in order that pegs can be placed at regular intervals along the centre line to enable earthwork quantities to be calculated (see Chapter 15). If the chainage of the intersection point, I, is known and the curve is then designed, the chainages of tangent points T and U, which both lie on the centre line, can be found as follows with reference to Figure 12.9:

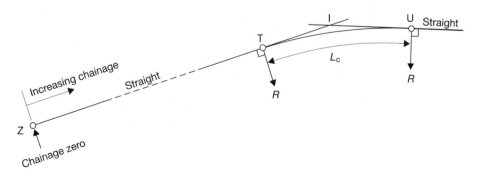

Figure 12.9 ● Chainage along a circular curve.

through chainage of T = through chainage of I − IT

through chainage of U = through chainage of T + L_C

A common mistake in the calculation of through chainage is to assume that (TI + IU) = L_C. This is not correct. Similarly, the chainage of U does not equal the chainage of I + IU. To avoid such errors the following rule must be obeyed:

When calculating through chainage from a point which does not lie on the centre line (for example, point I in Figure 12.9) it is necessary first to calculate the chainage of a point which lies further back on the centre line (that is, in the direction of zero chainage) before proceeding in a forward direction along the centre line.

Reflective summary

With reference to through chainage remember:

— *Through chainage* or *chainage* is a distance and is usually given in metres.

— It is used to reference points on the centre line of a project.

— It is a measure of the distance from the starting point of the project to the point in question.

— When calculating through chainage from a point which does not lie on the centre line you should first calculate the chainage of a point which lies further back on the centre line before proceeding in a forward direction along the centre line.

12.4 Designing circular curves

After studying this section you should be able to design circular curves to join two or more intersecting straights and you should appreciate that the design process is normally undertaken using computer software packages. You should understand that circular curves are just one of a variety of horizontal curves that can be used to solve the basic problem of enabling a vehicle to travel safely from one straight section of road or railway to another. You should also be aware that circular curves cannot always be used in isolation and may have to be combined with transition curves or even omitted altogether in favour of computer generated curves.

This section includes the following topics:

● A design method for circular curves

● The use of computer software packages in the design procedure

A design method for circular curves

In circular curve design there are three main variables: the deflection angle θ, the radius of curvature R and the design speed v.

All new roads are designed for a particular speed and the chosen value depends on the road type and location of the proposed road. Design speeds for particular classes of roads in the United Kingdom are specified in a Department of Transport publication entitled the *Design Manual for Roads and Bridges*. Volume 6 of this manual includes a traffic standard *TD 9/93 – Amendment No. 1 – Highway Link Design*, which gives guidelines for design speeds (see Section 13.2 for further discussion of *TD 9/93*). Hence v will be known, which leaves θ and R to be determined.

When *designing new roads*, there is usually a specific area (often referred to as a *band of interest*) within which the proposed road must fall to avoid certain areas of land and unnecessary demolition. When *improving existing roads*, this band of interest is very clearly defined and is often limited to the immediate area next to the road being improved. In both cases, there will be a limited range of values for both θ and R in order that the finished road will fall within this band of interest.

Nowadays, roads are almost invariably designed using software packages with values of θ and R being input and amended as necessary until a suitable design is finalised. The software contains a map of the area through which the road is to pass and a variety of different combinations of intersecting straights can be tested by superimposing them over the digital map. Once a solution is found, which ensures that the horizontal alignment fits into the band of interest and meets any other criteria specified in the design, θ is given by the software. Occasionally, hand designs are still undertaken. In these, the straights are drawn on a plan of the area and an initial value for θ is obtained by measuring it from the drawing using a protractor. This initial value is ideal for a preliminary design to ensure that the road will fit adequately into the area. However, at some stage, because of the inaccuracy of the protractor reading, θ must be measured on site in order that the design can be refined.

R is chosen with reference to design values again stipulated in *TD 9/93*. Basically, the manual limits the value of the minimum radius that can be used at a particular speed for a wholly circular curve. If a radius below the minimum is used it is necessary to incorporate transition curves into the design.

During the design process, an initial radius value, greater than the minimum without transitions is chosen. If a software package is being used, it automatically calculates the tangent lengths and fits them onto a map. If a hand design is being performed, the tangent lengths should be calculated using equation (12.2) and then fitted on the drawing by measuring along the straights from the intersection point. In both cases, if there are no problems with this fit, this initial design can be used; otherwise a new radius value should be chosen and a new fit obtained. Eventually, a suitable R value would be selected. The design is completed by calculating the *superelevation* required for the curve. This is fully discussed in Section 13.1.

This trial and error method is suitable if any value of R above the minimum without transitions can be used and literally thousands of designs are possible and will all be perfectly acceptable. However, if a curve has to have particular tangent lengths, then since θ will be known, R can be determined from $R\tan(\theta/2)$. R should then be checked against the values specified in *TD 9/93* to ensure that it is greater than the minimum without transitions. If it is not, either θ will have to be reduced or transition curves must be incorporated.

Once θ and R have been finalised the setting out of the centre line on site can begin. This is discussed in sections 12.5, 12.6 and 12.7.

Of all the various types of horizontal curve available, those with a constant radius are the easiest to design and have the simplest setting out calculations. However, they cannot always be used alone owing to limitations on their minimum radii, as specified in *TD 9/93*. If they cannot be used in isolation, circular curves can be combined with transition curves to form *composite curves*. It is usually possible to design a composite curve to fit any reasonable combination of deflection angle and radius. Transition curves, composite curves and the restrictions on radii specified in *TD 9/93* are discussed in detail in Section 13.2.

Increasingly, however, because of the use of highway design software packages, there is a tendency to eliminate circular curves and transition curves from the design altogether and instead to use curves of constantly changing radius. These are known as *polynominals* because their equations take the form of cubic polynomials, an example of such a curve being a *cubic spline*. These are true computer-based curves, which are generated by the highway design software once any design constraints have been input such as the positions of intersection points, the coordinates of points through which the curve must pass and the locations of points that must be avoided. Further information on these types of curves is given in Section 13.5.

The use of computer software packages in the design procedure

As discussed above, the design procedures for circular curves are nowadays often done using software packages. A wide range of such software is available with many different manufacturers offering complete suites of highway design and volume analysis packages. These cover all aspects of the design process from the initial choice and subsequent refining of the horizontal and vertical alignments to the calculation of volumes and the planning of the movement of earthworks necessary for the final designed route.

Although a detailed description of the ways in which such packages work is not included here, they are all based on the fundamental principles of curve design and earthwork calculations discussed in this chapter and in Chapters 13, 14 and 15. Consequently, throughout these chapters, references are made to software packages where appropriate. Sections 13.5 and 14.4, in particular, discuss their applications in highway design in greater detail.

Highway design and volume analysis packages cannot function without data and this is usually provided by a basic mapping package onto which the highway design and volume analysis software can be added as *modules* to extend the capabilities of the system. The basic package usually consists of a series of land surveying modules in which control is established and the detail and the contours are located. These enable a contoured site plan to be produced and a three-dimensional DTM of the existing ground surface to be generated. Such DTMs provide the basic data required for the highway design and volume analysis modules. Further information on DTMs and other aspects of computer-aided surveying and mapping are given in Section 10.9 in the *Detail Surveying and Mapping* chapter.

Reflective summary

With reference to designing circular curves remember:

—   In circular curve design there are three main variables: the deflection angle θ, the radius of curvature R and the design speed v.

—   Design speeds for particular classes of roads are specified in *TD 9/93*.

— θ is chosen such that the horizontal alignment will fit into the *band of interest* through which the centre line must pass and also meet any other criteria specified in the design.

— R is chosen with reference to *TD 9/93*, which limits the value of the minimum radius that can be used for a particular design speed for a wholly circular curve.

—   Circular curves cannot always be used in isolation and may have to be combined with transition curves.

—   Horizontal alignments are normally designed using software packages.

12.5 Introduction to setting out horizontal curves on site

After studying this section you should understand the importance of the centre line on site and you should have an appreciation of the factors to be considered when planning to set it out. You should be aware of the main categories of setting out methods that can be employed and you should be able to choose the most appropriate method for different situations where horizontal curves are used and in cases where the setting out operations may be temporarily impeded.

This section includes the following topics:

- The importance of the centre line
- The different setting out methods that are available
- Obstructions to setting out

The importance of the centre line

The centre line provides an important reference line on site. Once it has been pegged out, other features such as channels, verges, tops and bottoms of embankments, edges of cuttings and so on, can be fixed from it. Consequently it is vital that:

- The centre line pegs are established to a high degree of accuracy.

- They are protected and marked in such a way that site traffic can clearly see them and avoid disturbing them accidentally.

- In the event of them being disturbed, they can be relocated easily and quickly to the same accuracy to which they were initially set out.

The different setting out methods that are available

There are a number of methods by which the centre line can be set out, all of which fall into one of the following two categories.

- *Traditional methods*, which involve working along the centre line itself using the straights, intersection points and tangent points for reference. The equipment required to carry out these methods can include tapes, theodolites and total stations.

- *Coordinate methods*, which use control networks as reference. These networks take the form of control points located on site some distance away from the centre line for use with theodolites, total stations or GPS receivers.

Although both of the above categories are still used, coordinate methods have virtually superseded traditional ones for all major curve setting out operations for a number of reasons:

- There is now widespread use of total stations on site and GPS equipment is increasingly being used for setting out purposes.

- The almost universal adoption of highway design software packages, which are invariably based on coordinate methods, has eliminated the tedious nature of the calculations involved in such methods and enables setting out data to be produced in a form ready for immediate use by total stations and GPS receivers.

- Coordinate methods have the advantage that relocating centre line pegs, which have been disturbed, is much easier to carry out than by traditional methods.

However, coordinate methods are not always the most appropriate and traditional techniques are still used for less important curves, for example, those used on roads for housing estates, on minor roads, for kerb lines, for boundary walls and so on, where they are often more convenient and quicker to use than coordinate methods. They also represent the only possibility in cases where no nearby control points are available. The relative merits of traditional and coordinate methods are discussed further in Section 13.6.

If traditional methods are to be used, the first setting out operation is normally to fix the position of the intersection and tangent points on site. Points on the centre line can then be set out from the tangent points by one of a number of different methods. If coordinate methods are to be used, it is not normally necessary to set out the intersection or tangent points. Instead, coordinates of points on the curve are determined at the design stage by the software and presented in a form suitable for use by total stations or GPS receivers, which are then used to set them out from a control network. Traditional methods and coordinate methods of setting out circular curves are discussed in Sections 12.6 and 12.7, respectively, and those for transition curves are discussed in Section 13.6.

Obstructions to setting out

If care has been taken at the design stage with the route location and the choice of a suitable radius then the setting out of the centre line should proceed smoothly, no matter whether traditional or coordinate methods are being used. Occasionally, however, depending on the terrain and the existing use of the land over which the centre line is to be constructed, there may be temporary obstructions to the setting out operations. Should these be foreseen then it is advisable to opt for one of the coordinate methods of setting out, since these enable the centre line sections on either side of the obstruction to be set out thereby allowing the Works to proceed. Once the obstruction has been removed, the same coordinate method can be employed to establish the missing section of the centre line.

Reflective summary

With reference to setting out horizontal curves on site remember:

— The centre line provides an important reference line on site, which can be used to establish other features such as channels, verges and embankments.

— Centre lines can be established either by *traditional methods* or *coordinate methods* – coordinate methods are mainly adopted nowadays due to the widespread use of total stations and the increasing use of GPS receivers on site.

— It is sometimes more appropriate to use traditional methods rather than coordinate methods.

> — In cases where the proposed route of the centre line may be tempo-
> rarily obstructed, it is advisable to adopt one of the coordinate
> methods of setting out.

12.6 Setting out circular curves on site by traditional methods

After studying this section you should be able to locate the positions of intersection points and tangent points on site. You should know three traditional methods by which pegs on the centre lines of circular curves can be set out on site from their tangent points, which are *the tangential angles method, offsets from the tangent lengths* and *offsets from the long chord*.

This section includes the following topics:

● Locating the intersection point and tangent points

● Setting out using the tangential angles method

● Setting out using offsets from the tangent lengths

● Setting out using offsets from the long chord

Locating the intersection point and tangent points

When traditional methods are being used to set out the centre line, it is first necessary to locate the intersection and tangent points on site. When doing this, it is not
sufficiently accurate to scale the position of the intersection and tangent points from
a plan; they must be set out precisely on site. With reference to Figure 12.10, the
procedure is as follows:

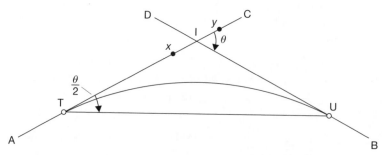

Figure 12.10 ● Location of intersection and tangent points.

- Locate the two straights AC and BD and define them on the ground using at least two pegs on each straight. Use nails in the tops of the pegs to define the straights precisely.

- Set up a theodolite over the nail in a peg on one of the straights (say AC) and sight the nail in another peg on this so that the theodolite telescope is pointing in the direction of the intersection point I.

- Drive in two additional pegs x and y on the straight AC such that straight BD will intersect the line xy. Again, use nails in the tops of pegs x and y to define the straight AC precisely.

- Join the nails in the tops of pegs x and y using a string line.

- Move the theodolite and set it up over the nail in the top of one of the pegs on straight BD. Sight the nail in the top of another peg on BD so that the telescope is again pointing towards I.

- Fix the position of I by driving in a peg where the line of sight from the theodolite on BD intersects the string line xy. Again, define I precisely using a nail in the top of the peg.

- Move the theodolite to I and measure angle $A\hat{I}B$. Calculate the deflection angle, θ, from $\theta = 180° -$ angle $A\hat{I}B$.

- Calculate tangent lengths IT and IU using $R\tan(\theta/2)$.

- Fix tangent points T and U by measuring back along the straights from I, either using tapes or a total station, allowing for the slope of the ground. Drive in pegs and mark the exact positions of T and U using nails in the tops of the pegs.

- Check the setting out by measuring angle $I\hat{T}U$, which should equal $\theta/2$.

The use of two theodolites simplifies the procedure by eliminating the need to set out the intermediate pegs x and y. In this case, point I would be fixed at the intersection of the lines of sight from each theodolite located on the two straights.

The method just described requires a theodolite to be set up at point I, which must be accessible if this method is to be successful. On occasions, however, it is impossible to use this method owing to the intersection point being inaccessible; for example, it may fall on a very steep hillside, in marshy ground or in a lake or a river. In such cases, the following procedure should be adopted in order that θ can be determined and the tangent points located. Consider Figure 12.11.

- Choose points A and B somewhere on the straights such that it is possible to sight from A to B and from B to A and also to measure AB.

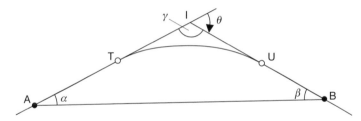

Figure 12.11 ● Location of tangent points when the intersection point is inaccessible.

- Measure AB.
- Measure angles α and β, calculate γ from $\gamma = 180° - (\alpha + \beta)$ and obtain θ, either from $\theta = (180° - \gamma)$ or $\theta = (\alpha + \beta)$.
- Use the Sine Rule to calculate IA and IB.
- Calculate IT and IU using $R\tan(\theta/2)$.
- Using AT = IA − IT and BU = IB − IU, set out T from A and U from B. If A and B were chosen to be on the other side of T and U then AT and BU will have negative values.
- If possible, measure angle $I\hat{T}U$ as a check – it should equal $(\theta/2)$.

Setting out using the tangential angles method

This is the most accurate of the traditional methods and it can be used for any circular curve. It can be carried out using either a *theodolite and a tape, two theodolites* or a *total station and a pole-mounted reflector*. The formula used for the tangential angles is derived as follows and uses the important relationships developed in Section 12.2. Consider Figure 12.12, in which tangential angles α_1, α_2 and α_3 are required. The assumption is made that arc TK = chord TK if the chord $\leq (R/20)$.

The length of the chord TK is given by

Chord TK = $R2\alpha_1$ (α_1 in radians)

Converting radians to degrees and rearranging gives

$$\alpha_1 = \left(\frac{TK}{2R}\right)\left(\frac{180}{\pi}\right) \text{ degrees}$$

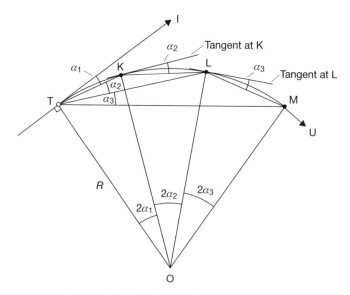

Figure 12.12 ● Tangential angles method.

Similarly

$$\alpha_2 = \left(\frac{KL}{2R}\right)\left(\frac{180}{\pi}\right) \text{ degrees}$$

and

$$\alpha_3 = \left(\frac{LM}{2R}\right)\left(\frac{180}{\pi}\right) \text{ degrees}$$

Note that the chord for α_2 is KL not TL and that for α_3 is LM not TM.
   In general

$$\alpha = \left(\frac{90}{\pi}\right)\left(\frac{\text{chord length}}{R}\right) \text{ degrees} \tag{12.8}$$

Tangential angles method using a theodolite and a tape

In this method, a theodolite is set up at the tangent point and used to turn the tangential angles to define the directions to each centre line peg. The exact positions of the pegs on these directions are fixed by measuring with a tape from peg to peg in turn. The calculation and setting out procedures are as follows.

Calculation procedure

● Determine the total length of the curve.

● Select a suitable chord length ≤ (R/20), for example, 10 m or 20 m. This will leave a sub-chord at the end and it is usually necessary to have an initial sub-chord in order to ensure that pegs are placed on the centre line at exact multiples of through chainage. This is done to help in subsequent earthwork calculations; for example, pegs would be required at chainages 0 m, 20 m, 40 m, 60 m and so on if a 20 m chord had been selected. The chord must be ≤ (R/20) in order that the assumptions made in the derivation of the tangential angles formula still apply.

● A series of tangential angles is obtained from equation (12.8) to give $\alpha_1, \alpha_2, \alpha_3$ and so on, corresponding to chords TK, KL, LM and so on.
   In practice, $\alpha_2 = \alpha_3 = \alpha_4$ and so on, since all chords except the first and the last will be equal. Therefore, only three tangential angles need to be calculated – one for the initial sub-chord, one for a general chord length (say 20 m) and one for a final sub-chord.

● The results are normally tabulated before setting out the curve on site. This would show cumulative tangential angles and individual chord lengths against chainage.

   Worked example 12.1 in Section 12.9 illustrates the calculation method outlined above.

Setting out procedure

- Using the methods described in the previous section, the tangent points are fixed and the theodolite is set up at one of them, preferably the one from which the curve swings to the right. This ensures that the tangential angles set on the horizontal circle will increase from 0°. This is shown in Figure 12.13, where the theodolite has been set up at point T.

- With reference to Figure 12.13, the intersection point I is sighted and the horizontal circle is set to read zero.

- The theodolite is rotated so that the tangential angle α_1 for the first chord TK is set on the horizontal circle.

- The first chord TK is then set out by lining in the tape with the theodolite along this direction and marking off the length of the chord from the tangent point. The chord lengths derived in the calculations are considered in the horizontal plane, therefore the chord length being set out must either be corrected for any slope or stepped. Once the first point has been located it is marked using a peg with a nail driven in its top to define the exact position of K.

- With reference to Figure 12.14, the telescope is then turned until the horizontal circle gives a reading of $(\alpha_1 + \alpha_2)$ – it is now pointing from T towards L. With one end of the tape hooked over the nail in the peg at point K, the length of the second chord KL is 'swung' from K until it meets the direction TL defined by the theodolite. As before, this length should either be corrected for slope or stepped as required. A peg and nail are then located at the position where the 'swung' chord meets the direction; this fixes the second point L on the curve.

- The procedure is repeated for all the other points to be fixed on the curve until point U is set out. In each case the *cumulative* angle from the tangent point T with reference to the tangent line TI is set on the theodolite, but the chord length swung is the *individual* length from the previous peg fixed on the curve. Since an error in setting out one peg will be carried forward to the next peg, this method has the advantage that any error will soon become apparent as the curve begins to take the wrong shape or when there is a misclosure at U.

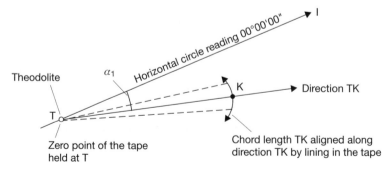

Figure 12.13 ● Setting out the first point.

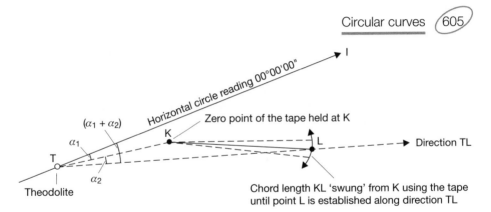

Figure 12.14 ● Setting out the second point.

● Once point U has been fixed, the theodolite should be moved to U so that tangential angle IÛT can be measured – it should equal ($\theta/2$).

Tangential angles method using two theodolites

This variation is used when the ground between the tangent points is of such a character that taping proves difficult, with very steep slopes, in undulating ground and across ploughed fields or if the curve is partly over marshy ground. The method is as follows and is shown in Figure 12.15.

Two theodolites are used, one being set at each tangent point. One disadvantage of the method is that two of everything are required, which means that two engineers, two instruments and, preferably, two assistants are needed to locate the pegs.

The method is one of intersecting points on the curve with the theodolites. For example, to fix point Z in Figure 12.15, α_1 is set out from T relative to TI and ($360° - (\theta/2) + \alpha_1$) is set out from U relative to UI. The two lines of sight intersect at Z where an assistant drives in a peg and a nail. Good liaison between the instrument operators and the assistant is vital and, for large curves, two-way radios are essential.

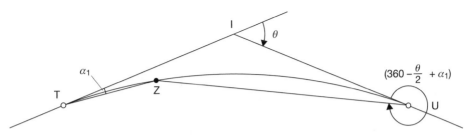

Figure 12.15 ● Tangential angles method using two theodolites.

Tangential angles method using a total station and a pole mounted reflector

In this method, a total station is set up at the tangent point and used to turn the tangential angles as in the method described earlier using a theodolite and a tape.

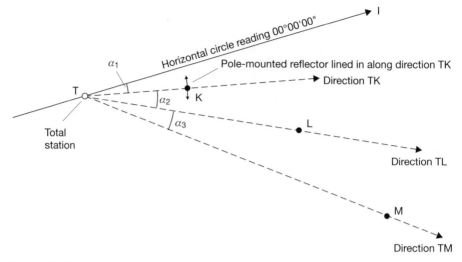

Figure 12.16 ● Setting out a circular curve using a total station and a pole mounted reflector.

However, instead of measuring lengths from one peg to another, the distance measuring component of the total station is used to set out the length to each peg directly from the tangent point.

From Figure 12.5 and Section 12.2, the length of the long chord TU was derived as $2R\sin(\theta/2)$; hence, considering Figure 12.12:

line TK = long chord of the curve of arc TK = $2R\sin(\alpha_1)$

line TL = long chord of the curve of arc TL = $2R\sin(\alpha_1 + \alpha_2)$

line TM = long chord of the curve of arc TM = $2R\sin(\alpha_1 + \alpha_2 + \alpha_3)$

This leads to the following calculation and setting out procedures.

Calculation procedure
● The tangential angles are obtained in exactly the same way as that described earlier for the method involving a theodolite and a tape and α_1, α_2, α_3 and so on, will be known.

● Using the formula for the long chords given above, the lengths of the long chords required to fix each point are determined to give the horizontal lengths TK, TL, TM and so on.

● The tangential angles and long chord lengths are tabulated for use on site.

Setting out procedure
● With reference to the method described earlier when using a theodolite and tape, the theodolite at T is replaced by a total station and the tape is replaced by a pole-mounted reflector, as shown in Figure 12.16.

- With reference to Figure 12.16, the tangential angle, α_1, for long chord TK is turned from TI to establish the direction TK. The total station is set to read horizontal distances directly and the horizontal distance TK is set out along this direction using the pole-mounted reflector. Point K is fixed using a nail in the top of a peg.
- The tangential angle $(\alpha_1 + \alpha_2)$ for the long chord TL is turned from TI to establish the direction TL, horizontal distance TL is set out along this direction using the pole mounted reflector and point L is fixed using a nail in the top of a peg.
- The procedure continues for point M and any other points on the curve until the second tangent point U is set out. Since each peg is always set out from the entry tangent point, this method has the advantage that an error in setting out one peg will not be carried forward to another. However, although a large error in the setting out of a peg should be easy to spot on site, a small error could go unnoticed so great care should be taken with this method.
- Finally, as a check, the total station should be moved to U so that tangential angle IÛT can be measured – it should equal $(\theta/2)$.

The above method has the advantage that the total station can correct automatically for the slope of the ground enabling horizontal distances to be set out directly. Worked example 12.2 shows how this method is applied.

Setting out using offsets from the tangent lengths

This traditional method of setting out requires two tapes. It is suitable for short curves or curves of small radius such as kerbs. It can also be used to set out additional points between those previously established by the tangential angles method or by coordinate methods. This is often necessary to give a better definition of the centre line. Consider Figure 12.17.

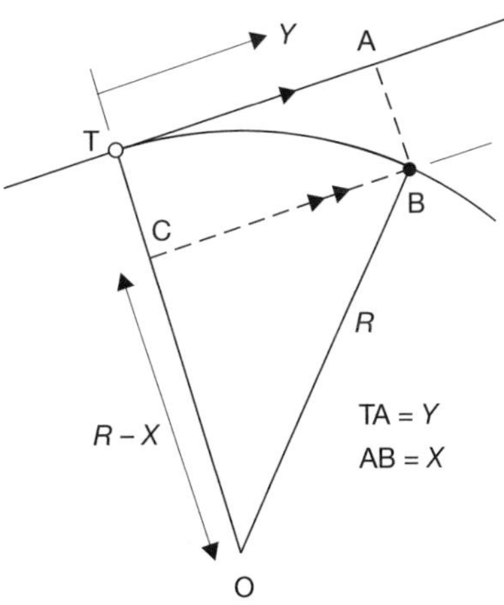

Required: The offset X, at a known distance Y along the tangent from T, to the curve.
In triangle OBC:

$$OB^2 = OC^2 + BC^2$$

Therefore

$$R^2 = (R - X)^2 + Y^2$$

From here there are two routes; *either*

$$R - X = \sqrt{R^2 - Y^2}, \text{ hence } X = R - \sqrt{R^2 - Y^2} \tag{12.9}$$

Figure 12.17 Setting out a circular curve from the tangent length.

or

$$R^2 = R^2 - 2RX + X^2 + Y^2$$

Dividing through by $2R$ gives

$$X = \left(\frac{Y^2}{2R}\right) + \left(\frac{X^2}{2R}\right)$$

but $(X^2/2R)$ will be very small since R is very large compared with X and it can be neglected to give

$$X = \left(\frac{Y^2}{2R}\right)$$

(12.10)

Equation (12.10) is accurate only for large radii curves and will give errors for small radii curves, where the effect of neglecting the second term cannot be justified.

Once the tangent points have been fixed, the lines of the tangents can be defined using a theodolite or ranging rods and the offsets (X) set off at right angles at distances (Y) from T and from U. Half the curve is set out from each tangent length. It is useful to tabulate the X and Y values before beginning the setting out. Worked example 12.4 at the end of this chapter shows this method.

Setting out using offsets from the long chord

This traditional method also requires two tapes. It is suitable for curves of small radius such as boundary walls and kerb lines at road intersections. Also, it is a very useful method when the tangent lengths are inaccessible and offsets from them cannot be used. Consider Figure 12.18.

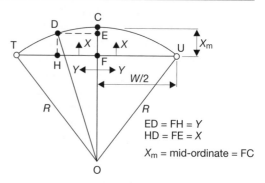

Required: The offset X from the long chord TU at a distance Y from F.

In this method all offsets are established from the mid-point F of the long chord TU. Let the length of chord TU = W.

In triangle TFO:

$$OT^2 = OF^2 + TF^2$$

Therefore

$$R^2 = (R - X_m)^2 + \left(\frac{W}{2}\right)^2$$

Figure 12.18 ● Setting out a circular curve from the long chord.

ED = FH = Y
HD = FE = X

X_m = mid-ordinate = FC

From which

$$(R - X_m) = \sqrt{R^2 - \left(\frac{W}{2}\right)^2}$$

and

$$X_m = R - \sqrt{R^2 - \left(\frac{W}{2}\right)^2} \tag{12.11}$$

In triangle ODE

$$OD^2 = OE^2 + DE^2$$

Therefore

$$R^2 = (OF + X)^2 + Y^2$$

from which

$$OF + X = \sqrt{R^2 - Y^2} \tag{12.12}$$

But

$$OF = (R - X_m)$$

Substituting X_m from equation (12.11)

$$OF = R - \left[R - \sqrt{R^2 - \left(\frac{W}{2}\right)^2}\right]$$

or

$$OF = \sqrt{R^2 - \left(\frac{W}{2}\right)^2}$$

Substituting this in equation (12.12) gives

$$X = \sqrt{R^2 - Y^2} - \sqrt{R^2 - \left(\frac{W}{2}\right)^2} \tag{12.13}$$

Once the tangent points have been fixed, the long chord can be defined and point F established. The offsets are then calculated at regular intervals from point F, both along FT and along FU. Again, it is useful to tabulate the offsets from FT and FU before beginning the setting out.

With reference to Figure 12.18, to set out a point D on the curve, its Y distance is measured from F towards T to fix point H where its offset X is set out at right angles. Worked example 12.4 at the end of this chapter shows this method.

Reflective summary

With reference to setting out circular curves on site by traditional methods remember:

— The first step is usually to set out the intersection points and the tangents points.

— The tangential angles method can be undertaken either using a theodolite and a tape, two theodolites or a total station and a pole-mounted reflector.

— The methods involving offsets from the tangent length or long chord can be undertaken using tapes alone.

— The tangential angles method is the most accurate of the traditional methods and is suitable for any circular curve.

— When using a theodolite and tape for the tangential angles method, an error in locating one peg will be carried forward to the next which should make any mistakes easy to spot.

— When using a total station and a pole-mounted reflector for the tangential angles method, an error in locating one peg is not carried forward to another. A large error in the location of a peg should be easy to spot, but small errors may go undetected so great care should be taken.

— The offset methods are best restricted to short curves or curves of small radius.

12.7 Setting out circular curves on site by coordinate methods

After studying this section you should know how to calculate coordinates of points on the centre lines of circular curves and you should be aware of methods by which these points can be set out from control networks.

This section includes the following topics:

● Coordinate methods

● Calculating the coordinates of points on the centre line

● Deriving the setting out data

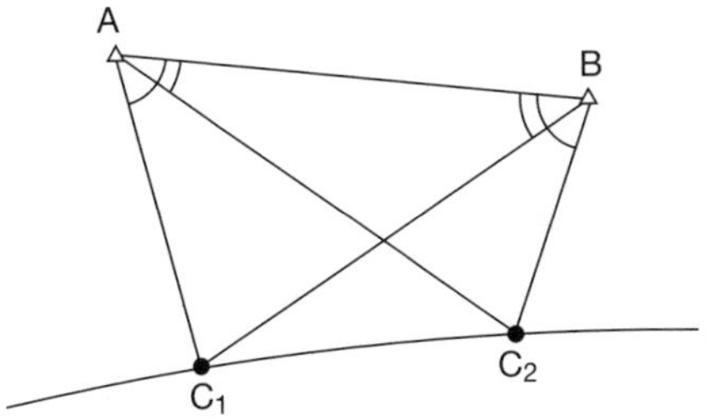

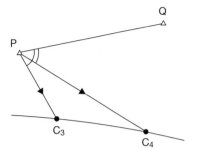

C$_1$ and C$_2$ fixed by intersection
from control points A and B

C$_3$ and C$_4$ fixed by bearing and
distance from control point P

Figure 12.19 ● Setting out a circular curve
by intersection.

Figure 12.20 ● Setting out a circular
curve by bearing and distance.

Coordinate methods

As discussed in Section 12.5, these methods are used nowadays in preference to traditional techniques. In such methods, which are suitable for all horizontal curves, not just circular ones, the coordinates of points on the curve centre line are calculated and then fixed from a control network using *either*

● *Intersection* or *angle (bearing) and distance* methods, as shown in Figures 12.19 and 12.20. These use total stations, theodolites and tapes to set out from two or more points in the network surrounding the scheme.

or

● *GPS* methods.

No matter which of these is used, the coordinates of points to be fixed on the centre line and the coordinates of the control network being used must be based on the same coordinate system.

Calculating the coordinates of points on the centre line

In all cases, it is first necessary to obtain the coordinates of the points that are to be set out on the curve centre line. Consider Figure 12.21 in which P and Q are ground control points and the coordinates of points C_1, C_2, C_3, C_4, C_5 and C_6 are required.

● To begin the calculations, the coordinates of I, T and U are required. Normally, where the design has been done using a software package, these will be determined by the software during the design process. In cases where a hand design has been undertaken, I, T and U will have to be fixed on the ground, as described in Section 12.6, and their coordinates measured either by intersection or bearing and distance methods from P and Q, or by including them in a traverse with P and Q.

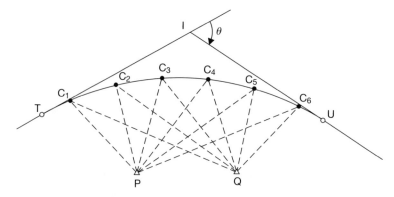

Figure 12.21 ● Setting out a complete curve using coordinates.

● Design the curve as if it was to be set out by the tangential angles method using a total station and a pole mounted reflector as discussed in Section 12.6. This will give, with reference to Figure 12.12, all the tangential angles (α values) and the lengths of the long chords from T (TC_1, TC_2, TC_3, TC_4, TC_5, TC_6 and TU in Figure 12.21).

● From the coordinates of T and I, calculate the bearing TI and use it with the tangential angles to calculate the bearings of all the long chords from T.

● Using the bearings of the long chords together with their lengths, calculate their coordinates from those of point T. Take care with this since there is no check on the calculations.

Some software packages will derive the coordinates of specified centre line points automatically during the design process by using a calculation similar to that just outlined above.

Deriving the setting out data

When setting out curves using total stations and GPS methods, the centre line coordinates will usually be pre-computed and the coordinate file for the curve will be downloaded into the total station or GPS receiver ready for setting out. In such a case, the operator does not have to do any calculations and the setting out follows procedures given in Section 11.3 for total station and GPS methods.

When setting out by theodolite and tape, *intersection* or *bearing and distance* methods are used to set out the curve and it is necessary to derive the angles and distances from control points to the centre line before this can be done. Sometimes, this information is already computed and available for use on site, but it may be necessary to compute the setting out data by hand. If this is required, then with reference to Figure 12.21, the procedure is as follows:

● Knowing the coordinates of P, Q, C_1, C_2, C_3, C_4, C_5 and C_6, derive the bearings and horizontal lengths from P and Q to each point by using rectangular → polar conversions.

- Set out pegs on the centre line by either
    - *intersection* from P and Q using bearings PC_1 and QC_1, PC_2 and QC_2, and so on

    or

    - *bearing and distance* from P and Q, using the bearings and horizontal lengths from P to each point and then checking them using the bearings and horizontal lengths from Q.

Worked example 12.3 at the end of this chapter shows how a circular curve can be set out by coordinate methods.

Reflective summary

With reference to setting out circular curves on site by coordinate methods remember:

- It is not always necessary to set out the intersection points and the tangent points.

- The coordinates of points on the curve centre line are calculated and the centre line is fixed using total stations, theodolites and tapes or GPS methods.

- The coordinates of points to be fixed on the centre line and the coordinates of the control network being used must be based on the same coordinate system.

12.8 Plotting the centre lines of circular curves on drawings

After studying this section you should know how to draw the centre line of a proposed circular curve on a drawing in order to see if your design is acceptable.

Hand drawing

Most highway alignment drawings are now produced in conjunction with highway design software packages on one of the wide range of plotters currently available, and further information on these is given in Section 10.9. However, there are still occasions during the initial design process when it is necessary to undertake a hand drawing of the centre line to see if a particular design is acceptable and falls correctly within the band of interest. In order to do this, the following procedure is recommended for each of the circular curves used in the design. It assumes that there is an existing survey plan of the area available.

- Draw the intersecting straights in their correct relative positions on a sheet of tracing paper.

- Calculate the length of each tangent length using $R\tan(\theta/2)$.

- Plot the tangent points by measuring this distance along each straight on either side of the intersection point at the same scale as the existing plan.

- Using either equation (12.9) for the *offsets from the tangent lengths* method or equation (12.13) for the *offsets from the long chord* method, draw up a table of offset X values for suitable Y values using the appropriate formula. Ensure that the Y values chosen will provide a good definition of the centre line.

- At the scale of the existing plan, plot the X and Y values on the tracing paper to establish points on the centre line and carefully join these to define the curve. A set of French curves is useful for this purpose, although with care a flexicurve can be used.

- Superimpose the tracing paper on the existing survey plan and decide whether or not the design is acceptable. If it is not, change the design and repeat the plotting procedure.

Reflective summary

With reference to plotting the centre lines of circular curves on drawings remember:

- Most highway alignment drawings are now produced in conjunction with highway design software packages on one of the wide range of machine plotters currently available.

- A hand drawing may be required to check a design to see whether it is acceptable and falls within the band of interest.

- Hand drawings are best done on tracing paper so that the design can be superimposed on the survey plan of the area.

- The centre line can be drawn on the tracing paper using either offsets from the tangent lengths or offsets from the long chord.

12.9 Circular curve worked examples

After studying this section you should understand how to calculate data to set out circular curves in a number of different ways. You will become familiar with the concept of through chainage and you will be able to calculate through chainage values for tangent points and other points on the curve. You will appreciate the differences between setting out by the tangential angles method using a theodolite and tape compared to using a total station and a pole-mounted reflector. You will understand how to compute coordinates of points on the curve and how these can be used to set out the curve using intersection techniques. You will appreciate how minor curves can be set out using offset methods, either from their tangent lengths or their long chords. You will become aware of the need to tabulate setting out data for ease of use and you will appreciate the importance of checking your calculations.

This section includes the following topics:

- Setting out a circular curve by the tangential angles method using a theodolite and a tape
- Setting out a circular curve by the tangential angles method using a total station and a pole-mounted reflector
- Setting out a circular curve by intersection from two nearby control points
- Setting out a circular curve by offsets from the tangent lengths and offsets from the long chord

Worked example 12.1: Setting out a circular curve by the tangential angles method using a theodolite and a tape

Question

It is required to connect two intersecting straights whose deflection angle is 13°16'00" by a circular curve of radius 600 m. The through chainage of the intersection point is 2745.72 m and pegs are required on the centre line of the curve at exact 25 m multiples of through chainage.

Tabulate the data necessary to set out the curve by the tangential angles method using a theodolite and a tape.

Solution

Consider Figure 12.22, in which it has been assumed that chainage is increasing from left to right. From equation (12.2)

$$tangent\ length = \text{IT} = R\tan\left(\frac{\theta}{2}\right) = 600 \tan 06°38'00" = 69.78 \text{ m}$$

Therefore

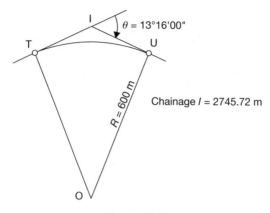

Figure 12.22

through chainage of T = chainage I – IT = 2745.72 – 69.78 = 2675.94 m

To fix the first point on the curve at chainage 2700 m (the next multiple of 25 m) an initial sub-chord is required where

length of initial sub-chord = 2700 – 2675.94 = 24.06 m

From equation (12.6)

length of circular curve = L_C = $R\theta$ (θ in radians)

In this case θ = 13°16'00" = 13.2667°.
   Therefore

$$L_C = \left(\frac{600 \times 13.2667 \times \pi}{180}\right) = 138.93 \text{ m}$$

From which

through chainage of U = chainage T + L_C = 2675.94 + 138.93 = 2814.87 m

Hence a final sub-chord is also required since 25 m chords can only be used up to chainage 2800 m. Therefore

length of final sub-chord = 2814.87 – 2800 = 14.87 m

Hence three chords are necessary:

initial sub-chord of 24.06 m
general chord of 25.00 m
final sub-chord of 14.87 m

The tangential angles for these chords are obtained from equation (12.8) as follows for initial sub-chord $= \dfrac{90}{\pi} \times \left(\dfrac{24.06}{600}\right) = 01°08'56''$

for general chord $= \dfrac{90}{\pi} \times \left(\dfrac{25.00}{600}\right) = 01°11'37''$

for final sub-chord $= \dfrac{90}{\pi} \times \left(\dfrac{14.87}{600}\right) = 00°42'36''$

These are applied to the whole curve to give the tabulated results shown in Table 12.1. The points on the centre line are designated C_1, C_2, C_3, C_4 and C_5 for use in worked examples 12.2 and 12.3.

Table 12.1

Point	Chainage (m)	Chord length (m)	Individual tangential angle				Cumulative tangential angle		
			°	'	''		°	'	''
T	2675.94	0	00	00	00		00	00	00
C_1	2700.00	24.06	01	08	56	(α_1)	01	08	56
C_2	2725.00	25.00	01	11	37	(α_2)	02	20	33
C_3	2750.00	25.00	01	11	37	(α_3)	03	32	10
C_4	2775.00	25.00	01	11	37	(α_4)	04	43	47
C_5	2800.00	25.00	01	11	37	(α_5)	05	55	24
U	2814.87	14.87	00	42	36	(α_6)	06	38	00
		$\Sigma138.93$ (checks)					$\theta/2 = 06° 38'00''$ (checks)		

As a check, the final cumulative tangential angle shown in Table 12.1 should equal $\theta/2$ within a few seconds. Also, the sum of the chords should equal the total length of the circular curve.

Note that α is proportional to the chord length. Any chords of equal length will have the same tangential angle and this is simply added to the cumulative total.

Worked example 12.2: Setting out a circular curve by the tangential angles method using a total station and a pole-mounted reflector

Question

The circular curve designed in Worked example 12.1 is to be set out by the tangential angles method using a total station and a pole-mounted reflector.

Tabulate the data necessary to set out the curve from entry tangent point T.

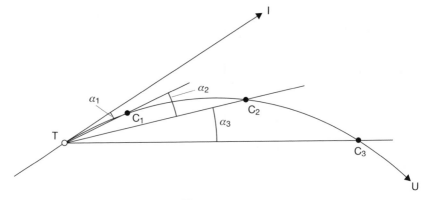

Figure 12.23

Solution

Figure 12.23 shows part of the curve designed in Worked example 12.1. In order to set this out using a total station and a pole-mounted reflector, the tangential angles (α values) are turned on the total station and the long chord lengths (TC_1, TC_2, TC_3 and so on) are set out from T using the pole-mounted reflector. The tangential angles were calculated in Worked example 12.1 and using these together with $R = 600$ m enables the long chord lengths to be calculated as follows using equation (12.5)

$$TC_1 = 2R \sin\alpha_1 \qquad = 1200 \sin 01°08'56" = 24.06 \text{ m}$$

$$TC_2 = 2R \sin(\alpha_1+\alpha_2) \qquad = 1200 \sin 02°20'33" = 49.05 \text{ m}$$

$$TC_3 = 2R \sin(\alpha_1+\alpha_2+\alpha_3) \quad = 1200 \sin 03°32'10" = 74.01 \text{ m}$$

and so on …

The data is tabulated as shown in Table 12.2, ready to be set out on site.

Table 12.2

Point being set out	Cumulative tangential angle to be turned from T relative to the line TI	Long chord (m) to be set out from T
T	00°00'00"	0
C_1	01°08'56"	24.06
C_2	02°20'33"	49.05
C_3	03°32'10"	74.01
C_4	04°43'47"	98.95
C_5	05°55'24"	123.84
U	06°38'00"	138.62

Worked example 12.3: Setting out a circular curve by intersection from two nearby control points

Question

The circular curve designed in Worked examples 12.1 and 12.2 is to be set out by intersection from two nearby traverse stations A and B. The whole circle bearing of straight TI is obtained from the design as 63°27'14", as are the coordinates of the entry tangent point T. The coordinates of A, B and T are as follows:

   A   829.17 mE, 724.43 mN
   B   915.73 mE, 691.77 mN
   T   798.32 mE, 666.29 mN

Using the relevant data from Worked examples 12.1 and 12.2, calculate

- The coordinates of all the points on the centre line of the curve, which lie at exact 25 m multiples of through chainage.
- The bearing AB and the bearings from A required to set out these points.
- The bearing BA and the bearings from B required to set out these points.

Solution

Figure 12.24 shows all the points to be set out together with traverse stations A and B.

The coordinates of all the points on the centre line of the curve, which lie at exact 25 m multiples of through chainage
Coordinates of C_1
With reference to Figure 12.25 and Table 12.2:

      bearing TC_1 = bearing TI + α_1
                              = 63°27'14" + 01°08'56"
                              = 64°36'10"
      horizontal length TC_1 = 24.06 m

Therefore

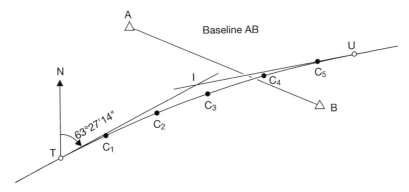

Figure 12.24

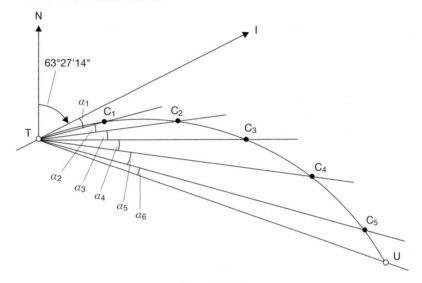

Figure 12.25

$$\Delta E_{TC_1} = 24.06 \sin 64°36'10" = +21.735 \text{ m}$$

$$\Delta N_{TC_1} = 24.06 \cos 64°36'10" = +10.319 \text{ m}$$

Hence

$$\mathbf{E_{C_1}} = E_T + \Delta E_{TC_1} = \mathbf{820.055 \text{ m}}$$

$$\mathbf{N_{C_1}} = N_T + \Delta N_{TC_1} = \mathbf{676.609 \text{ m}}$$

Coordinates of C_2

Again with reference to Figure 12.25 and Table 12.2:

$$
\begin{aligned}
\text{bearing TC}_2 \quad &= \text{ bearing TI} + (\alpha_1 + \alpha_2) \\
&= 63°27'14" + (02°20'33") \\
&= 65°47'47"
\end{aligned}
$$

horizontal length TC_2 = 49.05 m

Therefore

$$\Delta E_{TC_2} = 49.05 \sin 65°47'47" = +44.738 \text{ m}$$

$$\Delta N_{TC_2} = 49.05 \cos 65°47'47" = +20.110 \text{ m}$$

Hence

$$\mathbf{E_{C_2}} = E_T + \Delta E_{TC_2} = \mathbf{843.058 \text{ m}}$$

$$\mathbf{N_{C_2}} = N_T + \Delta N_{TC_2} = \mathbf{686.400 \text{ m}}$$

Coordinates of C_3, C_4 and C_5

These are calculated by repeating the procedures used to calculate the coordinates of C_1 and C_2. The values obtained are:

$$C_3 = 866.442 \text{ mE}, 695.220 \text{ mN}$$
$$C_4 = 890.183 \text{ mE}, 703.063 \text{ mN}$$
$$C_5 = 914.224 \text{ mE}, 709.908 \text{ mN}$$

Coordinates of U

These are calculated twice to provide a check.

Firstly, they are calculated by repeating the procedures used to calculate the coordinates of points C_1 to C_5. The values obtained are:

$$U = 928.652 \text{ mE}, 713.502 \text{ mN}$$

Secondly, they are calculated by working along the straights from T to I to U as follows:

bearing TI = 63°27'14"
horizontal length TI = 69.78 m (see Worked example 12.1)

Hence

$$\Delta E_{TI} = 69.78 \sin 63°27'14" = +62.423 \text{ m}$$

$$\Delta N_{TI} = 69.78 \cos 63°27'14" = +31.186 \text{ m}$$

Therefore

$$E_I = E_T + \Delta E_{TI} = 798.32 + 62.423$$
$$= 860.743 \text{ m}$$
$$N_I = N_T + \Delta N_{TI} = 666.29 + 31.186$$
$$= 697.476 \text{ m}$$

From Worked example 12.1, $\theta = 13°16'00"$; hence

bearing IU = bearing TI + θ
$$= 63°27'14" + 13°16'00" = 76°43'14"$$
horizontal length IU = 69.78 m

Therefore

$$\Delta E_{IU} = 69.78 \sin 76°43'14" = +67.914 \text{ m}$$
$$\Delta N_{IU} = 69.78 \cos 76°43'14" = +16.029 \text{ m}$$

from which

$$E_U = E_I + \Delta E_{IU} = 860.743 + 67.914$$
$$= 928.657 \text{ m}$$
$$N_U = N_I + \Delta N_{IU} = 697.476 + 16.029$$
$$= 713.505 \text{ m}$$

These check, within a few millimetres, the values obtained for the coordinates of U calculated from the long chord TU.

All the coordinates are listed in Table 12.3 and have been rounded to two decimal places to agree with the precision of the coordinate data given for points A, B and T.

Table 12.3

Point	Chainage (m)	Coordinates		Bearing from A			Bearing from B		
		mE	mN	°	'	"	°	'	"
T	2675.94	798.32	666.29	207	57	04	257	45	20
C$_1$	2700.00	820.05(5)	676.61	190	47	30	260	59	46
C$_2$	2725.00	843.06	686.40	159	56	09	265	46	26
C$_3$	2750.00	866.44	695.22	128	05	14	274	00	14
C$_4$	2775.00	890.18	703.06	109	18	14	293	50	23
C$_5$	2800.00	914.22	709.91	99	41	18	355	14	30
U	2814.87	928.65	713.50	96	16	12	30	44	04

Bearing AB = 110°40'19"   Bearing BA = 290°40'19"

Bearing AB and the bearings to the points on the centre line from A
These are calculated from the coordinates of the points either using the quadrants method of using rectangular → polar conversions as discussed in Section 6.2. The bearings are listed in Table 12.3.

Bearing BA and the bearings to the points on the centre line from B
Again, one of the methods discussed in Section 6.2 is used. The bearings are listed in Table 12.3.

Worked example 12.4: Setting out a circular curve by offsets from the tangent lengths and offsets from the long chord

Question
A kerb line is to be set out between two straights, which deflect through an angle of 75°00'00", such that it forms a circular curve of radius 20 m.

- Tabulate the data required to set out the centre line of the curve by offsets taken at exact 5 m intervals along the tangent lengths. The mid-point of the curve must also be fixed.

- Tabulate the data required to set out the centre line of the curve by offsets taken at exact 5 m intervals from the mid-point of the long chord. The mid-point of the curve must also be fixed.

Solution

Offsets from the tangent lengths
Figure 12.26 shows the curve to be set out and the setting out data is obtained from equation (12.9) which was derived in Section 12.6 as

$$X = R - \sqrt{R^2 - Y^2}$$

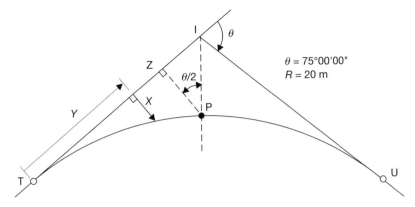

Figure 12.26

For this curve, from equations (12.2) and (12.3), respectively:

$$IT = R\tan\left(\frac{\theta}{2}\right) = 20\tan(37°30'00") = 15.35 \text{ m}$$

$$PI = R\left[\sec\left(\frac{\theta}{2}\right) - 1\right] = 20[\sec(37°30'00") - 1] = 5.21 \text{ m}$$

But from triangle IPZ in Figure 12.26:

$$IZ = PI\sin\left(\frac{\theta}{2}\right) = 5.21\sin(37°30'00") = 3.17 \text{ m}$$

Hence the length along the tangent for the offset to the mid-point P is given by

$$TZ = IT - IZ = 15.35 - 3.17 = 12.18 \text{ m}$$

Therefore, offsets are required at 5 m, 10 m and 12.18 m along the tangent length from T. Using equation (12.9), the setting out data given in Table 12.4 can be calculated. This data is used to set out half the curve along TI and half along UI. Point P is set out twice to provide a check. As a further check, the last offset in Table 12.4, to P, should equal $PI\cos(\theta/2)$, within a few millimetres. In this case, $PI\cos(\theta/2) = 5.21(\cos 37°30'00") = 4.13$ m, which checks.

Table 12.4

Length along tangent from T or U (Y values)	R	$\sqrt{R^2 - Y^2}$	Offset from the tangent lengths (X values)
0.00 m (T or U)	20.00 m	20.00 m	0.00 m (T or U)
5.00 m	20.00 m	19.36 m	0.64 m
10.00 m	20.00 m	17.32 m	2.68 m
12.18 m (P)	20.00 m	15.86 m	4.14 m (P)

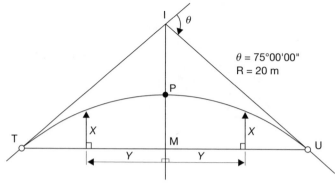

$$\theta = 75°00'00"$$
$$R = 20 \text{ m}$$

Figure 12.27

Offsets from the long chord

Figure 12.27 shows the curve to be set out and the setting out data is obtained from equation (12.13), which was derived in Section 12.6, that is

$$X = \sqrt{R^2 - Y^2} - \sqrt{R^2 - \left(\frac{W}{2}\right)^2}$$

For this curve, from equation (12.5) for the long chord

$$TU = W = 2R \sin\left(\frac{\theta}{2}\right) = 40 \sin 37°30'00" = 24.35 \text{ m}$$

Therefore

$$\frac{W}{2} = 12.18 \text{ m}$$

Hence offsets are required at 0 m, 5 m, 10 m and 12.18 m from the mid-point, M, of the long chord TU. Using equation (12.13), the setting out data given in Table 12.5 can be calculated.

Table 12.5

Length along the long chord from M (Y values)	$\sqrt{R^2 - Y^2}$	$\sqrt{R^2 - \left(\frac{W}{2}\right)^2}$	Offsets from the long chord (X values)
0.00 m (M)	20.00 m	15.87 m	4.13 m (P)
5.00 m	19.36 m	15.87 m	3.49 m
10.00 m	17.32 m	15.87 m	1.45 m
12.17(5) m (T or U)	15.87 m	15.87 m	0.00 m (T or U)

Reflective summary

With reference to circular curve worked examples remember:

— It is always advisable to sketch the straights and the curve and to add all the known data to the sketch before beginning the calculations.

— When calculating the positions of points on the centre line you will usually require an initial sub-chord, a general chord and a final sub-chord to ensure that pegs are located at specific multiples of through chainage.

— The data for setting out the tangential angles method using a theodolite and a tape is derived by working from point to point along the curve.

— The data for setting out the tangential angles method using a total station and a pole-mounted reflector is derived by always working from the starting tangent point.

— When calculating the coordinates of points on the curve, the coordinates of the exit tangent point should be calculated twice as a check.

— Data calculated for setting out a curve using one of the offset methods applies to each half of the curve.

— You should tabulate the calculations wherever possible so that they can be easily followed for checking purposes.

Exercises

12.1 A 4.56° circular curve is to be designed to fit between two intersecting straights. What is the radius of this curve?

12.2 Two straights, which deflect through an angle of 60°00'00", are to be connected by a circular curve of radius 80 m. The curve is to be set out by offsets from its tangent lengths. Calculate the data required:

(i)   to set out the mid point of the curve

(ii)   to set out pegs on the centre line of the curve by offsets taken at exact 10 m intervals along the tangent lengths

12.3 A circular curve of radius 900 m is to be constructed between two straights of a proposed highway. The deflection angle between the straights is 14°28'06" and the curve is to be set out by the tangential angles method using a theodolite and a tape. The through chainage of the intersection point is 1345.82 m and pegs are required on the centre line at exact 20 m multiples of through chainage.

(i)   Calculate the tangent lengths

(ii) Calculate the length of the circular curve

(iii) Calculate the through chainages of the two tangent points

(iv) Tabulate the data required to set out the curve

12.4 The bearings of three successive intersecting straights AB, BC and CD along the centre line of a proposed highway are 103°29'24", 125°43'22" and 116°12'54", respectively. The horizontal distance BC is 708.32 m.

It is proposed to connect AB and BC by a 1500 m radius curve and BC and CD by a 900 m radius curve such that there is an intervening straight on BC between the end of one curve and the start of the other. The through chainage of intersection point B is 1097.65 m and chainage increases from A to D.

Calculate the through chainages of the four tangent points on the two curves.

12.5 Two straights, which meet at an intersection angle of 135°00'00", are to be connected by a circular curve of radius 60 m. The curve is to be set out by offsets from its long chord. Calculate the data required:

(i)   to set out the mid point of the curve

(ii) to set out pegs on the centre line of the curve by offsets taken at exact 5 m intervals along its long chord

12.6 A circular curve of radius 1100 m is to be set out between two successive straights TI and IU on a proposed road.

When the straights are set out on site, it is found that the intersection point, I, is inaccessible. The distance from F, a point on TI, to G, a point on UI, is 197.36 m. The horizontal angle at F measured clockwise from FG to FT is 165°12'34" and the horizontal angle at G measured clockwise from GU to GF is 173°22'48". The bearing of TI is 64°16'26"and the through chainage of F is 895.23 m.

Calculate the through chainages of the entry tangent point T and the exit tangent point U.

12.7 A circular curve of radius 750 m is to connect two straights, which deflect through an angle of 14°36'12". The through chainage of the intersection point is 2319.87 m. The centre line of the curve is to be set out by the tangential angles method using a total station and a pole-mounted reflector. Pegs are required on the centre line at exact 25 m multiples of through chainage.

(i)   Calculate the through chainage of the two tangent points

(ii) Draw up a table listing the data required to set out the curve from the entry tangent point

(iii) Briefly describe how the centre line is set out

12.8 The circular curve in Exercise 12.7 is to be set out by bearing and distance methods from two control points. During the design of the curve, the coordinates of the entry tangent point T and the bearing of the tangent length TI were obtained as follows:

Bearing TI   110°17'16"
T   1359.67 mE, 2165.44 mN

Calculate the coordinates of the points on the centre line of the curve at exact 25 m multiples of through chainage.

Further reading and sources of information

For the latest information on the design of circular curves, consult

Design Manual for Roads and Bridges, Volume 6 Road Geometry, Section 1 Links, Part 1 TD9/93 – incorporating Amendment No. 1 dated Feb 2002 – Highway Link Design. Jointly published by the Overseeing Organisations of England, Scotland Wales and Northern Ireland, that is, The Highways Agency, the Scottish Executive Development Department, The National Assembly for Wales (Cynulliad Cenedlaethol Cymru) and The Department for Regional Development Northern Ireland. The manual can be accessed and printed from the Internet at: `http://www.official-documents.co.uk/document/deps/ha/dmrb/index.htm`. It is also available from The Stationery Office at `http://www.tso.co.uk/`.

chapter 13

Transition curves

 Aims

After studying this chapter you should be able to:

- Appreciate the limitations of circular curves
- Understand how transition curves can be used to improve the safety and comfort of passengers in vehicles travelling around horizontal curves
- Understand the purpose of superelevation and the limitations on its values
- Use the standards specified in the United Kingdom Department of Transport publication *Design Manual for Roads and Bridges* to choose suitable radii for horizontal curves designed for particular speeds
- Understand the terminology and geometry of transition curves
- Design horizontal curves, which incorporate transition curves
- Appreciate the use of computer software packages in the design of horizontal alignments
- Set out the centre lines of horizontal curves, which incorporate transition curves, using a variety of different methods
- Choose the most appropriate method of setting out for different situations where horizontal curves are used
- Plot the centre lines of horizontal curves, which incorporate transition curves, on drawings

This chapter contains the following sections:

13.1 The need for transition curves

13.2 Current United Kingdom Department of Transport design standards

13.3 Type of transition curve to be used

13.4 The geometry of transition curves

13.5 Designing composite and wholly transitional curves

13.1 The need for transition curves

After studying this section you should be aware of the limitations of circular curves due to the forces that act on vehicles as they travel around the curves. You should understand that transition curves can be used to introduce these forces gradually and uniformly, thereby increasing the safety of passengers in the vehicles. You should know what composite curves and wholly transitional curves are and you should be familiar with superelevation and appreciate how it can be used to further improve the safety of passengers.

This section includes the following topics:

- Radial force and design speed
- Use of transition curves
- Superelevation

Radial force and design speed

A transition curve differs from a circular curve in that its radius is continuously changing. As may be expected, such curves involve more complex formulae than curves of constant radius and their design is more complicated. Because circular curves are unquestionably easier to design than transition curves and they are easily set out on site, the questions naturally arise as to why are transition curves necessary and why is it not possible to use circular curves to join all intersecting straights?

The reason why transition curves are necessary is due to the *radial force* acting on a vehicle as it travels round the curve.

A vehicle travelling with a constant speed v along a curve of radius r is subjected to a radial force P such that $P = (mv^2/r)$, where m is the mass of the vehicle. This force is, in effect, trying to push the vehicle back on to a straight course.

On a straight road, r = infinity, therefore $P = 0$

On a circular curve of radius R, $r = R$, therefore $P = (mv^2/R)$

Roads are designed for particular speeds and v, the *design speed*, is constant for any given road. The design speed is determined from consideration of the alignment and layout of the road and is generally selected to be equivalent to the 85 percentile speed. This is the speed not normally exceeded by 85 per cent of the vehicles using the road. Similarly, the mass of the vehicle can be assumed to be constant. Consequently $P \propto 1/r$ and the smaller the radius of a road curve, the greater the force acting on a vehicle travelling along it.

Any vehicle leaving a straight section of road and entering a circular curve section of radius R will experience the full radial force (mv^2/R) instantaneously. If R is small and the vehicle is travelling too fast, the practical effect of this is for the vehicle to skid sideways, away from the centre of curvature, as the full radial force is applied. In severe cases the vehicle could overturn.

To counteract this, the Department of Transport in the UK has published a *Design Manual for Roads and Bridges*. Volume 6 of this manual includes a traffic standard *TD 9/93 – Amendment No. 1 – Highway Link Design*, which gives guidelines for highway design that specify minimum radii for wholly circular curves. These are discussed in Section 13.2. If it is necessary to use radii below those specified for wholly circular curves for the road speed in question then transition curves must be incorporated into the design.

Transition curves are curves in which the radius changes from infinity to a particular value. The effect of this is to gradually increase the radial force P from zero to its highest value and thereby reduce its effect. Consider a road curve consisting of two transitions and a circular curve as shown in Figure 13.1. For vehicles travelling from T to U, the force P gradually increases from zero along the first transition curve to its maximum value on the circular curve and then decreases to zero again along the second transition curve. This greatly reduces both the tendency to skid and the discomfort experienced by passengers in the vehicles. This is one of the purposes of

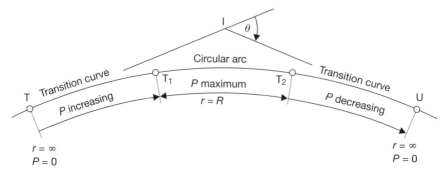

Figure 13.1 ● A composite curve.

transition curves – by introducing the radial force gradually and uniformly they minimise passenger discomfort. However, to achieve this, they must have a certain property.

For a constant speed v, the force P acting on the vehicle is (mv^2/r). Since any given curve is designed for a particular speed and the mass m of a vehicle can be assumed to be constant, it follows that $P \propto 1/r$.

However, if the force is to be introduced uniformly along the curve, it also follows that P must be proportional to l, where l is the length along the curve from the entry tangent point to the point in question.

Combining these two requirements gives $l \propto 1/r$ or

$$rl = K \tag{13.1}$$

where K is a constant. If L_T is the total length of each transition and R the radius of the circular curve, then at the end of the end of each transition curve, K can be evaluated using $RL_T = K$.

If the transition curve is to introduce the radial force in a gradual and uniform manner, equation (13.1) suggests that the product of the radius of curvature at any point on the curve and the length of the curve up to that point is a constant value. This is the definition of a spiral, and because of this, transition curves are also known as *transition spirals*. The types of curve used as transitions are discussed further in Section 13.3.

A further purpose of transition curves is to gradually introduce superelevation, and this is discussed later in this section.

Use of transition curves

Transition curves can be used to join intersecting straights in one of two ways, either in conjunction with circular curves to form *composite curves* or alone in pairs to form *wholly transitional curves*.

Composite curves

Figure 13.1 shows an example of a composite curve. In these, transition curves of equal length are used on either side of a central circular arc of radius R. A design method for composite curves is discussed in Section 13.5.

Although this type of design has widespread use, it has the disadvantage that the radius and hence the radial force are constant throughout the circular arc and, if the force is large, the length of the circular arc represents a danger length over which the maximum force applies. For this reason the use of wholly transitional curves is sometimes preferred.

Wholly transitional curves

These curves consist of two transition curves of equal length with no intervening central circular arc, as shown in Figure 13.2. Each of the transitions in this curve has a continuously changing radius and hence a continuously changing force. Therefore

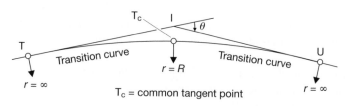

Figure 13.2 ● A wholly transitional curve.

there is only one point (the *common tangent point* T_C in Figure 13.2) where the force is a maximum. This means that wholly transitional curves are safer than composite curves. However, it is not always possible to fit this type of curve between the two intersecting straights owing to limitations on the minimum radii values that can be used. This is discussed in Section 13.5, where a design method for wholly transitional curves is considered.

Superelevation

Although transition curves can be used to introduce the radial force gradually in an attempt to minimise its effect, this can also be greatly reduced and even eliminated by raising one side of the roadway relative to the other. This procedure is shown in Figure 13.3 and the difference in height between the road channels is known as the *superelevation*.

In theory, by applying sufficient superelevation, the resultant force acting on a vehicle travelling around a curve can be made to act perpendicularly to the road surface, thereby forcing the vehicle down on to the road surface rather than throwing it off.

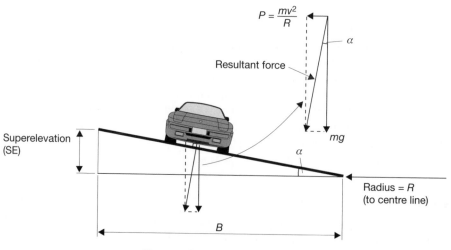

Figure 13.3 ● Superelevation.

The maximum superelevation (SE) occurs when $r = R$ along the central circular arc of a composite curve or at the common tangent point T_C of a wholly transitional curve. With reference to Figure 13.3

$$\tan \alpha = \frac{mv^2/R}{mg} = \frac{v^2}{gR}$$

But

$$SE = B \tan \alpha$$

Therefore

$$maximum\ theoretical\ SE = \left(\frac{Bv^2}{gR}\right) \text{ where } v \text{ is in m s}^{-1} \tag{13.2}$$

In practice, for high design speeds, wide carriageways and small radii, the maximum superelevation could be very large and if actually constructed would be alarming to drivers approaching the curve. Also, any vehicle travelling below the design speed would tend to slip down the road surface towards the centre of curvature and the driver would have to understeer to compensate. Interestingly, should the *maximum* SE actually be constructed then any vehicle travelling at the design speed should travel round the curve without the driver needing to adjust the steering wheel.

Because of these aesthetic effects, *TD 9/93* gives a number of recommendations for the *maximum and minimum allowable values of superelevation* that should be designed into new roads. Some of these are listed below, but considerably more information about superelevation can be found in *TD9/93* itself. The recommendations are

- Superelevation shall normally balance out only 45% of the radial force *P*.
- In rural areas superelevation shall not exceed 7% (approximately 1 in 14.5) and, wherever possible, radii should be chosen such that superelevation is kept within the desirable value of 5% (1 in 20).
- In urban areas with at-grade junctions and side accesses, superelevation shall be limited to 5%.
- The *minimum allowable* SE, to allow for drainage, is 2.5% (1 in 40).

Although the *maximum theoretical* SE $= (Bv^2/gR)$, in practice the *maximum allowable* SE $= 0.45(Bv^2/gR)$, since superelevation can only balance out 45% of the radial force. Expressing v in kph and R in metres, and substituting for $g = 9.81$ m s^{-2} gives

$$maximum\ allowable\ SE = \left(\frac{Bv^2}{282.5R}\right) \text{ m}$$

and, for convenience, this is written in *TD 9/93* as

$$maximum\ allowable\ SE = \left(\frac{Bv^2}{200\sqrt{2}R}\right) = \left(\frac{Bv^2}{282.8R}\right) \text{ m} \tag{13.3}$$

Expressing the maximum allowable superelevation as a percentage, s, such that $s = 100(SE/B)$ gives

$$s\% = \frac{v^2}{(2\sqrt{2})R} = \frac{v^2}{2.828R} \tag{13.4}$$

These expressions for maximum allowable superelevation hold for values of R down to two steps below the desirable minimum values only (see Section 13.2).

Once the maximum allowable value has been calculated, if it gives a cross slope on rural roads greater than 7% then only 7% should be used. If, for example, a design requires a superelevation of 10%, only 7% should be used. However, it is not advisable to do this, and if the maximum SE is exceeded the road should, if possible, be redesigned to meet standards. Because the SE only balances out 45% of the radial force, the other 55% and any extra superelevation not accounted for in the final design are assumed to be taken by the friction between the road surface and the tyres of the vehicle. This is the reason why some vehicles skid in wet or greasy conditions on roads containing lower R values.

Once the allowable superelevation has been determined it is built into the curve. In the case of a *composite curve* it is constant along the circular section and gradually introduced on the entry transition curve and gradually removed on the exit transition curve. For a *wholly transitional curve* it is gradually introduced and removed on the entry and exit transition curves, reaching its maximum value at one point only, where the two curves meet. If a design consists only of a *circular curve* then between one-half and two-thirds of the maximum allowable superelevation should be introduced on the entry straight leading into the curve and the remainder just after the beginning of the curve. The superelevation should be run out into the exit straight in a similar manner – *TD9/93* gives further information on this.

TD9/93 states that superelevation shall not be introduced so gradually as to create large almost flat areas of carriageway nor so sharply as to cause discomfort or to kink the edges of carriageways. In order to ensure that this is not the case it recommends that the superelevation should be introduced at a rate of no more than about 1% (1 in 100) on all-purpose roads and 0.5% (1 in 200) on motorways. As an example, if on an all-purpose road the maximum allowable superelevation is calculated as 0.36 m, a transition curve of length of at least $0.36 \times 100 = 36$ m is required to enable the superelevation to be introduced at a rate of no more that 1%.

Because of the restrictions in *TD9/93* it is important that in cases where transition curves are incorporated into the design, the lengths of the transitions are sufficient to allow the superelevation to be introduced. Consequently, when designing such curves, the lengths of the transitions obtained from the design process must be checked to ensure that they are long enough to allow the superelevation to be introduced. If they are not then their lengths should be increased to the values required for them to do so. This is discussed further in Section 13.5 and Worked example 13.1 at the end of this chapter illustrates this point.

Reflective summary

With reference to the need for transition curves remember:

— Vehicles travelling along a circular curve are subjected to a radial force which is trying to make them travel in a straight line.

— The radial force is inversely proportional to the radius of curvature. This means that the smaller the radius the greater the force and the greater the passenger discomfort.

— The radial force is directly proportional the square of the design speed. This means that the faster the speed the greater the force.

— The Department of Transport publication *TD9/93* specifies minimum radii for a range of design speeds below which it is necessary to incorporate transition curves.

— Transition curves are curves in which the radius of curvature is either decreasing from infinity or increasing to infinity – a true transition curve is a spiral.

— Transition curves serve two purposes: to introduce the radial force gradually and uniformly in order to minimise passenger discomfort and to gradually introduce superelevation.

— Superelevation involves raising one of the road channels relative to the other and enables the effect of the radial force to be reduced.

— *TD9/93* specifies maximum allowable values of superelevation for aesthetic effects and minimum allowable values of superelevation to ensure that water can drain from the road surface.

— Transition curves can be used in conjunction with circular curves to form composite curves or in pairs to form wholly transitional curves.

13.2 Current United Kingdom Department of Transport design standards

After studying this section, you should be aware of the highway design standards currently used in the United Kingdom. You should know the choice of design speeds that is available and you should appreciate the limiting radii values that should be considered during the design process.

This section includes the following topics:

- *TD9/93 – Amendment No. 1 – Highway Link Design*
- Using *TD9/93* to choose a suitable radius

TD9/93 – Amendment No. 1 – Highway Link Design

In the United Kingdom, the Department of Transport stipulates allowable radii for particular design speeds. These are discussed in a design standard entitled *TD 9/93 – Amendment No. 1 – Highway Link Design*, which is contained in Volume 6 of the *Design Manual for Road and Bridges*. This standard completely replaces a number of earlier documents which have been withdrawn – details of these withdrawn publications are given in the foreword to *TD 9/93*.

Six design speeds ranging from 120 kph to 50 kph are specified as shown in Table 13.1, which has been reproduced from *TD9/93*. Values are listed in the table for each design speed against four parameters: *STOPPING SIGHT DISTANCE, HORIZONTAL CURVATURE, VERTICAL CURVATURE* and *OVERTAKING SIGHT DISTANCES*. The use of *TD9/93* in HORIZONTAL CURVATURE is discussed in the following section and the applications of the other three parameters in Table 13.1 to the design process are discussed in Chapter 14.

Using *TD9/93* to choose a suitable radius

It is strongly recommended that *TD9/93* be studied in great detail before the commencement of any highway link design and the following example is included only to give a brief overview of the use of the standard when selecting a radius for use in *HORIZONTAL CURVATURE* design.

Consider two intersecting straights on a proposed dual carriageway to be designed for a speed of 100 kph, which are to be joined by a horizontal curve. With reference to the section headed HORIZONTAL CURVATURE in Table 13.1, the various radii choices are as follows.

If only a *circular curve* is to be designed, the minimum R value that can be used is 2040 m. If this radius is too large to meet the design requirements then transition curves must be incorporated into the design.

If *transition curves* are to be incorporated then the following radii are permitted for various superelevation (*SE*) values:

For $SE = 2.5\%$, R must be ≥ 1440 m
For $SE \leq 3.5\%$, R must be ≥ 1020 m
For $SE \leq 5\%$, R must be ≥ 720 m   (*Desirable Minimum*)
For $SE \leq 7\%$, R must be ≥ 510 m   (One Step below *Desirable Minimum*)
For $SE \leq 7\%$, R must be ≥ 360 m   (Two Steps below *Desirable Minimum*)

Table 13.1 ● Current Department of Transport highway design standards (published here by permission of The Stationery Office).

DESIGN SPEED kph	120	100	85	70	60	50	V^2/R
STOPPING SIGHT DISTANCE m							
Desirable Minimum	295	215	160	120	90	70	
One Step below Desirable Minimum	215	160	120	90	70	50	
HORIZONTAL CURVATURE m							
Minimum R^* without elimination of Adverse Camber and Transitions	2880	2040	1440	1020	720	520	5
Minimum R^* with Superelevation of 2.5%	2040	1440	1020	720	510	360	7.07
Minimum R^* with Superelevation of 3.5%	1440	1020	720	510	360	255	10
Desirable Minimum R^* with Superelevation of 5%	1020	720	510	360	255	180	14.14
One Step below Desirable Minimum R with Superelevation of 7%	720	510	360	255	180	127	20
Two Steps below Desirable Minimum R with Superelevation of 7%	510	360	255	180	127	90	28.28
VERTICAL CURVATURE m							
Desirable Minimum* Crest K Value	182	100	55	30	17	10	
One Step below Desirable Min Crest K Value	100	55	30	17	10	6.5	
Absolute Minimum Sag K Value	37	26	20	20	13	9	
OVERTAKING SIGHT DISTANCES							
Full Overtaking Sight Distance FOSD m	*	580	490	410	345	290	
FOSD Overtaking Crest K Value	*	400	285	200	142	100	

*Not recommended for use in the design of single carriageways.

The V^2/R values simply represent a convenient means of identifying the relative levels of design parameters, irrespective of design speed.

This appears to give a wide choice, but in practice *TD9/93* recommends that in order to ensure a high standard of road safety, designs incorporating at least the Desirable Minimum radii values should be the initial objective. Desirable Minimum values represent the comfortable values dictated by the design speed, and in this case the designer should try to ensure that the final chosen R value is no less than 720 m. If this is not possible then a *relaxation* to a lower level design speed step may be introduced at the discretion of the designer, having regard to the advice given in *TD9/93* and all the relevant local factors.

As explained in *TD9/93*, relaxations may be introduced in cases where strict application of Desirable Minimum standards would lead to disproportionately high construction costs or severe environmental impact on people, properties or landscapes. Even if relaxations are introduced, the level of service provided by the road may remain generally satisfactory and a road may not become unsafe. Depending on the type of road being designed, relaxations up to one or more design speed steps below the Desirable Minimum values may be permitted. So in this case, depending on the situation, the designer may consider relaxing the radius value either by one step to 510 m or two steps to 360 m, if permitted. If only a one step relaxation is permitted, down to 510 m, and the designer wishes to use a radius below this value, down to 360 m, then this becomes a *departure* from standard. Departures from standard may be allowed in situations of exceptional difficulty in order to lower the R value even further. Proposals to adopt departures from the standard must be submitted to the Department of Transport or its appointed agent for approval. Full details of the criteria covering relaxations and departures are given in *TD 9/93*.

The example given above was for a proposed dual carriageway. However, had the proposed road been a two-lane single carriageway there would have been much greater restrictions on the R values that could have been used, as indicated in Table 13.1 by the asterisks shown next to several of the radii to indicate that they are not recommended for use in the design of single carriageways. *TD9/93* recommends that two-lane single carriageways should be designed such that there are clear overtaking sections and clear non-overtaking sections. The former require large radii and the latter require low radii. To accommodate this, *TD9/93* states that for clear overtaking sections the R values should be considerably greater than the minimum without transitions and for non-overtaking sections the radii should be between the Desirable Minimum value and one step below the Desirable Minimum value. In practice, the values for clear overtaking sections are raised by one design speed step and, for a road designed for 100 kph, the minimum recommended radius for clear overtaking is 2880 m, which is the limiting radius for no transitions for a design speed of 120 kph. Similarly, that for 85 kph is raised to 2040 m, the limiting radius for 100 kph. For a two-lane single carriageway designed for a speed of 100 kph, the radius should either be greater than 2880 m or between 510 m and 720 m. This restriction is set to avoid the use of radii (in this case between 720 m and 2880 m) which could produce long dubious overtaking conditions for vehicles travelling in the left-hand curve direction, and simply reduce the length of overtaking straight that could otherwise be achieved. Further information on the choice of radii for two-lane single carriageway roads is given in *TD9/93*.

Reflective summary

With reference to the current United Kingdom Department of Transport design standards remember:

- When designing horizontal curves in the United Kingdom, Department of Transport publication *TD9/93* should be consulted.

- For horizontal curvature design, *TD9/93* specifies a range of radii values for six different design speeds.

- If only a circular curve is to be designed, its radius of curvature must be greater than the minimum value without transitions specified in *TD9/93* for the design speed in question.

- If a curve involving transitions is to be designed then, wherever possible, the minimum radius of curvature must be at or greater than the Desirable Minimum value specified in *TD9/93* for the design speed in question.

- If it is not possible to use a minimum radius of curvature at or greater than the Desirable Minimum value then it may be possible to apply a relaxation of one or more design speed steps below the Desirable Minimum value.

- If the relaxation values are still not appropriate and a smaller radius is being considered then this is a departure from standard for which special permission must be given.

- In order to ensure that there are clear overtaking sections and clear non-overtaking sections, *TD9/93* imposes much greater restrictions on the minimum radii of curvature that can be used in the design of 2-lane single carriageways compared with one-way roads, dual carriageways and motorways.

13.3 Type of transition curve to be used

After studying this section you should understand the properties required of a transition curve and appreciate that the ideal transition curve is one which is a true spiral. You should know the equations of two different types of curves that are used for transitions, that is, the *clothoid* and the *cubic parabola,* and you should appreciate the differences between them. You should be able to determine which is the most appropriate type of transition curve to use in practice.

This section includes the following topics:

- Types of curve
- The clothoid

- The cubic parabola
- Choice of transition curve

Types of curve

Equation (13.1) shows that for a transition curve, the expression $rl = K$ must apply, which means that its radius of curvature must decrease in proportion to its length. This is the property of a spiral and one curve having this property is the *clothoid*. Because of this, it is often referred to as the *ideal transition curve* or the *ideal transition spiral*.

Another transition curve in common use is the *cubic parabola*, which is derived from the clothoid. The cubic parabola is not a spiral, in that it does not have the property that rl is always constant. However, it can be used over a certain range and the calculations involved in its design are simpler than those of the clothoid.

The clothoid and the cubic parabola are discussed in the following sections and it is recommended that the clothoid section be studied first since much of the cubic parabola theory is based on it.

The clothoid

This is a true transition curve in that it has the property that at any point on it rl is always a constant value. The equation of the clothoid can be derived by considering Figure 13.4, which shows two points M and N close together on a transition curve of length L_T. On this diagram

ϕ is the *deviation angle* between the tangent at M and the straight TI
δ is the *tangential angle* to M from T with reference to TI
x is the offset to M from the straight TI at a distance y from T
l is the length from T to any point M on the curve (not shown)
δl is the length along the curve from M to N
$\delta\phi$ is the angle subtended by the arc δl of radius r

The distance δl on the curve is considered short enough to assume that the radius of curvature r at both M and N is the same. Therefore

$$\delta l = r\delta\phi$$

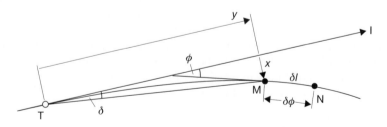

Figure 13.4 ● Clothoid geometry.

From equation (13.1), $rl = K$ is required and substituting $1/r = (l/K)$ gives

$$\delta\phi = \left(\frac{l}{K}\right)\delta l$$

Integration gives $\phi = (l^2/2K) + \text{constant}$, but when $l = 0$, $\phi = 0$, so the constant $= 0$. Therefore

$$\phi = \left(\frac{l^2}{2K}\right)$$

At the end of a transition, $r = R$ and $l = L_T$, giving $K = rl = RL_T$. This gives

$$\phi = \left(\frac{l^2}{2RL_T}\right) \text{ (in radians)} \tag{13.5}$$

This is the basic equation of the clothoid. If its conditions are satisfied and speed is constant, radial force will be introduced uniformly.

The maximum value of ϕ will occur at the common tangent point T_1 where the transition meets the circular arc when $l = L_T$. This gives

$$\phi_{\max} = \left(\frac{L_T}{2R}\right) \text{ (in radians)} \tag{13.6}$$

Derivation of setting out formulae for the clothoid

Using the basic equation of the clothoid it is possible to derive values to enable it to be set out on site. In Chapter 12, several methods of setting out were discussed for circular curves, including offsets from the tangent lengths and tangential angles. These methods can also be employed for transition curves but the formulae used are different from those for circular curves.

With reference to Figure 13.4, for the *offsets from the tangent lengths* method a formula is required linking the offset x to the length along the tangent y and for the *tangential angles* method a formula is required linking the tangential angle δ to the length l along the curve from T.

Consider Figure 13.5, which shows an enlarged section of Figure 13.4. Since M and N are close, it can be assumed that the curve length MN = chord length MN = δl and expressions for δx and δy can be derived as follows.

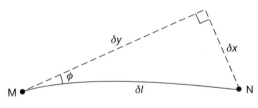

Figure 13.5

For x

$$\delta x = \delta l \sin \phi = \left[\phi - \frac{\phi^3}{3!} + \frac{\phi^5}{5!} - \ldots\right]\delta l$$

$$= \left[\left(\frac{l^2}{2K}\right) - \frac{(l^2/2K)^3}{3!} + \frac{(l^2/2K)^5}{5!} - \ldots\right]\delta l$$

Integration gives

$$x = \left(\frac{l^3}{6K}\right) - \left(\frac{l^7}{336K^3}\right) + \left(\frac{l^{11}}{42240K^5}\right) - \ldots \qquad (13.7)$$

For y

$$\delta y = \delta l \cos \phi = \left[1 - \frac{\phi^2}{2!} + \frac{\phi^4}{4!} - \ldots\right]\delta l$$

$$= \left[1 - \frac{(l^2/2K)^2}{2!} + \frac{(l^2/2K)^4}{4!} - \ldots\right]\delta l$$

Integration gives

$$y = l - \left(\frac{l^5}{40K^2}\right) + \left(\frac{l^9}{3546K^4}\right) - \ldots \qquad (13.8)$$

For x and y there are no constants of integration since $x = y = 0$ when $l = 0$.

The above formulae for x and y can be used to set out the clothoid using *offsets from the tangent lengths* as described in Section 13.6.

For the *tangential angles method,* a formula linking δ and l is required. With reference to Figure 13.4, tan $\delta = (x/y)$ and by calculating x and y values for particular l values along the curve, δ values can be calculated. The use of the tangential angles method for setting out transition curves is also described in Section 13.6.

Unfortunately, the clothoid formulae for x and y are both in the form of infinite series, which can make their exact calculation difficult. However, the third and subsequent terms in each expression tend to be very small and can be neglected. In the past, special tables were produced listing values for these series up to and including the second terms. Nowadays, however, they can easily be evaluated using hand calculators and even these have been superseded by the wide range of highway design software packages currently available. These usually incorporate clothoids in their design procedures together with the option to use another type of curve known as a cubic spline. The use of such packages is discussed in Section 13.5.

The cubic parabola

Unlike the clothoid, the cubic parabola is not a true spiral and because of this it cannot always be used. However, it approximates very closely to a spiral over a certain range and its advantage of having simpler formulae than those of the clothoid giving rise to easier calculations has led to it being widely adopted. In practice, for the lengths over which it tends to be used, it can be considered to be identical to the clothoid from which its formulae are derived by making certain assumptions. The validity of these assumptions is discussed later in this section.

Derivation of setting out formulae for the cubic parabola

In the previous section, formulae in the form of infinite series were developed for x and y offsets on a transition in order that the clothoid could be set out using these either from the tangent length or by the tangential angles method. These formulae are used to obtain the equation of the cubic parabola by assuming that the second and subsequent terms in equations (13.7) and (13.8) can be neglected. For the cubic parabola therefore

$$x = \left(\frac{l^3}{6K}\right) \quad \text{and} \quad y = l$$

Substituting $y = l$ in the expression for x gives

$$x = \left(\frac{y^3}{6K}\right)$$

Since

$$K = rl = RL_T$$

it follows that

$$x = \left(\frac{y^3}{6RL_T}\right) \qquad (13.9)$$

This is the basic equation of the cubic parabola and it can be used to set out the curve by *offsets from the tangent lengths*.

As with the clothoid, it is also possible to set out the cubic parabola by the *tangential angles method*. With reference to Figure 13.4, the formula for the tangential angle δ in terms of the length l along the curve is derived as follows:

$$\tan \delta = \left(\frac{x}{y}\right)$$

But, for the cubic parabola

$$x = \left(\frac{y^3}{6RL_T}\right)$$

Substituting for x gives

$$\tan \delta = \frac{y^3/6RL_T}{y} = \left(\frac{y^2}{6RL_T}\right)$$

Here, another assumption is made in that only small angles are considered so that tan $\delta = \delta$ radians. This gives

$$\delta = \left(\frac{y^2}{6RL_T}\right) \text{ (in radians)} \qquad (13.10)$$

and this formula can be used to set out the cubic parabola by the *tangential angles method*. However, in order that it can be set out using a theodolite or total station, an expression in terms of degrees is preferred where

$$\delta = \left(\frac{y^2}{6RL_T}\right)\left(\frac{180}{\pi}\right) \text{ degrees}$$

or

$$\delta = \left(\frac{l^2}{6RL_T}\right)\left(\frac{180}{\pi}\right) \text{ degrees} \qquad (13.11)$$

since y is assumed to be equal to l for the cubic parabola.

A useful relationship can be developed for the cubic parabola between δ, the tangential angle, and ϕ, the deviation angle, as follows.

Since $\phi = (l^2/2RL_T)$ is the basic equation of the clothoid and $y = l$ for the cubic parabola, it follows that $\phi = (y^2/2RL_T)$ for the cubic parabola.

However, equation (13.10) gives

$$\delta = \left(\frac{y^2}{6RL_T}\right)$$

and it follows that for the cubic parabola

$$\delta = \left(\frac{\phi}{3}\right) \qquad (13.12)$$

In addition

$$\delta_{max} = \left(\frac{\phi_{max}}{3}\right) \qquad (13.13)$$

These relationships are shown in Figure 13.6 and are used to check the setting out of a cubic parabola (see Section 13.6).

Validity of the assumptions used in the derivation of the cubic parabola formulae

Three assumptions are made during the derivation of the cubic parabola formulae:

- The second and subsequent terms in the expansion of sin ϕ and cos ϕ are neglected as being too small – the validity of this will depend on the value of ϕ.

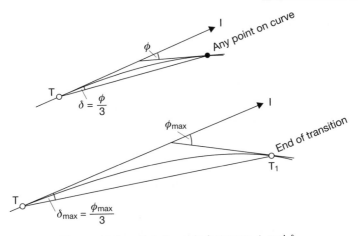

Figure 13.6 ● Relationship between ϕ and δ.

● Tan δ is assumed to equal δ radians. Since $\delta = \phi/3$ the validity of this will also depend on the value of ϕ.

● The length along the tangent is assumed to equal the length along the curve or $y = l$. Again, the value of ϕ will be critical to the validity of this, since the greater the deviation, the less likely is this assumption to be true.

All of these assumptions are only valid and the cubic parabola can only be used as a transition curve if the deviation angle, ϕ, is below some acceptable value.

It can be shown that if the deviation angle remains below approximately 12°, there is no difference between the clothoid and the cubic parabola. However, beyond 12° the assumptions made in the derivation of the cubic parabola formulae begin to break down and, to maintain accuracy, further terms must be added to equations (13.9) and (13.11), thereby losing the advantage offered by the simple equations. In fact, as shown in Figure 13.7, once the deviation angle reaches 24°06′, even if the formulae are expressed as infinite series, the cubic parabola becomes useless as a transition curve because its curvature begins to increase with its length; that is, rl is no longer a constant. This shows that, in theory, the cubic parabola can only be used as

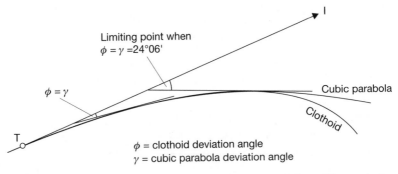

Figure 13.7 ● Relationship between the clothoid and the cubic parabola.

a transition curve if ϕ_{max} is less than approximately 24°, but in practice it tends to be restricted to curves where ϕ_{max} is less than approximately 12° and δ_{max} is less than approximately 4° (since $\delta_{max} = \phi_{max}/3$) in order that the simple formulae can be used.

Choice of transition curve

In practice both the clothoid and the cubic parabola are used. The angles involved are usually well below the limiting values for the cubic parabola and hence the final choice is usually one of convenience or habit. The clothoid can be used in any situation but has more complex formulae – the cubic parabola is easier to calculate by hand but cannot always be used. Nowadays, if highway design software packages are used, the question of choosing between these two becomes irrelevant, since the clothoid, being the ideal transition curve, would always take precedence over the cubic parabola. However, another type of curve, a *cubic spline*, is normally offered as an alternative to the clothoid in such packages. This is discussed further in Section 13.5.

The remainder of this chapter is devoted to the cubic parabola simply because its formulae and calculations are easier to show in written form.

Reflective summary

With reference to the type of transition curve to be used remember:

— The clothoid is sometimes referred to as the ideal transition curve because it is a true spiral.

— The calculations required to set out the clothoid on site are complicated by the fact that some of its formulae involve infinite series.

— A curve known as the cubic parabola, which is derived by making certain assumptions in the clothoid formulae, is often used instead of the clothoid.

— The calculations required to set out the cubic parabola on site are more straightforward for hand computation since its formulae do not involve infinite series.

— The cubic parabola is not a true spiral and should only be used where the deviation angle is less than approximately 12°.

— In practice both the clothoid and the cubic parabola are used, but if one of the many commercially available highway design software packages is available, the clothoid will always be preferred since it is a true spiral.

13.4 The geometry of transition curves

After studying this section you should know how to calculate the length of transition curve required to ensure that passenger discomfort is minimised. You should understand the geometry of transition curves and appreciate that much of the terminology used when referring to circular curves also applies to transition curves. You should also know how to calculate the tangent lengths and the lengths of curves used in composite and wholly transitional curves.

This section includes the following topics:

- The length of a transition curve required to minimise passenger discomfort
- The shift of a cubic parabola
- Tangent lengths and total curve lengths

The length of a transition curve required to minimise passenger discomfort

In order that transition curves can be designed and set out on site it is necessary to consider their geometry. The choice of the length to be used for transition curves is very important, since it must ensure that passenger discomfort is minimised and that the maximum allowable superelevation can be introduced gradually. In addition, it is necessary to determine the tangent lengths and, in the case of composite curves, the length of the central circular arc. Much of the terminology used in transition curve geometry is the same as that used in circular curve geometry and it is strongly recommended that Section 12.2 is reviewed before continuing.

As discussed in Section 13.1, one of the purposes of transition curves is to introduce the radial force gradually and uniformly such that passenger discomfort is minimised. It is essential, therefore, that transition curves are long enough to allow this to be achieved. Consider Figure 13.8, which shows part of a composite curve having transition curves of length L_T and a circular curve of radius R. The radial force at any

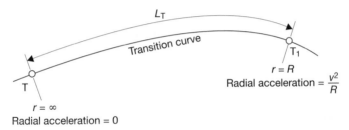

Figure 13.8 ● Rate of change of radial acceleration.

point on the curve is given by $P = mv^2/r$ but since *force = mass × acceleration*, it follows that the *radial acceleration* at any point on the curve is given by v^2/r.

Since v is constant for any given curve, the radial acceleration is inversely proportional to the radius. Therefore the rate at which the radial acceleration changes is inversely proportional to the rate at which the radius changes. The faster the change in radius, the greater the rate of change of radial acceleration and hence the faster the introduction of the radial force, resulting in greater passenger discomfort. In effect, the shorter the transition curve, the greater the potential danger.

The transition curve must therefore be long enough to ensure that the radius can be changed at a slow enough rate in order that the radial force can change at a rate which is acceptable to passengers.

As can be seen, the rate at which the radial acceleration changes is a very important parameter and it can be used to calculate the value of L_T. It is known as the *rate of change of radial acceleration* (c) and should be considered as a safety or comfort factor, the value of which has an upper limit beyond which discomfort is too great. *TD9/93 recommends a maximum value for c of 0.3 m s^{-3}*, although it may be necessary to increase this value up to 0.6 m s^{-3} in difficult cases. In practice, wherever possible, transition curves with c values at 0.3 m s^{-3} should be used to ensure that they achieve the comfort requirement without being excessively long. A summary of the design method together with the final choice of a c value is given in Section 13.5.

The length L_T of a transition curve can be obtained by consideration of the rate of change of radial acceleration c as follows. In Figure 13.8

the radial acceleration at $T_1 = \left(\dfrac{v^2}{R}\right)$ and

the radial acceleration at T = zero

The difference between these gives

the change in radial acceleration from T to $T_1 = \left(\dfrac{v^2}{R}\right)$

But

the time taken to travel along the transition curve $= \left(\dfrac{L_T}{v}\right)$

Combining these gives

the rate of change of radial acceleration $= c = \left[\dfrac{(v^2/R)}{(L_T/v)}\right]$

or

$$c = \left(\frac{v^3}{L_T R}\right)$$

Rearranging gives the length of the transition as

$$L_T = \left(\frac{v^3}{cR}\right) \text{ where } v \text{ is in m s}^{-1}$$

If v is in kph

$$L_T = \left(\frac{v^3}{3.6^3 cR}\right) \text{ metres} \tag{13.14}$$

This is the formula used to calculate the lengths of transition curves in highway design.

The shift of a cubic parabola

In order that the tangent lengths can be calculated, a parameter known as the shift must be calculated.

Figure 13.9 shows a typical composite curve arrangement. The dotted arc between V and W represents a circular curve of radius $(R + S)$, which has been replaced by a circular arc T_1T_2 of radius R plus two transition curves, entry TT_1 and exit T_2U. By doing this, the original curve VW has been shifted inwards a distance S, where $S = VG = WK$. Hence, this distance S is known as the *shift*.

It must be noted that the composite curve TT_1T_2U and the original circular curve VW are completely different since the radius of the new circular arc T_1T_2 is R and the

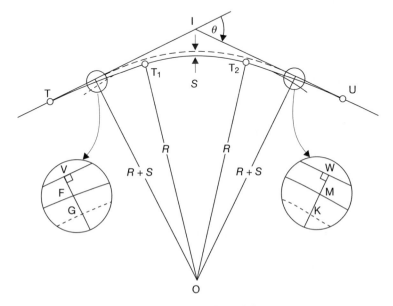

Figure 13.9 ● The shift.

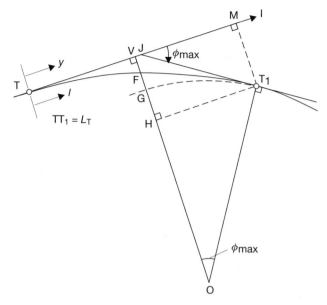

Figure 13.10

radius of the old circular curve VW is $(R + S)$. Also, the tangent points and the lengths of the circular arcs are not the same.

Figure 13.10 shows an enlargement of the left-hand side of Figure 13.9, and with reference to this a formula for the shift can be obtained as follows.

In quadrilateral VJT_1O

$$\text{angle } O\hat{V}J = \text{angle } J\hat{T}_1O = 90°$$

Because angle $V\hat{J}T_1 = 180° - \phi_{max}$

$$\text{angle } T_1\hat{O}V = \phi_{max}$$

The shift S is

$$S = VG = (VH - GH) = [MT_1 - (GO - HO)]$$

From the cubic parabola equation (13.9)

$$x = \left(\frac{y^3}{6RL_T}\right)$$

When $y = L_T$, $x = MT_1$ and equation (13.9) becomes

$$MT_1 = \left(\frac{L_T^3}{6RL_T}\right) = \frac{L_T^2}{6R}$$

With $GO = R$ and $HO = R \cos \phi_{max}$

$$S = \left(\frac{L_T^2}{6R}\right) - (R - R\cos\phi_{max})$$

$$= \left(\frac{L_T^2}{6R}\right) - R\left[1 - \left(1 - \frac{\phi_{max}^2}{2!} + \frac{\phi_{max}^4}{4!} - \ldots\right)\right]$$

This expression for S involves an infinite series but, again assuming small deviation angles, terms greater than ϕ_{max}^2 can be neglected as being too small. With $\phi_{max} = L_T/2R$

$$S = \left(\frac{L_T^2}{6R}\right) - R\left(\frac{\phi_{max}^2}{2!}\right) = \left(\frac{L_T^2}{6R}\right) - \left(\frac{R}{2}\right)\left(\frac{L_T}{2R}\right)^2$$

Simplifying gives the formula for the shift of a cubic parabola transition curve as

$$S = \left(\frac{L_T^2}{24R}\right) \tag{13.15}$$

The shift is an important parameter in the design and setting out of composite and wholly transitional curves and, using equation (13.15), part of the tangent lengths can be calculated as shown later in this section. However, to enable the whole of the tangent lengths to be determined another property of the shift is required.

Consider Figure 13.10 again, in which F is the point where the shift and the transition curve cross each other. Since the angles involved are small, it is assumed that

$$FT_1 = GT_1$$

Since GT_1 lies on the extended central circular arc, it is equal to $R\phi_{max}$. This gives

$$FT_1 = R\phi_{max}$$

Since $\phi_{max} = L_T/2R$, FT_1 can be expressed as

$$FT_1 = R\left(\frac{L_T}{2R}\right) = \frac{L_T}{2}$$

Because $TT_1 = L_T$ this shows that

$$FT = \frac{L_T}{2} \tag{13.16}$$

Since $y = l$ for a cubic parabola, it follows that

$$VT = FT = \frac{L_T}{2} \tag{13.17}$$

From Figure 13.10, when $x = VF$, $y = L_T/2$ and substituting these values into equation (13.9) gives

$$VF = \frac{(L_T/2)^3}{6RL_T} = \left(\frac{L_T^2}{48R}\right)$$

However

$$VG = S = \left(\frac{L_T^2}{24R}\right)$$

And comparison of these shows that

$$VF = \left(\frac{shift}{2}\right) = FG$$

This gives the property that the shift at VG is bisected by the transition curve and the transition curve is bisected by the shift. This property, together with the formula for the shift, enables the tangent lengths of composite and wholly transitional curves to be calculated as shown in the next section.

Tangent lengths and total curve lengths

Figure 13.11 shows that, in order for the composite curve to move vehicles through the deflection angle θ from one straight to the next, each transition curve moves them through the angle ϕ_{max} and the central circular arc moves them through the remaining $(\theta - 2\phi_{max})$. With reference to Figure 13.11, the *tangent lengths* IT and IU are given by

$$IT = IV + VT = IW + WU = IU$$

Because IV is the tangent length for circular curve VW, from equation (12.2)

$$IV = (R + S)\tan\left(\frac{\theta}{2}\right)$$

From equation (13.17)

$$VT = \frac{L_T}{2} = WU$$

This gives the tangent length as

$$IT = (R + S)\tan\left(\frac{\theta}{2}\right) + \frac{L_T}{2} \tag{13.18}$$

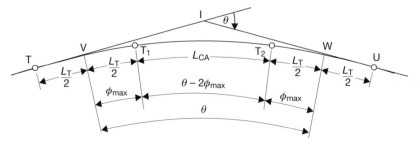

Figure 13.11 ● Tangent lengths and curve lengths.

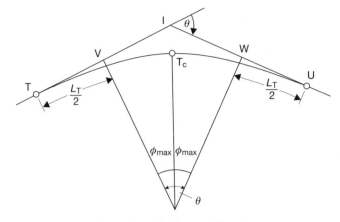

Figure 13.12 ● Wholly transitional curve.

This formula applies not only to composite curves but also to wholly transitional curves, as inspection of Figure 13.12 will show.

If the total length of the *composite curve* (L_{total}) in Figure 13.11 is required it can be obtained from

$$L_{total} = TV + VW + WU = \frac{L_T}{2} + R\theta + \frac{L_T}{2} \tag{13.19}$$

Alternatively, L_{total} can be obtained from the fact that the composite curve consists of a transition curve TT_1, a circular arc T_1T_2 and a transition curve T_2U. The length of the circular arc (L_{CA}) is given by

$$L_{CA} = R(\theta - 2\phi_{max}) \tag{13.20}$$

and the total length of the curve will be

$$L_{total} = TT_1 + T_1T_2 + T_2U = L_T + R(\theta - 2\phi_{max}) + L_T \tag{13.21}$$

In the case of a *wholly transitional curve* (Figure 13.12), since it does not contain a central circular arc, the length is simply given by

$$L_{total} = 2L_T \tag{13.22}$$

Reflective summary

With reference to the geometry of transition curves remember:

— Much of the terminology used in transition curve geometry is the same as that used in circular curve geometry.

— The choice of the length of transition curve to be used is very important. It must be long enough not only to ensure that the radial force is introduced gradually and uniformly so that passenger discomfort is minimised but also to enable the maximum allowable superelevation to be introduced gradually.

— The rate of change of radial acceleration is a measure of how quickly the radial force is introduced. The greater the rate of change of radial acceleration the quicker the force is introduced and the greater is the potential passenger discomfort.

— The greater the rate of change of radial acceleration the shorter the transition curve and the greater the potential danger.

— TD9/93 recommends a maximum value of 0.3 m s^{-3} for the rate of change of radial acceleration although in may be necessary to increase this value up to 0.6 m s^{-3} in difficult cases.

— A parameter known as the shift is required to enable the tangent lengths to be calculated. It is a measure of how far an equivalent circular curve has been shifted away from the tangent lengths by the introduction of transition curves.

— In any composite or wholly transitional curve, the shift bisects the transition curve and the transition curve bisects the shift.

— In a composite curve consisting of two equal length transition curves and a central circular arc, which is designed to connect two intersecting straights having a deflection angle of θ, each transition takes ϕ_{max} and the circular arc takes the remaining $(\theta - 2\phi_{max})$.

13.5 Designing composite and wholly transitional curves

After studying this section you should know the steps involved in the design of composite and wholly transitional curves. You should be able to undertake the design of such curves and you should appreciate the need to consider the vertical alignment before finalising the design of the horizontal alignment. You should be aware of the widespread use of computer software packages in the design of roads and you should have an appreciation of the general concepts on which these packages are based.

This section includes the following topics:

● Overview of horizontal alignment design
● A design method for composite curves

- A design method for wholly transitional curves
- Phasing of horizontal and vertical alignments
- Summary of horizontal curve design
- An introduction to computer-aided road design

Overview of horizontal alignment design

The main purpose of any horizontal alignment design is to ensure that vehicles travel safely and comfortably from one intersecting straight to another. How this is achieved will depend on the layout of the straights and any limitations that apply to the area through which they pass. Often, however, there is more than one solution to the problem in that it may be possible to join the straights using only circular curves, only transition curves or using composite curves.

Section 12.4 described the approach used when designing circular curves and it is recommended that this section be studied before continuing since many of the concepts discussed there are relevant to the design of transition curves. In particular, the concept of the *band of interest* within which the road being designed must fall is very important as this has considerable effect on the value of the deflection angle θ, which in turn governs to a large extent the decision as to whether or not transition curves should be included in the design. In addition, it is important to appreciate that the design procedure is usually undertaken with the aid of computer software packages, the principles of which are discussed later in this section.

A design method for composite curves

Composite curves were defined in Section 13.1. Their design is based on the fact that the composite curve must deflect the road through angle θ. Of this, the circular arc takes $(\theta - 2\phi_{max})$ and each transition takes ϕ_{max}, as shown in Figure 13.11.

Given: Deflection angle θ, design speed v and road type.

Problem: To calculate a suitable composite curve to fit between the straights TI and IU which have been drawn on a plan of the area in which the road is to be located.

Solution: Before detailing the design method, it must be noted that there are many solutions to this problem, all of them perfectly acceptable. Consequently, the following method can only be taken as a guide to design from which a suitable rather than a unique solution can be found. The procedure is based on the Department of Transport design standards given in Table 13.1 and is as follows.

- In this case, the deflection angle θ is given. If θ is unknown then it must be determined. If a computer package is being used, θ will be given by the software, otherwise it is measured accurately on site. How this can be done is described in Sections 12.6 and 13.6.

- Select a value of R greater than the Desirable Minimum given in Table 13.1 for the design speed in question and choose a c value of 0.3 m s^{-3} so that the design starts with the recommended limiting values for both R and c. If the design does not work they can be amended later.

- Calculate the length of each transition curve from equation (13.14), $L_T = v^3/3.6^3 cR$.

- Calculate the shift, S, from equation (13.15), $S = L_T^2/24R$.

- Calculate the tangent lengths IT and IU from equation (13.18)

$$IT = (R + S)\tan\left(\frac{\theta}{2}\right) + \frac{L_T}{2}$$

- The calculated lengths IT and IU should now be fitted to the straights drawn on the plan of the area to see if they are acceptable. Due to the band of interest, it may be necessary to alter the lengths of IT and IU to obtain a suitable fit. Assuming that θ cannot be changed, this can be done by altering R and/or c. Ideally, R should be maintained at a value greater than the Desirable Minimum value and c should normally not exceed 0.3 m s^{-3}. Wherever possible, the value of c should be at or just below 0.3 m s^{-3} in order to give transition curves that are long enough to ensure safe designs. Excessively long transition curves should be avoided due to their higher construction costs. If necessary, it may be possible to relax R to the One Step below Desirable Minimum value and even in cases of exceptional difficulty to the Two Steps below Desirable value. Equally, it may be possible to raise the value of c although a value of 0.6 m s^{-3} is the maximum that may be permitted. The process is an iterative one and ends when the tangent lengths are of an acceptable length to fit the given situation.

- Calculate the maximum allowable superelevation using either equation (13.3) or equation (13.4), $Bv^2/282.8R$ m or $v^2/2.828R\%$. At this point it is necessary to check that the length of each transition curve obtained earlier in the design process is long enough to enable the superelevation to be introduced. If it is not then the length of the transition will have to be increased and the new R value determined.

- Once R has been finalised, calculate ϕ_{max} from equation (13.6), $\phi_{max} = L_T/2R$.

- Calculate $(\theta - 2\phi_{max})$, and hence obtain the length of the circular arc, L_{CA}, using equation (13.20), $L_{CA} = R(\theta - 2\phi_{max})$.

Worked examples 13.2 and 13.3 at the end of this chapter show some of the calculations discussed above.

A design method for a wholly transitional curve

As defined in Section 13.1, these are curves which consist only of transitions. They can be considered as composite curves which have central circular arcs of zero length. Figure 13.12 shows such a curve. Wholly transitional curves have the advantage that there is only one point at which the radial force is a maximum, and therefore the safety is increased. Unfortunately, it is not always possible to fit a wholly transitional curve into a given situation.

The design procedure described below considers wholly transitional curves with equal tangent lengths only. However, although it is possible to design wholly transitional curves with unequal tangent lengths by using a different rate of change of radial acceleration for each half of the curve, they are rarely used.

Wholly transitional curves with equal tangent lengths have a very interesting property, which helps greatly with their design. With reference to Figure 13.12, since the circular arc is missing, it follows that

$$\theta = 2\phi_{max}$$

But

$$\phi_{max} = \left(\frac{L_T}{2R}\right)$$

Therefore, for a wholly transitional curve

$$\theta = 2\left(\frac{L_T}{2R}\right) = \frac{L_T}{R} \tag{13.23}$$

In addition, all the other transition curve equations still apply. Consequently, equation (13.14) for the length of a transition that gives comfort and safety must still apply. This equation is

$$L_T = \left(\frac{v^3}{3.6^3 cR}\right)$$

However, from equation (13.23), $L_T = R\theta$ and substituting for this gives

$$R\theta = \left(\frac{v^3}{3.6^3 cR}\right)$$

Rearranging this gives

$$R = \sqrt{\left(\frac{v^3}{3.6^3 c\theta}\right)} \text{ metres} \tag{13.24}$$

This leads to the property of wholly transitional curves that for any given two straights there is only one symmetrical wholly transitional curve that will fit between them for a given design speed if the rate of change of radial acceleration is maintained at a particular value. In other words, since v and θ are usually fixed, R has a unique value if c is maintained at, say, 0.3 m s^{-3}.

This is, in fact, the method of designing such curves and it is summarised as follows.

Given: Deflection angle θ, design speed v, and road type.

Problem: With reference to Figure 13.12, to calculate a suitable wholly transitional curve to fit between the straights TI and IU, which have been
drawn on a plan of the area in which the road is to be located.

Solution:

- Choose a value for c of 0.3 m s^{-3}, as discussed earlier in this section for the design
of composite curves.

- Substitute this into equation (13.24), $R = \sqrt{(v^3/3.6^3 c\theta)}$ and calculate the
minimum radius of curvature, R.

- This R value must be checked against the values given in Table 13.1 for the design
speed in question. Ideally, it should be greater than the Desirable Minimum
value given in the table. If this is not the case then, if possible, it should be greater
than the One Step below Desirable Minimum value and in exceptional cases it
may be permitted to use it providing it is above the Two Steps below Desirable
Minimum value.

  If R is acceptable when measured against these criteria then the length of transition for comfort and safety can be calculated using $L_T = v^3/3.6^3 cR$.

  If R cannot meet these criteria then the value of c must be reduced and the
calculations repeated.

  Worked example 13.4 at the end of this chapter shows the way in which the
radius value is checked.

- Once a suitable R for comfort and safety has been determined, calculate the
maximum allowable superelevation using either $Bv^2/282.8R$ m or $v^2/2.828R$%.
Now check that the length of each transition curve obtained is long enough to
enable the superelevation to be introduced. If it is not, then the length of the
transition will have to be increased and a new R value determined and checked
against the values given in Table 13.1.

- Once L_T has been determined for comfort, safety and superelevation, it is necessary to ensure that the curve will fit within the band of interest. For a quick check,
the assumption can be made with wholly transitional curves that the tangent
length is equal to the curve length. However, when the design is finalised, the
tangent lengths IT and IU should be obtained from IT $= (R + S)\tan(\theta/2) + L_T/2$. IT
and IU should then be checked for fit on the plan of the area. If they do not fit, it is
necessary to return to the start of the calculations and change one or more of the
variables, either v, θ or c. However, a solution may not be found, since it is not
always possible to fit a wholly transitional curve between intersecting straights
within the limits stipulated by the Department of Transport.

Phasing of horizontal and vertical alignments

It is very unusual for a horizontal alignment to be designed in isolation since allowance must also be made for the change in height of the ground surface along the
proposed centre line. This requires consideration of the vertical shape of the centre
line and leads to the design of the *vertical alignment*. Just as horizontal curves are used

to join intersecting straights in the design of the horizontal alignment, *vertical curves* are used to join intersecting *gradients* in the design of the vertical alignment.

A full description of the design and setting out of vertical curves is given in Chapter 14, but because of the interdependence of the horizontal and vertical alignments, each must be considered during the design of the other. In practice, they should be correctly *phased*, that is, their tangent points should coincide to ensure that they are the same length. If this is the case, it will avoid the creation of apparent kinks or optical illusions in the road surface, which could distract drivers.

Before finalising the design of the horizontal alignment, the total length of each composite or wholly transitional curve involved should be calculated and compared with the length of its equivalent vertical curve, as appropriate. The length of either the horizontal or the vertical curve should then be changed to ensure that the two curves are equal. Normally, it is the vertical curve length that is changed since this is usually easier to do. This need to equate the horizontal and vertical alignments is discussed further in Section 14.3.

Summary of horizontal curve design

In Section 12.4 and earlier in this section, methods for designing wholly circular, composite and wholly transitional curves have been discussed. Usually, these three techniques are combined into one general design and considered as possible solutions to the same problem – to design the best curve to fit a particular set of conditions.

Often, only the design speed v and the type of road are known (motorway, dual carriageway, 2-lane carriageway) and the problem becomes one of choosing the ideal combination of θ, R and c to fit into the band of interest concerned while maintaining current design standards.

If a vertical curve is designed in conjunction with the horizontal curve, which is normally the case, then the problem is further complicated by the need to phase the two curves correctly.

Because of all the different parameters involved, the design can be tedious and time consuming when done manually. Fortunately, the iterative processes involved are ideal for solution by computer. In theory, a computer program is written, which consists of the basic design steps and the design standards. The known curve parameters, for example, v and θ, are input together with chainage values, reduced levels and any external constraints. On running the program, the software calculates suitable values for the curve parameters which were not known, for example, c, R and superelevation and also produces the setting out data, if required.

In practice, there is now a wide range of commercially available highway design software which will undertake far more than just the basic design steps. They cover all stages of the design procedure, starting with trial horizontal and vertical alignments and continuing to the production of longitudinal and cross-sectional drawings, the listing of setting out data, the calculation of volumes, the planning of the movement of materials and, in some cases, even the preparation of computer graphics to show how the proposed design will look on completion.

These packages have revolutionised the design process. They have not only freed engineers from the tedious calculations and allowed them to concentrate on the important design concepts, but have also speeded up the design procedure to such an extent that different parameters can be tried and different designs compared in a very short period of time. The general concepts on which such packages are based are discussed in the following section.

An introduction to computer-aided road design

In highway alignment design, many factors such as design standards, topography, environment and the visual impact of the road have to be considered. This creates a demand for a number of alternative routes to be studied for any given road scheme and, for each route, the ability to produce a visual representation or model of the proposed road is highly desirable as a means of checking design work and for presentation at public hearings.

The preparation of different alignments by hand methods involves much work and the production of the various drawings for each design manually is an almost impossible task. However, by using computer systems in road design, these problems can be overcome to such an extent that many trial designs can be studied and presented with relative ease.

As a consequence, considerable emphasis has been placed on the development of highway design software packages that are capable of undertaking complex horizontal and vertical alignment designs. There is now a wide range of these available from commercial surveying and software companies for use with workstations, PCs, total stations and GPS receivers and, given the wide choice, it is not possible to give a detailed review of such individual packages in a general textbook such as this. For further information on their capabilities and costs, however, the reader is recommended to study surveying journals such as *Civil Engineering Surveyor*, *Engineering Surveying Showcase* and *Geomatics World*, which produce annual reviews of the latest software packages.

Although individual packages are not reviewed here, the general concepts on which they are based tend to be similar and the block diagram of Figure 13.13 shows the various stages involved. These are briefly described as follows.

Initially, a DTM is produced of the area covered by the corridor or band of interest. The DTM is formed using air or ground survey methods as described in Section 10.9 and is essentially a map of the area stored digitally in a computer. In addition to surface information, the results of any site investigations can also be stored in the DTM. Such data may include ground-water conditions, geotechnical characteristics of the area and any other properties which may affect the design.

After the DTM has been completed, many trial alignments can be studied by the computer. For horizontal alignments, two methods are used by the computer: the conventional method, in which straight sections of road are joined by circular and/or transition curves (see Chapter 12 and previous sections of this chapter) or a technique based on curves known as *cubic splines*. A cubic spline is a curve of continually changing radius, the equation of which takes the form of a cubic polynomial. Cubic

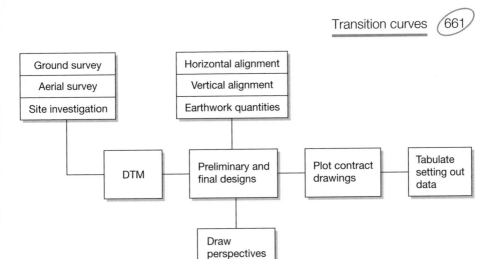

Figure 13.13 ● Stages involved in computer-aided road design.

splines are specified to fit between given location points, for example, straights, end points of a scheme, points the curve must pass through to avoid obstacles and so on.

For vertical alignment design (see Chapter 14), three methods are used by the computer: the traditional method based on intersecting gradients and parabolic curves, an extension of this traditional method in which parabolic curves are fitted to various fixed elements along the horizontal alignment such as sections of gradient, bridges, tie-ins to existing road junctions and so on and the cubic spline method mentioned above.

Each combined horizontal and vertical alignment, as designed by the computer, is passed through the DTM and the computer produces a longitudinal section, as many cross-sections as desired and an estimate of the earthwork quantities involved. This considerably shortens the time required to carry out these procedures by manual methods, details of which can be found in Chapter 15. In addition, for any alignment, the computer system can produce perspective drawings showing views along the proposed road. Such drawings can be used for visually checking the design and for preparing material for reports, exhibitions and public enquiries. The flexibility of highway design software packages enables any amount of design data to be combined with DTMs and it is possible to carry out a much more thorough preliminary design than that which could ever be undertaken by conventional methods.

As soon as the optimum alignment has been chosen, further data is entered into the DTM to enable a set of contract drawings to be produced by the computer interfaced with a suitable plotter. If all the relevant information for the optimum road alignment is in digital form, these drawings will consist of a series of plans showing all aspects of the road construction including longitudinal and cross-sections along the main alignment and also at interchanges, junctions, slip roads and so on. Based on these, schedules of earthwork quantities can be produced by the computer along any section of road and setting out tables can be computed giving angles and distances relative to existing survey stations.

The greatest benefits of using a computer system in road design are the ability to investigate different alignments and a reduction in the overall time taken for the design and production of contract drawings. If the design should change at any time, these changes can be entered reasonably quickly into the system and modified drawings produced.

The main drawback to the use of computer systems, namely the perceived high cost of purchasing the necessary hardware and software, has been eliminated. Most of the current packages are available for very reasonable prices and virtually all can run on PCs, the prices for which continue to fall despite their constantly improving specifications. With capabilities far greater than those which not so long ago would have required a mainframe or minicomputer for their operation, modern highway design software packages are now well within the budget of any engineering and surveying practice, no matter how small.

Reflective summary

With reference to the designing composite and wholly transitional curves remember:

— The main purpose of any horizontal design is to ensure that vehicles travel safely and comfortably from one intersecting straight to another through the deflection angle θ.

— There are often several alternative solutions that will meet this purpose perfectly satisfactorily.

— The design of composite curves is based on the fact that each transition curve takes ϕ_{max} of the deflection angle and the circular arc takes the remaining portion $(\theta - 2\phi_{max})$.

— When designing wholly transitional curves if the design speed v and the deflection angle θ are fixed (as is often the case) then there is only one minimum radius of curvature value R that will apply for a given rate of change of radial acceleration value, c.

— When designing any horizontal alignment containing transition curves, radii values greater then the desirable minimum values specified in *TD9/93* for the design speed in question should be used wherever possible.

— The rate of change of radial acceleration should not normally exceed 0.3 m s^{-3}, and wherever possible this figure of 0.3 m s^{-3} should be used rather than a smaller value to avoid long transition curves and higher construction costs.

- The length of transition curve required for safety and comfort should be compared to the length required to ensure that the maximum allowable superelevation can be introduced at a rate of no more than 1% for all-purpose roads and 0.5% for motorways.

- Before the horizontal alignment can be finalised it must be phased with any vertical alignment being undertaken at the same location to avoid the creation of optical illusions or apparent kinks in the road surface, which could distract drivers.

- Horizontal alignments are normally designed using software packages in conjunction with Digital Terrain Models.

13.6 Setting out composite and wholly transitional curves

After studying this section you should know two traditional methods by which pegs on the centre lines of transition curves can be set out on site from their tangent points. These are known as the *tangential angles method* and *offsets from the tangent lengths*. You should know how to calculate coordinates of points on the centre lines of composite and wholly transitional curves and you should be aware of methods by which these points can be set out from control networks. You should understand the relative merits of coordinate methods of setting out compared to traditional methods.

This section includes the following topics:

- The methods available to set out the centre line
- Setting out using the tangential angles method
- Setting out using offsets from the tangent lengths
- Setting out using coordinate methods
- Coordinate methods compared with traditional methods

The methods available to set out the centre line

As discussed in Section 12.5, the centre line provides an important reference on site from which other features can be established and it can be set out either by *traditional* or *coordinate* methods. Although these were defined in Chapter 12, they apply equally to composite and wholly transitional curves.

As discussed in Section 13.5, the initial requirement is to obtain an accurate value for the deflection angle θ in order that the design can be undertaken. To achieve this, it may first be necessary to set out the intersection point, as described in Section 12.6, so that θ can be measured. Once the design has been completed, the tangent points which lie on each straight can be pegged out on site.

If traditional methods are used, the centre line is then set out from these tangent points either by the tangential angles method or using offsets from the tangent lengths.

If coordinate methods are used, the coordinates of the tangent points are determined and then used to help calculate the coordinates of points at regular intervals along the centre line. These points are then pegged out on the centre line either by taking observations from nearby control points or by using GPS receivers.

Both traditional and coordinate methods are used nowadays, although coordinate techniques are normally preferred for all major curves. Traditional methods, however, still have their place and they are often more convenient and quicker to use when defining the centre lines of less important curves, for example, minor roads, boundaries, kerbs, housing estate roads and so on. If there are no control points nearby and GPS receivers are not available then only traditional methods can be used.

Setting out using the tangential angles method

The tangential angles method for circular curves was described in Section 12.6, and the method for transition curves is based on similar principles. It is the most accurate of the traditional methods and it can be used for any transition curve. It is undertaken using a theodolite and a tape and, as with the circular curves method, it is necessary to first set out the intersection point, I. The method by which this is done is identical to that described for circular curves. Once I has been fixed, tangent points T and U can be established as follows.

- Calculate the shift using $S = L_T^2/24R$.
- Calculate the tangent lengths using $IT = IU = (R + S)\tan(\theta/2) + L_T/2$.
- Measure back from I to locate T and forward from I along the other straight to locate U.

Once tangent points T and U have been located, the tangential angles method can be used to set out the entry and exit transition curves and also the central circular arc in the case of composite curves.

The entry and exit transition curves

The actual calculation and setting out procedures are as follows.

Setting out the first peg A on the entry transition curve
- In Figure 13.14, l_1 is chosen as a chord length such that it is $\leq R/20$, where R is the minimum radius of curvature.

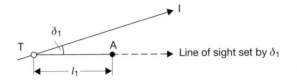

Measure l_1 from T along line of
sight to fix peg at A

Figure 13.14

- δ_1 is calculated from l_1 using $\delta_1 = (l_1^2/6RL_T)(180/\pi)$ degrees.
- A theodolite is set at T, aligned to I with a reading of $00°00'00"$ and δ_1 is turned off.
- A chord of length l_1 is measured from T and lined in at point A using the theodolite. A peg is driven into the ground at this point and a nail in this is used to locate A.

Setting out the second peg B and subsequent pegs on the entry transition curve
- In Figure 13.15, l_2 is the distance around the curve from T to B.

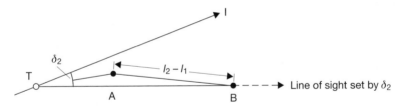

Measure $l_2 - l_1$ from A to line
of sight to fix peg at B

Figure 13.15

- δ_2 is calculated from l_2 using $\delta_2 = (l_2^2/6RL_T)(180/\pi)$ degrees
- δ_2 is set on the horizontal circle of the theodolite.
- A chord of length $(l_2 - l_1)$ is measured from A and lined in at point B using the theodolite. A peg with a nail is used to locate B.
- The procedure is repeated for all subsequent setting out points up to the common tangent point T_1 at the end of the entry transition curve. Usually, as with circular curves, a sub-chord is necessary at the beginning of the curve to maintain pegs at exact multiples of through chainage and a final sub-chord is normally required to set out the common tangent point T_1.

Setting out the pegs on the exit transition curve
- The exit transition curve is set out from U to T_2 with the theodolite set at U and aligned to I such that the horizontal circle is reading $00°00'00"$. The tangential angles are then subtracted from $360°$ to give the required directions.

- As for the entry transition curve, sub-chords are usually required at the beginning and end of the exit transition curve to ensure that pegs are placed at exact multiples of through chainage.

- If a wholly transitional curve is being set out, the common tangent point between the two transition curves is set out again, having already been fixed at the end of the entry transition curve. The difference between the two positions gives a measure of the accuracy of the setting out.

Setting out the central circular arc

This only applies to composite curves since wholly transitional curves have no central circular arc. The central circular arc is normally set out from T_1 to T_2 and it is first necessary to establish the line of the common tangent at T_1.

Figure 13.16 shows the entry transition curve and part of the central circular arc, in which the final tangential angle for T to T_1 will be $\delta_{max} = \phi_{max}/3$. With reference to Figure 13.16, the setting out procedure is as follows.

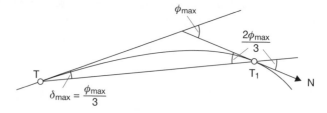

Figure 13.16

- Move the theodolite to T_1, align back to T with the horizontal circle reading $180°-(2\phi_{max}/3)$. The common tangent along T_1N now corresponds to a reading of $00°00'00"$ on the theodolite.

- Rotate the telescope in azimuth until a reading of $00°00'00"$ is obtained and set out pegs on the circular arc from T_1 to T_2 using the tangential angles method for circular curves described in Section 12.6. This involves the tangential angles formula for circular curves given in equation (12.8).

    Again, initial and final sub-chords are normally required to ensure that pegs are located on the centre line at exact multiples of through chainage.

- Finally, point T_2, the second common tangent point is established. Since T_2 is also fixed when setting out the exit transition curve from U, the difference between its two positions gives a measure of the accuracy of the setting out.

In practice, the tangential angle and chord data are tabulated ready for use on site. Worked example 13.2 at the end of this chapter shows how this is done.

Setting out using offsets from the tangent lengths

This method is similar to that described for circular curves and again requires that the tangent points have been set out. Two tapes are required and the method is best used for setting out short transition curves, since accurate taping becomes more difficult as the curve gets longer.

In the case of a *wholly transitional curve*, the entry transition curve is set out from the tangent point on the entry straight and the exit transition curve is set out from the tangent point on the exit straight.

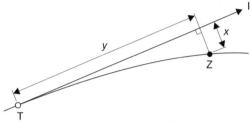

Consider Figure 13.17, which shows part of a cubic parabola transition curve. To set out any point Z on the curve, the method involves choosing y and calculating x using

Figure 13.17 ● Setting out a transition curve by offsets from the tangent length.

$x = y^3/6RL_T$. For a complete curve, x and y values should be tabulated for use on site.

In the case of a *composite curve*, the entry and exit transitions are set out in the same way as those for a wholly transitional curve and the central circular arc is then set out by offsets from the long chord (this is described for a circular curve in Section 12.6).

Worked example 13.4 shows the calculations involved when setting out a wholly transitional curve using offsets from the tangent lengths.

Setting out using coordinate methods

So far in this section, two traditional methods of establishing the centre lines of composite and wholly transitional curves have been described. Although these methods are still used, they have been virtually superseded for all major curves by *coordinate* methods that use control networks. In such methods, which are equally applicable to transition curves and circular curves, the coordinates of points at regular intervals along the centre line are calculated with reference to a site control network. The points are then pegged out on site either using a total station set at points in the ground control network surrounding the scheme as shown in Figure

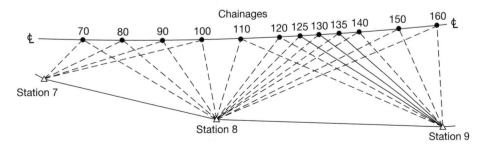

Figure 13.18

13.18 or by using a GPS receiver. In both cases the coordinates of points to be fixed on the centre line and the coordinates of the control network being used must be based on the same site coordinate system.

Nowadays, the coordinate calculations involved are usually done within computer software highway design packages and results of such computations are normally presented in the form of computer printouts ready for immediate setting out use on site. Table 13.2 shows a typical format for a printout and gives all the information required to set out the curve shown in Figure 13.18. The curve is to be set out by bearing and distance from control points 7, 8 and 9 with a total station, each centre line point being established from one control point and checked from another. The calculations required to produce Table 13.2 are summarised as follows.

- The coordinates of the control points are found from the control survey data.
- The horizontal alignment is designed and the coordinates of the intersection and tangent points are calculated.
- Assuming that the centre line is to be pegged at exact multiples of through chainage, chord lengths and tangential angles are calculated for the entry and exit transition curves and the central circular arc.
- The coordinates of the points to be established on the centre line are calculated using the chord lengths, tangential angles and the coordinates of the intersection and tangent points.
- Control points which are visible from and which will give a good intersection to the proposed centre line are found and the bearings/distances are calculated from the control points to the centre line points.
- A printout is obtained with a format similar to that shown in Table 13.2.

Worked example 13.3 at the end of this chapter shows the calculations involved when a composite curve is to be set out by bearing and distance methods.

The procedure described above is applicable to total stations. If GPS is to be used, the points can be set out directly but it is also possible to use a total station to set out using the coordinates directly. All of these are discussed further in Section 11.3.

Coordinate methods compared with traditional methods

The relative merits of coordinate methods compared with traditional methods for setting out horizontal curves were briefly considered in Section 12.5 and are further considered here. When compared with the traditional methods of setting out from the tangent points, coordinate methods have a number of important advantages. However, they are not always appropriate and some of the relative merits of the two categories of technique are listed below. Further comparisons are made in Section 11.3.

- Coordinate methods can be carried out by anyone who is capable of using a total station or a GPS receiver. Since the data is in the form of either bearings and distances or coordinates, no knowledge of curve design is necessary. This is not the case with traditional methods.

Table 13.2 ● Example computer printout format.

JOB REFERENCE J01777

PORTSMOUTH RAIL BRIDGE MAIN ALIGNMENT CH 70 TO CH 160

CENTRE LINE CHAINAGE TO BE SET OUT	STATION REFERENCE NUMBER	WHOLE CIRCLE BEARING DEG	MIN	SEC	HORIZONTAL DISTANCE FROM STATION TO CENTRE LINE (m)	STATION REFERENCE NUMBER	WHOLE CIRCLE BEARING DEG	MIN	SEC	HORIZONTAL DISTANCE FROM STATION TO CENTRE LINE (m)
70	7	24	10	57	13.695	8	276	02	31	38.734
80	7	41	38	08	22.183	8	285	31	19	30.504
90	7	49	00	38	31.585	8	301	13	45	23.726
100	7	52	53	24	41.281	8	325	47	54	19.939

WHOLE CIRCLE BEARING FROM STATION 7 TO STATION 8 = 79 DEG 12 MIN 09 SEC
WHOLE CIRCLE BEARING FROM STATION 8 TO STATION 7 = 259 DEG 12 MIN 09 SEC

110	8	354	04	27	20.844	9	270	04	27	54.792
120	8	15	22	10	25.954	9	275	26	51	45.969
125	8	22	43	14	29.480	9	278	59	41	41.768
130	8	28	24	28	33.392	9	283	19	50	37.772
135	8	32	51	00	37.571	9	288	40	14	34.052
140	8	36	22	09	41.937	9	295	15	48	30.709
150	8	41	30	43	51.034	9	313	00	40	25.732
160	8	45	01	39	60.436	9	335	51	19	24.166

WHOLE CIRCLE BEARING FROM STATION 8 TO STATION 9 = 68 DEG 34 MIN 09 SEC
WHOLE CIRCLE BEARING FROM STATION 9 TO STATION 8 = 248 DEG 34 MIN 09 SEC

- The increased use of highway design computer software packages in which the setting out data is presented ready for use in coordinate form has produced a corresponding increase in the adoption of such methods.

- The widespread use of computers has also greatly speeded up the calculation procedures associated with coordinate methods, which were always perceived to be more difficult to perform by hand when compared with those associated with the traditional methods.

- Traditional methods require tangent points to be occupied. This can delay the construction process. Coordinate methods enable the work to proceed unhindered.

- On any site involving centre lines, the pegs will inevitably be disturbed by the construction process and they will need to be re-established as each stage of the work is completed. Any disturbed centre line pegs can quickly be relocated from the control stations in coordinate methods since the control points will be well protected and located away from site traffic. In traditional methods, however, the tangent points will have to be reoccupied and these can often themselves be lost during construction, requiring extra work in their relocation.

- In coordinate methods, each peg on the centre line is fixed independently of all the other pegs on the centre line. This ensures that any error made when locating one peg is not carried forward to the next peg, as can occur in the tangential angles method, for example.

- Coordinate methods enable key sections of the centre line to be set out in isolation, such as a bridge centre line, in order that work can progress in more than one area of the site.

- Obstacles on the proposed centre line, which may be the subject of disputes, can easily be by-passed using coordinate methods to allow work to proceed while arbitration takes place. Once the obstacle is removed, it is an easy process to establish the missing section of the centre line. This is not usually possible with traditional methods.

- Coordinate methods have the disadvantage that there is very little check on the final setting out. Large errors will be noticed when the centre line does not take the designed shape, but small errors could pass unnoticed. In the tangential angles method, checks are provided by locating common tangent points from two different positions.

- Although the widespread use of total stations and the increasing use of GPS receivers on sites encourages the use of coordinate techniques, such equipment may not always be available and it may be simpler to use traditional methods that work along the centre line. This will particularly be the case where minor curves are being set out, such as those used for roads on housing estates, kerbs at road intersections, short curves and boundaries.

Reflective summary

With reference to setting out composite and wholly transitional curves remember:

— As for circular curves, they can be set out either by traditional methods or coordinate methods.

— Traditional methods normally require the intersection and tangent points to be located – this is often not required with coordinate methods.

— Two of the most commonly used traditional methods are those involving tangential angles and chords and offsets from the tangent lengths.

— The tangential angles method can be applied to any curve but the offsets method is best restricted to shorter curves.

— When using coordinate methods, the coordinates of points to be fixed on the centre line and the coordinates of the control network being used must be based on the same coordinate system.

— Theodolites and total stations are used with control points to set out the centre line using either *intersection* or *bearing and distance* methods.

— Pole-mounted GPS receivers are used to set out the centre line points directly.

— Coordinate methods have considerable advantages over traditional methods although they are not always appropriate.

13.7 Plotting the centre lines of composite and wholly transitional curves

After studying this section you should know how to draw the centre line of a proposed composite or wholly transitional curve on a drawing in order to see if your design is acceptable.

The plotting procedure

As discussed for circular curves, despite the widespread use of computer plotting facilities, there are still occasions during the initial horizontal alignment design when it is necessary to undertake a hand drawing of the proposed centre line. For composite and wholly transitional curves the following procedure is recommended. It assumes that there is an existing plan of the area available.

- Draw the intersecting straights in their correct relative positions on a sheet of tracing paper.

- Calculate the length of each tangent using $IT = IU = (R + S)\tan(\theta/2) + L_T/2$.

- Plot the tangent points by measuring this distance along each straight on either side of the intersection point at the same scale as the existing plan.

- To plot the *entry and exit transition curves*, use the offsets from the tangent lengths method as described in Section 13.6. Use $x = y^3/6RL_T$ to prepare a table of offset values x for suitable y values and ensure that the y values chosen will provide a good definition of the centre line.

- At the scale of the existing plan, plot the x and y values on the tracing paper from the tangent lengths to establish points on the entry and exit transition curves as shown in Figure 13.17.

- To plot the *central circular arc* (where appropriate), carefully join the plotted ends of the entry and exit transition curves. This is the long chord of the central circular arc.

- Using equation (12.13) for the offsets from the long chord method, prepare a table of offset X values for appropriate Y values. Again, ensure that the Y values chosen will provide a good definition of the centre line.

- At the scale of the existing plan, plot the X and Y values from the long chord to establish points on the centre line of the centre circular arc.

- Carefully join all the points plotted to define the complete centre line. A set of French curves is useful for this purpose, although with care a flexicurve can be used.

- Superimpose the tracing paper on the existing plan and decide whether or not the design is acceptable. If it is not, change the design and repeat the plotting procedure.

Reflective summary

With reference to plotting the centre lines of composite and wholly transitional curves remember:

- There are still occasions during the initial design process when it is necessary to undertake a hand drawing of the proposed centre line to see if it is acceptable and falls within the band of interest.

— Hand drawings are best done on tracing paper so that the design can be superimposed on the survey plan of the area.

— The centre lines of transition curves can be drawn on the tracing paper using offsets from the tangent lengths.

— The centre line of the central circular arc of a composite curve can be drawn on the tracing paper using offsets from the long chord.

— Most highway alignment drawings are now produced in conjunction with highway design software packages.

13.8 Transition curve worked examples

After studying this section you should understand how to calculate superelevation along a transition curve and how to check that the curve is long enough for it to be introduced. You will be able to determine through chainage values at regular intervals along composite curves and know how to derive and tabulate the data required to set out such curves by the tangential angles method. You will understand how to compute coordinates of points at regular chainage intervals along the centre line of a composite curve and how these can be used to set out the centre line using a total station. You will know how to compute the minimum radius of curvature of a wholly transitional curve and be aware of the procedures required to check whether the value obtained is acceptable. You will be able to calculate the data required to set out transition curves by offsets from their tangent lengths.

This section includes the following topics:

- Calculating and checking superelevation
- Setting out a composite curve by the tangential angles method
- Setting out a composite curve by coordinate methods
- Calculating the minimum radius of curvature for a wholly transitional curve and tabulating the data required to set it out by offsets from its tangent lengths

Worked example 13.1: Calculating and checking superelevation

Question

On a proposed road having a design speed of 100 kph and a carriageway width of 7.30 m, a composite curve consisting of two transition curves and a central circular arc of radius 750 m is to join two intersecting straights having a deflection angle of

09°34'28". The rate of change of radial acceleration for the road is to be 0.3 m s^{-3} and the superelevation should be introduced at a rate of no more than 1%.

- Calculate the amount of superelevation that must be built into the central circular arc.
- Check that the transition curves are long enough for the superelevation to be introduced.
- Calculate the amount of superelevation that should be constructed along the entry transition curve at 20 m intervals from the entry tangent point.

Solution

The amount of superelevation that must be built into the central circular arc
The maximum allowable superelevation is given by equation (13.3) as

$$\text{maximum allowable SE} = \left(\frac{Bv^2}{282.8R}\right)$$

$$= \left(\frac{7.30 \times 100^2}{282.2 \times R}\right)$$

$$= \mathbf{0.344 \ m}$$

Expressing this as a percentage using equation (13.4) gives

$$s\% = \frac{v^2}{2.828R} = \frac{100^2}{2.828 \times 750} = 4.71\%$$

With reference to Table 13.1, the radius of 750 m is greater than the desirable minimum value of 720 m for 100 kph and the superelevation of 4.71% is less than the value of 5% required if the desirable minimum radius is used. Hence **0.344 m (4.71%) superelevation** should be built into the central circular arc.

Checking that the transition curves are long enough
The length of each transition curve required for comfort and safety is obtained from equation (13.14) as

$$L_T = \left(\frac{v^3}{3.6^3 cR}\right) = \left(\frac{100^3}{3.6^3 \times 0.3 \times 750}\right) = 95.26 \ m$$

The superelevation value of 0.344 m must be introduced and removed over a length of 95.26 m, which represents a gradient of

$$\frac{0.344}{95.26} = 0.36\%$$

Since this is less that the maximum allowable rate of introduction of 1%, **the transitions are long enough**.

The amount of superelevation that should be constructed along the entry transition curve at 20 m intervals from the entry tangent point

For transition curves, from equation (13.1), $rl = K = RL_T$. This gives

$$K = 95.26 \times 750 = 71,445$$

At 20 m along the curve from the entry tangent point

$$r = \frac{K}{20} = \frac{71445}{20} = 3572.25 \text{ m}$$

Substituting this into equations (13.3) and (13.4) gives

$$SE \text{ at 20 m along the curve} = \frac{(7.30 \times 100^2)}{(282.8 \times 3572.25)} = 0.07 \text{ m}$$

$$s\% \text{ at 20 m along the curve} = \frac{100^2}{2.828 \times 3572.25} = 0.99\%$$

Because this is less that the minimum allowable value of 2.5% for drainage, a value of 2.5% must be used, therefore

$$SE \text{ built at 20 m along the curve} = 2.5\% \text{ of } B = 0.025 \times 7.30 = \mathbf{0.18 \text{ m}}$$

Repeating this procedure at 40 m, 60 m, 80 m and 95.26 m along the curve gives the superelevation values shown in Table 13.3.

Table 13.3

l	r	SE = $Bv^2/282.8R$		SE built	
		m	%	m	%
0	∞	0	0	0.183	2.50
20	3572.25	0.072	0.99	0.183	2.50
40	1786.13	0.145	1.98	0.183	2.50
60	1190.75	0.217	2.97	0.217	2.97
80	893.06	0.289	3.96	0.289	3.96
95.26	750	0.344	4.71	0.344	4.71

Worked example 13.2: Setting out a composite curve by the tangential angles method

Question

The deflection angle between two intersecting straights is measured as 14°28'26". The straights are to be joined by a composite horizontal curve consisting of a central circular arc and two cubic parabola transition curves of equal length.

The design speed of the road is 85 kph and the radius of the circular curve is 600 m. The rate of change of radial acceleration is to be 0.3 m s^{-3}.

If the through chainage of the intersection point is 461.34 m, prepare the setting out tables for the three curves such that pegs can be located on the centre line at exact 20 m multiples of through chainage using the tangential angles method.

Solution

Figure 13.19 shows the data and layout of the composite curve.

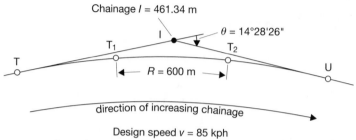

Chainage I = 461.34 m

$\theta = 14°28'26"$

R = 600 m

direction of increasing chainage

Design speed v = 85 kph
Rate of change of radial acceleration c = 0.3 m s^{-3}

Figure 13.19

Design of the entry transition from T to T_1

From equations (13.14), (13.15) and (13.18)

$$L_T = \left(\frac{v^3}{3.6^3 cR}\right) = \left(\frac{85^3}{3.6^3 \times 0.3 \times 600}\right) = 73.13 \text{ m}$$

$$S = \left(\frac{L_T^2}{24R}\right) = \frac{73.13^2}{24 \times 600} = 0.37 \text{ m}$$

$$IT = (R + S)\tan\left(\frac{\theta}{2}\right) + \frac{L_T}{2} = 76.24 + 36.56 = 112.80 \text{ m}$$

Hence

through chainage of T = 461.34 − 112.80 = 348.54 m

through chainage of T_1 = 348.54 + 73.13 = 421.67 m

To keep to exact 20 m through chainage values, the chord lengths for the entry transition curve are as follows.

initial sub-chord length = chainage of first point on transition − chainage T
= 360 − 348.54
= 11.46 m

general chord length = 20.00 m

final sub-chord length = chainage of T_1 − chainage of last point on transition
= 421.67 − 420
= 1.67 m

Table 13.4

Through chainage m	Chord length m	_l_ m	δ °	Clockwise tangential angle from T relative to TI			
				°	′	″	
348.54 (T)	0	0	0	00	00	00	
360.00 (C$_1$)	11.46	11.46	0.0286	00	01	43	(δ_1)
380.00 (C$_2$)	20.00	31.46	0.2154	00	12	55	(δ_2)
400.00 (C$_3$)	20.00	51.46	0.5763	00	34	35	(δ_3)
420.00 (C$_4$)	20.00	71.46	1.1113	01	06	41	(δ_4)
421.67 (T$_1$)	1.67	73.13	1.1639	01	09	50	(δ_{max})
	Σ73.13 (checks)						

Using the tangential angles formula for transition curves, equation (13.11), $\delta = (l^2/6RL_T)(180/\pi)°$, Table 13.4 is obtained. As a check on this, $\phi_{max}/3$ should be calculated and compared to δ_{max} since from equation (13.13), δ_{max} should equal $\phi_{max}/3$.

From equation (13.6)

$$\phi_{max} = \left(\frac{L_T}{2R}\right) = \left(\frac{73.13 \times 180}{2 \times 600 \times \pi}\right)^° = 03°29'30''$$

and

$$\delta_{max} = \frac{\phi_{max}}{3} = 01°09'50'' \text{ (checks with Table 13.4)}$$

Design of the central circular arc from T_1 to T_2
The circular arc takes $(\theta - 2\phi_{max})$. With $\phi_{max} = 03°29'30''$, $2\phi_{max} = 06°59'00''$ and

$$(\theta - 2\phi_{max}) = 14°28'26'' - 06°59'00'' = 07°29'26'' = 0.130735 \text{ rad}$$

From equation (13.20)

$$\text{length of circular arc} = L_{CA} = R(\theta - 2\phi_{max}) = 600(0.130735) = 78.44 \text{ m}$$

Hence

$$\text{through chainage of } T_2 = 421.67 + 78.44 = 500.11 \text{ m}$$

Using 20 m chords and keeping to exact 20 m multiples of through chainage, the chord lengths for the circular arc are as follows:

initial sub-chord length = 18.33 m
general chord length = 20.00 m
final sub-chord length = 0.11 m

Table 13.5

Through chainage m	Chord length m	Tangential angle for each chord				Cumulative clockwise tangential angle from T relative to the common tangent		
		°	′	″		°	′	″
421.67 (T₁)	0	00	00	00		00	00	00
440.00 (C₅)	18.33	00	52	31	(α_1)	00	52	31
460.00 (C₆)	20.00	00	57	18	(α_2)	01	49	49
480.00 (C₇)	20.00	00	57	18	(α_3)	02	47	07
500.00 (C₈)	20.00	00	57	18	(α_4)	03	44	25
500.11 (T₂)	0.11	00	00	19	(α_5)	03	44	44
	Σ78.44 (checks)							

Using the tangential angles formula for circular curves, equation (12.8),

$$\alpha = \left(\frac{90}{\pi}\right)\left(\frac{\text{chord length}}{R}\right),$$

Table 13.5 is obtained. As a check, the final cumulative tangential angle should equal $(\theta - 2\phi_{max})/2$ within a few seconds. In this case

$$\alpha = (\theta - 2\phi_{max})/2 = 03°44'43'' \text{ (checks with Table 13.5)}$$

Design of the exit transition curve from U to T₂ (in the opposite direction to increasing chainage)

Since the curve is symmetrical, the length of the exit transition curve again equals 73.13 m. Therefore

through chainage of U = 500.11 + 73.13 = 573.24 m

To keep to exact 20 m multiples of through chainage, the chord lengths for the exit transition curve, working from U to T₂, are as follows:

initial sub-chord length = 13.24 m
general chord length = 20.00 m
final sub-chord length = 19.89 m

Again using equation (13.11), the setting out data shown in Table 13.6 are obtained. The check that $\delta_{max} = 01°09'50''$ is again applied.

The last column of Table 13.6 shows the clockwise tangential angles from U relative to UI which have been obtained by subtracted the δ values from 360°. These will apply when an optical reading theodolite is being used to ensure that the curve is set out on the correct side of UI. However, if an electronic theodolite or a total station is being used in which the direction of the horizontal circle can be changed then the values given in the δ column of Table 13.6 will apply.

Table 13.6

Through chainage m	Chord length m	*l* m	*δ*			Clockwise tangential angle from U relative to UI		
			°	′	″	°	′	″
573.24 (U)	0	0	00	00	00	360	00	00
560.00 (C₁₁)	13.24	13.24	00	02	17 (δ_{11})	359	57	43
540.00 (C₁₀)	20.00	33.24	00	14	26 (δ_{10})	359	45	34
520.00 (C₉)	20.00	53.24	00	37	01 (δ_{9})	359	22	59
500.11 (T₂)	19.89	73.13	01	09	50 (δ_{max})	358	50	10
	Σ73.13 (checks)							

Worked example 13.3: Setting out a composite curve by coordinate methods

Question

The composite curve designed in Worked example 13.2 is to be set out by bearing and distance methods from two horizontal control points G and H. The intersection point I and the entry tangent point T have been set out on site and the coordinates of these, together with those of points G and H, are listed in Table 13.7. Using the data calculated in Worked example 13.2, calculate

- The coordinates of all the points on the centre line of the curve which lie at exact 20 m multiples of through chainage.
- The bearing GH and the bearings and horizontal lengths from G that are necessary to set out all the pegs on the centre line using a total station.

Solution

Figure 13.20 shows all the points to be set out together with points G and H. Their chainage values and the required tangential angles and chords are listed in Tables 13.4, 13.5 and 13.6.

Table 13.7

Point	m*E*	m*N*
G	727.61	893.83
H	940.57	886.28
I	789.14	863.72
T	704.95	788.64

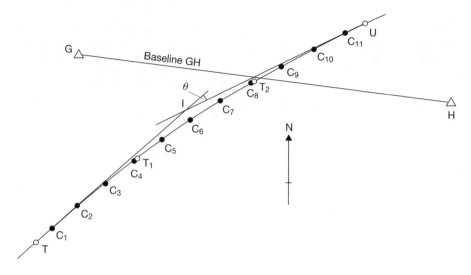

Figure 13.20

The coordinates of all the points on the centre line of the curve which lie at exact 20 m multiples of through chainage

Coordinates of C_1

With reference to Figure 13.21 and Table 13.4

$$\text{bearing } TC_1 = \text{bearing } TI + \delta_1$$

But

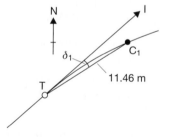

Figure 13.21

$$\Delta E_{TI} = E_I - E_T = 789.14 - 704.95 = 84.19 \text{ m}$$

$$\Delta N_{TI} = N_I - N_T = 863.72 - 788.64 = 75.08 \text{ m}$$

and, from a rectangular $\rightarrow$ polar conversion

$$\text{bearing } TI = 48°16'25''$$

This gives

$$\text{bearing } TC_1 = 48°16'25'' + 00°01'43'' = 48°18'08''$$

Since the horizontal length of $TC_1 = 11.46$ m

$$\Delta E_{TC_1} = 11.46 \sin 48°18'08'' = +8.557 \text{ m}$$

$$\Delta N_{TC_1} = 11.46 \cos 48°18'08'' = +7.623 \text{ m}$$

The *coordinates of C_1* are

$$E_{C_1} = E_T + \Delta E_{TC_1} = 704.95 + 8.557 = \textbf{713.507 m}$$

$$N_{C_1} = N_T + \Delta N_{TC_1} = 788.64 + 7.623 = \textbf{796.263 m}$$

Coordinates of C_2
With reference to Figure 13.22, application of the Sine Rule in triangle TC_1C_2 gives

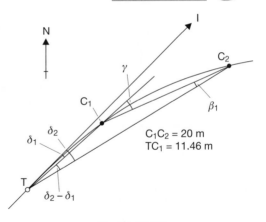

Figure 13.22

$$\frac{TC_1}{\sin \beta_1} = \frac{C_1C_2}{\sin(\delta_2 - \delta_1)}$$

Substituting values from Table 13.4 gives

$$\sin \beta_1 = \left(\frac{11.46}{20.00}\right) \sin 00°11'12"$$

From which

$$\beta_1 = 00°06'25"$$

Therefore

$$\gamma = \beta_1 + (\delta_2 - \delta_1) = 00°17'37"$$

and

$$\begin{aligned} \text{bearing } C_1C_2 &= \text{ bearing } TC_1 + \gamma \\ &= 48°18'08" + 00°17'37" = 48°35'45" \end{aligned}$$

With horizontal length $C_1C_2 = 20.00$ m, the *coordinates of C_2* are obtained as follows

$$\Delta E_{C_1C_2} = 20.00 \sin 48°35'45" = +15.001 \text{ m}$$

$$\Delta N_{C_1C_2} = 20.00 \cos 48°35'45" = +13.227 \text{ m}$$

$$\mathbf{E_{C_2}} = E_{C_1} + 15.001 = \mathbf{728.508 \text{ m}}$$

$$\mathbf{N_{C_2}} = N_{C_1} + 13.227 = \mathbf{809.490 \text{ m}}$$

Coordinates of C_3
With reference to Figure 13.23, the chord length TC_2 can be taken to equal the curve length TC_2, that is

$$TC_2 = TC_1 + C_1C_2 = 31.46 \text{ m}$$

In triangle TC_2C_3

$$\frac{TC_2}{\sin \beta_2} = \frac{C_2C_3}{\sin(\delta_3 - \delta_2)}$$

$$\sin \beta_2 = \left(\frac{31.46}{20.00}\right) \sin 00°21'40"$$

$$\beta_2 = 00°34'05"$$

Since $(\gamma + \beta_1) = (\delta_3 - \delta_2) + \beta_2$

$$\gamma = 00°21'40" + 00°34'05" - 00°06'25" = 00°49'20"$$

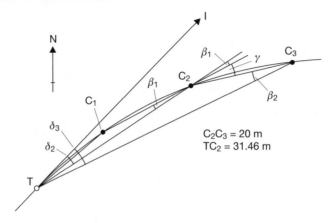

Figure 13.23

and

bearing C_2C_3 = bearing $C_1C_2 + \gamma$

= 48°35'45" + 00°49'20" = 49°25'05"

The *coordinates of C_3* are obtained as follows

$\Delta E_{C_2C_3}$ = 20.00 sin 49°25'05" = +15.190 m

$\Delta N_{C_2C_3}$ = 20.00 cos 49°25'05" = +13.011 m

E_{C_3} = E_{C_2} + 15.190 = **743.698 m**

N_{C_3} = N_{C_2} + 13.011 = **822.501 m**

Coordinates of C_4 and T_1
The coordinates of C_4 and T_1 are calculated from those of C_3 and C_4 by repeating the procedures used above. The values obtained are as follows:

C_4 = **759.188 mE, 835.152 mN**

T_1 = **760.498 mE, 836.187 mN**

Coordinates of C_5
Point C_5 lies on the central circular arc as shown in Figure 13.24. From this Figure and the data in Table 13.5

bearing T_1Z = bearing TI + ϕ_{max}

= 48°16'25" + 03°29'30" = 51°45'55"

and

bearing T_1C_5 = bearing $T_1Z + \alpha_1$

= 51°45'55" + 00°52'31" = 52°38'26"

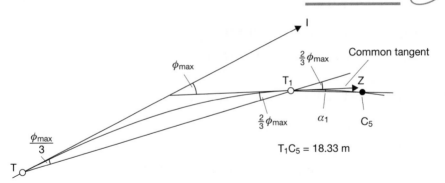

Figure 13.24

Therefore, the *coordinates of C₅* are obtained as follows:

$$\Delta E_{T_1 C_5} = 18.33 \sin 52°38'26" = +14.569 \text{ m}$$

$$\Delta N_{T_1 C_5} = 18.33 \cos 52°38'26" = +11.123 \text{ m}$$

$$E_{C_5} = E_{T_1} + 14.569 = \mathbf{775.067 \text{ m}}$$

$$N_{C_5} = N_{T_1} + 11.123 = \mathbf{847.310 \text{ m}}$$

Coordinates of C₆
With reference to Figure 13.25 and Table 13.5

$$\lambda_1 = (\alpha_1 + \alpha_2) = 01°49'49"$$

$$\text{bearing } C_5 C_6 = \text{bearing } T_1 C_5 + \lambda_1$$

$$= 52°38'26" + 01°49'49" = 54°28'15"$$

and the *coordinates of C₆* are obtained as follows

$$\Delta E_{C_5 C_6} = 20.00 \sin 54°28'15" = +16.276 \text{ m}$$

$$\Delta N_{C_5 C_6} = 20.00 \cos 54°28'15" = +11.622 \text{ m}$$

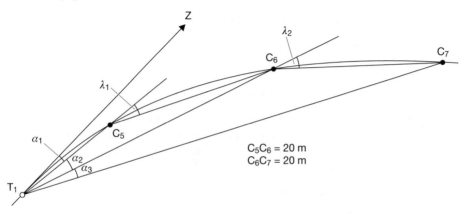

Figure 13.25

$$E_{C_6} \quad = E_{C_5} + 16.276 = \textbf{791.343 m}$$

$$N_{C_6} \quad = N_{C_5} + 11.622 = \textbf{858.932 m}$$

Coordinates of C_7

With reference to Figure 13.25 and Table 13.5

$$\lambda_2 = (\alpha_2 + \alpha_3) = 01°54'36''$$

bearing $C_6 C_7$ = bearing $C_5 C_6 + \lambda_2$

$$= \ 54°28'15'' + 01°54'36'' = 56°22'51''$$

and the *coordinates of C_7* are obtained as follows

$$\Delta E_{C_6 C_7} = \ 20.00 \ \sin 56°22'51'' = +16.655 \ m$$

$$\Delta N_{C_6 C_7} = \ 20.00 \ \cos 56°22'51'' = +11.073 \ m$$

$$E_{C_7} \quad = E_{C_6} + 16.655 = \textbf{807.998 m}$$

$$N_{C_7} \quad = N_{C_6} + 11.073 = \textbf{870.005 m}$$

Coordinates of C_8 and T_2

These coordinates are calculated using procedures similar to those used to calculate the coordinates of C_7 from C_6. The values obtained are as follows

$$\textbf{C}_8 \ = \ \textbf{825.013 m}E, \ \textbf{880.517 m}N$$

$$\textbf{T}_2 \ = \ \textbf{825.108 m}E, \ \textbf{880.573 m}N$$

Coordinates of U, C_{11}, C_{10}, C_9 and T_2

Using the deflection angle, bearing IU is calculated from

bearing IU $= \ $ bearing TI $+ \ \theta$

$$= \ 48°16'25'' + 14°28'26'' = 62°44'51''$$

Using the tangent length IU, the *coordinates of U* are calculated from those of I as follows

$$\Delta E_{IU} = \ 112.80 \ \sin 62°44'51'' = +100.279 \ m$$

$$\Delta N_{IU} = \ 112.80 \ \cos 62°44'51'' = +51.653 \ m$$

$$E_U \quad = E_I + 100.279 = \textbf{889.419 m}$$

$$N_U \quad = N_I + 51.653 = \textbf{915.373 m}$$

Starting from U and working back to T_2, the *coordinates of points C_{11}, C_{10}, C_9 and T_2* are calculated by repeating the procedures used to calculate the coordinates of points C_1, C_2, C_3, C_4 and T_1. The data required for this is given in Table 13.6. The coordinate values obtained are as follows

$$\textbf{C}_{11} = \ \textbf{877.653 m}E, \ \textbf{909.302 m}N$$

$$\textbf{C}_{10} = \ \textbf{859.933 m}E, \ \textbf{900.028 m}N$$

$$C_9 = 842.356 \text{ m}E, 890.486 \text{ m}N$$

$$T_2 = 825.109 \text{ m}E, 880.579 \text{ m}N$$

The coordinates of T_2 are calculated twice and this provides a check on the calculations. In this example, the two sets of coordinates for T_2 differ by 0.001 m in the eastings and 0.006 m in the northings, which is perfectly acceptable.

The coordinates of all the points on the curve are listed in Table 13.8 and have been rounded to two decimal places to agree with the precision of the chainage value of point I given in Worked example 13.2.

The bearing GH and the bearings and horizontal lengths from G that are necessary to set out all the pegs on the centre line using a total station

The bearings and horizontal lengths are calculated from the coordinates using one of the methods discussed in Section 6.2. The required bearings and horizontal lengths are listed in Table 13.8.

Table 13.8

Point	Through chainage	Coordinates		Bearing from G			Horizontal length from G
	m	mE	mN	°	'	"	m
T	348.54	704.95	788.64	192	09	25	107.60
C_1	360.00	713.51	796.26	188	13	23	98.58
C_2	380.00	728.51	809.49	179	23	19	84.34
C_3	400.00	743.70	822.50	167	17	18	73.12
C_4	420.00	759.19	835.15	151	42	43	66.64
T_1	421.67	760.50	836.19	150	17	26	66.36
C_5	440.00	775.07	847.31	134	25	37	66.46
C_6	460.00	791.34	858.93	118	42	22	72.66
C_7	480.00	808.00	870.01	106	30	17	83.84
C_8	500.00	825.01	880.52	97	46	53	98.31
T_2	500.11	825.11	880.58	97	44	20	98.40
C_9	520.00	842.36	890.49	91	40	02	114.80
C_{10}	540.00	859.93	900.03	87	19	02	132.47
C_{11}	560.00	877.65	909.30	84	06	48	150.84
U	573.24	889.42	915.37	82	25	03	163.24

Bearing GH = 92°01'50"

Worked example 13.4: Calculating the minimum radius of curvature for a wholly transitional curve and tabulating the data required to set it out by offsets from its tangent lengths

Question

Part of a proposed rural road consists of two straights, which intersect at an angle of 172°46'18". These are to be joined using a wholly transitional horizontal curve having equal tangent lengths. The transition curves are to be cubic parabolas. The design speed for the road is to be 85 kph and the rate of change of radial acceleration 0.3 m s^{-3}.

● Calculate the minimum radius of curvature of the curve and comment on its suitability.

● Tabulate the data required to set out the transition curves by offsets taken at exact 20 m intervals along their tangent lengths.

Solution

The minimum radius of curvature and its suitability
The curve is shown in Figure 13.26. From equation (13.24)

$$R = \sqrt{\left(\frac{v^3}{3.6^3 c\theta}\right)} \text{ metres}$$

But since the intersection angle is 172°46'18"

$$\theta = 180° - 172°46'18" = 07°13'42" = 0.12616 \text{ rad}$$

Therefore

$$R = \sqrt{\frac{85^3}{3.6^3 \times 0.3 \times 0.12616}} = \mathbf{589.73 \ m}$$

$$= \text{minimum radius of the curve at the common tangent point } T_C$$

From Table 13.1, the desirable minimum radius for this road is 510 m, the one step below desirable minimum radius value is 360 m and the two steps below desirable minimum radius value is 255 m. In this case, a radius of 589.73 m is perfectly acceptable since it is greater than the desirable minimum value. If it had been less than 510

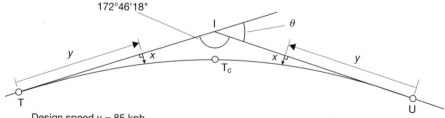

Design speed v = 85 kph
Rate of change of radial acceleration = c = 0.3 m s^{-3}

Figure 13.26

m but greater than 360 m, the designer could have used his or her judgement, depending on the prevailing site factors, to decide whether or not to use such a value if a one step relaxation was permitted. The same considerations would be applied if the radius had been less than 360 m and a two-step relaxation was permitted. If the radius had been less than 510 m and no relaxations were permitted then either a departure from standard would have to be requested or one or more of the variables would have to be changed, that is, v, θ or c, in order to try to increase the radius above the minimum desirable value.

The data required to set the curves out by offsets from their tangent lengths
From equation (13.14), the length of each transition curve is given by

$$L_T = \left(\frac{v^3}{3.6^3 cR}\right) = \left(\frac{85^3}{3.6^3 \times 0.3 \times 589.73}\right) = 74.40 \text{ m}$$

For cubic parabola transition curves, the offsets from the tangents are given by equation (13.9) as $x = y^3/6RL_T$. Substituting 20 m multiples of y into this equation enables the offset x values in Table 13.9 to be produced. These values apply to both transition curves with one being set out from tangent length TI and the other from tangent length UI as shown in Figure 13.26. Since, in transition curves, the length along the tangent is equal to the length along the curve, the final offset in Table 13.9, that is, $x = 1.56$ m at $y = L_T = 74.40$ m, will fix the common tangent point, T_C. This point is fixed twice as a check on the setting out.

Table 13.9

y m	$x = y^3/6RL_T$ m
0.00	0.00
20.00	0.03
40.00	0.24
60.00	0.82
74.40	1.56

Reflective summary

With reference to transition curve worked examples, remember:

— As with circular curve worked examples, it is advisable to sketch the straights and the curve and to add all the known data to the sketch before beginning the calculations.

— When calculating superelevation along a transition curve it is necessary to check whether the curve is long enough to enable it to be introduced in accordance with the Department of Transport traffic standard *TD 9/93*.

— A minimum superelevation of 2.5% must be maintained along a horizontal curve to allow for drainage.

— When calculating the positions of points on the centre line you will usually require an initial sub chord, a general chord and a final sub chord to ensure that pegs are located at specific multiples of through chainage.

— When calculating data for the tangential angles method, the type of theodolite to be used will influence the way in which the data is presented. With an optical reading instrument it may be necessary to subtract the tangential angles from 360° to ensure that the curve is set out on the correct side of the tangent length. In the case of an electronic instrument, the direction of the horizontal circle can usually be changed to facilitate this.

— When calculating the coordinates of points on a composite curve, the coordinates of the second common tangent point should be calculated twice as a check.

— It will be necessary to check the value obtained using the equation for the radius of a wholly transitional curve to ensure that it complies with the guidelines given in *TD 9/93*.

— Data calculated for setting out a wholly transitional curve using offsets from its tangent lengths applies to each half of the curve.

— You should tabulate the calculations wherever possible so that they can be easily followed for checking purposes.

Exercises

13.1 Two intersecting straights are to be joined by a horizontal curve, which is to be designed as part of a proposed dual carriageway having a design speed of 120 kph. With reference to the *Design Manual for Roads and Bridges* published by the Department of Transport, summarise the various choices of radii that are available.

13.2 A composite curve consisting of two equal length transition curves and a central circular arc is to be used to connect two intersecting straights TI and IU, which have a deflection angle of 12°24'46". The design speed of the road is to be 70 kph, the rate of change of radial acceleration 0.30 m s^{-3} and the radius of curvature of the circular arc 450 m. The transition curves are to be set out by offsets taken at exact 20 m intervals along the tangent lengths from T and U. The central circular curve is to be set out by offsets taken at exact 20 m intervals from the mid-point of its long chord.

Tabulate the data required to set out the three curves.

13.3 A composite curve consisting of entry and exit transition curves of equal length and a central circular arc is to connect two straights on a new road. The design speed for the road is 100 kph, the radius of the circular arc is 800 m and the rate of change of radial acceleration is 0.30 m s^{-3}. The through chainage of the intersection point is 3246.28 m and the deflection angle is 15°16'48".

(i)   Prepare tables for setting out all three curves by the tangential angles method if pegs are required at each 20 m multiples of through chainage

(ii)   Describe the procedure necessary to establish the common tangent between the entry transition curve and the central circular arc giving the values of any angles required

13.4 A road 7.30 m wide deflects through an angle of 18°47'26", the through chainage of the intersection point being 1659.47 m. A circular arc of radius 600 m and two equal length transition curves are to be designed for a speed of 85 kph with a rate of change of radial acceleration of 0.30 m s^{-3}.

Calculate:

(i)   The through chainages of the four tangent points

(ii)   The theoretical and actual values for the maximum superelevation on the circular arc, taking g as 9.81 m s^{-2}

(iii)   The data required to set out the entry transition curve and the exit transition curve by the tangential angles method if pegs are to be placed on the centre line at exact 50 m multiples of through chainage

13.5 A wholly transitional horizontal curve having equal tangent lengths is to join two straights TI and IU. The design data are:

Design speed	100 kph
Rate of change of radial acceleration	0.33 m s^{-2}
Intersection angle between the straights	169°18'38"
Through chainage of the intersection point	2618.22 m

Through chainage increases from T to U.

Calculate:

(i)   The minimum radius of curvature of the road curve

(ii)   The tangential angles and chords to be set out from T relative to TI and from U relative to UI to fix pegs on the entry and exit transition curves if pegs are to be set out at exact 30 m multiples of through chainage

13.6 A curve connecting two highway straights is to be wholly transitional, consisting of two identical transition curves. The distance from the intersection point of the straights to the intersection point of the two transitions is to be 5.35 m. The deflection angle between the straights is 12°18'22" and the maximum allowable superelevation built into the curve is 5.5%.

Calculate:

(i)   The length from each tangent point to the intersection point

(ii)   The minimum radius of curvature of the curve

(iii) The design speed for the curve, taking g as 9.81 m s^{-2}

(iv)   The rate of change of radial acceleration when vehicles travel at the design speed

13.7 On a section of an existing road, two straights that intersect at a through chainage of 1873.62 m and at an angle of 168°52'12" are joined by a wholly circular curve that has a dangerously small radius of curvature. It is proposed to replace this with a single carriageway composite horizontal curve consisting of entry and exit transition curves each 120.00 m in length and a central circular arc. The design speed for the curve is to be 100kph, the rate of change of radial acceleration is to be 0.20 m s^{-3} and the proposed carriageway width is 7.30 m.

Calculate:

(i)   The through chainages of the tangent points at the start and finish of the central circular arc

(ii)   The maximum allowable superelevation that must be built into the composite curve, taking g as 9.81 m s^{-2}

(iii) The data required to set out the entry transition curve by the tangential angles method if a peg is to be placed on the centre line at every exact 25 m multiple of through chainage

13.8 On a highway development scheme, the bearings of three successive intersecting straight sections of road AB, BC and CD are 32°41'23", 44°16'18" and 26°45'38", respectively.

It is proposed to connect AB and BC with a wholly transitional curve and BC and CD with a composite curve consisting of entry and exit transition curves of equal length and a central circular arc. The design data for the two curves are:

Design speed	70 kph
Rate of change of radial acceleration	0.20 m s^{-3}
Through chainage of the tangent point on EF	2913.85 m
Horizontal length FG	505.78 m
Radius of the circular arc on the composite curve	750 m

Through chainage increases from A to D

Tabulate the data required to set out the entry transition of the composite curve if pegs are to be placed on the centre line at exact 20 m multiples of through chainage using the tangential angles method

13.9 On a proposed highway, two intersecting straights AI and IB meet at a point I which has a through chainage of 7256.19 m. A horizontal curve, comprising a central circular arc and two equal length transitions, is to be used to connect these two straights. The design data are as follows:

Design speed	85 kph
Rate of change of radial acceleration	0.30 m s^{-3}
Minimum radius of curvature	600 m
Overall road width	10.00 m
Bearing IA	255°39'18"
Bearing IB	99°16'44"

The through chainage increases from A to B.

(i)   Calculate the setting out data for the entry and exit transition curves and the central circular arc if the tangential angles method is to be used and pegs are to be placed on the centre line at exact 25 m multiples of through chainage

(ii)   Calculate the amount of allowable superelevation that should be constructed along the entry transition curve at exact 25 m multiples of through chainage

(iii)   Using the appropriate data given above, calculate the chainage of the common tangent point if the curve design is changed to a wholly transitional type

13.10 A wholly transitional curve having equal tangent lengths is be designed to connect two intersecting straights which meet at a deflection angle of 07°34'56". If each tangent length is to be 95.38 m and the design speed for the road is 100 kph, calculate:

(i)   The minimum radius of curvature

(ii)   The total length of the curve

(iii)   The maximum rate of change of radial acceleration

13.11 A composite horizontal curve consisting of entry and exit transition curves of equal length and a central circular arc is to connect two straights on a new road, which intersect at a point I.

The design speed for the road is 70 kph, the radius of the central circular arc is 510 m and the rate of change of radial acceleration is 0.30 m s^{-3}. The through chainage of I is 3196.23 m and the deflection angle is 10°17'46".

The coordinates of point I are (596.39 mE, 622.45 mN) and the bearing from the entry tangent point T to point I is 102°35'28". Pegs are to be placed on the centre line at exact 20 m multiples of through chainage.

Calculate the through chainages and the coordinates of all the points that are to be set out on the curve.

Further reading and sources of information

For the latest information on the design of transition curves, consult

Design Manual for Roads and Bridges, Volume 6 Road Geometry, Section 1 Links, Part 1 TD9/93 – incorporating Amendment No. 1 dated Feb 2002 – Highway Link Design. Jointly published by the Overseeing Organisations of England, Scotland Wales and Northern Ireland, that is, The Highways Agency, the Scottish Executive Development Department, The National Assembly for Wales (Cynulliad Cenedlaethol Cymru) and The Department for Regional Development Northern Ireland. The manual can be accessed and printed from the Internet at: http://www.official-documents.co.uk/document/deps/ha/dmrb/index.htm. It is also available from The Stationery Office at http://www.tso.co.uk/.

For the latest information on highway design software packages, consult

Software Systems, Electronic Surveying Supplement, Civil Engineering Surveyor, pp. 44–50, published by the Institution of Civil Engineering Surveyors, Spring 2003. http://www.ices.org/

Fort, M. J. (2003) Surveying for geomatics software. *Engineering Surveying Showcase 2003*, No. 2, pp. 33–49. http://www.pvpubs.com/.

chapter 14

Vertical curves

 ## Aims

After studying this chapter you should be able to:

- Understand what gradients are and the limitations that are imposed on their values
- Appreciate how vertical curves can be used to improve the safety and comfort of passengers in vehicles travelling from one intersecting gradient to another
- Understand the terminology and geometry of vertical curves
- Use the standards specified in the United Kingdom Department of Transport publication *Design Manual for Roads and Bridges* to help with the calculation of the length of vertical curve required
- Design vertical curves to join intersecting gradients
- Appreciate the use of computer software packages in the design of vertical alignments
- Calculate reduced levels along the centre lines of vertical curves
- Calculate the reduced levels of the highest and lowest points on vertical curves
- Plot vertical alignments on drawings
- Set out vertical curves on site

This chapter contains the following sections:

14.1 The need for vertical curves

After studying this section you should know what gradients are and that vertical curves are required to enable passengers in vehicles to travel safely and comfortably from one gradient to another. You should be aware that in the UK the Department of Transport recommends maximum and minimum gradient values for all new highways. You should appreciate that vertical curves can either be crest curves or sag curves and you should know how to differentiate between them. You should understand that the main purposes of vertical curves are to provide adequate visibility, and passenger comfort and safety.

This section includes the following topics:

- Gradients
- Purposes of vertical curves

Gradients

In the same way that horizontal curves are used to connect intersecting straights in the horizontal plane, vertical curves are used to connect intersecting straights in the vertical plane. These straights are usually referred to as *gradients* and the combination of the gradients and vertical curve is knows as the *vertical alignment*.

Gradients are usually expressed as percentages, for example, 1 in 50 = 2%, 1 in 25 = 4%. In the UK, the Department of Transport publication *TD 9/93 – Amendment No. 1 – Highway Link Design*, which forms Volume 6 of the *Design Manual for Roads and Bridges*, recommends *desirable maximum* and *absolute maximum* gradient values for all new highways, and these are shown in Table 14.1. Wherever possible, the desirable

Table 14.1

Type of road	Desirable maximum gradient	Absolute maximum gradient
Motorways	3%	4%
Dual carriageways	4%	8%
Single carriageways	6%	8%

maximum values should not be exceeded. Any gradient steeper than 4% on motorways and 8% on all purpose roads is considered to be a departure from standard.

For drainage purposes, *TD 9/93* recommends that the road channels should have a minimum gradient of 0.5% (1 in 200) wherever possible. This is achieved on level sections of road by steepening the channels between drainage gullies while the road itself remains level.

Further information concerning gradients can be found in *TD9/93*, details of which are given in the Further reading and sources of information section at the end of this chapter.

In the design calculations, which are discussed in Section 14.4, the *algebraic difference A* between the gradients is used. This requires the introduction of the sign convention that gradients rising in the direction of increasing chainage are considered to be positive and those falling are considered to be negative.

This leads to *six* different combinations of gradient, which are shown in Figure 14.1 together with the value of their algebraic difference, *A*. In Figure 14.1, chainage has been assumed to increase from left to right and various entry gradients (*m*% values) are shown intersecting with various exit gradients (*n*% values). *A* is obtained from

$$A = (\text{entry gradient } \%) - (\text{exit gradient } \%) \tag{14.1}$$

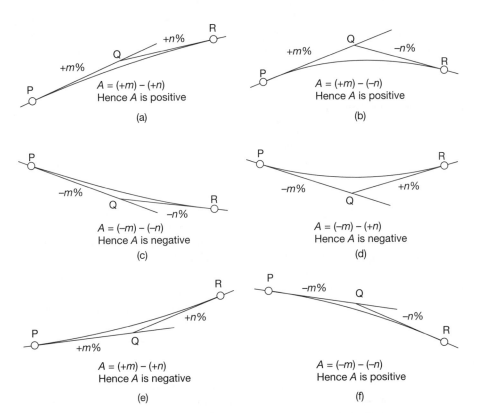

Figure 14.1 ● The six possible combinations of gradient.

The signs of the gradients are retained in the calculation of A. For example, in Figure 14.1(a)

$$A = (+m\%) - (+n\%)$$

Note that A is calculated in the direction of increasing chainage and can be either positive or negative.

Throughout the remainder of this chapter, reference will be made to the terms crest curve and sag curve and, in order to avoid confusion, these terms are defined as follows. A *crest curve*, which can also be referred to as a summit or hogging curve, is one for which the algebraic difference of the gradients is positive, and a *sag curve*, which can also be referred to as a valley or sagging curve, is one for which the algebraic difference of the gradients is negative.

Using these definitions, (a), (b) and (f) in Figure 14.1 are crest curves and (c), (d) and (e) are sag curves.

Purposes of vertical curves

Vertical curves have other similarities with horizontal curves in that they are designed for a particular speed, which is constant for each particular vertical curve, and one of their main functions is to ensure that passengers in vehicles travelling along the curves are transported *safely and comfortably*. In addition, they must ensure that there is *adequate visibility* to enable vehicles to be able either to stop safely or to overtake safely.

Passenger comfort and safety

As the vehicle travels along the vertical curve a radial force, similar to that which occurs in horizontal curves, acts on it in the vertical plane. This has the effect of trying to force the vehicle away from the centre of curvature of the vertical curve. In crest design, this could cause the vehicle to leave the road surface, as in the case of hump-back bridges, while in sag design the underside of the vehicle could come into contact with the road surface, particularly where the gradients are steep and opposed. Both scenarios result in discomfort and danger to passengers travelling in the vehicle and hence their possibility must be minimised. This is achieved firstly by restricting the gradients (see Table 14.1), which has the effect of reducing the force, and secondly by choosing a suitable type and length of curve such that this reduced force is introduced as gradually and uniformly as possible. The type of curve used is discussed in Section 14.2 and the length of curve required is discussed in Section 14.3.

Adequate visibility

In order that vehicles travelling at the design speed can stop or overtake safely, it is essential that oncoming vehicles or any obstructions in the road can be seen clearly and in good time. The length of the vertical curve used must ensure that any visibility requirement is met and this is achieved by the use of sight distances and K values, which are discussed in Section 14.3.

Reflective summary

With reference to the need for vertical curves remember:

— Vertical curves are used to connect intersecting gradients in the vertical plane.

— Gradients are usually expressed as percentages and *TD9/93* recommends *desirable maximum* and *absolute maximum* gradient values for all new highways. Any gradient steeper than 4% on motorways and 8% on all purpose roads is considered to be a departure from standard.

— There are six different possible combinations of gradient, three being *crest curves* and three being *sag curves*.

— The value of the *algebraic difference* between the gradients for crest curves is *positive* and for sag curves is *negative*.

— The two main requirements of vertical curves are that they provide acceptable levels of *passenger comfort and safety* and *adequate visibility*.

14.2 The type of vertical curve to be used and its geometry

After studying this section you should know that vertical curves are flat curves. You should understand that although circular or elliptical curves could be used as vertical curves, in practice parabolas are used since they have a uniform rate of change of gradient, which enables the vertical radial force to be introduced and removed gradually. You should know the equation of the vertical curve and be aware of the assumptions that are made during its derivation that simplify the design and setting out procedures. You should be able to calculate the reduced levels of points on the centre line of the vertical curve at regular through chainage intervals and also the through chainage and reduced level of the highest and lowest points on vertical curves.

This section includes the following topics:

- Vertical curve geometry
- Assumptions made in vertical curve calculations
- The equation of the vertical curve
- Calculating reduced levels along the centre line of a vertical curve
- The highest point of a crest and the lowest point of a sag

Vertical curve geometry

In practice, due to the restrictions placed on the gradients by the Department of Transport, vertical curves can be categorised as *flat* where the definition of a flat curve is that if its length is L_V and its radius is R then $L_V / R \leq 1/10$. This definition does assume that the vertical curve forms part of a circle of radius R but, again due to the restricted gradients, there is no appreciable difference between a circular arc, an ellipse or a parabola and the definition can be applied to all three types of curve by approximating the value of R.

The final choice of curve is governed by the requirement for passenger safety and comfort as discussed in Section 14.1. In practice, a *parabolic curve* is used to achieve a uniform rate of change of gradient and therefore a uniform introduction of the vertical radial force. This uniformity of the rate of change of gradient is given by the basic equation of a parabola as

$$x = cy^2 \qquad (14.2)$$

$$\frac{dx}{dy} = 2cy$$

$$\frac{d^2x}{dy^2} = 2c = \text{constant} = \text{rate of change of gradient}$$

Assumptions made in vertical curve calculations

The choice of a parabola simplifies the calculations and further simplifications are possible if certain assumptions are made. Consider Figure 14.2, which is greatly exaggerated for clarity and shows a parabolic vertical curve having equal tangent lengths joining two intersecting gradients PQ and QR. In general, vertical curves are

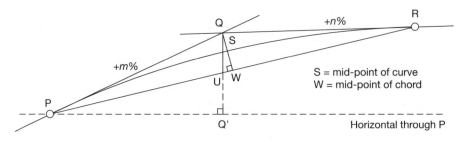

S = mid-point of curve
W = mid-point of chord

Horizontal through P

Figure 14.2

designed such that the two tangent lengths are equal, that is PQ = QR, but it is possible to design vertical curves with unequal tangent lengths, and these are discussed in Section 14.4.

The assumptions are as follows:

- Chord PWR = arc PSR = PQ + QR.
- Length along the tangents = horizontal length, that is PQ = PQ'.
- QU = QW, that is, there is no difference in dimensions measured either in the vertical plane or perpendicular to the entry tangent length.

The first two assumptions are very important since they are saying that the length is the same whether measured along the tangents, the chord, the horizontal or the curve itself. The third assumption is used when calculating reduced levels along the curve and when plotting it on a drawing.

All of these assumptions are only valid if the Department of Transport recommendations for gradients as listed in Table 14.1 are adhered to.

Equation of the vertical curve

Since the curve is to be parabolic, its equation will be of the form given in equation (14.2) as $x = cy^2$, y being measured along the tangent length and x being set off at right angles to it. In fact, from the assumptions, x can also be set off in a vertical direction without any introduction of appreciable error.

The basic equation (14.2) is usually modified to a general equation containing some of the parameters involved in the vertical curve design. This general equation will be developed for the equal tangent length crest curve shown in Figure 14.3, but the same equation can be derived for sags and applies to all six possible combinations of gradient.

Consider Figure 14.3. Let QS = e and let the total length of the curve = L_V. Using the assumptions

$$\text{Level of Q above P} = \left(\frac{m}{100}\right)\left(\frac{L_V}{2}\right) = \frac{mL_V}{200} \qquad (14.3)$$

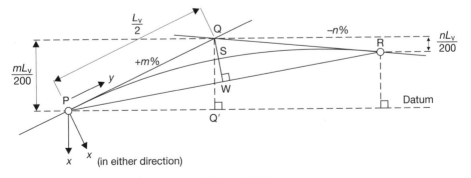

Figure 14.3

$$\text{Level of R below Q} = \left(\frac{n}{100}\right)\left(\frac{L_V}{2}\right) = \frac{nL_V}{200} \qquad (14.4)$$

Hence

$$\text{Level of R above P} = \left(\frac{mL_V}{200}\right) - \left(\frac{nL_V}{200}\right) = \frac{(m-n)L_V}{200}$$

But, from the assumptions, PW = WR, therefore

$$\text{Level of W above P} = \frac{(m-n)L_V}{400}$$

From the properties of the parabola

$$QS = \frac{QW}{2} = SW$$

Since QW = Level of Q above P − Level of W above P

$$QS = \frac{1}{2}\left[\frac{mL_V}{200} - \frac{(m-n)L_V}{400}\right] = \frac{(m+n)L_V}{800}$$

In this case the algebraic difference of the gradients $A = (+m) - (-n) = (m+n)$ and

$$QS = e = \frac{AL_V}{800} \qquad (14.5)$$

The basic equation of the parabola from equation (14.2) is $x = cy^2$. Therefore, at point Q, when $y = L_V/2$ and $x = e$

$$e = c\left(\frac{L_V}{2}\right)^2$$

This gives

$$c = \frac{e}{(L_V/2)^2}$$

and substituting this into equation (14.2) gives

$$x = \frac{ey^2}{(L_V/2)^2}$$

But, from equation (14.5)

$$e = \frac{AL_V}{800}$$

Therefore

$$x = \frac{Ay^2}{200L_V} \qquad (14.6)$$

Equation (14.6) is the general equation of parabolic vertical curves and is used in the calculation of reduced levels along the curve centre line.

Calculating reduced levels along the centre line of a vertical curve

Figure 14.4 shows a vertical curve having equal tangent lengths PQ and QR. Through chainage increases from tangent point P to tangent point R and the entry gradient is $+m\%$ and the exit gradient is $-n\%$. If P is taken as the datum level then the level of any point Z on the curve with respect to P is given by ΔH, where

$$\Delta H = \left[\frac{(m)y}{100} - \frac{(A)y^2}{200L_V} \right] \tag{14.7}$$

This is a general expression and ΔH can be either positive or negative, depending on the signs of m and A.

All ΔH values are related to the RL of tangent point P and should be added to or subtracted from this to obtain the reduced levels of points on the curve, which lie a general distance y from P.

Worked examples 14.1 and 14.2 in Section 14.6 show how reduced levels along the centre line of a vertical curve can be calculated using equation (14.7).

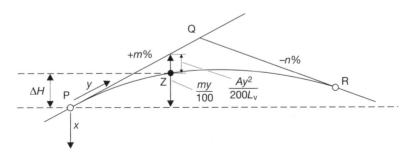

Figure 14.4

The highest point of a crest and the lowest point of a sag

In order that drainage gullies can be positioned effectively, it is necessary to know the through chainage and reduced level of the highest or lowest point on the vertical curve. The highest point occurs when ΔH in Figure 14.4 is a maximum.

For a maximum or minimum value of ΔH, $d(\Delta H)/dy = 0$. Applying this to equation (14.7)

$$\frac{d(\Delta H)}{dy} = \left(\frac{m}{100} \right) - \left(\frac{Ay}{100L_V} \right) = 0$$

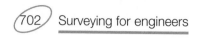

Rearranging gives

$$\left(\frac{m}{100}\right) = \left(\frac{Ay}{100L_V}\right)$$

And for a maximum or minimum value of ΔH

$$y = \frac{L_V m}{A} \tag{14.8}$$

This gives the point along the curve from tangent point P at which the maximum or minimum level occurs. To find the reduced level at this point it is necessary to substitute equation (14.8) back into equation (14.7) as follows

$$\Delta H_{max/min} = \left(\frac{m}{100}\right)\left(\frac{L_V m}{A}\right) - \left(\frac{A}{200L_V}\right)\left(\frac{L_V^2 m^2}{A^2}\right)$$

Simplifying this gives the maximum or minimum height on the curve above or below point P as

$$\Delta H_{max/min} = \left(\frac{L_V m^2}{200 A}\right) \tag{14.9}$$

Worked example 14.1 in Section 14.6 shows how the through chainage and reduced level of the highest point on the centre line of a vertical curve can be calculated.

Reflective summary

With reference to the type of vertical curve to be used and its geometry remember:

— Due to the gradient restrictions recommended in *TD9/93*, vertical curves are considered to be *flat* curves and because of this, circular arcs, ellipses or parabolas could all be used as vertical curves

— In practice, parabolas are used to achieve a uniform rate of change of gradient and therefore a uniform introduction of the vertical radial force to ensure passenger comfort and safety.

— Assumptions made during the derivation of the general parabolic vertical curve equation mean that when designing vertical curves the length is the same whether measured along the tangents, the chord, the horizontal or the curve itself, and there is no difference in dimensions measured either in the vertical plane or perpendicular to the entry tangent length.

- The basic equation of a parabolic vertical curve is $x = Ay^2/200L_v$.

- It is possible to calculate reduced levels of points at regular through chainage intervals along a vertical curve.

- It is also possible to calculate the through chainage and reduced level of the highest or lowest point on the curve in order that drainage gullies can be positioned appropriately.

14.3 The length of vertical curve to be used

After studying this section you should know that the length of vertical curve to be used to join two intersecting gradients depends on the distance of visibility, or sight distance, from one side of the carriageway on which the vertical curve is located to the other. You should understand the terms Stopping Sight Distance and Full Overtaking Sight Distance and you should know how they are determined. You should be familiar with K Values and you should be able to use them to determine the minimum length of vertical curve required. You should appreciate the need to phase the horizontal and vertical alignments before finalising the length of a vertical curve.

This section includes the following topics:

- Sight distances
- K values
- Final choice of vertical curve length

Sight distances

The length of vertical curve to be used in any given situation depends on the *sight distance*. This is the distance of visibility from one side of the carriageway on which the vertical curve is located to the other. There are two categories of sight distance

- *Stopping Sight Distance (SSD)*, which is the theoretical forward sight distance required by a driver in order to stop safely and comfortably when faced with an unexpected hazard on the carriageway

and

- *Full Overtaking Sight Distance (FOSD)*, which is the length of visibility required by drivers of vehicles to enable them to overtake vehicles on the road ahead in safety and comfort.

Since it requires a greater distance to overtake than to stop, the FOSD values are greater than the SSD values.

When designing vertical curves, it is essential to know whether safe overtaking is to be included in the design. If it is, the FOSD must be incorporated, if it is not then the SSD must be incorporated.

It is usually necessary to consider whether to design for overtaking only at crest curves on single carriageways, since overtaking should not be a problem on dual carriageways and visibility is usually more than adequate for overtaking at sag curves on single carriageways.

In *TD9/93*, recommended stopping sight distances and overtaking sight distances for six different design speeds are given and these are shown in the sections headed *STOPPING SIGHT DISTANCE* and *OVERTAKING SIGHT DISTANCE* in Table 13.1. These recommended values were obtained as follows.

The SSD ensures that there is an envelope of clear visibility such that, at one extreme, drivers in low vehicles are provided with sufficient visibility to see low objects, while, at the other extreme, drivers of high vehicles are provided with visibility to see a significant portion of other vehicles. This envelope of visibility is shown in Figure 14.5 in which 1.05 m represents the driver's eye height for low vehicles and 2.00 m that for high vehicles. A lower object height of 0.26 m is used to include the rear tail lights of other vehicles and an upper object height of 2.00 m ensures that a sufficient portion of a vehicle ahead can be seen to identify it as such.

The FOSD ensures that there is an envelope of clear visibility between the 1.05 m and 2.00 m driver's eye heights above the centre of the carriageway as shown in Figure 14.6.

When constructing vertical curves, the SSD and FOSD must be measured in both the horizontal and vertical planes and checked to ensure that they comply with the values given in Table 13.1 for the design speed in question. They are measured using the dimensions given in Figures 14.5 and 14.6 for the envelopes of visibility.

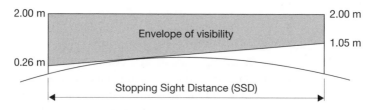

Figure 14.5 ● Measurement of SSD.

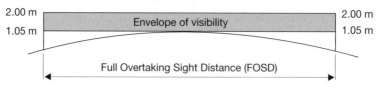

Figure 14.6 ● Measurement of FOSD.

K Values

In the past it was necessary to use the appropriate sight distance for the road type and design speed in question to calculate the minimum length of the vertical curve required. Nowadays, constants known as K values specified in *TD9/93* greatly simplify the calculations.

The minimum length of vertical curve (*min* L_V) for any given road is obtained from the equation

$$min\ L_V = KA \text{ metres} \tag{14.10}$$

where K is the constant obtained from *TD9/93* for the particular road type and design speed and A is the algebraic difference of the gradients, the absolute value (always positive) being used.

Table 13.1, which is taken from *TD9/93*, shows the current K Values for six different design speeds. The K Values ensure that the minimum length of vertical curve obtained from equation (14.10) contains adequate visibility and provides sufficient comfort. It must be noted that the length obtained is the *minimum* required and it is perfectly acceptable to increase the value obtained. This may be necessary when trying to phase the vertical alignment with the horizontal alignment as discussed later in this section.

The units of K are metres and their values have been derived from the sight distances discussed earlier. In Table 13.1, under the sections headed *VERTICAL CURVATURE* and *OVERTAKING SIGHT DISTANCES*, there are three categories of K values for crests and one category of K Values for sags. These are discussed below.

Crest K Values

If a full overtaking facility is to be included in the design of single carriageways then the *FOSD Overtaking Crest K Value* given in the bottom row of Table 13.1 should be used in equation (14.10). An overtaking requirement is not normally included in the design of motorways, dual carriageways and one-way carriageways since it is not critical in such cases.

If overtaking is not to be considered in the design then there are two possible K Values available for crests. With reference to the *VERTICAL CURVATURE* section of Table 13.1, these are the *Desirable Minimum Crest K Value* and the *One Step below Desirable Minimum Crest K Value*.

Examples of the use of the crest K Values are given below.

● *Example (1): Dual carriageway, design speed 85 kph, crest*
From Table 13.1

FOSD Overtaking Crest K Value	= 285 m
Desirable Minimum Crest K Value	= 55 m
One Step below Desirable Minimum Crest K Value	= 30 m

Since a dual carriageway is being designed, overtaking is not critical. Hence the FOSD Overtaking Crest K Value does not apply and, in such a case, *TD9/93* recommends that at least the Desirable Minimum Crest K Value should be used. Therefore from equation (14.10), if possible use

$L_V \geq 55A$ metres

If this is not possible then, depending on the site conditions, the designer may be permitted to apply a *relaxation* (see Section 13.2) to the One Step below Desirable Minimum Crest K Value and use $L_V \geq 30A$ metres.

● *Example (2): Single carriageway, design speed 60 kph, crest*
From Table 13.1

FOSD Overtaking Crest K Value	= 142 m
Desirable Minimum Crest K Value	= 17 m
One Step below Desirable Minimum Crest K Value	= 10 m

Since a single carriageway is being designed, a decision has to be made as to whether or not full overtaking is to be allowed for in the design.

If full overtaking is being designed then equation (14.10) gives

min $L_V = 142A$ metres

If full overtaking is not to be included, it would appear that

min $L_V = 17A$ metres

However, *TD9/93* states that for crests on single carriageways, unless FOSD Overtaking Crest K Values can be used, it is sufficient to use only the One Step below Desirable Minimum Crest K Value since the use of the Desirable Minimum Crest K Values may result in sections of road having dubious visibility for overtaking.

In summary, this means that on single carriageway crests, overtaking should either be easily achieved or not possible at all. Hence, in this example

If possible, use $L_V \geq 142A$ metres

Otherwise use $L_V = 10A$ metres

Further details on restrictions involved in the design of single carriageways can be found in *TD9/93*.

Sag K Values

Only one category of K Values is given in *TD9/93* for sag curves since overtaking visibility is usually unrestricted on this type of vertical curve. This is the set of *Absolute Minimum Sag K Values* shown in the *VERTICAL CURVATURE* section of Table 13.1, which ensure adequate stopping visibility and comfort. The use of these K Values is illustrated in the following example.

● *Example (3): Single carriageway, design speed 100 kph, sag*
From Table 13.1

Absolute Minimum Sag K Value = 26 m

Since *TD9/93* states that sag curves should normally be designed to the Absolute Minimum Sag K Values, the required length of vertical curve is given by equation (14.10) as

$$L_V = 26A \text{ metres}$$

Final choice of vertical curve length

Often the value for the minimum length of curve obtained from the K Values is not used, a greater length being chosen. This may be done for several reasons, for example, it may be necessary to fit the curve into particular site conditions. However, there is another factor, which must be considered before deciding on the final length of the vertical curve. Normally, vertical alignments are designed in conjunction with a horizontal alignment and, where this is the case, the two alignments must be considered together so that they can be *phased* correctly. To avoid the creation of optical illusions in the road surface, the tangent points of the vertical curve must, wherever possible, coincide exactly with the tangent points of the horizontal curve, where applicable. This is known as *phasing*.

If a vertical curve is started during a horizontal curve then to a driver travelling along the curve the road appears disjointed due to the vertical directional change of the vertical curve being introduced on the horizontal curve at a point where the horizontal radial force and superelevation may be severe. This can lead to driver error and must be avoided wherever possible.

In most cases, the required length of the horizontal curve will be greater than the minimum required length of the vertical curve and it will be necessary to increase the vertical curve length to that of the horizontal curve. Should the minimum vertical curve length be greater than the required length of the horizontal curve then the opposite will apply.

When phasing vertical and horizontal alignments, the curves should run between the start and finish tangent points, not between any two tangent points. This is shown in Figure 14.7. To introduce the two alignments at different tangent points would again create optical illusions.

Reflective summary

With reference to the length of vertical curve to be used remember:

— The length of curve to be used in any given situation depends on the *sight distance*.

— There are two types of sight distance, the *Stopping Sight Distance (SSD)*, which is the sight distance required by a drivers to stop safely and comfortably and the *Full Overtaking Sight Distance (FOSD)*, which

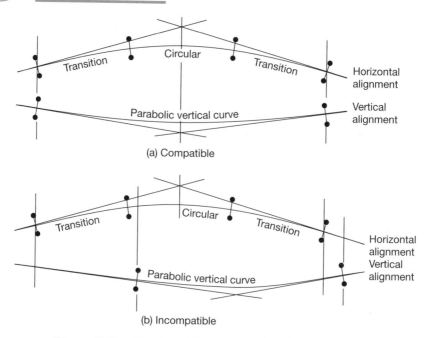

Figure 14.7 ● Phasing of horizontal and vertical alignments.

is the sight distance required by drivers to overtake vehicles ahead of them in safety and comfort.

— It is essential to know whether the length is to be such that safe overtaking is possible. If it is then the length must be based on the FOSD but if it is not then it should be based on the SSD.

— The minimum required length of a vertical curve for a particular design speed is obtained using K Values specified in *TD9/93*. These are derived from the sight distances and ensure that the minimum length of vertical curve obtained contains adequate visibility and provides sufficient comfort.

— The lengths obtained using K Values are minimum values and it is perfectly acceptable to increase them if site conditions and design standards permit.

— If the vertical curve is being designed in conjunction with a horizontal curve then it is important that the lengths of both curves are the same in order that they can be phased correctly to avoid the creation of optical illusions in the road surface.

14.4 Designing vertical curves

After studying this section you should understand the procedures involved in the design of vertical curves. You should know how to design vertical curves having either equal tangent lengths or unequal tangent lengths, and how to design a vertical curve to pass through a particular point. You should know how to plot vertical curves on longitudinal sections to see if your design is acceptable. You should appreciate that the design procedure is usually undertaken with the aid of commercially available highway alignment software design packages.

This section includes the following topics:

- Designing vertical curves having equal tangent lengths
- Designing vertical curves having unequal tangent lengths
- Designing a vertical curve to pass through a particular point
- Computer aided vertical alignment design

Designing vertical curves having equal tangent lengths

The first problem considered in this section is to design a suitable vertical curve having equal tangent lengths to fit between two intersecting gradients PQ and QR for a particular design speed. It is assumed that a longitudinal section has been plotted to show the existing ground profile along the proposed centre line as shown in Figure 14.8 (see also Section 15.3). The solution to the design is as follows.

- Choose suitable gradients by drawing them on the longitudinal section such that they balance out the cut and fill areas on the centre line as much as possible, as shown in Figure 14.8. The values chosen for the gradients must be within those stipulated in *TD9/93* for the road type in question as shown in Table 14.1.

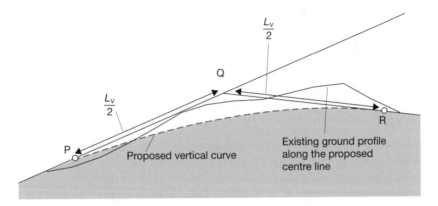

Figure 14.8

- From the drawing, measure the gradients and the reduced level of the point Q where the two gradients intersect.
- Calculate the algebraic difference of the gradients, A.
- Obtain the appropriate K Value for the design speed from Table 13.1. There are three categories of K Values for crests and one category of K Values for sags. If safe overtaking is to be included in the design the FOSD Overtaking Crest K Values must be used otherwise at least the Desirable Minimum Crest K Values should be adopted wherever possible for all roads except single carriageways. For crests on single carriageway roads where overtaking is not included in the design, it is sufficient to use the One Step below Desirable Minimum Crest K values to avoid sections of road having dubious visibility for overtaking. Sag curves are normally designed using the Absolute Minimum Sag K Values. Examples of the use of K Values to obtain the lengths of vertical curves are given in Section 14.3.
- Use the chosen K Value in equation (14.10), $min\ L_V = KA$, to calculate the minimum required length of the vertical curve. At this point it may be necessary to phase the vertical and horizontal alignments as described in Section 14.3 and an alteration in the gradients may be necessary to accommodate this. Should this be necessary then an attempt should again be made on the longitudinal section to balance out cut and fill areas on the centre line.
- On the drawing measure back along the entry and exit gradients from Q a distance of $L_V/2$ to fix the positions of the entry tangent point P and the exit tangent point R as shown in Figure 14.8.
- The reduced levels of P and R should be calculated from the reduced level of Q using equations (14.3) and (14.4), respectively,

$$RL_R = RL_Q \pm mL_V/200 \text{ and } RL_R = RL_Q \pm nL_V/200$$

- Equation (14.7)

$$\Delta H = \left[\frac{(m)y}{100} - \frac{(A)y^2}{200L_V} \right]$$

is used together with the reduced level of the entry tangent point P to calculate the reduced levels of points on the curve itself. In the case where vertical and horizontal alignments are being designed together, the points used will be the same as those defined for the horizontal alignment. If there is no corresponding horizontal alignment, points should be chosen at regular chainage intervals along the vertical curve centre line. As a check on the calculations, the reduced level of the exit tangent point R should be calculated from equation (14.7) and compared to the value obtained using equation (14.4).

- The curve is plotted on the longitudinal section – in Figure 14.8 this is shown as a broken line. This can be done in one of two ways, either by using the reduced levels of points already calculated for the curve or by using equation (14.6), $x = Ay^2/200L_V$, to calculate offset (x) values to be set off at regular distance (y) intervals along the entry gradient starting at point P. This latter method is shown in

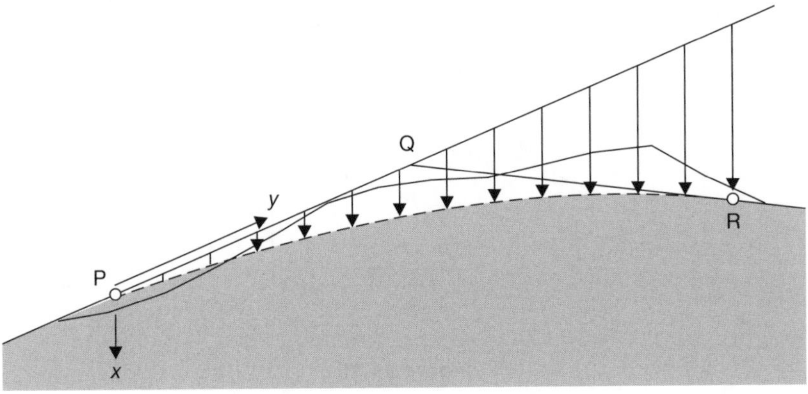

Figure 14.9

Figure 14.9, where it should be noted that the offset (*x*) values are measured from the entry gradient in the vertical direction since one of the assumptions made in Section 14.2 is that there is no difference in dimensions measured either in the vertical plane or perpendicular to the entry tangent length. The whole of the curve can be plotted from the entry gradient. Once the curve has been plotted on the longitudinal section it should be inspected to see if it is acceptable. If it is then is can be set out on site. If it is not then it will be necessary to change the gradients and repeat the whole design process.

- Once the design has been accepted the curve can be set out on site. This is described in Section 14.5.

Worked example 14.1 in Section 14.6 illustrates some of the steps involved when designing a vertical curve having equal tangent lengths.

Designing vertical curves having unequal tangent lengths

In the foregoing discussion, a vertical curve having equal tangent lengths (*symmetrical*) was considered. These are easy to design and can be fitted to the majority of cases but, occasionally, either to meet particular site conditions or to avoid large amounts of cut or fill, it becomes necessary to design a curve having unequal tangent lengths (*asymmetrical*).

With reference to Figure 14.10, the easiest method of designing such a curve is to introduce a third gradient BCD, which splits the total curve PR into two consecutive equal tangent length curves PC and CR. The common tangent line BCD is parallel to the chord PR and C is the common tangent point between the two curves.

The first curve PC is equal in length to the entry tangent length PQ and the second curve CR is equal in length to the exit tangent length QR. B is the mid-point of PQ and D is the mid-point of QR.

From Figure 14.10

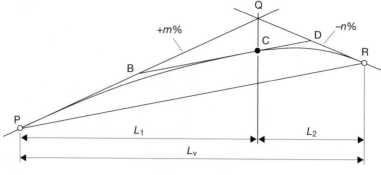

Figure 14.10

$PC = L_1$, $CR = L_2$

$L_V = L_1 + L_2$

$PB = BC = \dfrac{PQ}{2} = \dfrac{L_1}{2}$

$CD = DR = \dfrac{QR}{2} = \dfrac{L_2}{2}$

When calculating reduced levels at regular chainage intervals along the curves, each curve is treated as a separate equal tangent length vertical curve. Worked example 14.2 in Section 14.6 shows how this is done.

Designing a vertical curve to pass through a particular point

Occasionally when designing vertical curves, a specific design requirement has to be met which will greatly influence the design process, for example, where a vertical curve must have a minimum clearance over a culvert or under a bridge. In such cases, there will be a point of known chainage and reduced level through which the curve must pass. These requirements can be accommodated in the design and Worked example 14.3 in Section 14.6 shows the calculations involved.

Computer-aided vertical alignment design

Earlier in this section, a step-by-step guide to designing a vertical curve with equal tangent lengths has been given. This described how the design calculations could be performed by hand and how the curve could be plotted by hand on the longitudinal section to see if the design was acceptable. Although calculations and drawings for highway design are still sometimes undertaken by hand, it is much more usual nowadays for commercially available highway design software packages to be used. Many of these are available and their relatively low cost coupled with the falling prices of desktop and laptop computers makes them ideal for even the smallest engineering or surveying practice.

Such packages have a number of very significant advantages over hand methods for vertical alignment design. Their speed enables the calculations to be performed very quickly and their graphic capability gives an instantaneous on-screen view of the gradients and vertical curves. The need for phasing, as discussed in Section 14.3, is much easier to perform using the computer software. Editing facilities enable site constraints to be incorporated and different alignments to be tried until a suitable solution is achieved. Graphical, numerical and setting out data are easily provided, as required.

The principles on which these packages base their vertical alignment design are identical to those discussed in this chapter. Gradients are specified, sight distances and K Values are used and reduced levels along the curve are calculated. A range of curves is normally available; for example, symmetrical parabolas, asymmetrical parabolas and circular arcs are usually provided.

Since vertical curves are almost invariably designed in conjunction with horizontal curves, further information about currently available highway design software packages is given in Section 13.5.

Reflective summary

With reference to designing vertical curves remember:

— Vertical curves can have either equal tangent lengths or unequal tangent lengths and can be designed to pass through specific points, if required.

— At the start of the design process, when fitting the gradients to the longitudinal section, an attempt should be made to balance areas of cut and fill along the centre line.

— If safe overtaking is to be included in the design the FOSD Overtaking Crest K Values must be used otherwise at least the Desirable Minimum Crest K Values should be adopted wherever possible for all roads except single carriageways.

— For crests on single carriageway roads where overtaking is not included in the design, it is sufficient to use the One Step below Desirable Minimum Crest K Values to avoid sections of road having dubious visibility for overtaking.

— Sag curves are normally designed using the Absolute Minimum Sag K Values.

— If the vertical curve is being designed in conjunction with a horizontal curve then the designer must ensure that the vertical and horizontal alignments are correctly phased.

— Although calculations and drawings for highway design are still sometimes undertaken by hand, it is much more usual nowadays for commercially available highway design software packages to be used.

14.5 Setting out vertical curves on site

After studying this section you should understand why vertical curves are set out on site. You should be able to define the shape of a vertical curve on site using sight rails. You should appreciate that vertical curves can also be established on site using machine control systems.

This section includes the following topics:

● The purpose of defining vertical curves on site

● Defining vertical curves using sight rails

● Defining vertical curves using machine control systems

The purpose of defining vertical curves on site

Once a vertical curve has been designed, it must be set out on site. This involves defining its shape so that embankments and cuttings can be formed as required to recreate the shape of the curve on the ground surface. Once the embankments and cuttings have been constructed, they provide a foundation on which the road pavement can be laid. There are a number of ways in which vertical curves can be set out on site. Traditionally, sight rails have been used but increasingly machine control methods are being employed.

Defining vertical curves using sight rails

Sight rails are discussed in Section 11.3 and the procedure for using them to define vertical curves on site is as follows.

● The plan positions of the centre line points for which the reduced levels have been calculated at the design stage must be set out on site. If a horizontal alignment is being done in conjunction with the vertical alignment, these points should already have been pegged out.

- Wooden stakes to which the sight rails are to be attached should be driven into the ground, either at these points or offset a known horizontal distance from them to avoid them being disturbed.

- Sight rails are attached to these stakes such that their top edges are offset vertically some known height above or below the reduced levels of the points calculated along the centre line. Travellers are used in conjunction with the sight rails to monitor the construction work.

Once the sight rails have been fixed then if a string line were to be run across their top edges it would define a curve parallel to the designed vertical curve. As construction proceeds and the reduced levels of the ground surface change, it may be necessary to adjust the sight rails or set out new ones to maintain the definition of the shape of the vertical curve on site. Slope rails (see Section 11.3) are also used with sight rails during the construction process to define the side slopes of the embankments and cuttings and to monitor their construction.

Defining vertical curves using machine control systems

Although sight rails are still widely used on site to define vertical curves, there is an increasing use of machine control systems for this purpose. Several systems are available including those controlled by total stations, GPS and lasers. Each involves a computer mounted in the cab of the earthmover or digger to control the elevation and angle of the earthmover blade or digger bucket using information relayed to the computer by the total station, GPS receiver or laser beam. Such systems have many advantages over the sight rail method. They have been shown to improve productivity, reduce costs and produce more accurate results. In addition, the earthmoving plant can work anywhere on site without having to wait for sight rails to be located and there is no requirement for the work to be checked using travellers. Slope rails are also not needed when machine control systems are employed. Further information on these is given in Section 11.6.

Reflective summary

With reference to setting out vertical curves on site remember:

- It is important to define the shape of a vertical curve on site in order that embankments and cuttings can be formed on which the road pavement or railway tracks can be laid.

- They can be set out using sight rails or machine control systems.

— If sight rails are used they should be located on site, either at or offset from centre line points such that their top edges define a curve parallel to the designed vertical curve.

— As construction proceeds it may be necessary to adjust the sight rails or set out new ones to maintain the curve definition. Slope rails are used to define the side slopes of the embankments and cuttings and to monitor their construction.

— Machine control systems, such as those involving total stations, GPS receivers and laser instruments, are increasingly being used to define vertical curves , embankments and cuttings on site. They can improve productivity, reduce costs and produce more accurate results.

14.6 Vertical curve worked examples

After studying this section you should understand the calculation procedures involved in the design of vertical curves to meet a variety of different requirements. You will be able to use K Values to calculate the minimum lengths of vertical curves and you will understand that it may be necessary to use lengths greater than the minimum values. You will know how to calculate the position and reduced level of the highest or lowest point on a vertical curve and you will be able to design a vertical curve to pass through a particular point. You should appreciate the need to tabulate chainage and reduced level data and you will be aware of the importance of checking your calculations.

This section includes the following topics:

● Designing a vertical curve having equal tangent lengths

● Designing a vertical curve having unequal tangent lengths

● Designing a vertical curve to pass through a particular point

Worked example 14.1: Designing a vertical curve having equal tangent lengths

Question

The reduced level at the intersection of a rising gradient of 1.5% and a falling gradient of 1.0% on a proposed road is 93.60 m. Given that the K Value for this particular road is 55, the through chainage of the intersection point is 671.34 m and the vertical curve is to have equal tangent lengths, calculate

- The through chainages of the tangent points of the vertical curve if the minimum required length is to be used.
- The reduced levels of the tangent points and the reduced levels at exact 20 m multiples of through chainage along the curve.
- The position and reduced level of the highest point on the curve.

Solution

The through chainage of the tangent points of the vertical curve if the minimum required length is to be used

Figure 14.11 shows the curve in question.

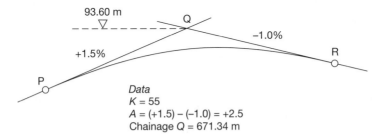

Data
K = 55
A = (+1.5) – (–1.0) = +2.5
Chainage Q = 671.34 m

Figure 14.11

From equation (14.1)

$$A = (\text{entry gradient } \%) - (\text{exit gradient } \%) = (+1.5) - (-1.0) = +2.5$$

From equation (14.10)

$$min\ L_V = KA = 55 \times 2.5 = 137.5\ \text{m}$$

Since this minimum length is to be used and using the assumption that the length along the gradients is the same as the length along the horizontal

$$\textit{through chainage of P} = 671.34 - \left(\frac{137.5}{2}\right) = \textbf{602.59 m}$$

$$\textit{through chainage of R} = 671.34 + \left(\frac{137.5}{2}\right) = \textbf{740.09 m}$$

The reduced levels of the tangent points and the reduced levels at exact 20 m multiples of through chainage along the curve

From Figure 14.11, it is obvious that P and R are both lower than Q. Therefore, using equations (14.3) and (14.4) and ignoring the signs of m and n

$$\textit{reduced level of P} = 93.60 - \left(\frac{mL_V}{200}\right) = 93.60 - 1.03 = \textbf{92.57 m}$$

$$\textit{reduced level of R} = 93.60 - \left(\frac{nL_V}{200}\right) = 93.60 - 0.69 = \textbf{92.91 m}$$

Table 14.2 (all quantities are in metres)

Chainage	y	(m)y/100	(A)y²/200Lᵥ	ΔH	RL
602.59 (P)	0	0	0	0	92.57
620.00	17.41	+0.26	+0.03	+0.23	92.80
640.00	37.41	+0.56	+0.13	+0.43	93.00
660.00	57.41	+0.86	+0.30	+0.56	93.13
680.00	77.41	+1.16	+0.54	+0.62	93.19
700.00	97.41	+1.46	+0.86	+0.60	93.17
720.00	117.41	+1.76	+1.25	+0.51	93.08
740.00	137.41	+2.06	+1.72	+0.34	92.91
740.09 (R)	137.50	+2.06	+1.72	+0.34	92.91

To keep to exact 20 m multiples of through chainage there will need to be an initial short value of y of 620.00 − 602.59 = 17.41 m and y will then increase in steps of 20 m such that the final y value will be equal to the length of the curve, that is, 137.5 m.

Working from P towards R, the reduced levels (RL) of the points on the curve are obtained using equation (14.7) as

$$RL = 92.57 + \Delta H = 92.57 + \left[\frac{(1.5)y}{100} - \frac{(2.5)y^2}{200L_V} \right] \qquad (14.11)$$

The results are tabulated in Table 14.2.

As a check on the calculations, the reduced level of R should equal that calculated earlier as is the case in this example.

For this curve, both *m* and *A* are positive and their positive signs have been retained in the RL calculations made using equation (14.11). When either *m* or *A* is negative, the negative sign should also be retained and taken into account when using equation (14.7) for ΔH in order that ΔH can have the correct sign.

The position and reduced level of the highest point on the curve
From equation (14.8), the highest point on a vertical curve occurs when

$$y = \frac{L_V m}{A} = \left(\frac{137.5 \times 1.5}{2.5} \right) = \textbf{82.50 m}$$

and using equation (14.9), the highest reduced level on a vertical curve can be calculated from the RL of the entry tangent point P as follows

$$\text{highest RL} = RL_P + \left(\frac{L_V m^2}{200 A} \right)$$

Therefore

$$\text{highest RL} = 92.57 + \left(\frac{137.5 \times 1.5^2}{200 \times 2.5}\right) = \textbf{93.19 m}$$

These values can be confirmed by inspection of Table 14.2.

Worked example 14.2: Designing a vertical curve having unequal tangent lengths

Question

A parabolic vertical curve is to connect a –2.50% gradient to a +3.50% gradient on a highway designed for a speed of 100 kph. The K Value for the highway is 26 and the minimum required length is to be used.

The reduced level and through chainage of the intersection point of the gradients are 59.34 m and 617.49 m respectively and, in order to meet particular site conditions, the through chainage of the entry tangent point is to be 553.17 m. Calculate

- The reduced levels of the tangent points.
- The reduced levels at exact 20 m multiples of through chainage along the curve.

Solution

From equation (14.1)

$$A = (-2.50) - (+3.50) = -6.00$$

From equation (14.10)

$$min\ L_V = KA = 26 \times 6.00 = 156.00 \text{ m}$$

Figure 14.12 shows the required curve and, from this, the tangent lengths are

$$L_1 = PQ = 617.49 - 553.17 = 64.32 \text{ m}$$

$$L_2 = QR = 156.00 - 64.32 = 91.68 \text{ m}$$

Since these are unequal, a third gradient BCD is introduced as discussed in Section 14.4.

The reduced levels of the tangent points, P and R

From Figure 14.12, it can be seen that

$$\textbf{RL}_\textbf{P} = RL_Q + \frac{(2.50 \times PQ)}{100} = 59.34 + \frac{(2.50 \times 64.32)}{100} = \textbf{60.95 m}$$

$$\textbf{RL}_\textbf{R} = RL_Q + \frac{(3.50 \times QR)}{100} = 59.34 + \frac{(3.50 \times 91.68)}{100} = \textbf{62.55 m}$$

The reduced levels at exact 20 m multiples of through chainage along the curve

For the curve

through chainage of P = 553.17 m

through chainage of R = through chainage of P + L_V
= 553.17 + 156.00 = 709.17 m

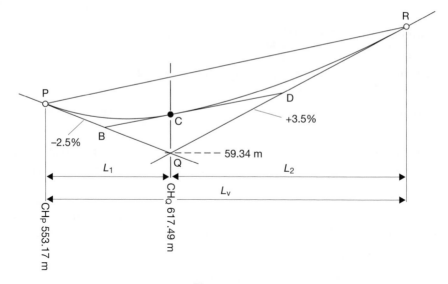

Figure 14.12

Also, from Figure 14.12, it can be seen that

gradient of BCD = gradient of PR

$$= \frac{[(-2.50)L_1 + (+3.50)L_2]}{L_V}\%$$

$$= \frac{[(-2.50)64.32 + (+3.50)91.68]}{156.00}\%$$

$$= +1.03\%$$

For the vertical curve PC, the reduced levels are calculated from P to C with reference to RL_P using equation (14.7) as follows

$$RL = RL_P + \left[\frac{(m)y}{100} - \frac{(A)y^2}{200L_1}\right]$$

where

$m = -2.50\%$
$L_1 = 64.32$ m
$A = (-2.50) - (+1.03) = -3.53$
through chainage of C = through chainage of Q = 617.49 m

The RLs calculated along curve PC are shown in Table 14.3.

For the vertical curve CR, the reduced levels are calculated from C to R with reference to RL_C using equation (14.7) as follows

$$RL = RL_C + \left[\frac{(m)y}{100} - \frac{(A)y^2}{200L_2}\right]$$

Table 14.3 (all quantities are in metres)

Chainage	y	(m)y/100	(A)y²/200L₁	ΔH	RL
553.17 (P)	0	0	0	0	60.95
560.00	6.83	−0.17	−0.01	−0.16	60.79
580.00	26.83	−0.67	−0.20	−0.47	60.48
600.00	46.83	−1.17	−0.60	−0.57	60.38
617.49 (C)	64.32	−1.61	−1.14	−0.47	60.48

where

$m = +1.03\%$
$L_2 = 91.68$ m
$A = (+1.03) - (+3.50) = -2.47$
through chainage of $R = 709.17$ m

The RLs calculated along curve PC are shown in Table 14.4. Note that the value obtained for the reduced level of tangent point R in Table 14.4 agrees with the value obtained in the solution to the first part of the question.

Table 14.4 (all quantities are in metres)

Chainage	y	(m)y/100	(A)y²/200L₂	ΔH	RL
617.49 (C)	0	0	0	0	60.48
620.00	2.51	+0.03	0.00	+0.03	60.51
640.00	22.51	+0.23	−0.07	+0.30	60.78
660.00	42.51	+0.44	−0.24	+0.68	61.16
680.00	62.51	+0.64	−0.53	+1.17	61.65
700.00	82.51	+0.85	−0.92	+1.77	62.25
709.17 (R)	91.68	+0.94	−1.13	+2.07	62.55

Worked example 14.3: Designing a vertical curve to pass through a particular point

Question

A parabolic vertical curve having equal tangent lengths is to connect a falling gradient of 2.40% to a rising gradient of 2.10% on a road designed for a speed of 100 kph. The length of curve used must be at least enough to ensure that the minimum K Value is 26.

In order to ensure that there will be sufficient clearance over a culvert, the curve must pass through a point Z of through chainage 2871.92 m and reduced level

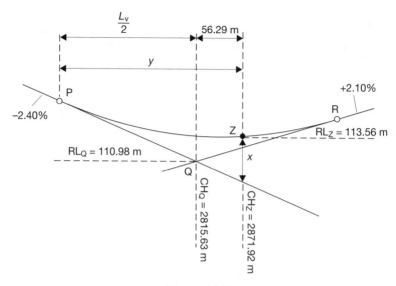

Figure 14.13

113.56 m. The reduced level and the through chainage of the point of intersection Q of the two gradients are 110.98 m and 2815.63 m, respectively.

Calculate the length of vertical curve that will meet these requirements.

Solution

Figure 14.13 shows the curve in question, in which point Z is located a horizontal distance y from the entry tangent point P and at a vertical offset distance x above the entry gradient.

From equation (14.1)

$$A = (-2.40) - (+2.10) = -4.50$$

Since point Z has a larger through chainage value than point Q, it must lie beyond the intersection point as shown in Figure 14.13. With reference to this figure, the distance y from the entry tangent point P to point Z is given by

$$y = \frac{L_V}{2} + (2871.92 - 2815.63) = \frac{L_V}{2} + 56.29$$

Therefore

$$L_V = 2y - 112.58 \qquad (14.12)$$

Also, since x is the vertical offset distance from the entry gradient to point Z, then

$$x = RL_Z - RL_Q + \frac{m(y - L_V/2)}{100}$$

$$= 113.56 - 110.98 + \frac{(2.40 \times 56.29)}{100} = 3.931 \text{ m}$$

But, from equation (14.6)

$$x = \frac{Ay^2}{200L_V} = \frac{4.50y^2}{200L_V}$$

Therefore

$$3.931 = \frac{4.50y^2}{200L_V}$$

Substituting for L_V in equation (14.12) gives

$$3.931 = \frac{4.50y^2}{200(2y - 112.58)}$$

Therefore

$$y^2 - 349.42y + 19668.98 = 0$$

Solving this quadratic equation gives

$$y = \frac{+349.42 \pm \sqrt{[349.42^2 - 4(1)(19,668.98)]}}{2(1)}$$

From which

$$y = 278.895 \text{ m or } 70.524 \text{ m}$$

Substitution into equation (14.12) gives

$$L_V = 445.21 \text{ m or } 28.47 \text{ m}$$

However, the minimum required length of vertical curve is given by equation (14.10) as

$$min\ L_V = KA = 26 \times 4.50 = 117.00 \text{ m}$$

Therefore $L_V = \textbf{445.21 m}$ will meet the requirements.

Reflective summary

With reference to vertical curve worked examples remember:

— It is always advisable to sketch the gradients and the curve and to add all the known data to the sketch before beginning the calculations.

— The length of vertical curve calculated from the equation $L_V = KA$ is a minimum one that can be increased as required to fit site conditions and other design criteria.

— If the reduced level of the intersection point Q is known, the reduced levels of the entry tangent point P and the exit tangent point R are obtained by applying the values of $mL_V/200$ and $nL_V/200$, respectively, to the reduced level of Q. It will be obvious from the sketch whether these values should be added or subtracted.

— When calculating reduced levels of points on the curve relative to the entry tangent point P, the signs of m and A in the formula

$$\left[\frac{(m)y}{100} - \frac{(A)y^2}{200L_V}\right]$$

should be retained in the calculations.

— You should tabulate the calculations wherever possible so that they can be easily followed for checking purposes.

— When designing curves to have unequal tangent lengths it is necessary to introduce a third gradient which is parallel to the line joining the entry and exit tangent points P and R. This creates two consecutive vertical curves, one equal in length to the entry tangent length PQ and one equal in length to the exit tangent length QR.

— In cases where the curve has to pass through a point of known change and reduced level, it is necessary to derive a quadratic equation and then compare the two solutions obtained to the minimum required length of the curve. Only one of the solutions will be acceptable.

Exercises

14.1 A parabolic vertical curve is to connect a +3.1% gradient to a −2.3% gradient on a single carriageway road having a design speed of 70 kph. With reference to the current UK Department of Transport design standards, calculate the minimum required length of curve if:

(i) The curve is to be designed for overtaking

(ii) The curve is to be designed for stopping only

14.2 A parabolic vertical curve is to connect a −2.2% gradient to a +1.9% gradient on a road having a design speed of 85 kph. Using the current UK Department of Transport design standards, calculate the minimum required length of the vertical curve.

14.3 With reference to the current UK Department of Transport design standards, discuss the options available for the minimum length of a vertical curve to be used to join a rising gradient of 2.9% to a falling gradient of 3.6% on a road designed for a speed of 85 kph, if the road is to be

(i) A dual carriageway

(ii) A two-lane single carriageway

14.4 A falling gradient of 2.5% meets a rising gradient of 3.2% at a reduced level of 235.60 m and a through chainage of 1172.45 m. A parabolic vertical curve having equal tangent lengths is to be used to connect the gradients and the K Value for the curve is 26. Calculate

(i) The reduced levels of the tangents points

(ii) The reduced levels of points on the curve at exact 30 m multiples of through chainage

(iii) The through chainage and the reduced level of the lowest point on the curve

14.5 On a road liable to flooding, a falling gradient of 3.09% meets a rising gradient of 2.87% at a reduced level of 3.16 m. A parabolic vertical curve having equal tangent lengths is to be used to connect the two gradients, which meet at a through chainage of 2619.52 m. A K Value of 20 was used to obtain the curve length.

Calculate the through chainage and the reduced level of the point at which a drainage gully should be located in order that it will be the most effective.

14.6 A parabolic vertical curve having equal tangent lengths is to be used to connect an upward gradient of 1.3% to a downward gradient of 1.5%. The reduced level of the entry tangent point on the 1.3% gradient is 189.46 m above datum. In order that the alignment is phased correctly, the tangent points for the vertical curve must coincide exactly with those of a horizontal curve of length 176.54 m. The K Value for the road is 55 and the through chainage of the entry tangent point is 2163.49 m.

(i) Show that the curve will meet the design criteria

(ii) Calculate the reduced level of the tangent point on the 1.5% downward gradient

(iii) Tabulate the reduced levels of points on the curve at exact 50 m multiples of through chainage

(iv) Calculate the through chainage and reduced level of the highest point on the curve

14.7 On a section of a proposed road, a falling gradient of 3.2% meets a rising gradient of 2.4% which in turn meets a falling gradient of 2.8%. The through chainages of the first and second intersection points are 3452.79 m and 3568.41 m, respectively, and the reduced level of the first intersection point is 229.86 m. The gradients are to be joined by two equal tangent length vertical curves consisting of a sag curve, AB, and a crest curve, BC, of the same length. Point B defines the end of the first curve and the start of the second.

Calculate

(i)   The through chainages of the tangent points of the two vertical curves

(ii) The reduced levels of the tangent points of the two vertical curves

(iii) The RLs of points on the sag curve AB at exact 25 m multiples of through chainage

(iv) The through chainage and reduced level of the highest point on the crest curve BC

14.8 A parabolic vertical curve having equal tangent lengths is to connect a rising gradient of 2.30% to a falling gradient of 1.80% on a road designed for a speed of 100 kph. The length of curve used must be at least enough to ensure that the minimum K Value is 100.

In order to ensure that there will be sufficient clearance over a culvert the curve must pass through a point Z of through chainage 1963.22 m and reduced level 113.95 m. The reduced level and the through chainage of the point of intersection of the two gradients are 117.27 m and 1938.67 m, respectively.

Calculate the length of the vertical curve that will meet these requirements.

14.9 A parabolic vertical curve having equal tangent lengths is to connect a rising gradient of 0.9% to a falling gradient of 1.2% on a road designed for a speed of 85 kph. The curve is to be designed for overtaking and its length must be at least sufficient to ensure that the FOSD Overtaking Crest K Value of 285 applies.

Particular site conditions dictate that the curve must pass through a point, which has a through chainage of 2634.21 m and a reduced level of 89.43 m. The reduced level and the through change of the intersection point of the two gradients are 92.68 m and 2698.37 m, respectively.

Calculate the length of vertical curve that will meet these requirements.

14.10 A parabolic vertical curve having equal tangent lengths is to connect a falling gradient of 2.1% to a rising gradient of 1.9% on a road designed for a speed of 100 kph. The through chainage and reduced level of the inter-section point of the two gradients are 1257.18 m and 102.23 m, respectively.

In order to ensure that there will be sufficient clearance under a bridge, the curve must pass through a point B of through chainage 1174.82 m and reduced level 103.59 m. The length of curve used must be enough to ensure that the K Value that applies has a value of at least 26.

Calculate the length of the vertical curve that will meet these requirements.

14.11 A parabolic vertical curve is to connect a +2.3% gradient to a −1.5% gradient on a highway designed for a speed of 85 kph. The K Value for the highway is 55 and the minimum required length is to be used.

The reduced level and the through chainage of the intersection point of the gradients are 94.18 m and 889.82 m, respectively, and in order to meet site conditions, the vertical curve is to have unequal tangent lengths such that the through chainage of the exit tangent point must be 976.54 m.

Calculate:

(i) The reduced levels of the tangent points

(ii) The reduced levels along the curve at exact 30 m multiples of through chainage

(iii) The through chainage and reduced levels of the highest point on the curve

Further reading and sources of information

For the latest information on the design of vertical curves, consult

Design Manual for Roads and Bridges, Volume 6 Road Geometry, Section 1 Links, Part 1 TD9/93 – incorporating Amendment No. 1 dated Feb 2002 – Highway Link Design. Jointly published by the Overseeing Organisations of England, Scotland Wales and Northern Ireland, that is, The Highways Agency, the Scottish Executive Development Department, The National Assembly for Wales (Cynulliad Cenedlaethol Cymru) and The Department for Regional Development Northern Ireland. The manual can be accessed and printed from the Internet at: http://www.official-documents.co.uk/document/deps/ha/dmrb/index.htm. It is also available from The Stationery Office at http://www.tso.co.uk/.

chapter 15

Earthwork quantities

 Aims

After studying this chapter you should be able to:

- Perform the calculations necessary to calculate the areas of parcels of land, which are enclosed by straight lines, irregular boundaries and a combination of both

- Collect the data required for and know how to produce longitudinal sections and cross-sections

- Calculate the areas of cross-sections and use them to estimate the volumes of embankments and cuttings in highway construction projects

- Use a regular grid of spot heights taken on the ground surface to estimate the volumes of material to be excavated when constructing foundations

- Estimate the volume of water contained within a proposed reservoir using contours

- Draw mass haul diagrams and use them to help plan the optimum movement of material during the construction of a new highway

- Appreciate the use of computer software packages in the determination of earthwork quantities

This chapter contains the following sections:

15.1 Types of earthwork quantities

15.2 Calculation of plan areas

15.3 Longitudinal sections and cross-sections

15.4 Calculating volumes from cross-sections

15.5 Calculating volumes from spot height and contours

15.6 Mass haul diagrams

15.1 Types of earthwork quantities

After studying this section, you should be aware of the various different types of quantities used in earthworks and the reasons why they have to be calculated. You should know the units that are used to represent these quantities and whether exact or estimated values can be obtained. In addition, you should understand why the movement of earthworks on a construction project needs to be carefully planned.

Earthwork quantities

In many engineering projects, large parcels of land are required for the site and huge amounts of material have to be moved in order to form the necessary embankments, cuttings, foundations, basements, lakes and so on, that have been specified in the design. Suitable land and materials can be very expensive and, if a project is to be profitable to the construction company involved, it is essential that its engineers make as accurate a measurement as possible of the earthwork quantities involved, that is the *areas of land* the project will occupy and the *volumes of material* required in its construction. The initial purchase of the land areas and the subsequent movement of the earthwork volumes can often form a considerable proportion of the overall cost of the project. Only by making accurate measurements of such quantities can appropriate estimates of the costs involved be obtained for inclusion in the tender documents. These are prepared during the planning of a project and include all the items relating to the construction of the proposed Works, including drawings, calculations, quantities and specifications. They are discussed further in Section 11.1.

Parcels of land are generally bounded either by straight sides, irregular sides or some combination of both. For these it is usual to calculate their *plan areas*, that is, the areas contained within their boundaries as they would appear on engineering plans ignoring any undulations in their ground surfaces. In the case of straight-sided areas it is possible to calculate very accurate measurements of their plan areas, whereas in the case of irregular-sided areas only estimated values can be obtained.

Cross-sections are often drawn to help with the volume calculations required for highway construction projects. They can take a number of different forms and they are normally based on *longitudinal sections*. When drawing cross-sections it is usual to approximate the shape of the ground surface to a series of straight lines. Hence, although accurate measurements of the *cross-sectional areas* contained within these straight sided boundaries can be made, the values obtained are only estimates of the true cross-sectional areas due to the approximation of the ground surface.

The unit used to represent plan areas and cross-sectional areas is usually that adopted in the *Système International* (SI), which is the *square metre* (m^2). However, when using square metres the figures involved can become very big if the area concerned is very large so other units are sometimes used. These are *ares*, *hectares* and *square kilometres* and they are related as follows:

$$100 \ m^2 \quad = \ 1 \ are$$
$$100 \ ares \quad = \ 1 \ hectare \qquad = \ 10 \ 000 \ m^2$$
$$100 \ hectares \ = \ 1 \ square \ kilometre \ = \ 10^6 \ m^2$$

Volumes of materials can be calculated in a number of ways, depending on the project concerned. The three major methods involve the use of cross-sections, spot heights and contours. Normally, only estimated values can be obtained for volumes due to the approximations that are usually made to the shape of the ground surface during their calculation. The unit used to represent volume is the *cubic metre* (m^3), which is used for all volumes, no matter how large or small. For certain projects, such as the construction of a new highway, where large volumes of material have to be excavated and moved around the site, careful planning of this movement is essential since charges may be levied not only on the volumes but also on the distances over which they are moved. One device for optimising the movement of such material is a *mass haul diagram*, which enables the most cost-effective use of earth-moving machinery to be planned in advance.

Reflective summary

With reference to the types of earthwork quantities remember:

— Accurate values of areas and volumes are required to enable the cost of a project to be determined as precisely as possible.

— Straight-sided areas can be determined exactly but only estimates can be made of irregular-sided ones.

— The standard unit of area is the square metre but units such as ares, hectares and square kilometres are sometimes used.

— It is unusual for exact volumes to be obtained due to the undulating nature of the ground surface.

— The standard unit of volume is the cubic metre.

— Careful planning of the movement of material on site can minimise the costs involved.

15.2 Calculation of plan areas

After studying this section, you should be aware of some of the common methods by which plans areas bounded by straight sides, irregular sides and a combination of both can be calculated. You should know the relative accuracies of each method and which is the most appropriate in certain situations.

This section includes the following topics:

● Areas enclosed by straight lines

● Areas enclosed by irregular lines

● Planimeters

Areas enclosed by straight lines

Into this category fall areas enclosed by traverse, network or detail survey lines. The results obtained for such areas will be exact since correct geometric equations and theorems can be applied.

Areas from triangles

Any straight-sided figure can be divided into well-conditioned triangles, the areas of which can be calculated using one of the following formulae

● $\text{Area} = \sqrt{[S(S-a)(S-b)(S-c)]}$ (15.1)

    where a, b and c are the lengths of the sides of the triangle and $S = (a + b + c)/2$

● $\text{Area} = \dfrac{(\text{base of triangle} \times \text{height of triangle})}{2}$ (15.2)

● $\text{Area} = \dfrac{ab \sin C}{2}$ (15.3)

    where C is the angle contained between side lengths a and b

The area of any straight-sided figure can be calculated by splitting it into triangles and summing the individual areas.

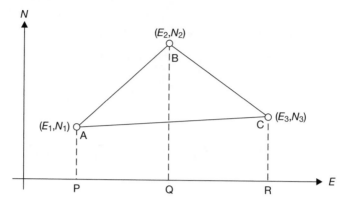

Figure 15.1 ● Cross coordinate method.

Areas from coordinates

In traverse and network calculations, the coordinates of the junctions of the sides of a straight-sided figure are calculated and it is possible to use them to calculate the area enclosed by the control network lines. This is achieved using the *cross coordinate method*.

Consider Figure 15.1, which shows a three-sided clockwise control network ABC in which the required area = ABC.

$$\text{area of ABC} = \text{area of ABQP} + \text{area of BCRQ} - \text{area of ACRP} \qquad (15.4)$$

These figures are trapezia for which the area is obtained from

$$\text{area of trapezium} = (\text{mean height} \times \text{width})$$

Therefore

$$\text{area of ABQP} = \frac{(N_1 + N_2)(E_2 - E_1)}{2}$$

Using similar expressions for areas BCRQ and ACRP, equation (15.4) becomes

$$\text{area ABC} = \frac{(N_1 + N_2)(E_2 - E_1)}{2} + \frac{(N_2 + N_3)(E_3 - E_2)}{2} - \frac{(N_1 + N_3)(E_3 - E_1)}{2}$$

and

$$2 \times \text{area ABC} = N_1 E_2 - N_1 E_1 + N_2 E_2 - N_2 E_1 + N_2 E_3 - N_2 E_2$$
$$+ N_3 E_3 - N_3 E_2 - N_1 E_3 + N_1 E_1 - N_3 E_3 + N_3 E_1$$

Rearranging this gives

$$2 \times \text{area ABC} = (N_1 E_2 + N_2 E_3 + N_3 E_1) - (E_1 N_2 + E_2 N_3 + E_3 N_1)$$

The similarity between the two brackets should be noted.

Although the example given is only for a three-sided figure, the formula can be applied to a figure containing N sides and the general formula for such a case is given by

$$2 \times \text{area} = (N_1E_2 + N_2E_3 + N_3E_4 + \ldots + N_{N-1}E_N + N_NE_1)$$
$$- (E_1N_2 + E_2N_3 + E_3N_4 + \ldots + E_{N-1}N_N + E_NN_1) \quad (15.5)$$

If the figure is numbered in the opposite direction, the signs of the two brackets are reversed.

The cross coordinate method can be used to subdivide straight-sided areas as shown in Worked example 15.1 below and can also be used to calculate the area of irregular cross-sections as discussed in Section 15.4.

Worked example 15.1: The division of an area using the cross coordinate method.

Question

The polygon traverse PQRSTP shown in Figure 15.2 is to be divided into two equal areas by a straight line that must pass through point R and which meets line TP at Z. The coordinates of the points are given in Table 15.1. Calculate the coordinates of point Z.

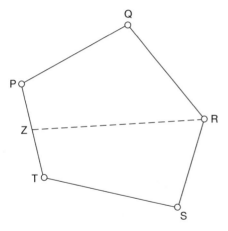

Table 15.1

Point	mE	mN
P	613.26	418.11
Q	806.71	523.16
R	942.17	366.84
S	901.89	203.18
T	652.08	259.26

Figure 15.2 ● Division of an area.

Solution

The traverse is lettered and specified in a clockwise direction. Using the clockwise version of the cross coordinate method from equation (15.5) gives

$$2 \times \text{area PQRSTP} = (1{,}452{,}532 - 1{,}314{,}662) = 68{,}935 \text{ m}^2$$

Therefore

$$\text{area PQRZP} = \text{area ZRSTZ} = \left(\frac{68{,}935}{2}\right) = 34{,}467.5 \text{ m}^2$$

Let point X have coordinates (E_Z, N_Z). Applying the clockwise version of the cross coordinate method to area PQRZP gives

$$68{,}935 = 213{,}432.58 - 51.27E_Z - 328.91N_Z$$

From which

$$E_Z = 2818.365 - 6.41525N_Z \qquad (15.6)$$

A similar application to area ZRSTZ gives

$$E_Z = 2.69651N_Z - 286.765 \qquad (15.7)$$

Solving equations (15.6) and (15.7) gives

$$\mathbf{E_Z = 632.16 \ m, \ N_Z = 340.78 \ m}$$

As a check, since Z lies on the line PT

$$\frac{(E_T - E_P)}{(N_T - N_P)} \ \text{should equal} \ \frac{(E_Z - E_P)}{(N_Z - N_P)}$$

Substituting the coordinates of P, T and Z gives

$$-0.2444 = -0.2444$$

which checks the coordinates of Z as calculated above.

Areas enclosed by irregular lines

For such cases only approximate results can be achieved. However, methods are adopted which will give the best approximations.

Give and take lines

In this method an irregular-sided figure is divided into triangles or trapezia, the irregular boundaries being replaced by straight lines such that any small areas excluded from the survey by the lines are balanced by other small areas outside the survey but included as shown in Figure 15.3.

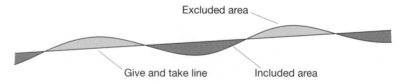

Figure 15.3 ● Give and take line.

The positions of these lines can be estimated by eye on a survey plan. The area is then calculated using one of the straight-sided methods.

Graphical method

This method involves the use of a transparent overlay of squared paper which is laid over the drawing or plan. The number of squares and parts of squares which are enclosed by the area is counted and, knowing the plan scale, the area represented by each square is known and the total area can be computed. This can be a very accurate method if a small grid is used.

Trapezoidal rule and Simpson's rule

These two methods make a mathematical attempt to calculate the area of an irregular-sided figure.

Figure 15.4 shows a control network contained inside an area having irregular sides. The shaded area is that remaining to be calculated after using one of the straight-sided methods to calculate the area enclosed by the control network lines.

Figure 15.5 shows an enlargement of Figure 15.4 along line AB. The offsets O_1, $O_2, O_3, ..., O_8$ are either measured directly in the field or can be scaled from a plan.

The *trapezoidal rule* assumes that if the interval L between the offsets is small, the boundary can be approximated by a series of straight lines as shown in Figure 15.6. This forms a series of trapezia, the areas of which are given by

$$A_1 = \left(\frac{L}{2}\right)(O_1 + O_2) \qquad A_2 = \left(\frac{L}{2}\right)(O_2 + O_3)$$

and so on

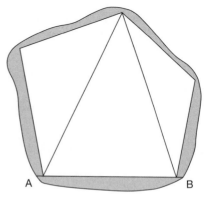

Figure 15.4

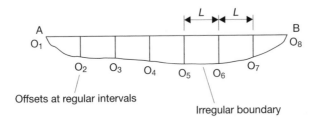

Figure 15.5

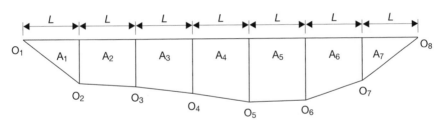

Figure 15.6 ● Trapezoidal rule.

For N offsets, the total area A is given by

$$A = \left(\frac{L}{2}\right)(O_1 + O_2) + \left(\frac{L}{2}\right)(O_2 + O_3) + \ldots + \left(\frac{L}{2}\right)(O_{N-1} + O_N)$$

which leads to the general trapezoidal rule shown below

$$A = \left(\frac{L}{2}\right)[O_1 + O_N + 2(O_2 + O_3 + O_4 + \ldots + O_{N-1}] \tag{15.8}$$

Putting this into words gives

$$\text{Total area} = \frac{L}{2} \text{ (first offset + last offset + twice all the other offsets)}$$

The trapezoidal rule applies to any number of offsets and Worked example 15.2 below shows how it is applied.

Worked example 15.2: Trapezoidal rule

Question

The following offsets, 8 m apart, were measured at right angles from a traverse line to an irregular boundary.

0 m   2.3 m   5.5 m   7.9 m   8.6 m   6.9 m   7.3 m   6.2 m   3.1 m   0 m

Calculate the area between the traverse line and the irregular boundary using the trapezoidal rule.

Solution

From equation (15.8)

$$\text{Area} = \left(\frac{8.0}{2}\right)[0 + 0 + 2(2.3 + 5.5 + 7.9 + 8.6 + 6.9 + 7.3 + 6.2 + 3.1)]$$

$$= 4 \times 2(47.8) = \mathbf{382.4 \ m^2}$$

Simpson's rule assumes that instead of being made up of a series of straight lines, the boundary consists of a series of parabolic arcs as shown in Figure 15.7. A more accurate result is obtained using this rule since a better approximation of the true shape of the irregular boundary is achieved.

Simpson's rule considers offsets in sets of three and it can be shown that the area between offset 1 and 3 is given by

$$A_1 + A_2 = \frac{L}{3}(O_1 + 4O_2 + O_3)$$

Similarly

$$A_3 + A_4 = \frac{L}{3}(O_3 + 4O_4 + O_5)$$

In general

$$\text{Total area} = \frac{L}{3}(O_1 + O_N + 4\Sigma \text{ even offsets} + 2\Sigma \text{ remaining odd offsets}) \quad (15.9)$$

Putting this into words gives

$$\text{Total area} = \frac{L}{3}(\text{first offset} + \text{last offset} + \text{four times the even offsets}$$

$$+ \text{ twice the remaining odd offsets})$$

However, the number of offsets **must** be an **odd** number for Simpson's rule to apply. When applying Simpson's rule to an even number of offsets, as in Figure 15.7, the first or last offset must be omitted, the rest of the area calculated using Simpson's rule and the first or last small area calculated as a trapezium using the trapezoidal rule.

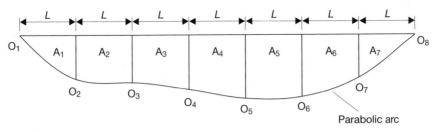

Figure 15.7 ● Simpson's rule.

Worked example 15.3 below shows how Simpsons's rule is applied.

Worked example 15.3: Simpson's rule

Question

Using the data given in Worked example 15.2, calculate the area between the traverse line and the irregular boundary using Simpson's rule wherever possible.

Solution

Since there is an even number of offsets, one solution is to calculate the area between 1 and 9 by Simpson's rule and the areas between 9 and 10 by the trapezoidal rule.

From equation (15.9)

$$\text{Area}_{1-9} = \left(\frac{8.0}{3}\right)[0 + 3.1 + 4(2.3 + 7.9 + 6.9 + 6.2) + 2(5.5 + 8.6 + 7.3)]$$

$$= \left(\frac{8.0}{3}\right)[3.1 + 4(23.3) + 2(1.4)]$$

$$= \left(\frac{8.0}{3}\right) \times 139.1 = 370.9 \text{ m}^2$$

From equation (15.8)

$$\text{Area}_{9-10} = \left(\frac{8.0}{2}\right)[3.1 + 0] = 12.4 \text{ m}^2$$

Therefore

Total Area = 370.9 + 12.4 = **383.3 m²**

Note the difference between this result and the value of 382.4 m² obtained using the trapezoidal rule in Worked example 15.2. Had Simpson's rule been applied between offsets 2 and 10 and the trapezoidal rule between offsets 1 and 2, a third slightly different result would have been obtained. This variation in answers illustrates that it is impossible to get an exact solution when measuring areas bounded by irregular sides.

Planimeters

Planimeters are instruments which automatically measure the area of any irregular-sided figure drawn on a map or a sheet of paper. Traditionally, *mechanical* devices were used, but these have now been superseded by *digital* instruments.

A high degree of accuracy can be achieved with a digital planimeter no matter how complex the shape of the area being measured. It consists of a magnifying eyepiece at the end of a tracing arm, which is attached to a small computer having a numeric keypad, an operations panel and a liquid crystal display as shown in Figure 15.8. Different models are available with different lengths of tracing arm to accommodate different paper sizes.

When being used to measure areas, their method of operation is very straightforward. The map or plan containing the area to be measured is placed on a smooth surface and the planimeter is placed on top such that the index mark on its magnifying eyepiece can be moved completely around the circumference of the area. The planimeter is activated and the scale of the map or plan is input to the on-board computer. The index mark is placed over a specific point on the circumference of the area and the planimeter reading is set to zero. The operator then carefully follows the circumference using the index mark returning to the original starting point and the area contained within the circumference is displayed automatically on the planimeter's display.

Figure 15.8 ● X-plan 460C III digital planimeter with printer (courtesy Ushikata Mfg. Co. Ltd).

Although originally designed for the measurement of areas, modern digital planimeters are very sophisticated instruments offering a wide range of facilities. These include the ability to measure the length of a line, the circumference of an area, the length of any segment or arc within a circumference, the radius of the arcs, the total length along a boundary, the rectangular coordinates of points based either on an arbitrary or an existing grid system, the position of the centroid of an area, the volumes of solids generated by rotating a section around a straight line and volumes contained between contour lines. In addition, some models can accommodate a detachable mini-printer (see Figure 15.8) and most have an internal memory in which data can be stored prior to downloading into a computer.

As well as being ideal for measuring plan areas, planimeters are extremely useful for measuring cross-sectional areas. They are often used for this purpose in preference to one of the special cross-sectional area formulae that are discussed in the following section and Worked example 15.4 below shows how a planimeter can be used to measure the area of a cross-section. They can also be used to help with mass haul calculations as shown in Worked example 15.7.

Worked example 15.4: Measuring an irregular cross-sectional area using a planimeter

Question

An irregular cross-sectional area was measured from a drawing using a digital planimeter, which gave readings directly in mm². The initial planimeter reading was set to zero and the final reading was 7362. If the horizontal scale of the cross-section was 1 in 200 and the vertical scale 1 in 100, calculate the true area represented by the cross-section.

Solution

The difference between the initial and final planimeter readings = 7362 mm². However, since the horizontal and vertical scales of the cross-section are 1 in 200 and 1 in 100, respectively, then on the drawing

$$1 \text{ mm}^2 \text{ represents an area of } 200 \text{ mm} \times 100 \text{ mm}$$

Hence

$$7362 \text{ mm}^2 = (7362 \times 200 \times 100) \text{ mm}^2$$
$$= \frac{(7362 \times 200 \times 100)}{(1000 \times 1000)} \text{ m}^2$$
$$= \mathbf{147.24 \text{ m}^2}$$

Reflective summary

With reference to the calculation of plan areas remember:

- The area of any straight-sided figure can be calculated by splitting it into triangles and summing the individual areas.

- If the coordinates of the points in a straight-sided figure are known, its area can be determined using the cross-coordinate method.

- Straight-sided areas can be determined exactly, but only estimates can be made of irregular-sided ones.

- The graphical method of measuring irregular-sided areas using a transparent square-grid overlay can be very accurate if small grid squares are used.

- Simpson's rule should be used if the irregular boundary is curved, whereas the trapezoidal rule should be used where the boundary is a series of straight lines.

- Digital planimeters can be used to determine the area of any closed shape drawn on a plan, no matter how complex its boundary, providing the scale of the plan is known.

15.3 Longitudinal sections and cross-sections

After studying this section, you should know what sectioning is and how longitudinal and cross-sections can be produced either from new data measured directly on site or from data collected from existing maps, aerial photographs and Digital Terrain Models. You should understand how longitudinal and cross-sections are used to represent the shape of the ground surface not only before any construction work has begun but also after the design stage has been completed.

This section includes the following topics:

- Sectioning
- Site methods of sectioning
- Office methods of sectioning

Sectioning

Before long narrow constructions such as roads can be designed, an accurate representation of the ground surface over which they are to be constructed must be

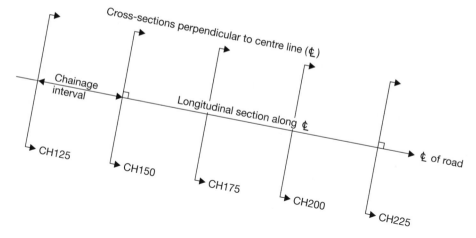

Figure 15.9 ● Longitudinal and cross-section layout.

obtained. One method of doing this is by *sectioning*, which involves two types of section: *longitudinal sections* and *cross-sections*. These are generally interdependent and Figure 15.9 shows their typical plan arrangement for a proposed road alignment.

A *longitudinal section* shows the shape or *profile* of the existing ground surface along the proposed centre line of the construction. It is produced by plotting distances to points along the centre line against their heights and it is prepared when determining the most economic formation level for a new construction, for example, a proposed road. *Formation level* is the term used to describe the level specified by the designer to which the existing ground surface is to be constructed or *formed*. In any construction work the aim is always to ensure that the project is constructed at its correct formation level. This is discussed further in Chapter 11.

Although it can be drawn by hand, a longitudinal section is usually produced nowadays with the aid of a computer-aided design package, which can also be used to determine the optimum position for the formation level. Longitudinal sections are also used in conjunction with mass haul diagrams to plan the movement of earth on a project as discussed in Section 15.6.

A longitudinal section defines the existing ground surface along a proposed centre line very precisely, but it gives no indication of the shape of the ground surface on either side of the centre line. While this may be sufficient for planning works such as sewers and pipelines, which only require narrow trenches, it is not sufficient for roads, where embankments and cuttings are invariably required. Hence additional surface information is normally required on each side of the proposed centre line and this is done using *cross-sections*, which indicate the shape of the existing ground surface perpendicular to the proposed centre line. They are produced by plotting distances from the centre line pegs to points at right angles to the centre line against their heights. As with longitudinal sections, although they can be drawn by hand, they are nowadays usually produced using computer-aided design packages for use in the design of embankments and cuttings for projects such as a new highway construction.

The data needed for sectioning can be obtained either directly on *site* using methods involving levelling, total stations or GPS surveys, or indirectly in an *office* using topographic maps, photogrammetric methods or computers interrogating Digital Terrain Models. In all cases, the first stage is to locate the position of the centre line so that levels can be taken along it at regular distance intervals. Regular distance intervals are normally required to help with subsequent volume calculations as discussed in Section 15.4.

Site methods of sectioning

Sections can be measured using techniques discussed in Chapter 2 (levelling), Chapter 5 (total stations) and Chapter 7 (GPS).

Sectioning using levelling techniques

In this method, the centre line of the proposed construction must first be physically defined on the existing ground surface. This is normally done using wooden pegs placed at regular horizontal distance (through chainage) intervals (typically 10 m, 20 m or 50 m) and techniques by which this can be achieved are discussed in Chapter 11 (setting out), Chapter 12 (circular curves) and Chapter 13 (transition curves). The *through chainage* of a peg on a road centre line is the horizontal distance along the centre line from the starting point of the road to the peg in question. Often the term through chainage is simply shortened to *chainage* and the starting point of the road is referred to as the position of *zero chainage*. Further information on this is given in Section 12.3.

Once the centre line has been defined, lines are established at right angles to this to define the positions of the cross-sections. Heights are then measured along the centre line and the lines at right angles to provide the height data for the longitudinal section and the cross-sections. Levelling techniques are discussed in detail in Chapter 2 and sectioning represents just one of their applications. If digital levelling equipment is being used then the readings will be recorded automatically. However, if optical levelling equipment is being used then it is usually sufficient to record readings to the nearest 0.01 m since the ground surface on which the levels are being taken is normally not so well defined to justify taking readings to millimetres.

For *longitudinal sections*, levels should be taken at the following points, the purpose being to survey the existing ground profile as accurately as possible in the time available.

- At the top and ground level of each centre line peg, noting the through chainage of each peg.
- At points on the centre line where the ground slope changes.
- Where features cross the centre line, such as fences, hedges, roads, pavements, ditches and so on. At points where roads or pavements cross the centre line, levels should be taken at the top and bottom of kerbs. At ditches and streams, the levels at the top and bottom of any banks as well as bed levels are required.

- Where necessary, the levels of underpasses and bridge soffits would be taken by levelling using an inverted staff.

When using optical levelling techniques for sectioning, it is advisable to reduce the levels using the *height of collimation method* since many intermediate sights will usually be required and this method is quicker to perform.

In order to be able to plot the longitudinal section, the chainage of each point where a level has been taken on the centre line is also required. Chainages will be known for the centre line pegs fixed at regular intervals on the centre line, but not for any additional points fixed between them. The most appropriate method of determining the chainages of these additional points is to measure the distances to them using a tape held horizontally between adjacent centre line pegs. The chainage and level of each point should be recorded simultaneously to avoid confusion at the plotting stage.

For *cross-sections*, the height data is obtained by taking levels along lines laid out at right angles to the centre line. Levels are taken on both sides of the centre line and, depending on the shape of the existing ground, they are taken either at regular distance intervals or at specific points where the slope of the ground surface changes. The purpose is again to obtain as true a representation of the ground profile as possible in the time available. For the best possible accuracy, a cross-section should be taken an every point where a level has been fixed on the centre line. However, since this would involve a considerable amount of fieldwork, this is generally not done and cross-sections are instead taken at regular intervals along the centre line, usually where the centre line pegs have been established. A right angle is set out at each peg either by eye for short lengths or by theodolite or total station for long distances or where greater accuracy is required. The line along which a cross-section is to be taken can be defined on site either using ranging rods or by laying a tape or string line along the ground surface. A tape is recommended, since it can also serve to measure the distances from the centre line to the points where levels are taken.

In practice, data for the longitudinal and cross-sections are usually collected at the same time. Starting at a TBM or OSBM, levels are taken at each centre line peg and at intervals along the cross-sections. These intervals may be regular on either side of the centre line peg or, where the ground is irregular, levels should be taken at all changes of slope such that a good representation of existing ground level is obtained over the full width of the proposed construction. The process is continued taking both longitudinal and cross-section levels in the one operation and the levelling is finally closed on another bench mark.

Such a line of levels can be very long and can involve many staff readings. Hence it is easy to make mistakes and it is possible for errors to occur at any stage in the process. The result is that if a large misclosure is found, all the levelling will have to be repeated, often a soul-destroying task. Consequently, great care is required and the following tips should be followed:

- It is good practice to include TBMs or OSBMs at regular intervals in the line of levels so that, if a large discrepancy is found, it can be isolated into a short stretch of the work.

- Be consistent in the way the data is collected, for example, always work along each cross-section in the same direction, and always work in the direction of increasing chainage when observing the points on the longitudinal section. Observe one cross-section at a time and do not mix them up.

- If using individual level sheets, record the levels, chainages and distances carefully, and make good use of the *remarks* column on the sheets. Most commercially available level books have a specific column in which chainages (distances) can be recorded. This is illustrated in Figures 15.10(a) and 15.11(a), which show level books for a longitudinal section and a cross-section, respectively.

- Prepare sketches of the sections on which to record relevant information – these can be particularly useful when complex sections are being measured.

When all the levelling fieldwork has been completed, the longitudinal section and the cross-sections can be plotted using the chainages, levels and distances measured on site. They are basically graphs of level against distance and examples of such sections obtained by levelling are shown in Figures 15.10(b) and 15.11(b). For *longitudinal sections* the origin of the graph is at the left-hand side with levels being plotted vertically and chainages horizontally. For *cross-sections* the point where the centre line meets the cross-section is taken as the origin. Distances from the centre line to points where the levels were taken are plotted horizontally to the left and right of the origin with the levels being plotted vertically. The points defining the levels are then joined together with straight lines to form the sections. Initially, the existing ground profile is plotted, but later, when the new horizontal and vertical alignments have been designed, the proposed formation levels are added as shown in Figures 15.12 and 15.13. Both the longitudinal section and the cross-sections are usually drawn with their horizontal and vertical scales at different values. For example

Scales for the longitudinal section

| Horizontal | same as the road layout drawings, for example, 1 in 500 |
| Vertical | exaggerated, for example 1 in 100 |

Scales for the cross-sections

| Horizontal | exaggerated, for example 1 in 200 |
| Vertical | exaggerated, for example 1 in 50 |

The reason for exaggerating the vertical scales in both sections and the horizontal scale of the cross-sections is to give a clear picture of the exact shapes of the ground surface along the sections. If the cross-sections have different horizontal and vertical scales then it is still possible to calculate their areas either by the graphical method discussed in Section 15.2 or by using a planimeter as shown in Worked example 15.4.

Although the methods discussed above provide an illustration of how to survey longitudinal and cross-sections by levelling, the procedures described are seldom used nowadays because they are slow and can be very tedious when a lot of data is required. Whilst levelling may be used for small projects where the terrain is relatively flat, sectional information on site is now normally obtained by using either total stations or GPS techniques.

(a)

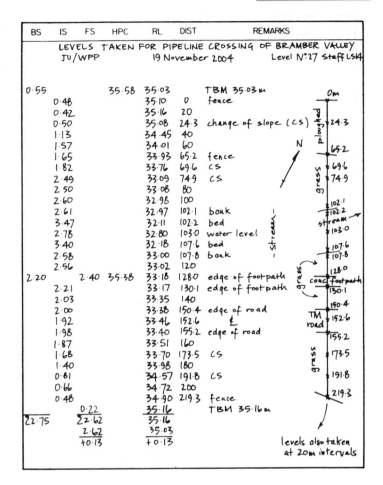

BS	IS	FS	HPC	RL	DIST	REMARKS
						LEVELS TAKEN FOR PIPELINE CROSSING OF BRAMBER VALLEY
						JU/WPP 19 November 2004 Level N°27 Staff LS14
0.55			35.58	35.03		TBM 35.03 m
	0.48			35.10	0	fence
	0.42			35.16	20	
	0.50			35.08	24.3	change of slope (CS)
	1.13			34.45	40	
	1.57			34.01	60	
	1.65			33.93	65.2	fence
	1.82			33.76	69.6	CS
	2.49			33.09	74.9	CS
	2.50			33.08	80	
	2.60			32.98	100	
	2.61			32.97	102.1	bank
	3.47			32.11	102.2	bed
	2.78			32.80	103.0	water level
	3.40			32.18	107.6	bed
	2.58			33.00	107.8	bank
	2.56			33.02	120	
2.20		2.40	35.38	33.18	128.0	edge of footpath
	2.21			33.17	130.1	edge of footpath
	2.03			33.35	140	
	2.00			33.38	150.4	edge of road
	1.92			33.46	152.6	₵
	1.98			33.40	155.2	edge of road
	1.87			33.51	160	
	1.68			33.70	173.5	CS
	1.40			33.98	180	
	0.81			34.57	191.8	CS
	0.66			34.72	200	
	0.48			34.90	219.3	fence
		0.22		35.16		TBM 35.16 m
Σ2.75		Σ2.62		35.16		
		2.62		35.03		
		+0.13		+0.13		

Remarks column notations: 0m, ploughed, 24.3, N, 65.2, grass, 69.6, 74.9, 102.1, 102.2, stream, 103.0, 107.6, 107.8, grass, conc footpath, 128.0, 130.1, 150.4, TM road, 152.6, 155.2, grass, 173.5, 191.8, 219.3.

levels also taken at 20 m intervals

LONGITUDINAL SECTION – BRAMBER VALLEY

(b)

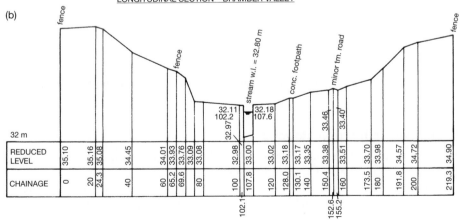

32 m																												
REDUCED LEVEL	35.10	35.16	35.08	34.45	34.01	33.93	33.76	33.09	33.08	32.98	33.00	33.02	33.18	33.17	33.35	33.38	33.51	33.70	33.98	34.57	34.72	34.90						
CHAINAGE	0	20	24.3	40	60	65.2	69.6	80	100	102.1	107.8	120	128.0	130.1	140	150.4	160	173.5	180	191.8	200	219.3						

stream w.l. = 32.80 m
32.11 / 102.2
32.18 / 107.6
32.97
32.98
33.46
33.40
152.6 / 155.2

Figure 15.10 ● Longitudinal section: (a) level book; (b) format for drawing section.

(a)

BS	IS	FS	HPC	RL	DIST	REMARKS

BADGER LANE – EXISTING LEVELS   CHAINAGE 420m
JU/WPP   26 November 2004   Level No.16 StaR LS B

1·32			150·47	149·15		TBM 149·15m ROADNAIL on ℄ AT
						CH 420
						LEFT
	1·43			149·04	2·3	channel
	1·26			149·21	2·3	kerb
	1·29			149·18	3·4	footpath edge
	1·13			149·34	4	⎫
	1·20			149·27	5	⎬ ground levels
	1·42			149·05	6	⎪
	1·68			148·79	7	⎪
	2·11			148·36	8	⎭
						RIGHT
	1·45			149·02	2·1	channel
	1·28			149·19	2·1	kerb
	1·32			149·15	3·2	footpath edge
	1·15			149·32	4	⎫
	1·10			149·37	5	⎬ ground levels
	1·01			149·46	6	⎪
	0·63			149·84	7	⎪
	0·46			150·01	8	⎭
		1·32		149·15		TBM 149·15m on ℄ at CH420

Note: LEFT and RIGHT viewed when
facing direction of increasing
chainage

(b)

BADGER LANE – EXISTING LEVELS AT CH 420 m

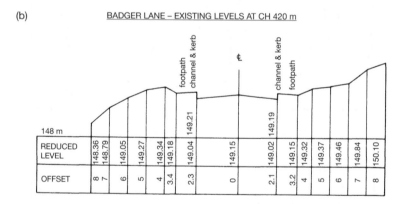

Figure 15.11 ● Cross-section drawing and associated levelling field book.

Sectioning using total stations

The techniques used for obtaining sectional information by levelling can be modified so that total stations could be used instead.

If the centre line has already been set out and the longitudinal section has been surveyed, the total station can be centred and levelled over a line peg, the height of the peg and instrument height entered into the instrument and a sighting taken to an adjacent line peg to define the centre line. By rotating the instrument through 90°, the line of the cross-section is defined exactly on a straight section of the centre line and

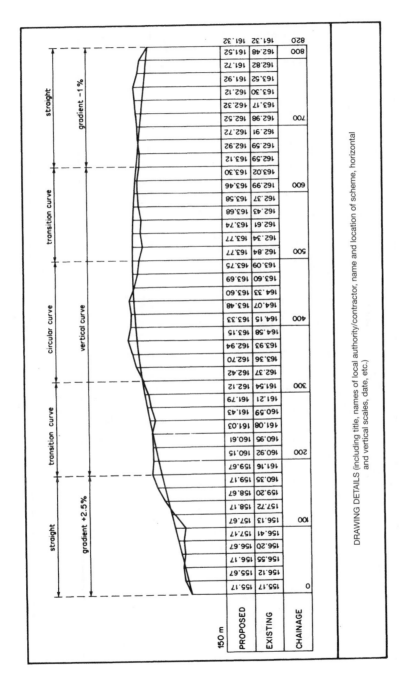

Figure 15.12 ● Example of a longitudinal section.

DRAWING DETAILS (including title, names of local authority/contractor, name and location of scheme, horizontal and vertical scales, date, etc.)

150 m			
PROPOSED			
EXISTING			
CHAINAGE			

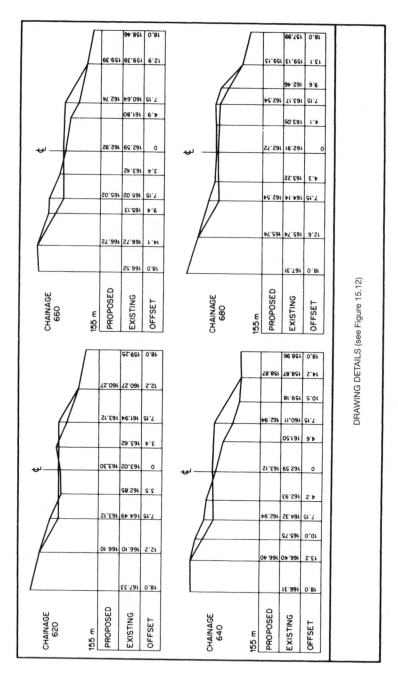

Figure 15.13 ● Example cross-sections.

with sufficient precision on a curved section. The detail pole is then moved along the line of sight of the total station and measurements are taken at changes of slope, where features cross the section and at regular offsets from the centre line as described for sectioning by levelling. When using a total station for this work, measurements are taken of distance and height along the cross-section and these are stored directly in the total station on data cards or in its data collector. A disadvantage of this method for sectioning is that it is time-consuming when a lot of sections have to be surveyed.

Measurement of sections can be made more efficient if the coordinates as well as the heights of each centre line peg are known. The total station is again set up over a line peg, the coordinates and height of which are entered into the instrument and it is orientated onto another line peg, whose coordinates have also been entered into the total station. Each cross-section within range of the total station is now measured. To do this, the person holding the detail pole estimates the position of each cross-section by eye and at each point sighted to define a section, the total station determines the coordinates and height using its coordinate measurement program. As many sections as possible are measured at this set up and to measure more sections, the total station is moved along the centre line to another line peg and the process repeated. As before, all data is recorded electronically. Because this method only gives the coordinates of points along each cross-section surveyed, this has to be processed by computer on site or at a later stage in the office to produce the required sections. Most of the major survey companies and producers of survey software have programs available for doing this. Although this method is much quicker at producing cross-sections than the first, it still requires the centre line to be set out and the coordinates and heights along this to be known.

The most efficient way of determining sectional data with a total station is to create a DTM of the area (DTMs are described in Section 10.9). The methods used for this are essentially a detail survey in which a series of control points are set up around the site from which measurements are taken to give the best representation of the ground surface along the proposed route covering the whole of the band of interest. This usually requires a regular grid of points to be surveyed as well as all changes of slope, and since large amounts of data can be obtained, all measurements are stored in the total station or a data collector. Like the previous method, this is an indirect method of determining sectional data, which are obtained later by processing the results in a computer. This technique has the advantage that it can determine both longitudinal and cross-sections before any setting out is done. Furthermore, the DTM can be interrogated as many times as desired to produce the optimum design for the earthworks.

Sectioning using GPS techniques

The techniques employed when using GPS to determine sectional data are similar to those used for detail surveys and mapping by GPS as described in Section 10.5. As with total stations, enough information is collected to form a reliable DTM of the site which is then used to obtain the required sections. For convenience and the best accuracy, an RTK system should be used in which a base station works with one or more rovers. The RTK rovers can be mounted on detail poles and these would be carried around the site and measurements could be taken in a stop-and-go method at discrete points on a grid and at changes of slope. However, it is also possible to use the RTK system in a continuous mode in which a data collection rate is set into the receiver. When using this method, the operator on the detail pole (rover) simply walks over the site without stopping and the RTK system records a position every few seconds, depending on the epoch or data collection rate entered into the receiver. To speed up the data collection process, the GPS antenna can be mounted on a moving vehicle of some sort and this can be driven over the site with data being collected at a pre-determined rate. This is obviously much faster than walking the site. A slight disadvantage of this is that data may not be collected at changes of slope and other significant points, but the density of data obtained by this method is usually so high that these can be interpolated with sufficient accuracy.

As an alternative to expensive RTK solutions, a mapping grade receiver could also be used for obtaining data, but for earthworks only. The advantages of using GPS in sectioning are that it is not necessary to set out the centre line. Instead, a DTM is produced, offering greater flexibility in design which is very effective where a lot of data has to be recorded, for example when working on large sites or when defining irregular ground surfaces.

Office methods of sectioning

These methods, which are undertaken in an office environment, enable longitudinal and cross-sections to be produced from ground surface data which already exists in a different form. Using these, sections can be obtained from an existing *topographic map* or from a *photogrammetric survey* stored in a digital plotting instrument or from a *Digital Terrain Model* held in a computer. As with the site methods discussed earlier, the centre line must again be defined before any data can be obtained. However, how it is defined will depend on the data measuring technique being employed.

Sectioning from topographic maps

In this method, the centre line can simply be drawn on the map with distances and heights being obtained from the scale of the map and the contours directly. Consider Figure 15.14, which shows part of a contoured topographic map of an area. The line XX is the proposed route for a straight section of a road centre line and relevant cross-sections are shown at chainages 525 m to 625 m. Using the contours, the approximate shape of the longitudinal and cross-sections can be obtained by scaling height and distance information from the map at points where the section lines cut contours, as shown in Figure 15.15.

Sectioning using photogrammetric surveys

Photogrammetry involves the use of photographs, usually aerial photographs, to provide measurement information in the form of drawings and three-dimensional coordinate data. The photographs are placed in instruments known as digital photogrammetric workstations and these are used to create digital stereomodels of

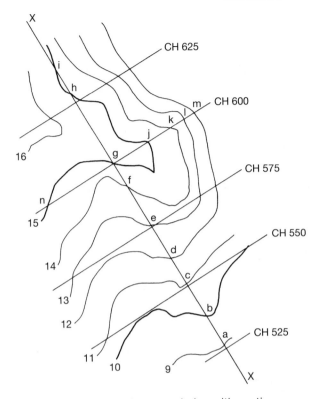

Figure 15.14 ● Contoured plan with sections.

the ground surface shown on the photographs. These stereomodels, which are stored in the controlling software of the workstation, can be used to produce either graphical information in the form of maps and plans or numerical information in the form of 3D coordinates. Longitudinal and cross-sections can be produced by using the software first to define the proposed centre line in the stereomodel and second to interrogate the stereomodel, both along the centre line and at right angles to it, to enable 3D data to be obtained. These can then be plotted as required in a format similar to those shown in Figures 15.12 and 15.13.

Sectioning using Digital Terrain Models

Digital Terrain Models are a combination of mathematical and geographical representations of parts of the Earth's surface. They can be derived using commercially available software packages from several data sources including levelling, total station observations, photogrammetric measurements, GPS surveys, laser scanning and LiDAR. They usually include both planimetric data (E, N) and relief data (RL, geographical elements and natural features such as rivers, ridge lines and so on). They are normally formed from the data sources either using a grid-based modelling system or a triangulation-based modelling system, both of which are discussed in Section 10.9.

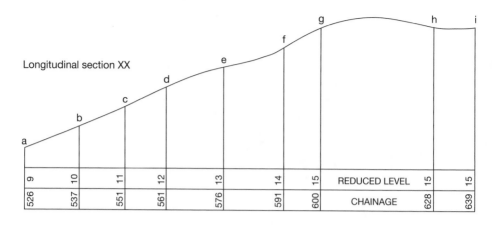

Longitudinal section XX

	9	10	11	12	13	14	15	REDUCED LEVEL	15	15
	526	537	551	561	576	591	600	CHAINAGE	628	639

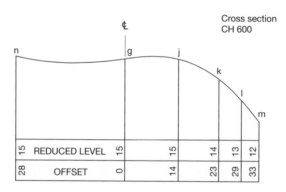

Cross section
CH 600

15	REDUCED LEVEL	15		15	14	13	12
28	OFFSET	0		14	23	29	33

Figure 15.15 ● Longitudinal and cross-sections from contours.

DTMs are essentially computer-generated surface models, which can be used for a variety of applications including the production of longitudinal and cross-sections. These can be produced in a similar manner to that described in the previous section for photogrammetric methods of sectioning – the software is first used to define the proposed centre line on the DTM and then used to interrogate the DTM both along and at right angles to the centre line to generate the 3D sectional data, which can then be used to plot the sections.

Reflective summary

With reference to longitudinal sections and cross-sections remember:

— Longitudinal sections and cross-sections are required to define the shape of the ground surface both at the design stage and during construction to help monitor the progress of construction work.

- The data required to produce the sections can be obtained either directly on site using methods involving levelling, total stations and GPS surveys or indirectly from existing information in the form of topographic maps, photogrammetric surveys and Digital Terrain Models (DTMs).

- Although they are still sometimes drawn by hand, longitudinal sections and cross-sections are usually produced using computer-aided design packages.

- They are usually drawn such that their vertical scales are larger than their horizontal scales to enable a clear picture of the exact shape of the ground surface along the sections to be obtained.

15.4 Calculating volumes from cross-sections

After studying this section, you should be aware of the various different types of cross-section that exist and how their dimensions and areas can be calculated. You should know how to draw the extent of embankments and cuttings on drawings in order to determine the amount of land required for a project. You should understand how volumes can be determined from cross-sections using the *end areas method* and the *prismoidal formula*. You should know the relative accuracies of these two techniques and which is the more appropriate in certain situations. You should understand how to improve the accuracy of volumes calculated over curved sections of embankments or cuttings.

This section includes the following topics:

- The types of cross-section that exist
- Determining the extent of embankments and cuttings
- Using the cross-sectional areas to calculate volumes
- The effect of curvature on volume
- Worked examples showing volume calculations from cross-sections

The types of cross-section that exist

Because the surface of the ground will always contain some degree of undulation, any cross-sectional shape produced on a drawing can only be an approximation of

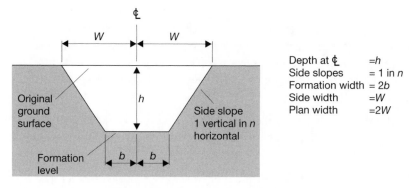

Figure 15.16 ● Level cross-section.

the true shape of the ground surface. Depending on the shape of the existing ground surface, there are several different types of cross-section that exist. These are discussed below.

Level cross-sections

These are the simplest types of cross-section and occur when the existing ground surface is horizontal as shown for the cutting in Figure 15.16, from which

$$\text{cross-sectional area} = A = h(2b + nh) \tag{15.10}$$

$$\text{plan width} = 2W = 2(b + nh) \tag{15.11}$$

For an embankment, the diagram is inverted and the same formulae apply.

The *side widths* (W values) are used to show the extent of the embankments and cuttings on the road drawings.

Two level cross-sections

These occur when the existing ground surface has a constant slope across the section. This is shown for a cutting in Figure 15.17, where

W_G = greater side width
W_L = lesser side width
h = depth of cut on the centre line from the existing to the proposed levels
1 in n = side slope
1 in s = ground or transverse slope

For two-level sections

$$W_G = \frac{s(b + nh)}{(s - n)} \tag{15.12}$$

$$W_L = \frac{s(b + nh)}{(s + n)} \tag{15.13}$$

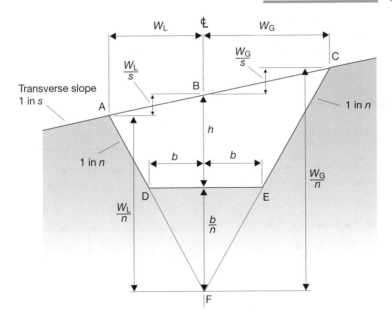

Figure 15.17 ● Two-level cross-section.

$$plan\ width = (W_G + W_L) \tag{15.14}$$

$$A = \frac{1}{2}\left[h + \left(\frac{b}{n}\right)\right](W_G + W_L) - \left(\frac{b^2}{n}\right) \tag{15.15}$$

Again, for embankments, Figure 15.17 is inverted and the same formulae apply.

The greater and lesser side widths (W_G and W_L) are used to show the extent of the embankments and cuttings on the road drawings.

Three-level cross-sections

These occur when the existing ground surface changes gradient at the centre line as it crosses the section. This is shown for a cutting in Figure 15.18, where

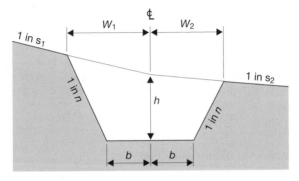

Figure 15.18 ● Three-level cross-section.

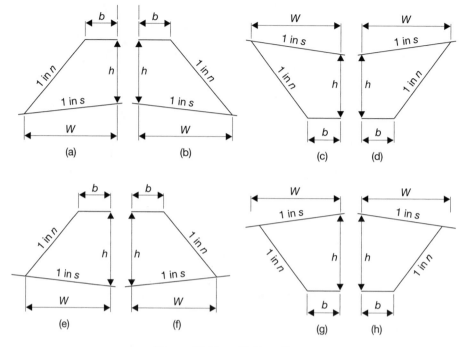

Figure 15.19 ● Half-sections.

W_1 and W_2 = side widths
h = depth of cut on the centre line from the existing to the proposed levels
1 in n = side slope
1 in s_1 and 1 in s_2 = transverse slopes

Cross-sections of this type are best considered as consisting of two separate half sections on either side of the centre line. There are eight possible types of half section, as shown in Figure 15.19, and it is possible to derive the following formulae for these half sections.

For the half sections shown in Figure 15.19(a–d)

$$W = \frac{s(b + nh)}{(s - n)} \tag{15.16}$$

For the half sections shown in Figure 15.19(e–h)

$$W = \frac{s(b + nh)}{(s + n)} \tag{15.17}$$

The *cross-sectional area* (A) of any combination of any two of the eight types of half section is given by

$$A = \frac{1}{2}\left[h + \left(\frac{b}{n}\right)\right](\text{sum of side widths}) - \left(\frac{b^2}{n}\right) \tag{15.18}$$

Cross-sections involving both cut and fill

These occur where the depth of cut or fill on the centre line is not great enough to give either a full cutting or a full embankment but instead gives a cross-section consisting partly of cut and partly of fill. Such a section can occur when a road is being built around the side of a hill and is used for economic reasons since the cut section can be used to provide the fill section and very little earth-moving distance is involved.

In practice, four types of section can occur as shown in Figure 15.20, where

h = depth of cut or fill on the centre line from the existing to the proposed levels
W_1 and W_2 = side widths
A_1 and A_2 = areas of cut or fill
1 in n and 1 in m = side slopes
1 in s = transverse slope

Two different side slopes are shown, since often a different side slope is used for cut compared to that used for fill.

The following formulae can be derived for W_1, W_2, A_1 and A_2.

For cross-sections similar to those shown in Figure 15.20(a) and (d)

$$W_1 = \frac{s(b - nh)}{(s - n)} \tag{15.19}$$

$$W_2 = \frac{s(b + mh)}{(s - m)} \tag{15.20}$$

$$A_1 = \frac{(b - sh)^2}{2(s - n)} \tag{15.21}$$

$$A_2 = \frac{(b + sh)^2}{2(s - m)} \tag{15.22}$$

For cross-sections similar to those shown in Figure 15.20(b) and (c)

$$W_1 = \frac{s(b + nh)}{(s - n)} \tag{15.23}$$

$$W_2 = \frac{s(b - mh)}{(s - m)} \tag{15.24}$$

$$A_1 = \frac{(b + sh)^2}{2(s - n)} \tag{15.25}$$

$$A_2 = \frac{(b - sh)^2}{2(s - m)} \tag{15.26}$$

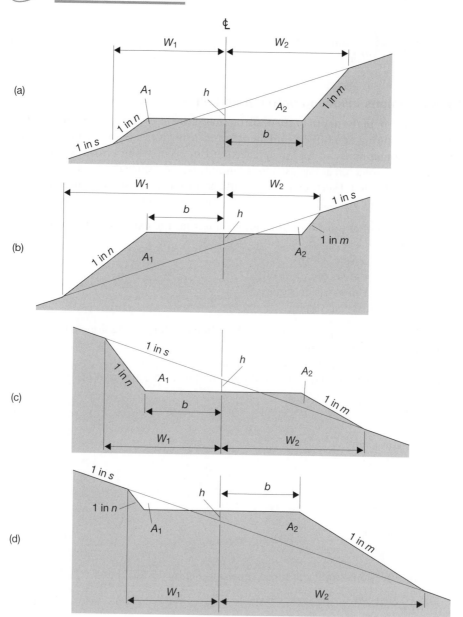

Figure 15.20 ● Sections involving cut and fill.

The side widths (W_1 and W_2) are again used to show the extent of the embankments and cuttings on the road drawings.

With any cross-section partly in cut and partly in fill, it is essential that a drawing be produced so that the correct formulae can be used.

Worked example 15.5 at the end of this section illustrates the application of these formulae.

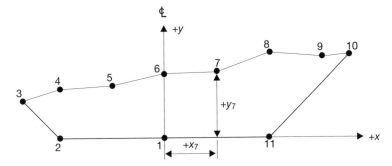

Figure 15.21 ● Irregular cross-section.

Irregular cross-sections

The different types of cross-section discussed on the previous pages all assumed that the ground surface was either flat or sloped uniformly either fully across the section or up to the surface position of the proposed centre line. In practice, however, it is more common for the ground surface to undulate in an irregular manner as it crosses the section, as shown in Figure 15.21. With such irregular sections it is not possible to derive formulae for side widths and areas. Instead, the side widths and plan width can be measured from the cross-sectional drawing and the cross-sectional area can be obtained using one of the techniques discussed in Section 15.2. Of these, the graphical method can give accurate results if a small grid is used, and a planimeter can be very effective.

Alternatively, the area of an irregular section could be found using the *cross coordinate method* discussed in Section 15.2. In order to apply this method, a coordinate system that has its origin at the intersection of the formation level and the centre line is used. Offset distances (x values) to the right of the centre line are taken as positive and those to the left of the centre line are taken as negative. Heights (y values) above the formation level are considered to be positive and those below the formation level are considered to be negative. In Figure 15.21, the points defining the section will have the coordinates given in Table 15.2. These coordinates are used to obtain the area of this section by substituting them into the cross coordinate formula given in equation (15.5) as follows

$$2 \times \text{Area} = (N_1E_2 + N_2E_3 + N_3E_4 + \ldots + N_{11}E_1)$$
$$- (E_1N_2 + E_2N_3 + E_3N_4 + \ldots + E_{11}N_1)$$

since a clockwise order is given in Figure 15.21 for the points.

Table 15.2

Point $n =$	1	2	3	4	5	6	7	8	9	10	11
E_n	0	$-b$	$-x_3$	$-x_4$	$-x_5$	0	x_7	x_8	x_9	x_{10}	b
N_n	0	0	y_3	y_4	y_5	y_6	y_7	y_8	y_9	y_{10}	0

A similar process can be applied to embankments and also to those sections involving both cut and fill. For the cut and fill sections, separate calculations are required.

In this type of calculation, careful attention must be paid to the algebraic signs involved.

Determining the extent of embankments and cuttings

The side widths (W values) of cross-sections can be determined either by using the formula appropriate to the type of cross-section, as discussed on the previous pages, or by scaling them directly from the cross-sectional drawings. Once they have been obtained, they can be used to mark the extent of the embankments and cuttings at each cross-section on the working drawings, as shown in Figure 15.22. Joining the end points of each cross-section with a curved line not only enables the area of land required for the construction to be defined but also allows the area of the site that must be cleared and stripped of topsoil before construction can begin to be determined. The extremities of the embankments and cuttings at each cross-section are usually joined with broken lines (as in Figure 15.22) since the actual points at which they meet the existing ground surface can only be approximated.

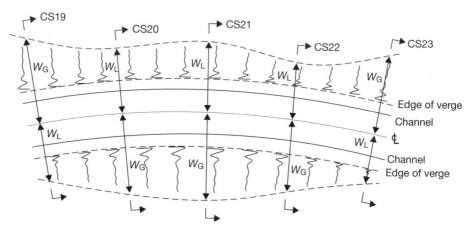

Figure 15.22 ● Use of side widths to determine the extent of embankments and cuttings.

Using the cross-sectional areas to calculate volumes

The cross-sectional areas are used to calculate the volumes and this is discussed below. However, before they can be used to obtain volumes it is necessary to modify their calculated values by making allowance for the depth of the road construction. The cross-sectional area of the road construction should be calculated or scaled from a plan of the construction and added to those cross-sectional areas that are in cut and subtracted from those cross-sectional areas that are in fill. Areas partly in cut and partly in fill should be inspected on the cross-sectional drawings and their areas modified accordingly.

Once the cross-sectional areas have been modified they are used to calculate the volume of material contained between them. These volume calculations can be greatly simplified if the cross-sections are taken at regular horizontal distance intervals along the centre line of the proposed construction. Because of this, pegs are normally set out on the centre line at regular intervals of through chainage.

Two methods are commonly used to determine volumes from cross-sectional areas: the *end areas method* and the *prismoidal formula*.

End areas method

This is comparable to the trapezoidal rule for areas. If two cross-sectional areas A_1 and A_2 are a horizontal distance d_1 apart, the volume contained between them V_1 is given by

$$V_1 = \left(\frac{d_1}{2}\right)(A_1 + A_2) \tag{15.27}$$

This leads to the *general end areas formula* for a series of N cross-sections, as follows

$$
\begin{aligned}
V_{total} &= V_1 + V_2 + V_3 + \ldots + V_{N-1} \\
&= \left(\frac{d_1}{2}\right)(A_1 + A_2) + \left(\frac{d_2}{2}\right)(A_2 + A_3) + \left(\frac{d_3}{2}\right)(A_3 + A_4) + \ldots + \left(\frac{d_{N-1}}{2}\right)(A_{N-1} + A_N)
\end{aligned}
$$

If $d_1 = d_2 = d_3 = d_{N-1} = d$

$$\text{total volume} = \left(\frac{d}{2}\right)[A_1 + A_N + 2(A_2 + A_3 + \ldots + A_{N-1})] \tag{15.28}$$

Putting this into words gives

$$\text{total volume} = \left(\frac{d}{2}\right)[\text{first area} + \text{last area} + \text{twice all other areas}]$$

The end areas method can be applied to any number of cross-sectional areas and will give accurate results if the cross-sectional areas are of the same order of magnitude. Worked example 15.6 at the end of this section shows how the end areas formula is applied.

Prismoidal formula

This is comparable to Simpson's rule for areas and is more accurate than the end areas method.

The volume contained between a series of cross-sections a constant horizontal distance apart can be approximated by the volume of a *prismoid*, which is a solid figure with plane parallel ends and plane sides. This is shown in Figure 15.23.

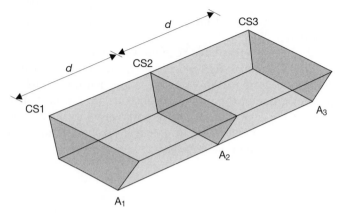

Figure 15.23 ● Prismoid for volume calculations.

It can be shown that for a series of three cross-sections, the volume, V_{1-3}, contained between them is given by

$$V_{1-3} = \left(\frac{d}{3}\right)(A_1 + 4A_2 + A_3) \tag{15.29}$$

This is the *prismoidal formula* and is used for earthwork calculations of cuttings and embankments and gives a true volume if *either*

> the transverse slopes at right angles to the centre line are straight and the longitudinal profile on the centre line is parabolic

or

> the transverse slopes are parabolic and the longitudinal profile is a straight line.

Taking account of these, unless the ground profile is regular both transversely and longitudinally it is likely that errors will be introduced in assuming that the figure is prismoidal over its entire length. These errors, however, are small and the volume obtained is a good approximation to the true volume.

If Figure 15.23 is extended to include cross-section 4 (A_4) and cross-section 5 (A_5), the volume from CS3 to CS5 (V_{3-5}) is given by equation (15.29) as

$$V_{3-5} = \left(\frac{d}{3}\right)(A_3 + 4A_4 + A_5)$$

Therefore, the total volume from CS1 to CS5 (V) is

$$V = \left(\frac{d}{3}\right)(A_1 + 4A_2 + 2A_3 + 4A_4 + A_5)$$

This leads to a *general prismoidal formula* for N cross-sections, where N **must** be **odd**, as follows

$$V = \left(\frac{d}{3}\right)(A_1 + A_N + 4\Sigma \text{ even areas} + 2\Sigma \text{ remaining odd areas}) \qquad (15.30)$$

Putting this into words gives

$$V = \left(\frac{d}{3}\right)(\text{first area + last area + four times the even areas}$$

$$+ \text{ twice the remaining odd areas})$$

This is often referred to as Simpson's rule for volumes and, although it can only be applied to an odd number of cross-sectional areas, it should be used wherever possible. Worked example 15.5 at the end of this section illustrates the application of the prismoidal formula.

The effect of curvature on volume

The foregoing discussion has assumed that the cross-sections are taken on a straight road or similar. However, where a horizontal curve occurs the cross-sections will no longer be parallel to each other and errors will result in volumes calculated either by the end areas method or the prismoidal formula.

To overcome this, *Pappus' theorem* must be used, which states that a volume swept out by a plane constant area revolving about a fixed axis is given by the product of the cross-sectional area and the distance moved by the centre of gravity of the section.

Using this, the volumes of cuttings that occur on circular curves can be calculated with a better degree of accuracy. Figure 15.24 shows an asymmetrical cross-section in which the centroid is situated at a horizontal distance c from the centre line, where c is referred to as the *eccentricity*. The centroid may be on either side of the centre line according to the transverse slope.

Figure 15.25 shows this cross-section occurring on a circular curve of radius R.

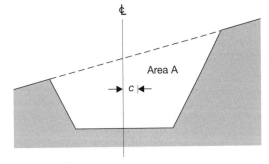

Figure 15.24 ● Asymmetrical cross-section with eccentricity c.

Length of path of centroid $= (R + c)\theta$ (where θ is in radians)

From Pappus' theorem, the volume V swept out is

$$V = A(R + c)\theta$$

But

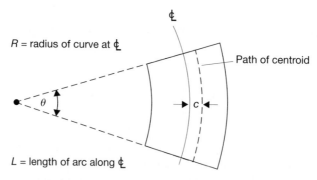

R = radius of curve at ℄

Path of centroid

θ

c

L = length of arc along ℄

Figure 15.25 ● Effect of eccentricity on volume calculation.

$$\theta \text{ (in radians)} = \frac{L}{R}$$

Hence

$$V = \frac{LA(R + c)}{R}$$

Therefore

$$V = L\left[A + \left(\frac{Ac}{R}\right)\right] = LA\left[1 + \left(\frac{c}{R}\right)\right] = LA' \qquad (15.31)$$

In equation (15.31), LA is the volume of a prismoid of length L and the term Ac/R can be regarded as the correction to be made to the cross-sectional area before calculating the volume as that of a normal prismoid. The corrected area can be expressed as

$$A' = A\left[1 \pm \left(\frac{c}{R}\right)\right] \qquad (15.32)$$

The ± sign is necessary since the centroid can lie on either side of the centre line. The negative sign is adopted if the centroid lies on the same side of the centre line as the centre of curvature and the positive sign if on the other side.

In practice, the shape of the cross-section will not be constant so that neither A nor c will be constant. However, the ratio c/R will usually be small and it is generally sufficiently accurate to calculate the correction for each cross-section and to use either the end areas method or the prismoidal formula to determine the volume.

The application of Pappus' theorem is normally only considered in those cases where the radius of curvature is small and where the curvature is to one side of the centre line only. In situations where the curvature alternates from one side of the centre line to the other, the effect of curvature on volume tends to be cancelled out, particularly on long projects such as motorways.

Worked examples showing volume calculations from cross-sections

The following two worked examples show the application of the formulae developed in this section. The first, Worked example 15.5, involves the calculation of cross-sectional areas and volumes for a series of sections that are partly in cut and partly in fill. The second, Worked example 15.6, illustrates how volumes should be calculated between cross-sections that are changing from all cut to all fill and vice versa.

Worked example 15.5: Combined cross-sectional area and volume calculations

Question

The centre line of a proposed road of formation width 12.00 m is to fall at a slope of 1 in 100 from chainage 50 m to chainage 150 m.

The existing ground levels on the centre line at chainages 50 m, 100 m and 150 m are 71.62 m, 72.34 m and 69.31 m, respectively, and the ground slopes at 1 in 3 at right angles to the proposed centre line.

If the centre line formation level at chainage 50 m is 71.22 m and the side slopes are to be 1 in 1 for cut and 1 in 2 for fill, calculate the volumes of cut and fill between chainages 50 m and 150 m.

Solution

Figure 15.26 shows the longitudinal section from chainage 50 m to chainage 150 m. From this

the centre line formation level at chainage 100 m = 71.22 − 0.50 = 70.72 m

the centre line formation level at chainage 150 m = 71.22 − 1.00 = 70.22 m

Figure 15.27 shows the three cross-sections.

Since all the cross-sections are partly in cut and partly in fill, equations (15.21), (15.22), (15.25) and (15.26) apply.

Cross-section 50 m

$s = 3$, $b = 6$ m, $n = 2$, $m = 1$

$h = 71.62 − 71.22 = +0.40$ m, that is, cut at the centre line

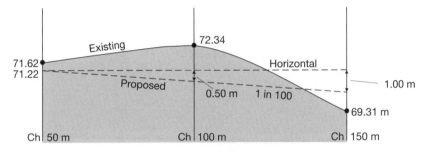

Figure 15.26

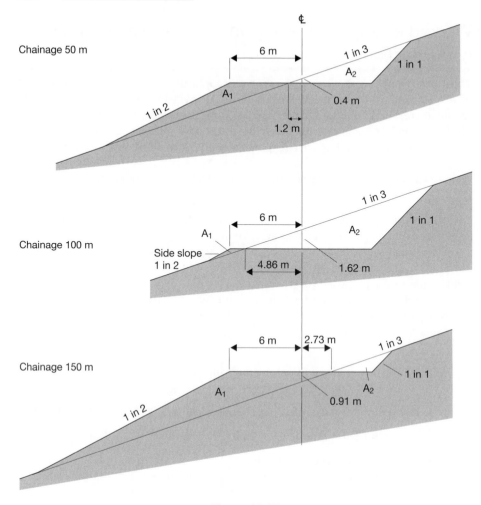

Figure 15.27

This cross-section is similar to that shown in Figure 15.20(a); hence from equations (15.22) and (15.21), respectively

$$area\ of\ cut = A_2 = \frac{(b + sh)^2}{2(s - m)}$$

$$= \frac{(6 + 3 \times 0.40)^2}{2(3 - 1)} = \textbf{12.96 m}^2$$

$$area\ of\ fill = A_1 = \frac{(b - sh)^2}{2(s - n)}$$

$$= \frac{(6 - 3 \times 0.40)^2}{2(3 - 2)} = \textbf{11.52 m}^2$$

Cross-section 100 m

    $s = 3$, $b = 6$ m, $n = 2$, $m = 1$

    $h = 72.34 - 70.72 = +1.62$ m, that is, cut at the centre line

This cross-section is similar to cross-section 50 m; hence again using equations (15.22) and (15.21)

$$area\ of\ cut = A_2 = \frac{(6 + 3 \times 1.62)^2}{2(3 - 1)} = \mathbf{29.48\ m^2}$$

$$area\ of\ fill = A_1 = \frac{(6 - 3 \times 1.62)^2}{2(3 - 2)} = \mathbf{0.65\ m^2}$$

Cross-section 150 m

    $s = 3$, $b = 6$ m, $n = 2$, $m = 1$

    $h = 69.31 - 70.22 = -0.91$ m, that is, fill at the centre line

This cross-section is similar to that shown in Figure 15.20(b); hence from equations (15.26) and (15.25), respectively

$$area\ of\ cut = A_2 = \frac{(b - sh)^2}{2(s - m)}$$

$$= \frac{(6 - 3 \times 0.91)^2}{2(3 - 1)} = \mathbf{2.67\ m^2}$$

$$area\ of\ fill = A_1 = \frac{(b + sh)^2}{2(s - n)}$$

$$= \frac{(6 + 3 \times 0.91)^2}{2(3 - 2)} = \mathbf{38.11\ m^2}$$

The prismoidal formula given in equation (15.30) can be used to calculate the volumes since the number of cross-sections is odd. This gives

$$volume\ of\ cut = \left(\frac{50}{3}\right)[12.96 + 2.67 + 4(29.48)] = \mathbf{2225.8\ m^3}$$

$$volume\ of\ fill = \left(\frac{50}{3}\right)[11.52 + 38.11 + 4(0.65)] = \mathbf{870.5\ m^3}$$

These figures would normally be rounded to at least the nearest cubic metre.

Worked example 15.6: Volume calculation from a series of cross-sections

Question

Figure 15.28 shows a longitudinal section along the proposed centre line of a road together with a series of six cross-sections taken at 20 m intervals. The areas of cut and/or fill at each section are indicated.

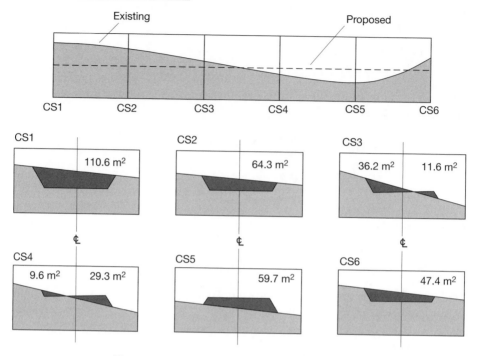

Figure 15.28 ● Data for Worked example 15.6.

Calculate the volumes of cut and fill contained between consecutive cross-sections from CS1 and CS6.

Solution

The volumes of cut and fill are calculated using either the end areas method or the prismoidal formula. If the cross-sections are all of the same type, one of the formulae, preferably the prismoidal, can be used. However, if the cross-sections are changing, as in Figure 15.28, it is best to work from one cross-section to the next using the end areas method as follows.

CS1 to CS2
From equation (15.27)

$$\text{volume of cut} = \left(\frac{20}{2}\right)(110.6 + 64.3) = \mathbf{1749\ m^3}$$

volume of fill = **zero**

CS2 to CS3
From equation (15.27)

$$\text{volume of cut} = \left(\frac{20}{2}\right)(64.3 + 36.2) = \mathbf{1005\ m^3}$$

The volume of fill presents a problem. Between CS2 and CS3 there is a point at which the fill begins. A good estimate of the position of this point must be made to enable accurate volume figures to be obtained. This is best done by assuming that the rate of increase of fill between CS2 and CS3 is the same as that between CS3 and CS4.

Between CS3 and CS4, the area of fill increases from 11.6 m² to 29.3 m² in a distance of 20 m. If this is extrapolated back as shown in Figure 15.29, the point at which the fill begins between CS2 and CS3 can be found.

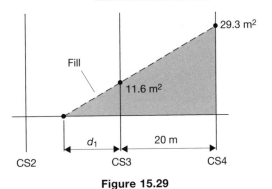

Figure 15.29

In Figure 15.29

$$\left(\frac{d_1}{11.6}\right) = \left(\frac{20}{29.3 - 11.6}\right)$$

Hence

$$d_1 = 13.1 \text{ m}$$

Therefore from equation (15.27)

$$\text{volume of fill} = \left(\frac{13.1}{2}\right)(0 + 11.6) = \textbf{76 m}^{\textbf{3}}$$

This extrapolation method will work only if the area of fill at CS4 is greater than twice the area of fill at CS3 otherwise a meaningless result will be obtained, that is $d_1 > d$. The only solution to such an occurrence is to inspect the cross-sectional drawings and the longitudinal section and to make a reasoned estimate of the position at which the fill begins.

The extrapolation method is suitable for both cut and fill, whether increasing or decreasing.

CS3 to CS4
From equation (15.27)

$$\text{volume of cut} = \left(\frac{20}{2}\right)(36.2 + 9.6) = \textbf{458 m}^{\textbf{3}}$$

$$\text{volume of fill} = \left(\frac{20}{2}\right)(11.6 + 29.3) = \textbf{409 m}^{\textbf{3}}$$

CS4 to CS5
The volume of cut must be calculated using the extrapolation method used above. From CS3 to CS4 the area of cut decreases from 36.2 m² to 9.5 m² in a distance of 20 m. Hence, the distance from CS4 towards CS5 at which the cut decreases (d_2) is given by

$$d_2 = \frac{(20 \times 9.6)}{(36.2 - 9.6)} = 7.2 \text{ m}$$

Therefore from equation (15.27)

$$\text{volume of cut} = \left(\frac{7.2}{2}\right)(9.6 + 0) = \mathbf{35 \ m^3}$$

$$\text{volume of fill} = \left(\frac{20}{2}\right)(29.3 + 59.7) = \mathbf{890 \ m^3}$$

CS5 to CS6
In this case the cross-section changes from
all fill to all cut. Again, for accuracy, it is
necessary to estimate the position where the
fill section ends and the cut section begins.
The relevant part of the longitudinal section
is shown in Figure 15.30.

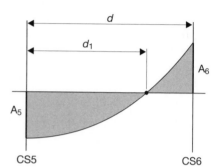

A linear relationship involving the cross-
sectional areas can be used as follows:

$$\frac{d_1}{A_5} = \frac{d}{(A_5 + A_6)}$$

Figure 15.30

Hence

$$d_1 = \frac{d \times A_5}{(A_5 + A_6)}$$

From equation (15.27), the volume of fill is obtained from

$$\left(\frac{d_1}{2}\right)(A_5 + 0)$$

and the volume of cut is obtained from

$$\left(\frac{d - d_1}{2}\right)(0 + A_6)$$

In this case

$$d_1 = \frac{(20 \times 59.7)}{(59.7 + 47.4)} = 11.1 \ m$$

Therefore from equation (15.27)

$$\text{volume of cut} = \left(\frac{20 - 11.1}{2}\right)(0 + 47.4) = \mathbf{211 \ m^3}$$

$$\text{volume of fill} = \left(\frac{11.1}{2}\right)(59.7 + 0) = \mathbf{331 \ m^3}$$

This linear method applies equally when the cross-section changes from all cut to all
fill.

15.5 Calculating volumes from spot heights and contours

After studying this section, you should know how to determine the volumes of earth to be removed when constructing underground features such as basements and underground tanks, and you should understand how volumes of large features such as reservoirs and earth dams can be determined from existing contoured maps.

This section includes the following topics:

● Using grids of spot heights to calculate volumes of excavations
● Determining volumes from contoured maps and plans

Using grids of spot heights to calculate volumes of excavations

Spot heights (see Section 10.6) can be used in the form of grids to obtain the volumes of earth to be removed during the construction of large deep excavations such as those required for basements, underground tanks and so on, where the formation levels can be sloping, horizontal or terraced.

A square, rectangular or triangular grid is established on the ground surface and spot heights are taken at each grid intersection. The smaller the grid interval, the greater will be the accuracy of the volume calculations, but the amount of fieldwork increases so a compromise is usually reached.

The formation level at each grid point must be known to enable the depth of cut from the existing to the proposed level at each grid intersection to be calculated.

Figure 15.31 shows a 10 m square grid with the depths of cut marked at each grid intersection. Consider the volume contained in grid square $h_1 h_2 h_6 h_5$; this is shown in Figure 15.32.

It is assumed that the surface slope is constant between grid intersections and the volume is given by

$$\text{volume} = \text{mean height} \times \text{plan area}$$
$$= \left(\frac{4.76 + 5.14 + 4.77 + 3.21}{4} \right) \times 100 = \textbf{447 m}^3$$

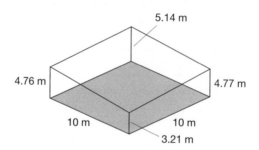

```
h₁          h₂          h₃          h₄
   4.76        5.14        6.72        8.10

h₅          h₆          h₇          h₈
   3.21        4.77        5.82        6.07

                                    Plan area of each
h₉          h₁₀         h₁₁          grid square = 100 m²
   1.98        2.31        3.55
```

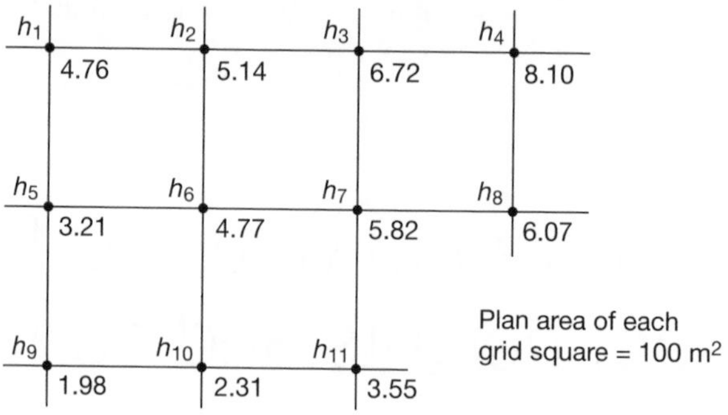

Figure 15.31 ● Grid heights for volume calculation.

5.14 m

4.76 m   4.77 m

10 m   10 m

3.21 m

Figure 15.32 ● Volume calculation for a square grid.

A similar method can be applied to each individual grid square and this leads to the following general formula for square and rectangular grids

$$\text{total volume} = \left(\frac{A}{4}\right)(\Sigma \text{ single depths} + 2\Sigma \text{ double depths}$$

$$+ 3\Sigma \text{ triple depths} + 4\Sigma \text{ quadruple depths}) + \delta V \qquad (15.33)$$

where

A	=	plan area of each grid square
single depths	=	depths such as h_1 and h_4, which are used once
double depths	=	depths such as h_2 and h_3, which are used twice
triple depths	=	depths such as h_7, which are used three times
quadruple depths	=	depths such as h_6, which are used four times
δV	=	the total volume outside the grid, which is calculated separately

Applying equation (15.33) to the example shown in Figure 15.31 gives

Volume contained within the grid area

$$= \left(\frac{100}{4}\right)[4.76 + 8.10 + 6.07 + 1.98 + 3.55$$

$$+ 2(5.14 + 6.72 + 3.21 + 2.31) + 3(5.82) + 4(4.77)]$$

$$= 25(24.46 + 34.76 + 17.46 + 19.08)$$

$$= \textbf{2394 m}^{\textbf{3}}$$

The result is only an approximation since it has been assumed that the surface slope is constant between spot heights.

If a triangular grid is used, the general formula must be modified as follows

● $A'/3$ must replace $A/4$, where A' = plan area of each triangle

and

● depths appearing in five and six triangles must be included.

Determining volumes from contoured maps and plans

Contours are discussed in Section 10.6, and height information in the form of existing contoured maps and plans can be very useful when it is required to determine the volumes of very large features such as reservoirs, earth dams, spoil heaps and so on.

The system adopted is to calculate the plan area enclosed by each contour and then treat this as a cross-sectional area. The contour interval provides the distance between cross-sections and either the prismoidal formula or the end areas method is used to calculate the volume. If the prismoidal formula is used, the number of contours must be odd.

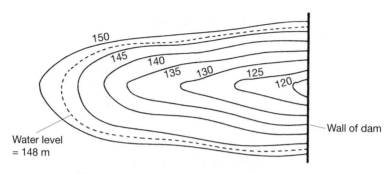

Figure 15.33 ● Plan of a proposed reservoir.

The plan area contained by each contour can be measured using a planimeter or one of the methods discussed in Section 15.2. The graphical method is particularly suitable in this case.

The accuracy of the result depends to a large extent on the contour interval, but normally great accuracy is not required and, in reservoir capacity calculations, volumes to the nearest 1000 m^3 are more than adequate. Consider the following example.

Figure 15.33 shows a plan of a proposed reservoir and dam wall. The vertical interval is 5 m and the water level of the reservoir is to be 148 m. The capacity of the reservoir is required.

The volume of water that can be stored between the contours can be found by reference to Figure 15.34, which shows a cross-section through the reservoir together with the plan areas enclosed by each contour and the dam wall.

$$\text{total volume} = \text{volume between the 148 m and 145 m contours}$$
$$+ \text{volume between the 145 m and 120 m contours}$$
$$+ \text{small volume below the 120 m contour}$$

The volume between the 148 m and 145 m contours is found by the end areas method using equation (15.27) to be

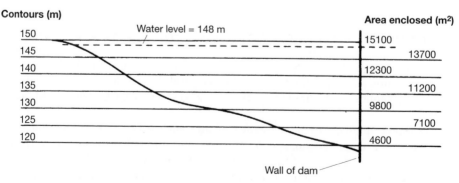

Figure 15.34 ● Cross-section through reservoir showing contour areas.

$$= \left[\frac{(15,100 + 13,700)}{2} \right] \times 3 = \mathbf{43,200 \ m^3}$$

The volume between the 145 m and 120 m contours is found by the end areas method using equation (15.28) to be

$$= \left(\frac{5}{2} \right) [13,700 + 4600 + 2(12,300 + 11,200 + 9800 + 7100)] = \mathbf{247,750 \ m^3}$$

The small volume below the 120 m contour can be found by decreasing the contour interval to, say, 1 m and using the end areas method or the prismoidal formula. Alternatively, if it is very small, it may be neglected. Let this volume = δV. Therefore

Total volume = 43,200 + 247,750 + δV
= $\mathbf{(290,950 + \delta V) m^3}$

This would usually be rounded to the nearest 1000 m³.

The second term in the above solution of 247,750 m³, was obtained by the end areas method applied between contours 145 m and 120 m. Alternatively, the prismoidal formula could have been used between the 145 m and 125 m contours (to keep the number of contours **odd**) and the end areas formula between the 125 m and 120 m contours. If this is done, the volume between the 155 m and 120 m contours is calculated to be **248,583 m³**.

As an alternative to the calculations just described, some digital planimeters have the facility to calculate volumes directly from areas contained within contour lines that have been measured by them and stored in their internal memories.

Reflective summary

With reference to calculating volumes from spot heights and contours remember:

— The volume values obtained will only be approximate.

— Volumes can be calculated from spot heights taken in the form of square, rectangular or triangular grids established on the ground surface. The smaller the length of the grid lines, the greater will be the accuracy of the volumes obtained but the amount of fieldwork required will be increased.

— Volumes can be obtained by applying either the end areas method or the prismoidal formula to areas enclosed by equally spaced contour lines.

— Some digital planimeters have the facility to measure areas contained within contour lines and then calculate the volume contained between them directly.

15.6 Mass haul diagrams

After studying this section, you should know how to plan the optimum movement of material required in the formation of embankments and cuttings using mass haul diagrams. You should understand how mass haul diagrams are produced, their terminology and their properties. You should also develop an appreciation of their uses, with particular reference to their application in determining the most economical method of moving the material involved in the construction.

This section includes the following topics:

- Overview of mass haul diagrams
- Drawing the diagram
- Terminology of mass haul diagrams
- Properties of the mass haul curve
- Economics of mass haul diagrams
- Choice of balancing line
- Uses of mass haul diagrams

Overview of mass haul diagrams

During the construction of long linear engineering projects such as roads, railways, and canals, there may be a considerable quantity of earth required to be brought on to the site to form embankments and to be removed from the site during the formation of cuttings. The earth brought to form embankments may come from another section of the site such as a tip formed from excavated material (known as a *spoil heap*) or may be imported on to the site from a nearby quarry. Any earth brought on to the site is said to have been *borrowed*. The earth excavated to form cuttings may be deposited in tips at regular intervals along the project to form spoil heaps for later use in embankment formation or may be *wasted* either by spreading the earth at right angles to the centre line to form verges or by carting it away from the site area and depositing it in suitable local areas.

This movement of earth throughout the site can be very expensive and, since the majority of the cost of such projects is usually given over to the earth-moving, it is essential that considerable care is taken when planning the way in which material is handled during the construction. The mass haul diagram is a graph of volume against chainage, which greatly helps in planning such earth-moving.

The *x*-axis represents the chainage along the project from the position of zero chainage.

The *y*-axis represents the aggregate volume of material up to any chainage from the position of zero chainage.

When preparing the mass haul diagram, volumes of cut are considered positive and volumes of fill are considered negative. The vertical and horizontal axes of the mass haul diagram are usually drawn at different scales to exaggerate the diagram and thereby facilitate its use.

The mass haul diagram considers only earth moved in a direction longitudinal to the direction of the centre line of the project and does not take into account any volume of material moved at right angles to the centre line.

Since the mass haul diagram is simply a graph of aggregate volume against chainage it will be noted that if the volume is continually decreasing with chainage, the project is all embankment and all the material will have to be imported on to the site, since there will be no fill material available for use. Such an occurrence will involve a great deal of earth-moving and is obviously not an ideal solution.

If a better attempt is made in the selection of a suitable formation level such that some areas of cut were balanced out by some areas of fill, a more economical solution will result. Because of this vital connection between the formation level and the mass haul diagram, the two are usually drawn together as shown in Figure 15.35, and the following section describes how this is done.

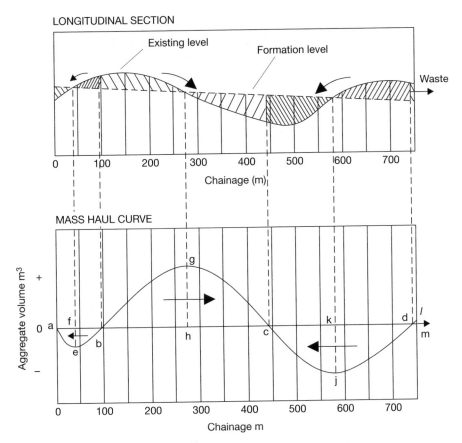

Figure 15.35 ● Mass haul diagram.

Drawing the diagram

Figure 15.35 was drawn as follows.

- The cross-sectional areas are calculated at regular horizontal distance intervals along the project, in this case every 50 m.
- The volumes between consecutive areas and the aggregate volume along the site are calculated, cut being positive and fill being negative.
- Before plotting can begin, a table should be drawn up similar to that shown in Table 15.3. One of the columns in Table 15.3 shows bulking and shrinkage factors. These are necessary owing to the fact that material usually occupies a different volume when it is used in a construction from that which it occupied in natural conditions. Very few soils can be compacted back to their original volume.

    If 100 m^3 of rock are excavated and then used for filling they may occupy 110 m^3 even after careful compaction, and the rock is said to have undergone *bulking* and has a *bulking factor* of 1.1.

Table 15.3

Chainage m	Individual volume m^3		Bulking/ shrinkage factors	Corrected individual volumes m^3		Aggregate volume m^3 Cut (+) Fill (−)
	Cut (+)	Fill (−)		Cut (+)	Fill (−)	
0	–	–	–	–	–	0
50	40	800	1.1	44	800	−756
100	730	–	1.1	803	–	+47
150	910	–	1.1	1001	–	+1048
200	760	–	1.1	836	–	+1884
250	450	–	1.1	495	–	+2379
300	80	110	1.1	88	110	+2357
350	–	520	–	–	520	+1837
400	–	900	–	–	900	+937
450	–	1120	–	–	1120	−183
500	–	970	–	–	970	−1153
550	–	620	–	–	620	−1773
600	200	200	0.8	160	200	−1813
650	590	–	0.8	472	–	−1341
700	850	–	0.8	680	–	−661
750	1120	–	0.8	896	–	+235

If 100 m^3 of clay are excavated and then used for filling they may occupy only 80 m^3 after careful compaction and the clay is said to have undergone *shrinkage* and has a *shrinkage factor* of 0.8.

Owing to the variable nature of the same material when found in different parts of the country, it is impossible to standardise bulking and shrinkage factors for different soil and rock types. Therefore a list of such factors has deliberately not been included since it would imply a uniformity that, in practice, does not exist. Instead, it is recommended that local knowledge of the materials in question should be considered together with tests on soil and rock samples from the area so that reliable bulking and shrinkage factors (which apply only to that particular site) can be determined.

As far as the mass haul diagram is concerned, it is the volumes of fill that are critical. For example, if a hole in the ground is 1000 m^3, the required volume is that amount of cut which will fill the hole. There are two methods that can be used to allow for such bulking and shrinkage. Either the calculated volumes of fill can be amended by dividing them by the factors applying to the type of material available for fill or the calculated volumes of cut can be amended by multiplying them by the factors applying to the type of material in the cut. In Table 15.3 the latter has been done.

- The longitudinal section along the proposed centre line is plotted, the proposed formation level being included.
- The axes of the mass haul diagram are drawn underneath the longitudinal profile such that chainage zero on the profile coincides with chainage zero on the diagram.
- The aggregate volume up to chainage 50 m is plotted at $x = 50$ m. The aggregate volume up to chainage 100 m is plotted at $x = 100$ m and so on for the rest of the diagram.
- The points are joined by curves or straight lines to obtain the finished mass haul diagram.

Terminology of mass haul diagrams

There is a considerable amount of terminology associated with mass haul diagrams and some of the terms used are defined below.

- *Haul distance* is the distance from the point of excavation to the point where the material is to be tipped.
- *Average haul distance* is the distance from the centre of gravity of the excavation to the centre of gravity of the tip.
- *Freehaul distance* is that distance, usually specified in the contract, over which a charge is levied only for the volume of earth excavated and not for its movement.

This is discussed further in the section headed *Economics of mass haul diagrams* on the following page.

- *Freehaul volume* is that volume of material which is moved through the free haul distance.

- *Overhaul distance* is that distance, in excess of the free haul distance over which it may be necessary to transport material. This is also discussed further in the section headed *Economics of mass haul diagrams*.

- *Overhaul volume* is that volume of material which is moved in excess of the free haul distance.

- *Haul.* This is the term used when calculating the costs involved in the earth-moving and is equal to the sum of the products of each volume of material and the distance through which it is moved. It is equal to the total volume of the excavation multiplied by the average haul distance and on the mass haul diagram is equal to the area contained between the curve and the balancing line (see the section headed *Properties of the mass haul curve* below).

- *Freehaul* is that part of the haul which is contained within the free haul distance.

- *Overhaul* is that part of the haul which remains after the freehaul has been removed. It is equal to the product of the overhaul volume and the overhaul distance.

- *Waste* is that volume of material which must be exported from a section of the site owing to a surplus or unsuitability.

- *Borrow* is that volume of material which must be imported into a section of the site owing to a deficiency of suitable material.

Properties of the mass haul curve

Consider Figure 15.35.

- When the curve is rising, the project is in *cut* since the aggregate volume is increasing, for example section *ebg*. When the curve is falling, the project is in *fill* since the aggregate volume is decreasing, for example section *gcj*. Hence, the *end of a section in cut* is shown by a *maximum point* on the curve, for example point *g*, and the *end of a section in fill* is shown by a *minimum point* on the curve, for example point *j*.

- The vertical distance between a maximum point and the next forward minimum represents the *volume of an embankment*, for example (*gh* + *kj*), and the vertical distance between a minimum point and the next forward maximum represents the *volume of a cutting*, for example (*ef* + *gh*).

- Any horizontal line which cuts the mass haul curve at two or more points balances cut and fill between those points and because of this is known as a *balancing line*.

   In Figure 15.35, the *x*-axis is a balancing line and the volumes between chainages *a* and *b*, *b* and *c*, and *c* and *d* are balanced out, that is, as long as the material is suitable, all the cut material between a and d can be used to provide the exact amount of fill required between a and d. The x-axis, however, does not always provide the best balancing line and this is discussed further in the section headed *Choice of balancing line* on p. 784.

- When a balancing line has been drawn on the curve, any area lying *above* the balancing line signifies that the material must be moved to the *right* and any area lying *below* the balancing line signifies that the material must be moved to the *left*. In Figure 15.35, the arrows on the longitudinal section and the mass haul diagram indicate these directions of haul.

- The length of balancing line between intersection points is the *maximum haul distance* in that section, for example the maximum haul distance in section bc is chainage c – chainage b.

- The area of the mass haul diagram contained between the curve and the balancing line is equal to the *haul* in that section, for example area *afbea*, area *bgchb* and area *ckdjc*.

    If the horizontal scale is 1 mm = R m and the vertical scale is 1 mm = S m³, then an area of T mm² represents a haul of TRS m³ m. This area could be measured using one of the methods discussed in Section 15.2. Note that the units of haul are m³ m (one cubic metre moved through a distance of one metre).

    Instead of calculating the centres of gravity of excavations and tips, which can be a difficult task, the *average haul distance* in each section can be easily found by dividing the *haul* in that section by the *volume* in that section, for example

$$\text{the } \textit{average haul distance} \text{ between } b \text{ and } c = \frac{(\text{area } bgchb)\text{m}^3 \text{ m}}{(gh) \text{ m}^3}$$

- If a surplus volume remains, this is *waste* which must be removed from the site, for example *lm*. If a deficiency of earth is found at the end of the project, this is *borrow* which must be imported on to the site. It is possible for waste and borrow to occur at any point along the site and this is also discussed in the section headed *Choice of balancing line*.

Economics of mass haul diagrams

When costing the earth-moving, there are four basic costs which are usually included in the contract for the project.

Cost of freehaul
Any earth moved over distances not greater than the free haul distance is costed only on the excavation of its volume, that is £A per m³.

Cost of overhaul
Any earth moved over distances greater than the free haul distance is charged both for its volume and for the distance in excess of the free haul distance over which it is

moved. This charge can be specified either for units of haul, that is, £B per m³ m, or for units of volume, that is £C per m³.

Cost of waste

Any surplus or unsuitable material which must be removed from the site and deposited in a tip is usually charged on units of volume, that is £D per m³. This charge can vary from one section of the site to another depending on the nearness of tips.

Cost of borrow

Any extra material which must be brought on to the site to make up a deficiency is also usually charged on units of volume, that is, £E per m³. This charge can also vary from one section of the site to another depending on the nearness of borrow pits or spoil heaps.

The following worked example illustrates how the costs of freehaul and overhaul can be calculated. Worked example 15.8 illustrates how the costs of borrowing and wasting can affect the final decision as to how the earth should be moved around the site.

Worked example 15.7: Costing using mass haul diagrams

Question

In a project for which a section of the mass haul diagram is shown in Figure 15.36, the free haul distance is specified as 100 m. Calculate the cost of earth-moving in the section between chainages 100 m and 400 m if the charge for moving the material within the freehaul distance is £A per m³ and that for moving any overhaul is £B per m3 m.

The x-axis should be taken as the balancing line and the areas between the curve and the balancing line in Figure 15.36 were measured with a digital planimeter and found to be as follows

$$\text{area of } (J + K + L + M) = 396{,}000 \text{ m}^3 \text{ m}$$

$$\text{area } J = 181{,}300 \text{ m}^3 \text{ m}$$

Solution

This type of problem can be solved in one of two ways.

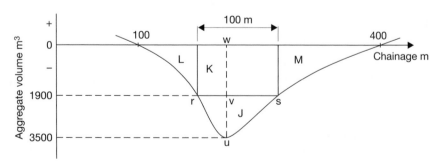

Figure 15.36

Solution 1 – Using planimeter areas only

Between chainages 100 m and 400 m, the *x*-axis balances cut and fill and the total volume to be moved in that section is given in Figure 15.36 as

$$uw = 3500 \text{ m}^3$$

The free haul distance of 100 m is fitted to Figure 15.36 so that it touches the curve at two points *r* and *s*. This means that the volume *uv* is the free haul volume and is therefore only charged for volume.

$$uv = (3500 - 1900) \text{ m}^3 = 1600 \text{ m}^3$$

Therefore area *J* can be removed since it is costed as **£1600*A***.

This leaves the volume *vw*, which is equal to 1900 m³, to be considered. This volume is the overhaul volume and has to be moved over a distance greater than the free haul distance. The distance through which it is moved has two components, the free haul distance and the overhaul distance, and this leads to two costs

- The overhaul volume moved through the free haul distance is costed on its volume only. This is area *K* in Figure 15.36 and the cost = **£1900*A***.
- The overhaul volume moved through the overhaul distance is the overhaul and is shown in Figure 15.36 as areas *L* and *M*. The cost is that involved in moving area *M* to area *L* and is obtained as follows:

$$\begin{aligned} \text{area contained in } L \text{ and } M &= (J + K + L + M) - (J + K) \\ &= 396{,}000 - [181{,}300 + (1900 \times 100)] \\ &= 24{,}700 \text{ m}^3 \text{ m} \end{aligned}$$

Hence

the cost of this overhaul = **£24,700*B***

Therefore

$$\begin{aligned} \textit{total cost} &= \text{free haul volume cost} + \text{overhaul volume costs} \\ &= \text{£1600}A + \text{£1900}A + \text{£24,700}B \\ &= \textbf{£3500}\boldsymbol{A} + \textbf{£24,700}\boldsymbol{B} \end{aligned}$$

Solution 2 – Using average haul distance and overhaul distance

Average haul distance between chainages 100 m and 400 m

$$\begin{aligned} &= (\text{haul between chainages 100 m and 400 m})/(\text{total volume between} \\ &\quad \text{chainages 100 m and 400 m}) \\ &= \frac{396{,}000}{3500} = 113 \text{ m} \end{aligned}$$

But the free haul distance = 100 m, hence

overhaul distance = 113 − 100 = 13 m

Therefore

$$\begin{aligned} \text{overhaul} &= \text{overhaul volume} \times \text{overhaul distance} \\ &= 1900 \times 13 = 24{,}700 \text{ m}^3 \text{ m} \end{aligned}$$

As for solution 1:

the cost of moving material over the free haul distance
= (free haul volume + overhaul volume) × £A
= £3500A (areas J and K)

the cost of overhaul = £24,700B (moving area M to area L)

Therefore

total cost = £3500A + £24,700B

Choice of balancing line

In Worked example 15.7 the x-axis was used as the balancing line. This is not always ideal. Figure 15.37 shows three possible balancing lines for the same mass haul diagram.

In Figure 15.37(a) the x-axis has been used and this results in waste near chainage 230 m.

In Figure 15.37(b), a balancing line in shown, which gives wastage near chainage 0 m. This may be better and cheaper if local conditions provide a suitable wasting point near chainage zero.

In Figure 15.37(c), two different balancing lines have been used, bc and de. This results in waste near chainage 0 m, where the curve is rising from a to b, borrow near chainage 125 m, where the curve is falling from c to d, and waste near chainage 210

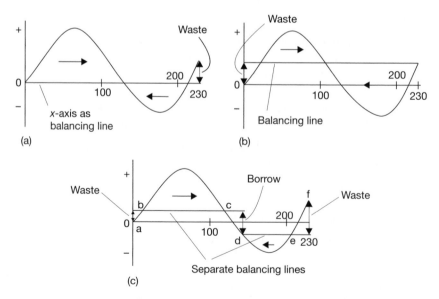

Figure 15.37 ● Balancing lines.

m, where the curve is rising from *e* to *f*. The two waste sections may be used to satisfy the central borrow requirement if economically viable.

Which choice is best depends on local conditions and on the proximity of borrow pits, quarries and suitable tipping sites. However, the following factors should be considered before a final choice is made.

- The use of more than one balancing line results in waste and borrow at intermediate points along the project, which will involve extra excavation and transportation of material.

- Short, unconnected balancing lines are often more economical than one long, continuous balancing line, especially where the balancing lines are shorter than the free haul distance since no overhaul costs will be involved.

- The direction of haul can be important. It is better to haul downhill to save power and, if long uphill hauls are involved, it may be better to waste at the lower points and borrow at the higher points.

- The main criterion should be one of economy. The free haul limit should be exceeded as little as possible in order that the amount of overhaul can be minimised.

- The haul is given by the area contained between the mass haul curve and the balancing line. Since the haul consists of freehaul and overhaul, if the haul area on the diagram can be minimised, the majority of it will be freehaul and hence overhaul will also be minimised. Therefore, the most economical solution from the haul aspect is to minimise the area between the curve and the balancing line. However, as shown in Figure 15.37(c), this can result in large amounts of waste and borrow at intermediate points along the project. The true economics can only be found by considering all the probable costs of hauling, wasting and borrowing.

- Where long haul distances are involved, it may be more economical to waste material from the excavation at some point within the free haul limit at one end of the site and to borrow material from a location within the free haul limit at the other end of the site rather than cart the material a great distance from one end of the site to the other. This possibility will become economical when the cost of excavating and hauling one cubic metre to fill from one end of the site to the other equals the cost of excavating and hauling one cubic metre to waste at one end of the site plus the cost of excavating and hauling one cubic metre to fill from a borrow pit at the other end of the site.

In practice, several different sets of balancing lines are tried and each costed separately with reference to the costs of wasting, borrowing and hauling. The most economical solution is usually adopted. The following worked example illustrates how this can be done.

Worked example 15.8: The use of balancing lines in costing

Question

In a project for which a section of the mass haul diagram is shown in Figure 15.38, the free haul distance is specified as 200 m. The earth-moving charges are as follows

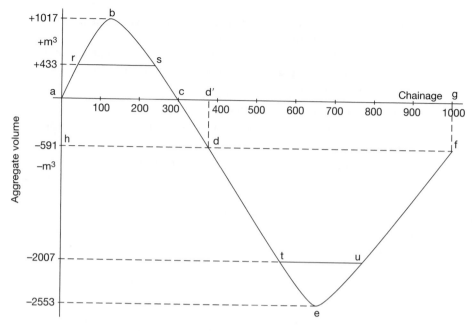

Figure 15.38

cost of free haul volume = £A per m³
cost of overhaul volume = £B per m³
cost of borrowing = £E per m³

Calculate the costs of each of the following alternatives

● Borrowing at chainage 1000 m only.

● Borrowing at chainage 0 m only.

● Borrowing at chainage 300 m only.

Solution

The 200 m free haul distance is added to Figure 15.38 as shown, that is

rs = *tu* = 200 m

The volumes corresponding to the horizontal lines *rs* and *tu* are interpolated from the curve to be +433 m³ and −2007 m³, respectively.

Borrowing at chainage 1000 m only

In this case, *acg* is used as a balancing line and borrow is required at *g* (chainage 1000 m) to close the loop *cefgc*.

free haul volume in section *ac* = 1017 − 433 = 584 m³
free haul volume in section *cg* = 2553 − 2007 = 546 m³

Hence

total free haul volume = 584 + 546 = 1130 m³

and

overhaul volume in section $ac = 433$ m^3
overhaul volume in section $cg = 2007$ m^3

Hence

total overhaul volume = 433 + 2007 = 2440 m^3

Also

borrow at g = 591 m^3

Therefore

cost of borrowing at chainage 1000 m only = £1130A + £2440B + £591E

Borrowing at chainage 0 m only
In this case, *hdf* is used as a balancing line and borrow is required at *h* (chainage 0 m) to close the loop *habdh*.
   As before

total free haul volume = 1130 m^3

However

overhaul volume in section $hd = 433 + 591 = 1024$ m^3
overhaul volume in section $df = 2007 - 591 = 1416$ m^3

Hence

total overhaul volume = 1024 + 1416 = 2440 m^3

Also

borrow at h = 591 m^3

Therefore

cost of borrowing at chainage 0 m only = £1130A + £2440B + £591E

This is the same as the cost of borrowing at chainage 1000 m provided that the cost of borrow is the same at chainages 0 m and 1000 m.

Borrowing at chainage 300 m only
In this case, two separate balancing lines *ac* and *df* are used and borrow is required at *c* (chainage 300 m) to fill the gap between *c* and *d*.
   As before

total free haul volume = 1130 m^3

However

overhaul volume in section $ac = 433$ m^3
overhaul volume in section $df = 2007 - 591 = 1416$ m^3

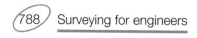

Hence

> *total overhaul volume = 433 + 1416 = 1849 m³*

Also

> *borrow at c = 591 m³*

Therefore

cost of borrowing at chainage 300 m only = £1130A + £1849B + £591E

This is the cheapest alternative assuming that the costs of borrow at chainages 0 m, 300 m and 1000 m are all equal. Considerably less overhaul is required when borrowing at chainage 300 m only.

Uses of mass haul diagrams

Mass haul diagrams can be used in several ways.

In design

Earlier in this section, the close link between the mass haul diagram and the formation level was discussed. If several formation levels are tried and a mass haul diagram is constructed for each, the one that gives the most economical results and maintains any stipulated standards such as gradient restrictions in vertical curve design can be used. Nowadays, mass haul diagrams tend to be produced using computer software packages and these greatly reduce the time required to obtain several different possible mass haul diagrams for comparison purposes. This is discussed further in Section 15.7.

In financing

Once the formation level has been designed, the mass haul diagram can be used to indicate the most economical method of moving the earth around the project and a good estimate of the overall cost of the earth-moving can be calculated.

In construction

By preparing a mass haul diagram, the required volumes of material are known before construction begins, enabling suitable plant and machinery to be chosen, sites for spoil heaps and borrow pits to be located and directions of haul to be established.

In forward planning

The mass haul diagram can be used to indicate the effect that other engineering works within the overall project, especially tunnels and bridges, will have on the earth-moving. Such constructions upset the pattern of the mass haul diagram by restricting the directions of haul, but since the volumes and the quantities of any waste and borrow will be known, suitable areas for spoil heaps and borrow pits can be located in advance of construction, enabling work to proceed smoothly.

Reflective summary

With reference to mass haul diagrams remember:

— Mass haul diagrams enable the most economical method of moving earth around long linear engineering projects such as roads, railways and canals to be determined.

— They are closely linked to longitudinal sections and the two are always drawn together.

— Bulking and shrinkage factors should be applied to the volumes to allow for the fact that excavated materials invariably change volume when they are compacted.

— Careful choice of balancing lines can enable waste or borrow to be moved over short distances that do not incur additional charges.

— Mass haul diagrams can be used at all phases of the construction process: during the design phase when different formation levels are being tested, during the procurement phase when the cost of the earthmoving is being determined, during the planning phase when types of equipment, locations of spoil heaps and borrow pits, and haul directions are being considered, and finally during the construction phase when the earth is actually being moved.

15.7 Computer-aided earthwork calculations

After studying this section, you should have an appreciation of the use of computer software packages in the determination of earthwork quantities.

Principles of computer-aided earthwork calculations

In simple terms, the main purpose of earthwork calculations is to determine the size of some geometrical figure as accurately as time and costs will allow. The figure can be two- or three-dimensional and its size can be measured as either an area or a volume. These can be stated as numerical quantities for costing purposes or, in the case of volumes, they can be presented graphically in the form of a mass haul diagram for planning purposes. With odd exceptions, such as a planimeter, the numerical values of area and volume are normally obtained from specific formulae and the mass haul diagram is basically a graph. All of this means that the whole nature of earthwork calculations can be broken down into two related elements: computing and plotting. While both of these can be easily performed by using hand calculators and manual drawing techniques, they are also ideal subjects for analysis by computer. Not surprisingly, then, there are now many computer software packages available which can be used for earthwork computations. Some are general graphics packages, which incorporate area and volume calculation routines. Others, however, have been produced specifically for land and engineering surveying purposes and many of the packages include software modules for the computation of various earthwork quantities. Given the wide choice, it is not possible to give a detailed review of such individual packages in a general textbook such as this. For further information on their capabilities and costs, however, the reader is recommended to study the surveying journals *Civil Engineering Surveyor*, *Engineering Surveying Showcase* and *Geomatics World*, which produce annual reviews of the latest software packages.

Although individual packages are not reviewed here, the general concepts on which they are based tend to be similar. The majority of packages involve the use of a DTM in conjunction with a database of all the points surveyed and computed for a construction project. DTMs and databases are discussed in Section 10.9 and, once they have been established, they can be used together with some of the techniques covered in this chapter to compute any required earthwork quantities.

The basic method by which the various packages compute areas and volumes relies on the fact that every point surveyed and computed is referenced by a unique number and has its own set of three-dimensional coordinates. In addition, if a DTM is formed, its triangular structure provides the computer with an interlocking network, which shows the interrelationship between the points. This enables the operator to interact with the computer and establish the geometrical figures that need to be measured. For example, areas can be defined by running straight lines and/or arcs between the points that define the boundary. The computer can then either apply the relevant formula to the points in question or it can break down the defined figure into a series of geometrical shapes (triangles, squares, segments and so on) from which it can calculate the area. These principles can be applied to obtain a wide range of area and volume information as outlined below.

- *Surface areas* and *plan areas* can be obtained by summing the individual surface areas and plan areas, respectively, of specific triangles in the DTM. The individual areas are calculated using one of the formulae given in Section 15.2.

Alternatively, for plan areas, if no DTM has been created, the cross coordinate method can be applied to the E,N values of the appropriate points contained in the database.

- *Cross-sectional areas* are represented in the computer by a series of linked points each being referenced by its elevation above a datum and its offset distance from the centre line. These can be used to establish a coordinate system from which the individual areas of cut and fill can be computed. This is similar to the technique described for irregular cross-sections in Section 15.4. Alternatively, the cross-section can be broken down into a series of figures made up of straight lines and arcs, the individual areas of which can be summed.

- *Volumes of road works* can be computed from the cross-sectional areas using the end areas method or the prismoidal formula. Bulking and shrinkage factors can be included as necessary.

- *Volumes of stockpiles* can be computed by specifying their base levels and summing the individual triangular prisms defined by each triangle in the DTM. This is identical to the method discussed in Section 15.5 for determining volumes from spot heights. The *volume between two contours* and the *volume of a void* (such as a hole in the ground or a lake) can be computed in a similar manner although, in these cases, two levels are specified and the final result is obtained by subtracting the volumes above or below each level.

- *Mass haul diagrams* and *longitudinal* and *cross-sections* can easily be computed and plotted once the highway design and the volume computations have been completed. Again, the computer uses techniques similar to those described in Sections 15.6 and 15.3.

The above examples by no means represent all the possibilities associated with computer-aided earthworks software systems. They are only intended to show some of their capabilities. However, the real advantage of such systems is their ability to speed up the calculations to such an extent that many different designs can be compared and an optimum solution found, with innovative designs that would have been impossible before the advent of computers becoming quite feasible.

In addition, complete interaction with the software is possible, enabling earthwork quantities to be specified before construction begins in order that the most appropriate geometrical figure or figures can be designed to meet the requirements. An example of this would be when dividing a parcel of land into a number of areas each required to have a different size and shape. Virtually any type of area or volumetric calculation can be performed by these packages and, should the reader wish to know more about the capabilities of specific modules, the reference material given at the end of this chapter is recommended.

Reflective summary

With reference to computer-aided earthwork calculations remember:

— The whole nature of earthwork calculations, being essentially a combination of computation and drawing, is ideal for analysis by computer software.

— There are many commercially available software packages from which plan areas, cross-sectional areas and volumes can be determined and longitudinal and cross-sectional drawings can be produced.

— The majority of packages involve the use of a *Digital Terrain Model* in conjunction with a database of all the points surveyed and computed.

— The great advantage of such software packages is that they speed up the calculations to such an extent that different designs can be compared and an optimum solution found very quickly.

— They allow innovative design solutions to be found that would have been impossible before the evolution of computers.

Exercises

15.1 State the units used when measuring plan areas and volumes.

15.2 A triangular piece of land has sides of length 67.39 m, 86.27 m and 44.42 m. It is proposed to construct a rectangular garage having dimensions of 6.5 m by 5.5 m on this land. What percentage of the total land area will the garage occupy?

15.3 The coordinates of a four-sided polygon traverse RSTUR as follows

Point	mE	mN
R	1761.32	1111.17
S	1959.77	1435.43
T	2168.32	1276.18
U	1994.81	972.36

Calculate the plan area contained within the traverse in hectares.

15.4 An area of land is bounded by straight sides in the shape of a five-sided clockwise polygon ABCDEA. It is required to split the area into two equal parts by erecting a fence, which is to run in a straight line from point C to a point F, which lies on side EA. A traverse is run around the area and the coordinates of A, B, C, D and E are calculated as follows

Point	mE	mN
A	679.11	775.87
B	826.93	807.32
C	1010.75	711.89
D	985.54	487.66
E	802.38	364.91

Calculate the length of the fence that should be erected.

15.5 With the aid of illustrations, describe **two** methods by which areas bounded by irregular lines can be measured.

15.6 An irregular shaped piece of land, which is known to occupy an area of 7956 m^2 is measured on a plan using a planimeter and a value of 5092 mm^2 is obtained. Calculate the scale of the plan.

15.7 A field is bounded by an irregular hedge running between points E and F and three straight fences FG, GH and HE. The following measurements are taken:

EF = 167.76 m, FG = 105.03 m, GH = 110.52 m, HE = 97.65 m and EG = 155.07 m

Offsets are taken to the irregular hedge from the line EF as follows. The hedge is situated entirely outside the quadrilateral EFGH.

E (0 m)	25 m	50 m	75 m	100 m	125 m	150 m	F(167.76 m)
0 m	2.13 m	4.67 m	9.54 m	9.28 m	6.39 m	3.21 m	0 m

Calculate the area of the field to the nearest m^2

– using the trapezoidal rule throughout

– using Simpson's rule wherever possible

15.8 Explain what is meant by the terms *longitudinal section* and *cross-section* and describe the fieldwork that must be undertaken in order that they can be obtained using optical levelling techniques.

15.9 With the aid of illustrations, discuss the various different types of cross-sections that can exist.

15.10 A cutting is to be formed as part of a proposed road having a formation width of 7.30 m and side slopes of 1 in 2. A cross-section is to be taken at chainage 2050 m, where the depth of dig to the proposed formation level is 10.89 m and the existing ground level is horizontal at right-angles to the proposed centre line. Calculate the plan width and the area of cut required at this cross-section.

15.11 A straight section of a proposed road having a formation width of 7.30 m is to be constructed as an embankment having side slopes of 1 in 3. At one particular cross-section, the transverse slope at right-angles to the proposed centre line is 1 in 20 and the height to the formation level at the centre line is 6.45 m. Calculate the area of fill required at this cross-section.

15.12 A straight section of a proposed road having a formation width of 10.00 m is to be constructed as a cutting having side slopes of 1 in 2. A cross-section is to be taken at chainage 1225 m, where the depth of dig to the proposed formation level is to be 4.82 m. At this cross-section, the transverse slope at right angles to the proposed centre line falls from left to right but changes at the proposed centre line from a fall of 1 in 11 to a fall of 1 in 17. Calculate the area of cut required at this cross-section.

15.13 A straight section of a proposed road having a formation width of 14.60 m is to be constructed with side slopes of 1 in 2 in cut and 1 in 3 in fill. A cross-section is to be taken at chainage 2350 m where the proposed depth of fill at the centre line is to be 0.26 m and the transverse slope at right angles to the proposed centre line rises from left to right at 1 in 5. Calculate the side widths and the areas of cut and fill at this section.

15.14 With the aid of illustrations, show how longitudinal and cross-sectional information can be obtained from contoured maps and plans.

15.15 Discuss how data for longitudinal sections and cross-sections can be obtained using levelling techniques.

15.16 A cutting is to be formed on a straight section of a proposed road. The values of eight consecutive cross-sectional areas taken at a constant horizontal distance interval of 30 m are 325 m^2, 398 m^2, 456 m^2, 580 m^2, 634 m^2, 546 m^2, 447 m^2, and 362 m^2, respectively. Calculate the volume contained between the first and last cross-sections using

– the end areas method throughout

– the prismoidal formula wherever possible

15.17 A cutting is to be formed on a section of a proposed road of formation width 10.00 m. The transverse slope at right-angles to the proposed centre line is 1 in 8 and the existing ground levels at chainages 1750 m, 1800 m and 1850 m are 176.32 m, 175.18 m and 174.87 m, respectively. The reduced level of the formation centre line at chainage 1750 m is 173.65 m and the formation is to have a falling gradient of 4% from chainage 1750 m to chainage 1850 m. The side slopes are to be 1 in 3. Calculate the volume of cut required to form the cutting between chainages 1750 m and 1850 m.

15.18 A straight section of a proposed road, between chainages 3450 m and 3550 m, is to be constructed on ground having a transverse slope, which rises at 1 in 7 from left to right, at right angles to the proposed centre line. The road is to be such that on any cross-section the formation level at the centre line is to be lower than the existing level and the area of cut is equal to the area of fill. The road design includes the following specifications:

– side slopes are to be 1 in 1 for cuttings and 1 in 3 for embankments

– formation width = 14.60 m

- the centre line of the road between chainage 3450 m and 3550 m is to be level

Calculate the volume of cut required to form the road between chainages 3450 m and 3550 m.

15.19 The junctions of the square grid shown below were levelled to determine the volume of excavation necessary in the construction of a basement floor. The reduced levels of the grid points are as shown.

x	x	x	x	x
185.67	186.22	187.34	187.45	188.00
x	x	x	x	x
186.33	187.03	187.22	187.56	187.90
x	x	x		
186.64	186.98	187.44		
x	x	x	x	x
187.08	187.35	187.89	187.34	187.43
x	x	x	x	x
187.24	187.46	188.02	187.93	187.26

The horizontal distance between the grid points is 20 m and the required formation level of the basement floor foundations is to be 178.00 m. Calculate the volume of the excavation within the grid area.

15.20 A planimeter was used to measure the plan areas contained between the proposed position of a dam wall and several contour lines on a 1 in 10000 scale map. The planimeter areas shown below were obtained. The proposed mean water level of the reservoir is 93 m and the volume below the 40 m contour can be neglected. Calculate the volume of the reservoir to the nearest 1000 m^3 using the prismoidal formula wherever possible.

Plan area mm^2	Contour m	Plan area mm^2	Contour m
15690	93	6020	65
12760	90	4120	60
11900	85	2100	55
10980	80	950	50
9750	75	240	45
7760	70	30	40

15.21 With the aid of illustrations, explain how mass haul diagrams are drawn.

15.22 Discuss the properties of mass haul diagrams with particular emphasis on how they indicate the directions in which the earth should be moved.

15.23 With reference to mass haul diagram, explain what is meant by the following terms: *Free haul distance, Overhaul distance, Haul, Waste* and *Borrow*.

15.24 Discuss the role of balancing lines in mass haul diagrams.

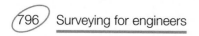

Further reading and sources of information

For the latest information on computer-aided earthwork calculations, consult

Software Systems, Electronic Surveying Supplement, Civil Engineering Surveyor, pp 58–63, published by the Institution of Civil Engineering Surveyors, Spring 2004; http://www.ices.org.uk/

Fort, M. J. (2003) Surveying for geomatics software. *Engineering Surveying Showcase 2003*, No. 2, pp. 33–49. http://www.pvpubs.com/.

Surveying calculations

The need for calculations

These days, engineers and surveyors use computers for performing calculations and for data processing. Although most of the survey information required on site is produced in this way, the ability to be able to compute by hand using a calculator of some sort is still needed from time to time, especially when setting-out data is missing and is urgently required on site. For example, when setting out the centre line of a road curve a computer will normally produce a printout of centre line coordinates defining the position of the curve plus the bearings and distances needed to set this out with a total station from nearby control points. Every now and then, the bearings and distances are not pre-calculated for some reason or the control points to which they refer are no longer accessible and a free station point has to be used instead. In these situations, the engineer or surveyor may have to compute the required bearings and distances by hand from the coordinates (see Section 6.2) or at least by using some propriety software if this is available. Nowadays, the coordinates of centre line and control points may already be downloaded into a total station or GPS controller ready for setting out. In this case, it always a good idea to check the data being generated by the total station or GPS equipment by calculating some random bearings and distances to see if they agree with the points being set out. Of course, any problems with setting out and other measurements on site should always be highlighted by independent checks, examples of which are given in previous chapters.

The important point to note here is that computers and instruments such as the total station and GPS have not completely replaced the need for some day-to-day calculations on site. For these reasons, some information about calculators and how to use them is covered in this appendix. Another reason for including this is that most students of surveying tend to use a calculator when learning the basics of surveying.

Like fieldwork, calculations should be carefully planned and carried out in a systematic manner and any data that is going to be used should be properly prepared before calculations start. Where possible, standardised tables or forms should be used to simplify calculations and if a calculation has not been checked, it is considered to be unreliable. Examples of the use of forms and tables in surveying calculations are shown throughout this book.

Anyone who is likely to use a calculator should study Table A.1 first, which lists those features of a calculator that are required for survey work.

Table A.1 ● Pocket calculator functions for engineering surveying.

Function or facility	Notes
Display	Should have at least 10 digits
Arithmetic	Basic functions required
Trigonometrical	sin, sin⁻¹, cos, cos⁻¹, tan, tan⁻¹ essential
Degrees, rad, gon	Facility for using trigonometrical functions in degree, rad and gon modes useful
Decimal degrees	Conversion between deg, min, sec and decimal degrees needed
Polar/rectangular	Conversion between polar (bearing and distance) and rectangular form (ΔE and ΔN) simplifies coordinate calculations greatly
Programmable	Useful (but not essential) for most calculations provided that program storage is available for repeat calculations
General purpose	$1/x$, x^3, $\sqrt{x}$, y^x occur frequently in engineering surveying
Logarithms	$\log x$, $10x$, $\ln x$, e^x sometimes used
Floating point	Essential when dealing with large or small numbers
Storage registers (memory)	Useful in complicated problems
Pre-programmed constants	π required
Statistical functions	$\bar{x}$ and s not essential but sometimes convenient

Significant figures

The way in which numbers are written or printed in surveying calculations and reports is important, whether these numbers represent measurements or are derived from measurements through a formula. As far as measurements are concerned, an indication of the precision achieved by a measurement is represented by the number of significant figures recorded. For example, a distance may be recorded as 15.342 m, implying a precision of 1 mm in the measurement, whereas the same distance recorded as 15.34 m implies a precision of 10 mm. In the same way, an angle written as 43°12'00" indicates it has been measured with a precision of 1", but when recorded as 43°12' it is assumed it has only been measured with a precision of 1'.

A number such as 15.342 has five significant figures, while 15.34 has four; the difference between these implies quite a difference in the equipment that might have been used to take the measurement. The position of the decimal point does not indicate the number of significant figures, as 0.0006521 has four significant figures and 0.098 has two. Care must be taken to ensure that the correct number of significant figures is used with a calculator (and displayed for a computer output) since both of these are capable of displaying many digits, some of which may not be significant. Something else to remember is that any quantity calculated from others cannot usually be quoted to a higher precision than that of the data supplied or that of any observations used in the calculations.

Various rules exist for determining significance for numbers resulting from a calculation. In general, it is the least precise component which gives the precision of the final result.

For *addition* and *subtraction*, an answer can only be quoted such that the number of figures shown after the decimal place does not exceed those of the number (or numbers) with the least significant decimal place.

For *multiplication* and *division*, an answer can only be quoted with the same number of significant figures as the least significant number used in the calculation.

These rules are illustrated in the following worked example.

Worked example A.1: Significant figures

Question
(a) Calculate the sum of 23.568, 1103.2, 0.3451 and 0.51
(b) Calculate the difference between 45.471 and 38.9
(c) Multiply 23.65 by 87.322
(d) Divide 112 by 22.699

Solution
(a) Listing each number gives

$$
\begin{array}{r}
23.568 \\
1103.2 \\
0.3451 \\
\underline{0.51} \\
1127.6231
\end{array} = \mathbf{1127.6}
$$

The result is quoted with five significant figures to agree with the number with the least significant decimal place, 1103.2, despite the fact that 0.3451 and 0.51 have fewer significant figures.

(b) The difference between 45.471 and 38.9 is

$$45.471 - 38.9 = 6.571 = \mathbf{6.6}$$

since 38.9 has the least significant place.

(c) Multiplying gives

$$23.65 \times 87.322 = 2065.1653 = \mathbf{2065}$$

since the least significant number, 23.65, has four significant figures.

(d) Dividing gives

$$\frac{112}{22.699} = \mathbf{4.93}$$

since 112 has three significant figures.

In some cases, results may be quoted to more significant figures than the above rules suggest. For example, the arithmetic mean (most probable value) of 14.56, 14.63, 14.59, 14.62 and 14.58 is

$$\frac{14.56 + 14.63 + 14.59 + 14.62 + 14.58}{5} = \frac{72.98}{5} = 14596$$

since the number 5 is an *exact number* and retaining the extra significant figure is justified for the mean value, which is more precise than a single value (however, this must be verified by calculating the standard error of each; see Section 9.2). Many circumstances occur in surveying in which numbers can be treated as exact, and these have to be carefully defined.

Units

Wherever possible throughout this book, *Système Internationale* (SI) units are used, although others are introduced as necessary. The units used in engineering surveying are as follows.

Length

millimetre mm
metre   m
kilometre   km

1 mm $= 10^{-3}$ m $= 10^{-6}$ km
10^3 mm $= 1$ m $= 10^{-3}$ km
10^6 mm $= 10^3$ m $= 1$ km

Area

Square metre m^2

Although not in the SI system, the hectare (ha) is often used to denote area where 1 ha = 100 m × 100 m = 10,000 m^2.

Volume

Cubic metre m^3

Angle

The SI unit of angle is the *radian*. However, most survey instruments and systems measure in degrees °, minutes ' and seconds " and some European countries use the gon ^{g} as the unit of angle. The relationships between all of these are as follows.

2π radians $= 360° = 400^g$

and taking π to be 3.141592654 gives

1 radian $= 57.295\ 779\ 513° = 57°17'44.806"$
1 radian $= 3437.746\ 771'$
1 radian $= 206\ 264.806"$

$90° = 1.570\ 796\ 327$ radians $= 100^g$
$1° = 0.017\ 453\ 293$ radians $= 1.111\ 111\ 111^g$

1' = 0.000 290 888 radians = 0.018 518 519^g
1" = 0.000 004 848 radians = 0.000 308 642^g

100^g = 90°
1^g = 0.9°
0.1^g = 5.4'
0.01^g = 32.4"

Worked example A.2: Angular conversions

Question
(a) Convert 36°11'39" to radians and 1.254 radians to degrees
(b) Determine the gon equivalent of 56°07' and the degree equivalent of 148.6524^g

Solution
(a) The decimal equivalent of 36°11'39" is 36.1942° since angles given to the nearest second should be calculated with four decimal places. This gives

$$36°11'39" = 36.1942\left(\frac{2\pi}{360}\right) = \textbf{0.631 707 radians}$$

since the 2 and 360 in the conversion are exact numbers.
The radian to degree conversion is

$$1.254 \text{ radians} = 1.254\left(\frac{360}{2\pi}\right) = 71.85° = \textbf{71°51'}$$

since angles recorded with two decimal places are known only to the nearest minute.

(b) The gon equivalent of 56°07' is

$$56°07' = 56.12\left(\frac{400}{360}\right) = \textbf{62.35}^g$$

and the degree equivalent of 148.6524^g is

$$148.6524^g = 148.6524\left(\frac{360}{400}\right) = 133.7872° = \textbf{133°47'14"}$$

Useful formulae

Some of the following are used in various chapters of the book and are listed here for reference.

Solution of right-angled triangle (Figure A.1)

$$\sin\theta = \frac{a}{c} \qquad \cos\theta = \frac{b}{c} \qquad \tan\theta = \frac{a}{b}$$

$$\operatorname{cosec} \theta = \frac{1}{\sin \theta} \qquad \sec \theta = \frac{1}{\cos \theta} \qquad \cot \theta = \frac{1}{\tan \theta}$$

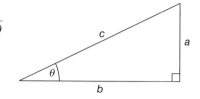

Figure A.1

Solution of oblique triangle ABC (Figure A.2)

Sine rule $\qquad \dfrac{a}{\sin A} = \dfrac{b}{\sin B} = \dfrac{c}{\sin C}$

Cosine rule $\qquad a^2 = b^2 + c^2 - 2bc \cos A$

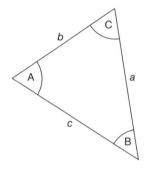

Figure A.2

Trigonometrical formulae

$$\sin(A \pm B) = \sin A \cos B \pm \cos A \sin B$$

$$\cos(A \pm B) = \cos A \cos B \mp \sin A \sin B$$

$$\tan(A \pm B) = \frac{\tan A \pm \tan B}{1 \mp \tan A \tan B}$$

$$2 \sin A \cos B = \sin(A + B) + \sin(A - B)$$

$$2 \cos A \sin B = \sin(A + B) - \sin(A - B)$$

$$2 \cos A \cos B = \cos(A + B) + \cos(A - B)$$

$$2 \sin A \sin B = \cos(A - B) - \cos(A + B)$$

$$\sin(-\theta) = -\sin \theta$$
$$\cos(-\theta) = \cos \theta$$
$$\tan(-\theta) = -\tan \theta$$

$$\cos 2\theta = \cos^2 \theta - \sin^2 \theta$$

$$\cos 2\theta = 2\cos^2 \theta - 1$$

$$\sin 2\theta = 2 \sin \theta \cos \theta$$

$$\tan 2\theta = \frac{2 \tan \theta}{1 - \tan^2 \theta}$$

$$\cos^2 \theta + \sin^2 \theta = 1$$

$$1 + \tan^2 \theta = \sec^2 \theta$$

$$\cot^2 \theta + 1 = \operatorname{cosec}^2 \theta$$

Series expansions for trigonometric functions

$$\sin x = x - \frac{x^3}{3!} + \frac{x^5}{5!} - \frac{x^7}{7!} + \dots \qquad x \text{ in radians}$$

$$\cos x = 1 - \frac{x^2}{2!} + \frac{x^4}{4!} - \frac{x^6}{6!} + \dots \qquad x \text{ in radians}$$

Derivatives of trigonometrical functions

y	$\dfrac{dy}{dx}$	y	$\dfrac{dy}{dx}$
$\sin(ax)$	$a\cos(ax)$	$\csc(ax)$	$-a\csc(ax)\cot(ax)$
$\cos(ax)$	$-a\sin(ax)$	$\sec(ax)$	$a\sec(ax)\tan(ax)$
$\tan(ax)$	$a\sec^2(ax)$	$\cot(ax)$	$-a\csc^2(ax)$

Derivatives of functions

Function of a function: If $y = f(u)$ and $u = f(x)$ then $\dfrac{dy}{dx} = \dfrac{dy}{du} \cdot \dfrac{du}{dx}$

Product: $\qquad \dfrac{d(uv)}{dx} = u\dfrac{dv}{dx} + v\dfrac{du}{dx}$

Quotient: $\qquad \dfrac{d}{dx}\left(\dfrac{u}{v}\right) = \dfrac{1}{v^2}\left[v\dfrac{du}{dx} - u\dfrac{dv}{dx}\right]$

Quadratic equation

If $ax^2 + bx + c = 0$ then

$$x = \frac{-b \pm \sqrt{b^2 - 4ac}}{2a}$$

Areas

Circle: $\quad Area = \pi r^2 \quad (r = \text{radius})$

Ellipse: $\quad Area = \pi ab \quad (a \text{ and } b \text{ are semi-major and semi-minor axes})$

Triangle: Referring to Figure A.2:

$$Area = \sqrt{s(s-a)(s-b)(s-c)} \text{ where } s = \frac{1}{2}(a + b + c)$$

$$Area = \frac{1}{2} bc \sin A$$

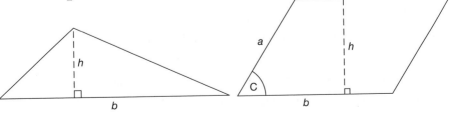

Figure A.3   Figure A.4

In Figure A.3:

$$Area = \frac{1}{2} bh$$

Parallelogram: Figure A.4
gives $Area = bh = ab \sin C$

Trapezoid: From Figure A.5

$$Area = \frac{1}{2}(a + b)h$$

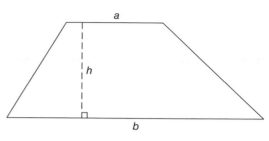

Figure A.5

Volumes

Sphere: $\frac{4}{3}\pi r^3$ (r = radius)

Cylinder: $\pi r^2 h$ (r = radius h = height)

Cone: $\frac{1}{3}\pi r^2 h$ (r = radius at base h = height from base to apex)

Pyramid: $\frac{1}{3} abh$ (ab = area of base h = height to apex)

Exercises

A.1 Calculate each of the following to the correct number of significant figures

- 21.587 140 + 0.058 + 8.9571 + 87.51
- 10.582 − 0.8524 − 0.652 − 0.2
- 0.6541 × 24.03 × 1.27 × 5.0006
- $\dfrac{2.593}{0.0006034}$
- 4529.015 cos 21°42'13"
- 12.036 sin 231°09'52"
- $\dfrac{38.365}{\cos 25°57'}$
- 301.22 tan 17°36'

A.2 Find the sine, cosine, tangent, secant, cosecant and cotangent of the following angles

     23°15'   142°19'35"   219°05'16"   348°47'

A.3 Convert the following angles into gons and radians

     117°01'29"   81°25'   12°20'55"   309°10'

A.4 Convert the following into degrees, minutes and seconds

     1.25617 radians   0.654002 radians   1.565 radians

     48.1256^g 184.07^g 12^g

Answers to numerical exercises

Chapter 2

2.11 357 m
2.12 0.024 m per 50 m
2.13 0.040 m per 40 m, 1.506 m, 1.462 m
2.14 A = 41.429 m, B = 41.529 m, C = 41.987 m, D = 41.941 m
2.15 A = 61.804 m, B = 61.767 m, C = 61.086 m, D = 62.177 m, E = 61.438 m, F = 61.578 m, G = 62.554 m, H = 61.063 m
2.16 Use HPC method as all readings are intermediate sights
200 m = 71.922 m, 210 m = 71.783 m, 220 m = 71.752 m, 230 m = 71.295 m, 240 m = 71.723 m, 250 m = 71.154 m, 260 m = 71.133 m, 270 m = 70.616 m, 280 m = 70.678 m
2.17 A = 31.275 m, B = 31.222 m, C = 31.217 m, D = 31.195 m, E = 31.286 m, soffit 1 = 34.489 m, soffit 2 = 34.384 m, soffit 3 = 34.268 m
2.18 Height change from R1 to R2 = 0.783 m, collimation error = 0.007 m per 70 m

Chapter 3

3.15 10"
3.16 1'43", 52", 21"
3.17 46", 21"
3.18 $A\hat{T}B$ = 38°04'26", $A\hat{T}C$ = 69°12'41", $A\hat{T}D$ = 136°52'20"
3.19 D1 = +02°37'24", D2 = −01°47'10", D3 = −05°18'51", D4 = +00°12'42"
Collimation error = 02'02"
3.20 Horizontal angles 135°35'04", 207°05'20", 309°57'33"
Vertical angles +02°53'24", −03°16'41", −01°18'04", +01°27'04"

Chapter 4

4.6 Slope corrections: −0.013 m, −0.050 m
Standardisation correction: +0.008 m
Tension corrections: −0.001 m, +0.002 m
Temperature corrections: −0.004 m, +0.002 m
Sag corrections: −0.024 m, −0.006 m

4.7 23.437 m

4.8 −0.022 m

4.9 32.217 m

4.10 21.049 m

4.11 −0.003m, −0.007 m

Chapter 5

5.18 ±3 mm, ±9 mm

5.19 45 mm

5.20 100m: 0.6 mm; 200 m: 2.5 mm; 300 m: 5.7 mm; 400 m: 10.0 mm; 500 m: 15.7 mm

5.21 8.357 m

Chapter 6

6.15 AS = 159.134 m, ST = 142.285 m, TA = 231.341 m, $\hat{A}$ = 37°15'51", $\hat{S}$ = 100°06'41", $\hat{T}$ = 42°37'28"

6.16 (i)   −60"

(ii)   Bearing HA = 147°06'10", Bearing HR = 174°38'10"

(iii)   Bearing HT = 268°25'30", Bearing TP = 176°52'40"
Bearing PE = 308°59'40", Bearing EW = 156°02'50"
Bearing WR = 77°20'20", Bearing RH = 354°38'10"

(iv)   HT: ΔE = −91.280 m ΔN = −2.509 m
TP: ΔE = +3.541 m ΔN = −64.927 m
PE: ΔE = −56.057 m ΔN = +45.385 m
EW: ΔE = +51.391 m ΔN = −115.685 m
WR: ΔE = +103.163 m ΔN = +23.177 m
RH: ΔE = −10.758 m ΔN = +114.559 m

(v)   1 in 47,200

(vi)   T = 224.170 mE, 172.608 mN   P = 227.711 mE, 107.681 mN
E = 171.654 mE, 153.066 mN   W = 223.045 mE, 37.381 mN
R = 326.208 mE, 60.558 mN

6.17 First traverse
B = 393.125 mE, 326.426 mN   A = 470.590 mE, 335.699 mN
H = 375.319 mE, 405.696 mN   X = 303.449 mE, 355.001 mN
P = 319.998 mE, 290.381 mN

Second traverse
T = 126.546 mE, 121.309 mN   A = 189.330 mE, 114.867 mN
F = 166.575 mE, 44.111 mN   M = 271.052 mE, 142.176 mN

6.18 (i)   −25"
   (ii)   Bearing MA = 142°00'45", Bearing AB = 59°10'00"
           Bearing BC = 347°32'25", Bearing CP = 82°13'45"
           Bearing PQ = 153°00'55"
   (iii)   MA: ΔE = +77.188 m ΔN = −98.841 m
            AB: ΔE = +154.209 m ΔN = +92.058 m
            BC: ΔE = −32.356 m ΔN = +146.419 m
            CP: ΔE = +244.847 m ΔN = +33.421 m
   (iv)   1 in 21,900
   (v)   A = 284.308 mE, 427.222 mN   B = 438.517 mE, 519.280 mN
           C = 406.161 mE, 665.699 mN

6.19 P = 787.792 mE, 494.904 mN   Q = 739.624 mE, 422.246 mN
      R = 631.572 mE, 412.103 mN   S = 544.625 mE, 426.011 mN

6.20 (i) a = 0.8070470; b = −0.5904976; c = −218.216; d = 57.888
   (ii)   B = 426.423 mE, 219.382 mN   D = 46.798 mE, 23.639 mN

6.21 T_1 = 302.284 mE, 394.778 mN T_2 = 337.955 mE, 295.391 mN

6.22 P = 372.160 mE, 355.672 mN

6.23 P = 237.877 mE, 687.427 mN

Chapter 8

8.19 At 10 m: 81.554 m, at 1000 m: 81.542 m

8.20 70.253 m

Chapter 9

9.13 *Primary first order*
      Distances 500 m: ±16.8 mm, 1 km: ±23.7 mm, 5 km: ±53.0 mm
      Angles 500 m: ±7.2", 1 km: ±5.1", 5 km: ±2.3"

      Primary second order
      Distances 500 m: ±33.5 mm, 1 km: ±47.4 mm, 5 km: ±106.1 mm
      Angles 500 m: ±14.5", 1 km: ±10.2", 5 km: ±4.6"
      Secondary control
      Distances 100 m: ±15.0 mm, 200 m: ±21.2 mm, 300 m: ±26.0 mm, 400 m:
      ±30.0 mm, 500 m: ±33.5 mm
      Angles 100 m: ±32.4", 200 m: ±22.9", 300 m: ±18.7", 400 m: ±16.2", 500 m:
      ±14.5"

9.14 *Primary first order*
      Distances 500 m: ±11.2 mm, 1 km: ±15.8 mm, 5 km: ±35.4 mm
      Angles 500 m: ±4.0", 1 km: ±2.8", 5 km: ±1.3"

Primary second order
Distances 500 m: ±16.8 mm, 1 km: ±23.7 mm, 5 km: ±53.0 mm
Angles 500 m: ±5.2", 1 km: ±3.6", 5 km: ±1.6"
Secondary control
Distances 100 m: ±15.0 mm, 200 m: ±21.2 mm, 300 m: ±26.0 mm, 400 m: ±30.0 mm, 500 m: ±33.5 mm
Angles 100m: ±32.4", 200 m: ±22.9", 300 m: ±18.7", 400 m: ±16.2", 500 m: ±14.5"
Tertiary control
Category 1
Distances 25 m: ±8 mm, 50 m: ±11 mm, 100 m: ±15 mm
Angles 25 m: ±01'05", 50 m: ±46", 100 m: ±32"
Category 2
Distances 25 m: ±25 mm, 50 m: ±35 mm, 100 m: ±50 mm
Angles 25 m: ±01'48", 50 m: ±01'16", 100 m: ±54"
Category 3
Distances 25 m: ±38 mm, 50 m: ±53 mm, 100 m: ±75 mm
Angles 25 m: ±02'24", 50 m: ±01'42", 100 m: ±01'12"
Category 4
Distances 25 m: ±50 mm, 50 m: ±71 mm, 100 m: ±100 mm
Angles 25 m: ±03'36", 50 m: ±02'33", 100 m: ±01'48"

9.17 42.2498 m ±0.0007 m
PD = 10 mm giving standard error = 4.0 mm. Since standard error of individual measurement = ±2.3 mm, observers meet specification.

9.18 53°51'34.5" ±5.4"

9.19 0.7367 m ±0.0008 m

9.20 Observer 3
Observer 1: 5 times; Observer 2: 7 times; Observer 3: 3 times
38°41'59.0" ± 1.1"

9.21 Reject 12°34'11", Angle = 12°34'16.8" ± 0.6"

9.22 ±14.14"

9.23 58.7135 m ± 0.010 m

9.24 ±01'12", ±6.25"

9.25 3.035 m ± 0.016 m. Improve measurement of vertical angle as this contributes most towards standard error in height of B.

9.26 1347.63 m^2 ± 0.21 m^2

9.27 Equation (9.8) gives sighting distance = 200 m. Because a digital level is used with high accuracy an impractical answer is obtained. Levelling could be done by setting up halfway between bench marks but this is not recommended as measurements are taken at the limit of the range of the level. It would be better to use two setups with 50 m sighting distances.

9.28 50 m: 1 round required but two should taken to check for gross errors, 100 m: 2 rounds, 200 m: 3 rounds

Chapter 11

11.2 At stake E: 0.49 m below; at stake F: 0.23 m below; at stake G: 0.06 m below; at stake H: 0.12 m below

11.3 Height of sight rail above cover level = 1.13 m; height above peg at A = 0.93 m; height above peg at B = 1.16 m; height above peg at C = 0.97 m

11.7 At stake A: 0.58 m below; at stake B: 0.72 m below; at stake C: 0.48 m below; at stake D: 0.12 m below

11.9 (i) Slope distance to R = 20.16 m, slope distance to S = 19.15 m, slope distance to T = 39.72 m, slope distance to U = 40.72 m

(ii) At stake R: 0.46 m below; at stake S: 0.16 m below; at stake T: 0.21 m below; at stake U: 0.27 m above

11.11 (i) Slope distance to A = 13.10 m, slope distance to B = 12.08 m, slope distance to C = 24.49 m, slope distance to D = 25.51 m

(ii) At stake A: 0.22 m below; at stake B: 0.77 m below; at stake C: 0.13 m above; at stake D: 0.18 m below

11.14 At stake V: 0.41 m below; at stake W: 0.27 m below; at stake X: 0.45 m below; at stake Y: 0.22 m below

11.16 (i) Clockwise angle from A to E relative to the line AB = 337°11'58"
Clockwise angle from A to F relative to the line AB = 314°08'14"
Clockwise angle from B to E relative to the line BA = 65°39'01"
Clockwise angle from B to F relative to the line BA = 34°25'04"

(ii) surface distance AE = 102.56 m; surface distance AF = 64.55 m

11.18 (i) Angle $A\hat{K}L$ = 27°41'31"; angle $A\hat{L}K$ = 69°58'48"

(ii) Slope length KA = 240.91 m; slope length LA = 119.17 m

(iii) Error in diagonal AC = −0.002 m; error in diagonal BD = +0.004 m

11.20 (i) 6241.54 mE, 5847.35 mN

(ii) 70.77 m

11.22 Angle $Y\hat{X}E$ = 36°43'46"; angle $Y\hat{X}F$ = 43°01'23"; angle $X\hat{Y}E$ = 70°55'22"; angle $X\hat{Y}F$ = 68°40'24"

11.26 (i) 259°16'55"

(ii) 99°45'44"

(iii) 741.02 m

11.28 (i) Angle $J\hat{H}R$ = 315°05'17", distance HR = 75.39 m; angle $J\hat{H}S$ = 294°55'10", distance HS = 40.36 m

(ii) Angle $H\hat{J}R$ = 29°57'20", distance JR = 106.60 m; angle $H\hat{J}S$ = 15°52'22", distance JS = 133.84 m

Chapter 12

12.1 1256.49 m

12.2 (i) 40 m, 10.72 m (ii) 0 m, 0 m; 10 m, 0.63 m; 20 m, 2.54 m; 30 m, 5.84 m

12.3 chainage (m), chord length (m), cumulative tangential angle (° ' "): 1231.58 (T), 0.00, 00 00 00; 1240.00, 8.42, 00 16 05; 1260.00, 20.00, 00 54 17; 1280.00, 20.00, 01 32 29; 1300.00, 20.00, 02 10 41; 1320.00, 20.00, 02 48

53; 1340.00, 20.00, 03 27 05; 1360.00, 20.00, 04 05 17; 1380.00, 20.00, 04
43 29; 1400.00, 20.00, 05 21 41; 1420.00, 20.00, 05 59 53; 1440.00, 20.00,
06 38 05; 1458.85 (U), 18.85, 07 14 05

12.4 802.52 m, 1384.57 m, 1723.31 m, 1872.66 m

12.5 (i) 0 m, 4.57 m; (ii) 5 m, 4.36 m; 10 m, 3.73 m, 15 m, 2.66 m; 20 m, 1.14 m

12.6 T: 749.70 m, U: 1160.75 m

12.7 (i) T = 2223.77 m, U = 2414.93 m; (ii) (point: cumulative tangential angle
(° ' "), long chord) T: 00 00 00, 0.00 m 1: 00 02 49, 1.23 m 2: 01 00 07,
26.23 m 3: 01 57 25, 51.22 m 4: 02 54 43, 76.20 m 5: 03 52 01, 101.16 m 6:
04 49 19, 126.09 m 7: 05 46 37, 150.98 m 8: 06 43 55, 175.84 m U: 07 18
08, 190.65 m

12.8 1: 1360.82 mE, 2165.01 mN 2: 1384.11 mE, 2155.92 mN 3: 1407.08 mE,
2146.05 mN 4: 1429.71 mE, 2135.42 mN 5: 1451.97 mE, 2124.05 mN 6:
1473.84 mE, 2111.93 mN 7: 1495.30 mE, 2099.10 mN 8: 1516.32 mE,
2085.56 mN U: 1528.64 mE, 2077.15 mE

Chapter 13

13.2 Entry and exit transition curves (starting at T and U, respectively): y = 0.00
m, x = 0.00 m; y = 20.00 m, x = 0.05 m; y = 40.00 m, x = 0.44 m; y = 54.46
m, x = 1.10 m

Central circular arc (measuring along the long chord from its mid-point in
both directions): y = 0.00 m, x = 0.51 m; y = 20.00 m, x = 0.07 m; y = 21.505
m, x = 0.00 m

13.3 Entry transition curve (chainage, chord, tangential angle from TI): 3094.26
m (T), 0.00 m, 00°00'00"; 3100.00 m, 5.74 m, 00°00'16"; 3120.00 m, 20 m,
00°05'19"; 3140.00 m, 20.00 m, 00°16'47"; 3160.00 m, 20.00 m, 00°34'39";
3180.00 m, 20.00 m, 00°58'57"; 3183.57 m (T_1), 3.57 m, 01°03'58"
Central circular arc (chainage, chord, tangential angle from common
tangent): 3183.57 m (T_1), 0.00 m, 00°00'00"; 3200.00 m, 16.43 m,
00°35'18"; 3220.00 m, 20.00 m, 01°18'16"; 3240.00 m, 20.00 m, 02°01'14";
3260.00 m, 20.00 m, 02°44'12"; 3280.00 m, 20.00 m, 03°27'10"; 3300.00 m,
20.00 m, 04°10'08"; 3307.61 m (T_2), 7.61 m, 04°26'29"
Exit transition curve (chainage, chord, tangential angle from UI): 3396.92 m
(U), 0.00 m, 360°00'00"; 3380.00 m, 16.92 m, 359°57'42"; 3360.00 m,
20.00 m, 359°49'04"; 3340.00 m, 20.00 m, 359°34'01"; 3320.00 m, 20.00 m,
359°12'33"; 3307.61 m (T_2), 12.39 m, 358°56'02"

13.4 (i)   T = 1523.57 m T_1 = 1596.70 m T_2 = 1720.34 m U = 1793.47 m
(ii) Theoretical = 0.69 m; maximum allowable = 0.31 m
(iii) Entry transition curve (chainage, chord, tangential angle from TI):
1523.57 m (T), 0.00 m, 00°00'00"; 1550.00 m, 26.43 m, 00°09'07"; 1596.70
m (T_1), 46.70 m, 01°09'50"
Exit transition curve (chainage, chord, tangential angle from UI): 1793.47 m
(U), 0.00 m, 360°00'00"; 1750.00 m, 43.47 m, 359°35'20"; 1720.34 (T_2),
29.66 m, 358°50'10"

13.5 (i) $R = 590.03$ m

(ii) Entry transition curve (chainage, chord, tangential angle from TI): 2507.90 m (T), 0.00 m, 00°00'00"; 2520.00 m, 12.10 m, 00°01'17"; 2550.00 m, 30.00 m, 00°15'38"; 2580.00 m, 30.00 m, 00°45'51"; 2610.00 m, 30.00 m, 01°31'58"; 2617.98 m (T_C), 7.98 m, 01°46'54"

Exit transition curve (chainage, chord, tangential angle from UI): 2728.06 m (U), 0.00 m, 360°00'00"; 2700.00 m, 28.06 m, 359°53'03"; 2670.00 m, 30.00 m, 359°30'16"; 2640.00 m, 30.00 m, 358°51'36"; 2617.98 m (T_C), 22.02 m, 358°13'06"

13.6 (i) Tangent length = 149.12 m

(ii) Radius = 691.63 m

(iii) Design speed = 103.72 kph

(iv) Rate of change of radial acceleration = 0.233 m s^{-3}

13.7 (i) $T_1 = 1846.54$ m $T_2 = 1900.82$ m;

(ii) Maximum allowable superelevation = 0.29 m

(iii) Entry transition curve (chainage, chord, tangential angle from TI): 1726.54 m (T), 0.00 m, 00°00'00"; 1750.00 m, 23.46 m, 00°02'57"; 1775.00 m, 25.00 m, 00°12'33"; 1800.00 m, 25.00 m, 00°28'51"; 1825.00 m, 25.00 m, 00°51'50"; 1846.54 m (T_1), 21.54 m, 01°16'59"

13.8 Entry transition on the composite curve (chainage, chord, tangential angle): 3200.21 m (T_4), 0.00 m, 00°00'00"; 3220.00 m, 19.79 m, 00°06'06"; 3240.00 m, 20.00 m, 00°24'41"; 3249.22 m (T_5), 9.22 m, 00°37'26"

13.9 (i) Entry transition curve (chainage, chord, tangential angle from TI): 7094.07 m (T), 0.00 m, 00°00'00"; 7100.00 m, 5.93 m, 00°00'28"; 7125.00 m, 25.00 m, 00°12'30"; 7150.00 m, 25.00 m, 00°40'51"; 7167.20 m (T_1), 17.20 m, 01°09'50"

Central circular arc (chainage, chord, tangential angle from the common tangent): 7167.20 m (T_1), 0.00 m, 00°00'00"; 7175.00 m, 7.80 m, 00°22'21"; 7200.00 m, 25.00 m, 01°33'58"; 7225.00 m, 25.00 m, 02°45'35"; 7250.00 m, 25.00 m, 03°57'12"; 7275.00 m, 25.00 m, 05°08'49"; 7300.00 m, 25.00 m, 06°20'26"; 7325.00 m, 25.00 m, 07°32'03"; 7341.46 m (T_2), 16.46 m, 08°19'12"

Exit transition curve (chainage, chord, tangential angle from UI): 7414.59 m (U), 0.00 m, 360°00'00"; 7400.00 m, 14.59 m, 359°57'13"; 7375.00 m, 25.00 m, 359°39'32"; 7350.00 m, 25.00 m, 359°05'31"; 7341.46 m (T_2), 8.54 m, 358°50'10"

(ii) Superelevation (chainage, amount): 7094.07 m (T), 0.25 m (minimum for drainage); 7100.00 m, 0.25 m (minimum for drainage); 7125.00, 0.25 m (minimum for drainage); 7150.00 m, 0.33 m; 7167.20 m (T_1), 0.43 m

(iii) Chainage of the common tangent point = 7179.75 m

13.10 (i) $R = 719.96$ m

(ii) Total length of the curve = 190.55 m

(iii) Maximum rate of change of radial acceleration = 0.31 m s^{-3}

13.11 Chainage, Easting, Northing: 3126.24 m (T), 528.08 mE, 637.71 mN; 3140.00 m (C_1), 541.51 mE, 634.69 mN; 3160.00 m (C_2), 560.97 mE,

630.09 mN; 3174.29 m (T_1), 574.80 mE, 626.50 mN; 3180.00 m (C_3), 580.30 mE, 624.96 mN; 3200.00 m (C_4), 599.42 mE, 619.10 mN; 3217.89 (T_2), 616.32 mE, 613.22 mN; 3220.00 m (C_5), 618.30 mE, 612.49 mN; 3240.00 m (C_6), 636.93 mE, 605.21 mN; 3260.00 m (C_7), 655.40 mE, 597.54 mN; 3265.94 m (U), 660.87 mE, 595.23 mN

Chapter 14

14.1 (i)   1080 m
     (ii)   If possible use 162 m; if not, use 91.8 m
14.2 82 m
14.4 (i) RL_P = 237.45 m, RL_R = 237.97 m
     (ii)   (Chainage, RL): 1098.35 m (P), 237.45 m; 1110.00 m, 237.19 m; 1140.00 m, 236.74 m; 1170.00 m, 236.65 m; 1200.00 m, 236.90 m; 1230.00 m, 237.41 m; 1246.55 m (R), 237.97 m
     (iii)   Chainage = 1163.35 m, RL = 236.64 m
14.5 Through chainage = 2621.72 m, RL = 4.05 m
14.6 (ii) RL_R = 189.28 m
     (iii)   (Chainage, RL): 2163.49 m (P), 189.46 m; 2200.00 m, 189.82 m; 2250.00 m, 189.99 m; 2300.00 m, 189.75 m; 2340.03 m (R), 189.28 m
     (iv)   Chainage = 2245.45(5)m, RL = 189.99 m
14.7 (i)   Chainage A = 3394.98 m, chainage B = 3510.60 m, chainage C = 3626.22 m
     (ii) RL_A = 231.71 m, RL_B = 231.25 m, RL_C = 231.02 m
     (iii)   (Chainage, RL): 3394.98 m (A), 231.71 m; 3400.00 m, 231.55(5)m; 3425.00 m, 230.97 m; 3450.00 m, 230.68 m; 3475.00 m, 230.70 m; 3500.00 m, 231.02 m; 3510.60 m (B), 231.25 m
     (iv)   Chainage = 3563.96 m, RL = 231.89 m
14.8 656.17 m
14.9 1177.73 m
14.10 318.16 m
14.11 (i) RL_P = 91.37 m, RL_R = 92.88 m
     (ii)   (Chainage, RL): 767.54 m (P), 91.37 m; 780.00 m, 91.65 m; 810.00 m, 92.23 m; 840.00 m, 92.70 m; 870.00 m, 93.05 m; 889.82 m (C), 93.22 m; 900.00 m, 93.28 m; 930.00 m, 93.30 m; 960.00 m, 93.10 m; 976.54 m (R), 92.89 m
     (iii)   Chainage = 917.40 m, RL = 93.32 m

Chapter 15

15.2 2.42%
15.3 9.7126 hectares
15.4 312.02 m
15.6 1 : 1250

15.7 (i) 14,824 m^2 (ii) 14,845 m^2
15.10 plan width = 50.86 m; area of cut = 316.68 m^2
15.11 175.96 m^2
15.12 100.92 m^2
15.13 side widths = 11.30 m (cut), 20.20 m (fill); areas = 6.00 m^2 (cut), 18.49 m^2 (fill)
15.16 (i) 102,135 m^3 (ii) 102,615 m^3 using prismoidal between 1 and 7 and end areas between 7 and 8 or 102,445 m^3 using end areas between 1 and 2 and prismoidal between 2 and 8
15.17 9305 m^3
15.18 1077 m^3
15.19 44,389 m^3
15.20 34,342,000 m^3 or 34,343,000 m^3

Appendix

A.1 118.11, 8.9, 99.8, 4297, 4207.95, −9.3754, 42.67, 95.55
A.2 0.3947, 0.9188, 0.4296, 1.088, 2.533, 2.328
0.6111626, −0.7915051, −0.7721524, −1.263416, 1.636226, −1.295081
−0.6305102, −0.7761809, 0.8123238, −1.288359, −1.586017, 1.231036
−0.19452, 0.98090, −0.19831, 1.0195, −5.1409, −5.0427
A.3 130.0275, 2.042467, 90.46, 1.421, 13.7207, 0.215524, 343.52, 5.3960
A.4 71°58'24", 37°28'18", 89°40', 43°18'47", 165°40', 11°

Index